MASS SPECTROMETRY IN BIOPHYSICS

WILEY-INTERSCIENCE SERIES IN MASS SPECTROMETRY

Series Editors:

Dominic M. Desiderio
Departments of Neurology and Biochemistry
University of Tennessee Health Science Center

Nico M. M. Nibbering
Vrije Universiteit Amsterdam, The Netherlands

John R. de Laeter • *Applications of Inorganic Mass Spectrometry*
Michael Kinter and Nicholas E. Sherman • *Protein Sequencing and Identification Using Tandem Mass Spectrometry*
Chhabil Dass, *Principles and Practice of Biological Mass Spectrometry*
Mike S. Lee • *LC/MS Applications in Drug Development*
Jerzy Silberring and Rolf Eckman • *Mass Spectrometry and Hyphenated Techniques in Neuropeptide Research*
J. Wayne Rabalais • *Principles and Applications of Ion Scattering Spectrometry: Surface Chemical and Structural Analysis*
Mahmoud Hamdan and Pier Giorgio Righetti • *Proteomics Today: Protein Assessment and Biomarkers Using Mass Spectrometry, 2D Electrophoresis, and Microarray Technology*
Igor A. Kaltashov and Stephen J. Eyles • *Mass Spectrometry in Biophysics: Conformation and Dynamics of Biomolecules*

MASS SPECTROMETRY IN BIOPHYSICS

Conformation and Dynamics of Biomolecules

Igor A. Kaltashov
Stephen J. Eyles
University of Massachusetts at Amherst

WILEY-INTERSCIENCE

A JOHN WILEY & SONS, INC., PUBLICATION

Published by John Wiley & Sons, Inc., Hoboken, New Jersey.
Published simultaneously in Canada.

Library of Congress Cataloging-in-Publication Data:

Kaltashov, Igor A.
Mass spectrometry in biophysics : conformation and dynamics of biomolecules / Igor A. Kaltashov, Stephen J. Eyles.
p. cm.
Includes bibliographical references and index.
ISBN 0-471-45602-0 (Cloth)
1. Mass spectrometry. 2. Biophysics. 3. Biomolecules—Spectra. I. Eyles, Stephen J. II. Title.
QP519.9.M3K35 2005
572′.33—dc22 2004012532

Printed in the United States of America.

10 9 8 7 6 5 4 3 2 1

CONTENTS

PREFACE

Strictly speaking, the term *biophysics* refers to the application of the theories and methods of physics to answer questions in the biological arena. This obviously now vast field began with studies of how electrical impulses are transmitted in biological systems and how the shapes of biomolecules enable them to perform complex biological functions. Over time, biophysicists have added a wide variety of methodologies to their experimental toolkit, one of the more recent additions being mass spectrometry (MS). Traditionally limited to the analysis of small molecules, recent technological advances have enabled the field of MS to expand into the biophysical laboratory, catalyzed by the 2002 Nobel prize winning work of John Fenn and Koichi Tanaka. MS is a rapidly developing field whose applications are constantly changing: this text represents only a snapshot of current techniques and methodologies.

The aim of this book is to present a detailed and systematic coverage of the current state of biophysical MS with special emphasis on experimental techniques that are used to study protein higher order structure and dynamics. No longer an exotic novelty, various MS-based methods are rapidly gaining acceptance in the biophysical community as powerful experimental tools to probe various aspects of biomolecular behavior both in vitro and in vivo. Although this field is now experiencing an explosive growth, there is no single text that focuses solely on applications of MS in molecular biophysics and provides a thorough summary of the plethora of MS experimental techniques and strategies that can be used to address a wide variety of problems related to biomolecular dynamics and higher order structure. The aim of this book is to close that gap.

We intended to target two distinct audiences: mass spectrometrists who are working in various fields of life sciences (but are not necessarily experts in biophysics) and experimental biophysicists (who are less familiar with recent developments in MS technology but would like to add it to their experimental arsenal). In order to make the book equally useful for both groups, the presentation of the MS-based techniques in biophysics is preceded by a discussion of general biophysical concepts related to structure and dynamics of biological macromolecules (Chapter 1). Although it is not meant to provide an exhaustive coverage of the entire field of molecular biophysics, the fundamental concepts are explained in some detail to enable anyone not directly involved with the field to understand the important aspects and terminology. Chapter 2 provides a brief overview of "traditional" biophysical techniques with special emphasis on those that are complementary to mass spectrometry and that are mentioned elsewhere in the book. These introductory chapters are followed by an in-depth discussion of modern mass spectrometric hardware used in experimental studies of biomolecular structure and dynamics. The purpose of Chapter 3 is to provide readers who are less familiar with MS with concise background material on modern MS instrumentation and techniques that will be referred to in the later chapters (the book is structured in such a way that no prior familiarity with biological MS is required of the reader).

Chapters 4 through 7 deal with various aspects of protein higher order structure and dynamics as probed by various MS-based methods. Chapter 4 focuses on "static structures," by considering various approaches to evaluate higher order structure of proteins at various levels of spatial resolution when crystallographic and nuclear magnetic resonance (NMR) data are either unavailable or insufficient. The major emphasis is on methods that are used to probe biomolecular topology and solvent accessibility (i.e., chemical cross-linking and selective chemical modification). In addition, the use of hydrogen–deuterium exchange for mapping protein–protein interfaces is briefly discussed. Chapter 5 presents a concise introduction to an array of techniques that are used to study structure and behavior of non-native protein states that become populated under denaturing conditions. The chapter begins with consideration of protein ion charge state distributions in electrospray ionization mass spectra as indicators of protein unfolding and concludes with a detailed discussion of hydrogen exchange, arguably one of the most widely used methods to probe the structure and dynamics of non-native protein states under equilibrium conditions. The kinetic aspects of protein folding and enzyme catalysis are considered in Chapter 6. Chapter 7 focuses on MS-based methods that are used to extract quantitative information on protein–ligand interactions (i.e., indirect methods of assessment of binding energy). The remainder of this chapter is devoted to advanced uses of mass spectrometry to characterize dynamics of multiprotein assemblies and its role in modulating protein function.

Complementarity of MS-based techniques to other experimental tools is emphasized throughout the book and is also addressed specifically in Chapter 8. Two examples presented in this chapter are considered in sufficient detail to illustrate the power of synergy of multiple biophysical techniques, where some methods provide overlapping information to confirm the evidence, while others provide completely unique details. Chapter 9 presents a discussion of MS-based methods to study higher-order structure and dynamics of biopolymers that are not proteins (oligonucleotides, polysaccharides, as well as polymers of nonbiotic origin). Chapter 10 provides a brief discussion of biomolecular properties in the gas phase, focusing primarily on the relevance of in vacuo measurements to biomolecular properties in solution.

The book concludes with a discussion of the current challenges facing biomolecular MS, as well as important new developments in the field that are not yet ready for routine use. Chapter 11 focuses on several areas where MS is currently making a debut. It begins with a discussion of novel uses of MS aimed at understanding "orderly" protein oligomerization processes, followed by consideration of "catastrophic" oligomerization, such as amyloidosis. This chapter also considers other challenging tasks facing modern MS, such as the detection and characterization of very large macromolecular assemblies (e.g., intact ribosomes and viral particles), as well as applications of various MS-based techniques to study the behavior of a notoriously difficult class of biopolymers—membrane proteins. The chapter concludes with a general discussion of the relevance of in vitro studies and reductionist models to processes occurring in vivo.

Throughout the entire book an effort has been made to present the material in a systematic fashion. Both the theoretical background and technical aspects of each technique are discussed in detail, followed by an outline of its advantages and limitations, so that the reader can get a clear sense of both current capabilities and potential future uses of various MS-based experimental methodologies. Furthermore, this book was conceived as a combination of a textbook, a good reference source, and a practical guide. With that in mind, a large amount of material (practical information) has been included throughout. An effort has also been made to provide the reader with a large reference base to the original research papers, so that the details of experimental work omitted in the book can easily be found. Because of space limitations and the vastness of the field, a significant volume of very interesting and important research could not be physically cited. It is hoped, however, that no important experimental techniques and methodologies have been overlooked. The authors will be grateful for any comments from the readers on the material presented in the book (Chapters 1, 3, 4, 5, 7, 10 and 11 were written mostly by I.K. and Chapters 6, 8 and 9 by S.E.; both authors contributed equally to Chapter 2). The comments can be emailed directly to the authors at kaltashov@chem.umass.edu and eyles@polysci.umass.edu.

We are grateful to Professors David L. Smith, Michael L. Gross, Max Deinzer, Lars Konermann, Joseph A. Loo, and Richard W. Vachet for helpful discussions over the past several years that have had direct impact on this book. We would also like to thank many other colleagues, collaborators, and friends for their support and encouragement during various stages of this challenging project. We are also indebted to many people who have made contributions to this book in the form of original graphics from research articles (the credits are given in the relevant parts of the text). We also thank the current and past members of our research group, who in many cases contributed original unpublished data for the illustrative material presented throughout. Finally, we would like to acknowledge the National Institutes of Health and the National Science Foundation for their generous support of our own research efforts at the interface of biophysics and mass spectrometry.

IGOR A. KALTASHOV
STEPHEN J. EYLES

University of Massachusetts at Amherst

1

GENERAL OVERVIEW OF BASIC CONCEPTS IN MOLECULAR BIOPHYSICS

This chapter provides a brief overview of the basic concepts and current questions facing biophysicists in terms of the structural characterization of proteins, protein folding, and protein–ligand interactions. Although this chapter is not meant to provide an exhaustive coverage of the entire field of molecular biophysics, the fundamental concepts are explained in some detail to enable anyone not directly involved with the field to understand the important aspects and terminology.

1.1. COVALENT STRUCTURE OF BIOPOLYMERS

Biopolymers are a class of polymeric materials that are manufactured in nature. Depending on the building blocks (or *repeat units* using polymer terminology), biopolymers are usually divided into three large classes. These are polynucleotides (built of nucleotides), peptides and proteins (built of amino acids), and polysaccharides (built of various saccharide units). In this chapter we only consider the general properties of biopolymers using peptides and proteins as examples; questions related to polynucleotides and polysaccharides will be discussed in some detail in Chapter 9.

All polypeptides are linear chains built of small organic molecules called *amino acids*. There are 20 amino acids that are commonly considered *canonical* or *natural*. This assignment is based on the fact that these 20 amino acids correspond to 61 (out of a total 64) codons within the triplet code with three remaining codons functioning as terminators of protein synthesis (Table 1.1) (1, 2), although there

Mass Spectrometry in Biophysics: Conformation and Dynamics of Biomolecules
By Igor A. Kaltashov and Stephen J. Eyles
ISBN 0-471-45602-0

TABLE 1.1. Chemical Structure and Masses of Natural (Canonical) Amino Acids

Symbol	Name	Molecular Formula (Residue)	Chemical Structure	Side Chain Character	Monoisotopic Mass[a] (Residue)	Average Mass (Residue)
Ala (A)	Alanine	C_3H_5NO	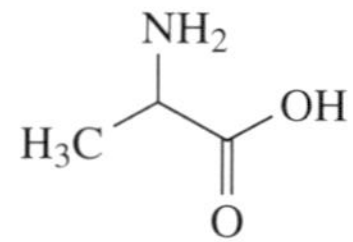	Nonpolar	71.037	71.079
Arg (R)	Arginine	$C_6H_{12}N_4O$	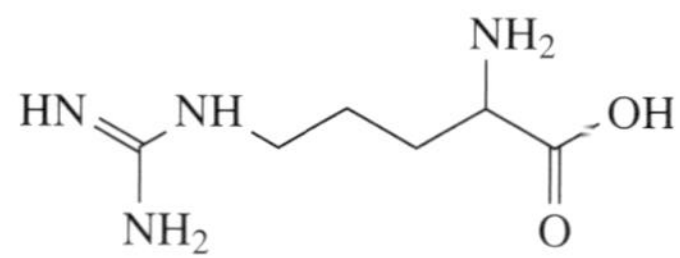	Basic	156.101	156.188
Asn (N)	Asparagine	$C_4H_6N_2O_2$	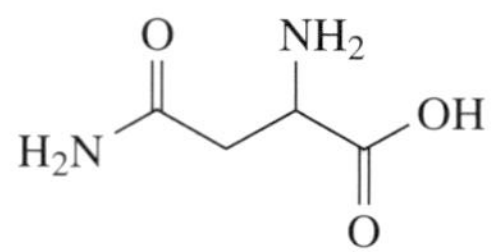	Polar	114.043	114.104
Asp (D)	Aspartic acid	$C_4H_5NO_3$	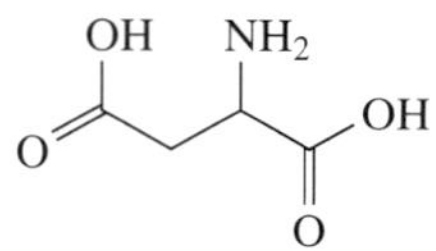	Acidic	115.027	115.089
Cys (C)	Cysteine	C_3H_5NOS	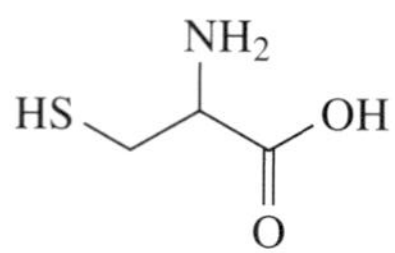	Polar/acidic	103.009	103.145

Gln (Q)	Glutamine	$C_5H_8N_2O_2$	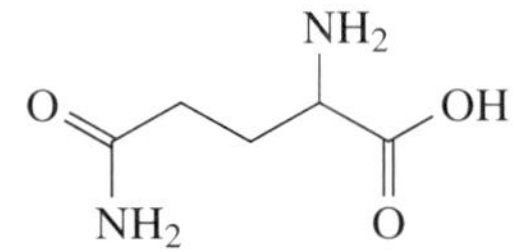	Polar	128.059	128.131
Glu (E)	Glutamic acid	$C_5H_7NO_3$	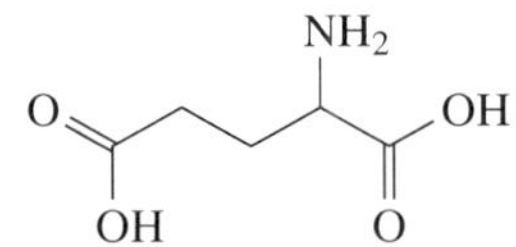	Acidic	129.043	129.116
Gly (G)	Glycine	C_2H_3NO	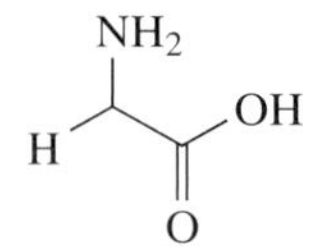	Nonpolar	57.021	57.052
His (H)	Histidine	$C_6H_7N_3O$	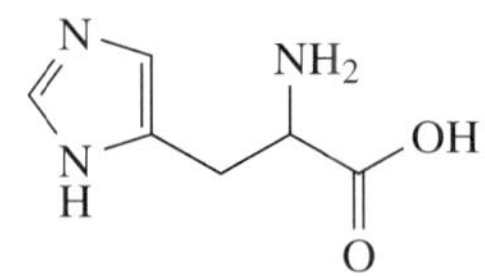	Basic	137.059	137.141
Ile (I)	Isoleucine	$C_6H_{11}NO$	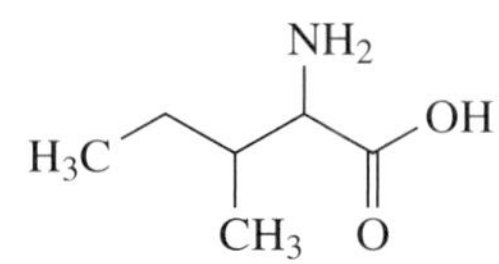	Nonpolar	113.084	113.160

TABLE 1.1 (*Continued*)

Symbol	Name	Molecular Formula (Residue)	Chemical Structure	Side Chain Character	Monoisotopic Mass[a] (Residue)	Average Mass (Residue)
Leu (L)	Leucine	$C_6H_{11}NO$	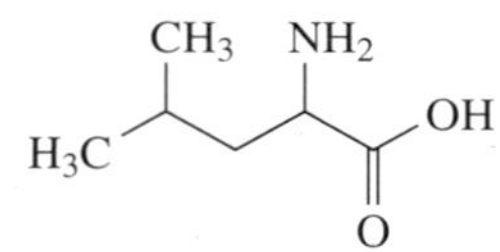	Nonpolar	113.084	113.160
Lys (K)	Lysine	$C_6H_{12}N_2O$	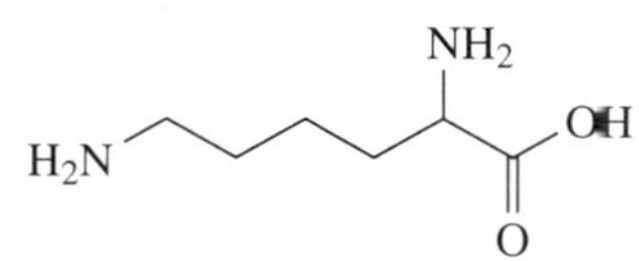	Basic	128.095	128.174
Met (M)	Methionine	C_5H_9NOS	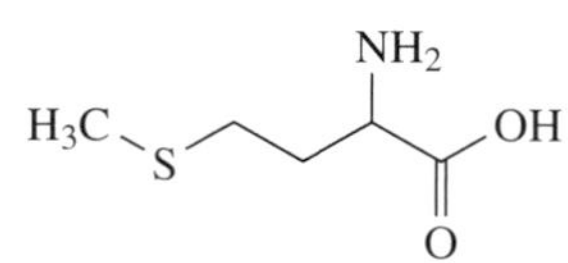	Nonpolar/amphipathic	131.040	131.199
Phe (F)	Phenylalanine	C_9H_9NO	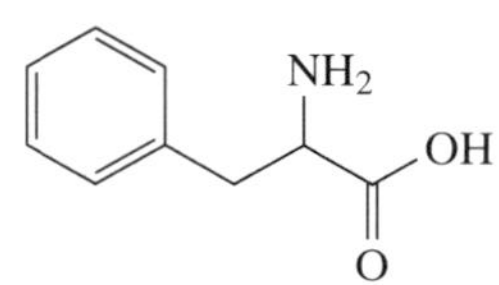	Nonpolar	147.068	147.177
Pro (P)	Proline	C_5H_7NO	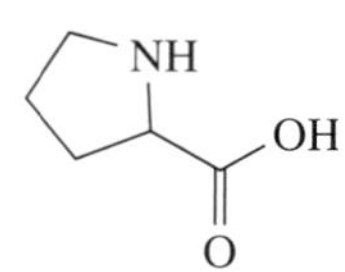	Nonpolar	97.053	97.117

Ser (S)	Serine	$C_3H_5NO_2$	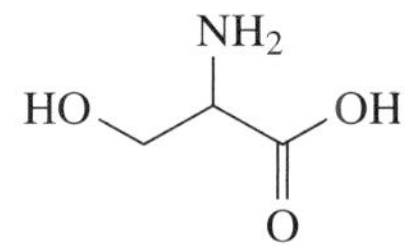	Polar	87.032	87.078
Thr (T)	Threonine	$C_4H_7NO_2$	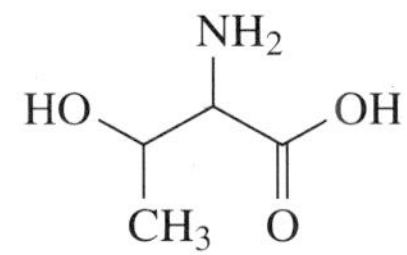	Polar/amphipathic	101.048	101.105
Trp (W)	Tryptophan	$C_{11}H_{10}N_2O$	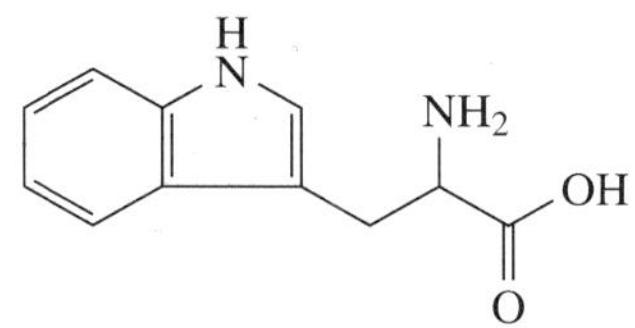	Amphipathic	186.079	186.213
Tyr (Y)	Tyrosine	$C_9H_9NO_2$	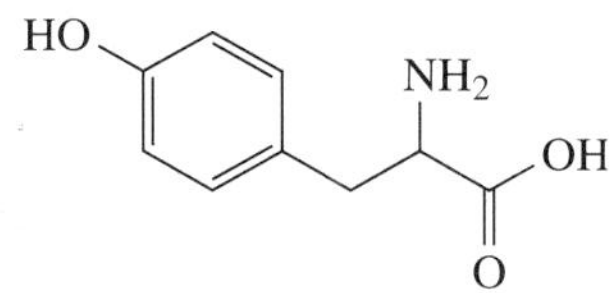	Amphipathic	163.063	163.176
Val (V)	Valine	C_5H_9NO	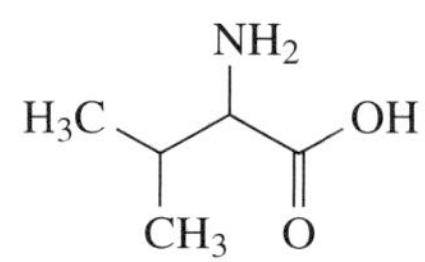	Nonpolar	99.068	99.133

[a] See Chapter 3 for a definition of monoisotopic and average masses.

are at least as many other amino acids that occur less frequently in living organisms (Table 1.2). Noncanonical amino acids are usually produced by chemical modification of a related canonical amino acid (e.g., oxidation of proline produces hydroxyproline), although at least two of them (selenocysteine and pyrrolysine) should be considered canonical based on the way they are produced and utilized in protein synthesis in vivo by some organisms [the UGA codon that was originally considered as a termination codon is now known to serve also as a Sec (selenocysteine) codon] (3, 4). Furthermore, new components can be added to the protein biosynthetic machinery of both prokaryotes and eukaryotes, which makes it possible to genetically encode unnatural amino acids in vivo (5, 6).

A peculiar structural feature of all canonical (with the exception of glycine) and most noncanonical amino acids is the presence of an asymmetric carbon atom (C^{α}), which should give rise to two different enantiomeric forms. Remarkably, all canonical amino acids are of the L-type. (D-Forms of amino acids can also be synthesized in vivo and are particularly abundant in fungi; however, these amino acids do not have access to the genetic code.) The rise and persistence of homochirality in the living world throughout the entire evolution of life remains one of the greatest puzzles in biology (7, 8). Examples of homochirality at the molecular level also include almost exclusive occurrence of the D-forms of sugars in the nucleotides. Manifestations of homochirality at the macroscopic level range from specific helical patterns of snail shells to chewing motions of cows.

Unlike most synthetic polymers and structural biopolymers (several examples of which will be presented in Chapter 9), peptides and proteins have a very specific sequence of monomer units. Therefore, even though polypeptides can be considered simply as highly functionalized linear polymers constituting a nylon-2 backbone, these functional groups, or *side chains*, are arranged in a highly specific order. All naturally occurring proteins consist of an exact sequence of amino acid residues linked by peptide bonds (Figure 1.1A), which is usually referred to as the *primary structure*. Some amino acids can be modified after translation, for instance, by phosphorylation or glycosylation. Among these modifications, formation of the covalent bonds between two cysteine residues is particularly interesting, since such *disulfide bridges* can stabilize protein geometry, in which the residues that are distant in the primary structure are held in close proximity to each other in three-dimensional space. A highly specific spatial organization of many (but not all) proteins under certain conditions is often referred to as *higher order structure* and is another point of distinction between them (as well as most biological macromolecules) and the synthetic polymers. Although the disulfide bridges are often important contributors to the stability of the higher order structure, correct protein folding does not necessarily require such covalent "stitches." In fact, cysteine is one of the less abundant amino acids, and many proteins lack it altogether. As it turns out, relatively weak noncovalent interactions between the functional groups of the amino acid side chains and the polypeptide backbone are much more important for the highly specific arrangement of the protein in three-dimensional space. The following section provides a brief overview of such interactions.

TABLE 1.2. Chemical Structure and Masses of Some Less Frequently Occurring Natural (Noncanonical) Amino Acids

Symbol	Name	Molecular Formula (Residue)	Chemical Structure	Side Chain Character	Monoisotopic Mass (Residue)	Average Mass (Residue)
Abu	2-Aminobutyric acid	C_4H_7NO	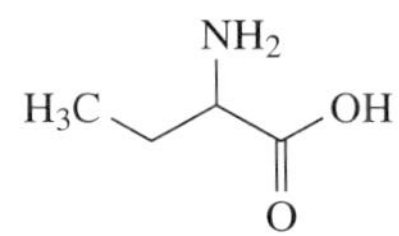	Nonpolar	85.053	85.106
Dha	Dehydroalanine	C_3H_3NO	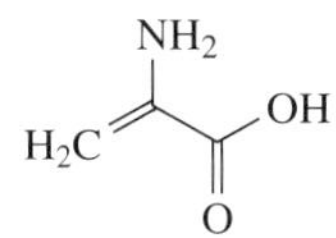	Nonpolar	69.021	69.063
Hse	Homoserine	$C_4H_7NO_2$	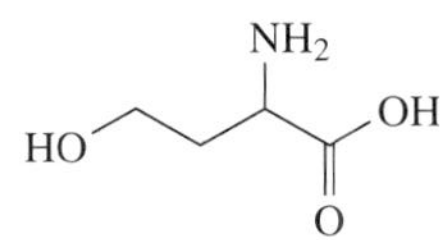	Polar	101.048	101.105
Hyp	Hydroxyproline	$C_5H_7NO_2$	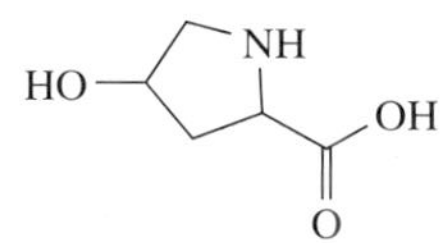	Polar	113.048	113.116
Nle	Norleucine	$C_6H_{11}NO$	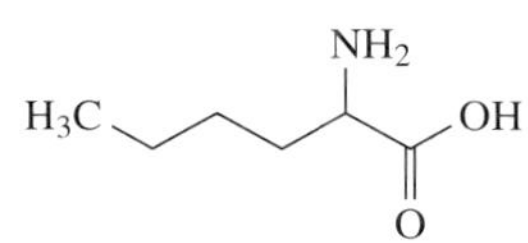	Nonpolar	113.084	113.160

TABLE 1.2 ***(Continued)***

Symbol	Name	Molecular Formula (Residue)	Chemical Structure	Side Chain Character	Monoisotopic Mass (Residue)	Average Mass (Residue)
Orn	Ornithine	$C_5H_{10}N_2O$	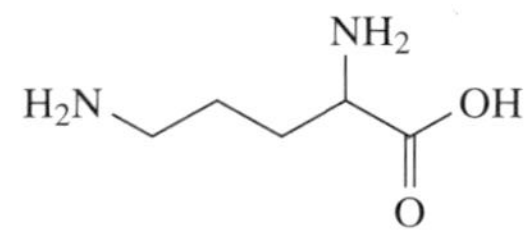	Basic	114.079	114.147
Pyr	Pyroglutamic acid	$C_5H_5NO_2$	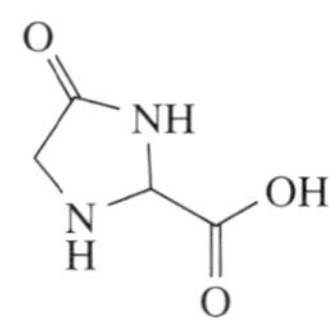	Moderately polar	111.032	111.100
Pyl	Pyrrolysine	$C_{11}H_{16}N_3O_2$ + R (NH_2, OH, or CH_3)	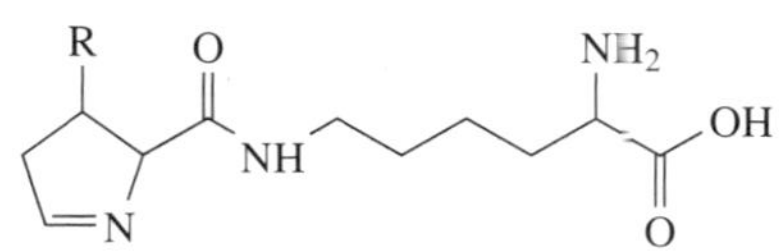	Polar		
Sec	Selenocysteine	C_3H_5NOSe	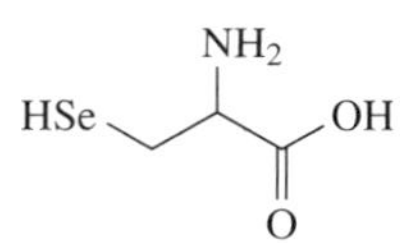	Polar/acidic	144.960 (150.954[a])	150.039

[a] Most abundant.

```
MTTASTSQVR QNYHQDSEAA INAQINLELY ASYVYLSMSY YFDRDDVALK NFAKYFLHQS
HEEREHAEKL MKLQNQRGGR IFLQDIKKPD CDDWESGLNA MECALHLEKN VNQSLLELHK
LATDKNDPHL CDFIETHYLN EQVKAIKELG DHVTNLRKMG APESGLAEYL FDKHTLGDSD NES
```

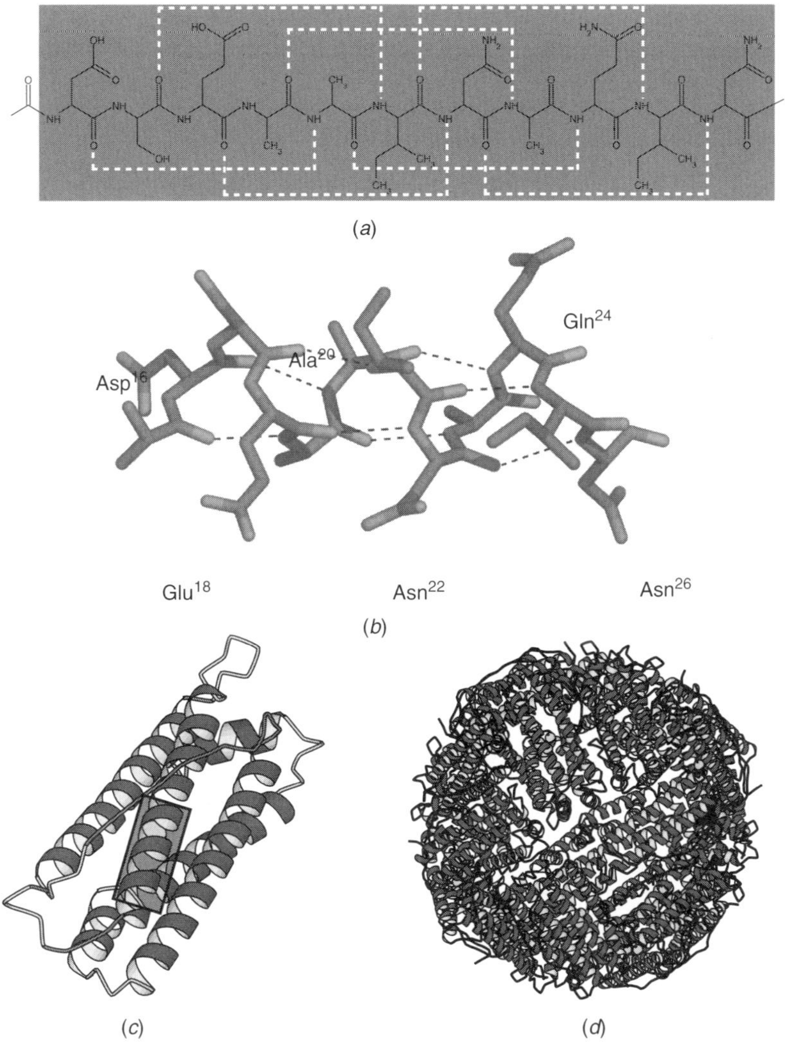

FIGURE 1.1. Hierarchy of structural organization of a protein (H-form of human ferritin). Amino acid sequence determines the primary structure (**A**). Covalent structure of the 11 amino acid residue long segment of the protein ($Glu^{16} \rightarrow Asn^{26}$) is shown in the shaded box. A highly organized network of hydrogen bonds along the polypeptide backbone (shown with dotted lines) gives rise to a secondary structure, an α-helix (**B**). A unique spatial arrangement of the elements of the secondary structure gives rise to the tertiary structure, with the shaded box indicating the position of the ($Glu^{16} \rightarrow Asn^{26}$) segment (**C**). Specific association of several folded polypeptide chains (24 in the case of ferritin) produces the quaternary structure (**D**).

1.2. NONCOVALENT INTERACTIONS AND HIGHER-ORDER STRUCTURE

Just like all chemical forces, the inter- and intramolecular interactions (both covalent and noncovalent) involving biological macromolecules are electrical in nature and can be described generally by the superposition of Coulombic potentials. In practice, however, the noncovalent interactions are subdivided into several categories, each being characterized by a set of unique features.

1.2.1. Electrostatic Interaction

The term *electrostatic interaction* broadly refers to a range of forces exerted among a set of stationary charges and/or dipoles. The interaction between two fixed charges q_1 and q_2 separated by a distance r is given by the Coulomb law:

$$E = \frac{q_1 q_2}{4\pi\varepsilon_0\varepsilon r}, \tag{1-2-1}$$

where ε_0 (defined in SI to have the numerical value of $8.85 \cdot 10^{-12}$ $C^2/N{\cdot}m$) is the absolute permittivity of vacuum and ε is the dielectric constant of the medium. Although the numerical values of the dielectric constants of most homogeneous media are readily available, the use of this concept at the microscopic level is not very straightforward (9, 10). The dielectric constant is a measure of the screening of the electrostatic interaction due to the polarization of the medium, hence the difficulty in defining a single constant for a protein, where such screening depends on the exact location of the charges, their environment, and so on. Although in some cases the values of the "effective" dielectric constants for specific protein systems can be estimated based on the experimental measurements of the electrostatic interactions, such an approach has been disfavored by many for a long time (11). In this book we will follow the example set by Daune (12) and will write all expressions with $\varepsilon = 1$.

Interaction between a charge q and a permanent dipole p separated by distance r is given by

$$E = -\frac{qp \cdot \cos\theta}{4\pi\varepsilon_0 r^2}, \tag{1-2-2}$$

where θ is the angle between the direction of the dipole and the vector connecting it with the charge q. If the dipole is not fixed directionally, it will align itself to minimize the energy (1-2-2); that is, $\theta = 0$. However, if such energy is small compared to thermal energy, Brownian motion will result in the averaging of all values of θ with only a small preference for those that minimize the electrostatic energy, resulting in a much weaker overall interaction:

$$E = -\frac{q^2 p^2}{(4\pi\varepsilon_0)^2 \cdot 3k_B T r^4}, \tag{1-2-3}$$

Interaction between two dipoles p_1 and p_2 separated by a distance r in this approximation is given by

$$E = -\frac{2p_1^2 p_2^2}{(4\pi\varepsilon_0)^2 \cdot 3k_B T r^6}, \tag{1-2-4}$$

while the interaction between the two fixed dipoles will be significantly stronger ($\sim 1/r^3$).

Polarization of a molecule can also be viewed in terms of electrostatic interaction using a concept of *induced dipoles* (12). Such interaction is, of course, always an attractive force, which is inversely proportional to r^4 (for a charge–induced dipole interaction) or r^6 (for a permanent dipole–induced dipole interaction). Finally, interaction between two polarizable molecules can be described in terms of a weak induced dipole–induced dipole interaction.

1.2.2. Hydrogen Bonding

The electrostatic interactions considered in the preceding sections can be treated using classical physics. *Hydrogen bonding* is an example of a specific noncovalent interaction that cannot be treated within the framework of classical electrostatics. It refers to an interaction occurring between a proton donor group (e.g., —OH, —NH_3^+) and a proton acceptor atom that has an unshared pair of electrons. Although hydrogen bond formation (e.g., R=Ö: ••• H—NR_2) may look like a simple electrostatic attraction of the permanent dipole–induced dipole type, the actual interaction is more complex and involves charge transfer within the proton donor–acceptor complex. The accurate description of such exchange interaction requires the use of sophisticated apparatus of quantum mechanics.

The importance of hydrogen bonding as a major determinant and a stabilizing factor for the higher-order structure of proteins was recognized nearly seventy years ago by Mirsky and Pauling, who wrote in 1936: "the [native protein] molecule consists of one polypeptide chain which continues without interruption throughout the molecule... this chain is folded into a uniquely defined configuration, in which it is held by hydrogen bonds between the peptide nitrogen and oxygen atoms..." (13). Considerations of the spatial arrangements that maximize the amount of hydrogen bonds within a polypeptide chain later led Pauling to the prediction of the existence of the α-helix, one of the most commonly occurring local motifs of higher order structure in proteins (14). Hydrogen bonds can be formed not only within the macromolecule itself, but also between biopolymers and water molecules (the latter act as both proton donors and acceptors). Hydrogen bonding is also central for understanding the physical properties of water, as well as other protic solvents.

1.2.3. Steric Clashes and Allowed Conformations of the Peptide Backbone: Secondary Structure

Both electrostatic and hydrogen bonding interactions within a flexible macromolecule would favor three-dimensional arrangements of its atoms that minimize

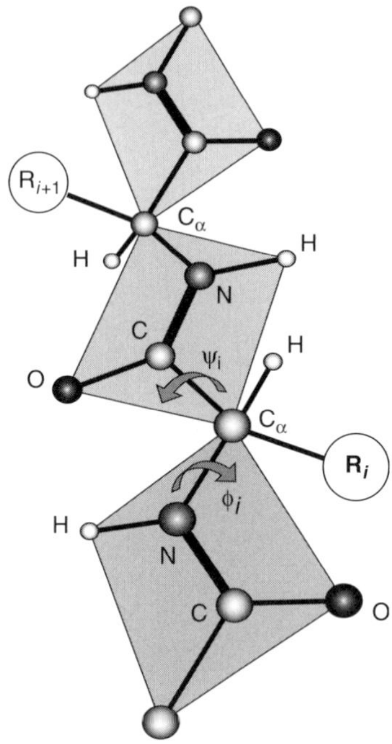

FIGURE 1.2. Peptide bond and the degrees of freedom determining the polypeptide backbone conformation.

the overall potential energy. However, there are two fundamental restrictions that limit the conformational freedom of the macromolecule. One is, of course, the limitation imposed by covalent bonding. The second limitation is steric hindrance, which also restricts the volume of conformational space available to the biopolymer. In this section we consider the limits imposed by steric clashes on the conformational freedom of the polypeptide backbone.

The peptide amide bond is represented in Figure 1.1A as a single bond (i.e., C—N); however, it actually has a partial double bond character in a polypeptide chain. The double bond character of the C—N linkage, as well as the strong preference for the *trans* configuration of the amide hydrogen and carbonyl oxygen atoms,* result in four atoms lying in one plane. A slight deformation of this configuration does occur in many cases, but it is rather insignificant. Figure 1.2 shows two successive planes linked by a C^α atom of the ith amino acid residue. The two degrees of freedom at this junction are usually referred to as ϕ_i and ψ_i angles and the backbone conformation of the polypeptide composed of n amino acid residues can be described using $n - 1$ parameters (pairs of ϕ_i and ψ_i). The

*The exception to this rule is offered by proline, which, as an imino acid, has its side chain also bonded to the nitrogen atom. Thus, the cis- and trans-forms are almost isoenergetic, leading to the possibility of *cis*-Xaa-Pro bonds (where Xaa is any amino acid residue) in folded proteins, and statistically at the level of 5–30% in unstructured polypeptides.

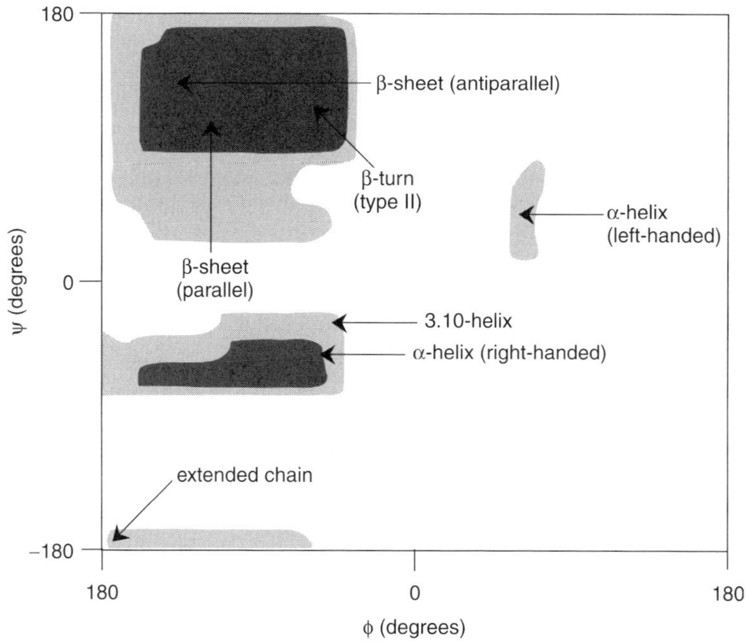

FIGURE 1.3. A schematic representation of the Ramachandran plot.

steric restrictions limit the conformational volume accessible to polypeptides, which is usually represented graphically on the (ϕ, ψ) plane using *conformational maps* or *Ramachandran plots* (15). An example of such a diagram shown in Figure 1.3 clearly indicates that only a very limited number of configurations of the polypeptide backbone are allowed sterically.

Several regions within the accessible conformational volume are of particular interest, since they represent the structures that are stabilized by highly organized networks of hydrogen bonds. The α-helix is one such structure, where the carbonyl oxygen atom of the ith residue is hydrogen bonded to the amide of the $(i+4)$th residue (Figure 1.1B). This local motif, or spatial arrangement of a segment of the polypeptide backbone, is an example of a *secondary structure*, which is considered the first stage of macromolecular organization to form higher-order structure. Another commonly occurring element of the secondary structure is located within a larger "island" of sterically allowed conformations on the Ramachandran plot. Such conformations [upper left corner on the (ϕ, ψ) plane in Figure 1.3] are rather close to the fully extended configuration of the chain and, therefore, cannot be stabilized by local hydrogen bonds. Nevertheless, formation of strong stabilizing networks of hydrogen bonds becomes possible if two strands are placed parallel or antiparallel to each other, forming the so-called *β-pleated sheets*.

The third important local structural motif is the *turn*, which causes a change in the chain direction within a folded protein. Whereas loops are generally flexible sections of chain, turn structures tend to be more rigid and are stabilized by hydrogen bonding or specific side chain interactions. These turn structures can be highly important particularly in antiparallel β-sheet structures, where a complete reversal of the chain is required to enable packing of adjacent strands. Other less frequently occurring elements of secondary structure (such as 3.10 or π helices) can also be identified on the Ramachandran plot.

So far, we have largely ignored the contributions of the amino acid side chains to the protein conformation. One obvious consequence of the existence of a variety of different side chains is the dependence of the Ramachandran plots for each particular (ϕ_i, ψ_i) pair on the identity of the ith amino acid residue. For example, a significantly larger conformational volume is available to glycine, as compared to amino acid residues with bulky side chains. Furthermore, different side chains placed at "strategic" locations may exert a significant influence on the stability of the secondary structural elements. We will illustrate this point using the α-helix as an example. All hydrogen bonds in a helix are nearly parallel to each other (and to the axis of the helix). This highly ordered pattern of hydrogen bonding results in a noticeable dipole moment, with the N-terminal end of the helix being a positive pole. Obviously, the presence of a positively charged residue at or near the N-terminal end of the helix will destabilize it due to the unfavorable charge–permanent dipole interaction (1-2-2). On the other hand, the presence of a negatively charged residue will be energetically favorable and increase the stability of the helix. Likewise, the presence of charged residues at or near the C-terminal end of the helix will also have a significant influence on the stability of this element of secondary structure.

It is important to note, however, that uncharged side chains may also be very important determinants of the higher-order structure of proteins and polypeptides due to the *hydrophobic interactions*. These will be considered in the following section.

1.2.4. Solvent–Solute Interactions, Hydrophobic Effect, Side Chain Packing, and Tertiary Structure

The term *hydrophobic effect* (16–19) refers to a tendency of nonpolar compounds (such as apolar amino acid side chains, Table 1.1) to be sequestered from the polar solutions (such as aqueous solution) to an organic phase. Such behavior is ubiquitous in nature and has been observed and described at least two millennia ago, although the term hydrophobic was coined only in 1915 (18). The initial view of the hydrophobic interaction was rather simplistic and implied attraction between the like media (e.g., oil–oil attraction). A very different view (which is now commonly accepted) was proposed in the mid-1930s by Hartley, who suggested that the nonpolar species are excluded from polar solvent because of their inability to compete with the strong interaction between the polar molecules themselves (20). In Tanford's words, "antipathy between hydrocarbon and water rests on the strong attraction of water for itself" (21). An intriguing aspect of the hydrophobic interaction is that the placement of a hydrocarbon molecule in water may be enthalpically favorable. This fact was the basis for a widespread skepticism over the concept of hydrophobic interactions, although such views did not prevail (22). It is now understood that the solvent—solute affinity is determined by the free energy (not the enthalpy alone), and it is the unfavorable free energy that leads to the observed disaffinity of water and nonpolar solutes.

Various microscopic explanations of the hydrophobic effect are usually based on the *frozen water patches* or *microscopic iceberg* model proposed originally by Frank and Evans (23), who suggested that placing a nonpolar solute in water creates a loose "cage" of first-shell water molecules around it. The creation of such a cage has a significant entropic price (due to the "forced" water ordering), hence the overall unfavorable free energy (despite favorable enthalpic contribution). A reader interested in a more detailed account of the physics of hydrophobicity and related phenomena is referred to a recent tutorial by Southall, Dill, and Haymet (18).

Although the initial work on the hydrophobic effect was focused on hydrocarbons, its main results and conclusions can easily be extended to nonpolar side chains of polypeptides and proteins, which are buried in a *hydrophobic core* of a folded or collapsed protein molecule in order to eliminate or at least minimize any contacts with the polar solvent. A very interesting historical account on the elucidation of the nature of hydrophobic interaction and its role in protein folding can be found in an excellent review by Tanford (24). While the hydrophobic side chains are generally more stable if sequestered away from solvent in the central core of the protein,* the hydrophilic residues usually decorate the solvent-exposed surface of the protein. This is achieved by combining the elements of secondary structure (α-helices, β-sheets, and turns) in a unique three-dimensional arrangement, or *tertiary structure*. It is the tertiary structure that affords proteins their unique biological function, whether it be purely structural, the precise spatial organization of side chains to effect catalysis of a reaction, presentation of a surface or loop for signaling or inhibition, creating a cavity or groove to bind ligand, or any of the other vast range of functions that proteins can perform.

Hydrophobic interaction is, of course, not the only driving force giving rise to a unique tertiary structure. Additional stabilization is afforded by close proximity of acidic and basic residues, which is frequently observed in the folded structure, enabling the formation of salt bridges. These can be viewed as charge–charge interactions (1-2-1). We have already mentioned that certain elements of secondary structure have intrinsic (permanent) dipole moments. Favorable arrangement of such dipoles with respect to one another (e.g., in the so-called helical bundles) may also become a stabilizing factor (1-2-2) in addition to the hydrophobic interaction. It is probably worth mentioning that in the vast majority of proteins the interactions stabilizing the tertiary structure are cooperative. In other words, significant enthalpic gains are achieved only if several segments of the protein are in close proximity and interact with each other. All such factors have been evolutionarily optimized for each protein but the important thing to realize is

*Proteins tend to be very well-packed molecules so the side chain atoms sequestered from the solvent must come into close contact with each other, hence the term *hydrophobic packing*.

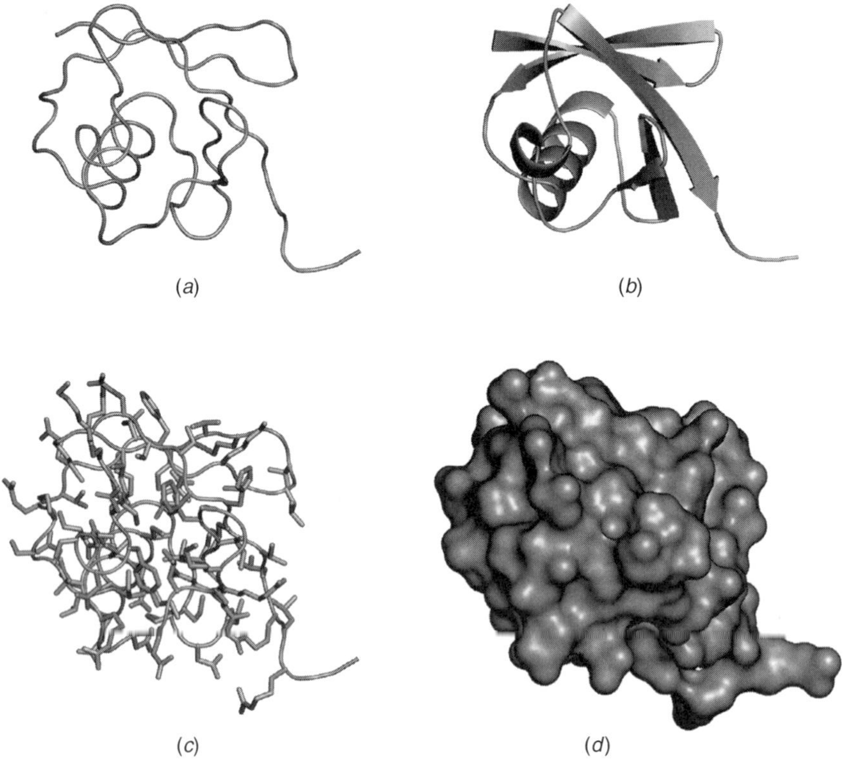

FIGURE 1.4. Different representations of the higher order structure of natively folded proteins.

that any one natural protein sequence has only a single most stable conformation, and the genetically encoded primary sequence alone is necessary and sufficient to define the final folded structure of the protein (Figure 1.4).

Many proteins adopt similar common structural motifs resulting from combinations of secondary structure elements such as the alternating $\beta\alpha\beta$ structure, 4-helix bundles, or β-barrels. As more and more protein structures are solved,* the number of protein architectures increases, although it has been predicted there are only a limited number of fold motifs (25–29). Such conclusion is based on the observations that (i) topological arrangements of the elements of secondary structure are highly skewed by favoring very few common connectivities and (ii) folds can accommodate unrelated sequences [as a general rule, structure is more robust than sequence (30, 31)]. Therefore, the fold universe appears to be dominated by a relatively small number of giant attractors,† each accommodating

*So far, structures have been solved for only about 1% of approximately 1,000,000 proteins of known sequences.

†The total number of folds is estimated to be less than 2000, of which 500 have already been characterized.

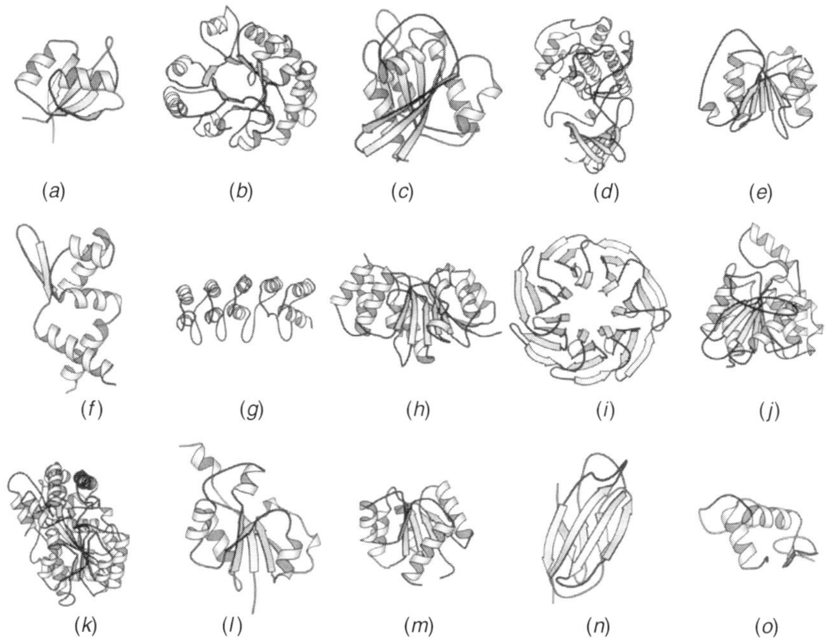

FIGURE 1.5. The 15 most populated folds selected on the basis of a structural annotation of proteins from completely sequenced genomes of 20 bacteria, five Archaea, and three eukaryotes. From left to right and top to bottom: ferredoxin-like (4.45%) (**A**), TIM-barrel (3.94%) (**B**), P-loop containing nucleotide triphosphate hydrolase (3.71%) (**C**), protein kinases (PK) catalytic domain (3.14%) (**D**), NAD(P)-binding Rossmann-fold domains (2.80%) (**E**), DNA:RNA-binding 3-helical bundle (2.60%) (**F**), α-α superhelix (1.95%) (**G**), *S*-adenosyl-L-methionine-dependent methyltransferase (1.92%) (**H**), 7-bladed β-propeller (1.85%) (**I**), α/β-hydrolases (1.84%) (**J**), PLP-dependent transferase (1.61%) (**K**), adenine nucleotide α-hydrolase (1.59%) (**L**), flavodoxin-like (1.49%) (**M**), immunoglobulin-like β-sandwich (1.38%) (**N**), and glucocorticoid receptor-like (0.97%) (**O**). The values in parentheses are the percentages of annotated proteins adopting the respective folds. Reproduced with permission from (32). © 2001 Springer-Verlag.

a large number of unrelated sequences. Figure 1.5 represents the 15 most populated folds selected on the basis of a structural annotation of proteins from completely sequenced genomes of 20 bacteria, five Archaea, and three eukaryotes (32).

The existence of a "finite set of natural forms" in the protein world has inspired some to invoke the notion of Platonic forms that are "determined by natural law" (33), a suggestion that seems more poetic than explanatory. What has become clear though is that very similar tertiary structures can be adopted by quite dissimilar primary sequences (32). Protein primary sequences can be aligned and regions identified that are identical or homologous (meaning the chemical nature of the amino acid side chain is similar—e.g., polar, nonpolar, acidic, basic). However, even sequences with quite low homology can have a very similar overall fold, depending on the tertiary interactions that stabilize them. Although tertiary structure is sometimes viewed as the highest level of spatial organization of the single-chain (i.e., monomeric) proteins, an even higher level of organization is often seen in larger proteins (generally more than 150 amino acid residues). Such proteins form clearly recognizable *domains*, which tend to be contiguous in primary structure and often enjoy certain autonomy from one another.

1.2.5. Intermolecular Interactions and Association: Quaternary Structure

Above and beyond the folding of monomeric chains, many protein chains can also assemble to form multisubunit complexes, ranging from relatively simple homodimers (such as hemoglobin molecules of primitive vertebrates, e.g. lamprey, hagfish) to large homo-oligomers (such as an iron-storage protein ferritin, assembled of 24 identical subunits) to indeed assemblies of different proteins (such as ribosomes). Such assemblies are usually considered to be the highest level of molecular organization at the microscopic level, which is usually referred to as *quaternary structure*. Although covalent links are sometimes formed between the monomeric constituents of a multimeric protein assembly (e.g., in a form of the disulfide bonds), the noncovalent interactions (discussed in the preceding sections) are usually much more important players.

The archetype of quaternary structure is the mammalian hemoglobin, which is a noncovalent tetramer ($\alpha_2\beta_2$) consisting of two pairs of similar monomeric chains (α- and β-globins). The arrangement of the monomers in the tetramer (which is in fact a dimer composed of two heterodimers) is crucial for the function of hemoglobin as an oxygen transporter. A tetramer composed of four identical globins (β_4) can also be formed and is indeed present in the blood of people suffering from some forms of thalassemia. However, this homotetramer (termed hemoglobin H, or Hb H), lacks the most important characteristic of the "normal" hemoglobin (Hb A), namely, high cooperativity of oxygen binding. We will revisit quaternary structure formation and functional characteristics in the closing chapter of this book.

1.3. THE PROTEIN FOLDING PROBLEM

1.3.1. What Is Protein Folding?

Polymers can adopt different conformations in solution depending on functionality and the interaction with neighboring chains, other parts of the same chain, and the bulk solvent. However, almost all synthetic copolymers (polymers consisting of more than one type of repeat units) consist of a range of different length chains and, in many cases, a nonspecific arrangement of monomer groups. On the other hand, the primary structure of a given protein is always the same, creating a homogeneous and highly monodisperse copolymer. Protein sequences are generally optimized to prevent nonspecific intermolecular interactions and individual molecules will fold to adopt a unique stable conformation governed solely by the primary sequence of amino acids. The ability of proteins to attain a unique higher-order structure sets them apart from most random copolymers. Most proteins can fold reversibly in vitro, without being aided by any sophisticated cellular machinery,* suggesting that the folding mechanism is solely determined by the primary structure of the protein, as well as the nature of the solvent. Folded proteins may remain stable indefinitely in most cases, suggesting that the native structures represent the global free energy minima among all kinetically accessible states (34).

Two classic puzzles are usually considered in connection with protein folding: (i) the *Blind Watchmaker's paradox* and (ii) the *Levinthal paradox*. The former is named after a classic book by Dawkins (35), an outspoken critic of the *intelligent design* concept (36). It states that biological (function-competent) proteins could not have originated from random sequences. The Levinthal paradox states that the folded state of a protein cannot be found by a random search (37). Both paradoxes have been historically framed in terms of a random search through vast spaces (sequence space in the *Blind Watchmaker's paradox* and conformational space in the *Levinthal paradox*), and the vastness of the searched spaces is equated with physical impossibility. Both paradoxes are elegantly solved within the framework of the *energy landscape* description of the folding process by invoking the notion of a guided search (38). The concept of protein energy landscapes and its relevance to the protein folding problem will be considered in some detail in the following sections of this chapter.

1.3.2. Why Is Protein Folding So Important

In the postgenomic era, structure determination has become of paramount importance since it leads to a three-dimensional picture of each gene product and, in many cases, gives hints as to the function of the protein. However, the static structure only represents the end point of the chemical reaction of protein folding. Polypeptide chains are translated as extended structures from RNA on the ribosome of cells, but how does this unstructured sequence fold into its final biologically active structure? Are specific local structures present in the newly translated chain? Is there a specific pathway or reaction coordinate of protein folding? The principles that govern the transitions of biopolymers from totally unstructured to highly ordered states (which often include several subunits assembled in a highly organized fashion) remain one of the greatest mysteries in structural biology (39, 40). Deciphering this code is key to understanding a variety of biological processes at the molecular level (e.g., recognition, transport, signaling, and biosynthesis), since the specificity of biological activity in proteins (as well as other biomolecules) is dictated by their higher-order structure.

*In the cell a number of helper proteins, or chaperones, assist efficient folding in the crowded cytosolic environment. However, it is generally agreed that chaperones do not in themselves direct folding, instead acting as gatekeepers to prevent misfolding and aggregation.

Aside from the obvious academic interest to biophysicists in discovering exactly how these biological machines work, there are many more practical implications. Only if we understand all of the processes that are involved in producing a biologically active protein can we hope to harness this power by designing proteins with specific functions. It may already be possible computationally to model an ideal binding site or even optimal arrangement of side chains to catalyze a chemical reaction, but without a thorough knowledge of how this site can be placed into an intact protein molecule, we cannot take advantage of the cellular machinery for the design of therapeutic protein drugs, or even molecules that can catalyze otherwise difficult chemical reactions. For instance, there are many enzymes in nature that catalyze reactions with extremely high specificity and efficiency, whereas chemists lag far behind. Hydrogenase enzymes, for example, catalyze the reduction of protons to produce diatomic hydrogen, a reaction that in a laboratory environment requires application of harsh reactants at elevated temperature or pressure, but which within the catalytic center of the protein occurs at physiological temperatures and with remarkably small energy requirements. Obviously, biological organisms have had a much longer time to optimize these processes relative to the chemical industry. If one can understand in detail the roles of each residue in a protein chain for both the folding and dynamics of the molecule, then the possibilities for protein engineering are boundless. Interestingly, artificial sequences quite often lead to proteins that either do not fold at all or are only marginally stable. This clearly demonstrates the extremely fine balance of forces present, which can be destroyed by just a single amino acid residue substitution, deletion, or insertion.

Another important aspect of understanding protein folding is to find ways of preventing the process from going awry (41–44). An ever-increasing number of pathological conditions that result from misfolding of proteins in the cell are being identified (45–49). Amyloid plaques actually result from the undesirable formation of quaternary structure when a normally monomeric peptide folds incorrectly and self-assembles to form long proteinaceous fibers. Similarly, other proteins that are not correctly folded may not present the correct binding surface for interaction with their physiological partners. Thus, not only correct folding but also the correct assembly of proteins are key to their correct biological function. Even relatively few mutations within a protein sequence may prevent folding to the native structure and hence prove pathological. In other cases, mutation can reduce the efficiency of folding or favor an alternative mode of folding that leads to aggregation and deposition of insoluble amyloid plaques within cells. We will consider the issues related to misfolding and aggregation in the following sections of this chapter.

Finally, one more fundamental problem related to protein folding that has become a focal point of extensive research efforts is the prediction of the native structure and function of a protein based on its primary structure. Since the sequence of each natural protein effectively encodes a single tertiary structure, prediction of the latter is, in essence, a global optimization problem, which is similar to one encountered in crystallography and the physics of clusters (50). The complication that arises when such a global optimization methodology is applied to determine the position of the global energy minimum for a protein is the vastness of the system that precludes calculations based on the first principles. So far, the most successful methods of structure prediction rely on the identification of a *template protein* of known structure, whose sequence is highly homologous to that of the protein in question. If no template structure can be identified, de novo prediction methods can be used, although it remains to be seen if such methods can predict structures to a resolution useful for biochemical applications (51). Prediction of protein function based on its sequence and structure is an even more challenging task, since homologous proteins often have different functions (52).

1.3.3. What Is the Natively Folded Protein and How Do We Define a Protein Conformation?

Before proceeding further with a description of protein folding, it would be useful to define some terms commonly used in the field in order to avoid confusion. The *native state* of a protein is defined as the fully folded biologically active form of the molecule. This has generally been considered as a single state with a well-defined tertiary structure, as determined by crystallography or NMR spectroscopy. More recently, researchers have come to appreciate the importance of dynamics within the protein structure. Even the native state is not a static single structure but may in fact, depending on the protein, have small or even large degrees of flexibility that are important for its physiological function.

Unfortunately, the use of the term *protein conformation* in literature has become rather inconsistent and often results in confusion. Historically, protein conformation referred to a specific "three-dimensional arrangement of its constituent atoms" (53). This definition, however, is rather narrow, since it does not reflect adequately the dynamic nature of proteins. One particularly annoying complication that arises when conformation is defined using only microscopic terms (e.g., atomic coordinates) is due to the fact that a majority of proteins have segments lacking any stable structure even under native conditions. These could be either the terminal segments that are often invisible in the X-ray structures or flexible loops whose conformational freedom is often required for a variety of functions ranging from recognition to catalysis. In general, it is more than likely that any two randomly selected natively folded protein molecules will *not* have identical sets of atomic coordinates and, as a result, will not be assigned to one conformation if the geometry-based definition is strictly applied. Therefore, it seems that the thermodynamics-based definition of a protein conformation is a better choice. Throughout this book, we will refer to the protein conformation not as a specific microstate, but as a *macrostate*, which can be envisioned as a collection of microstates separated from each other by low-energy barriers ($\leq k_B T$). In other words, if one microstate is accessible from another at room temperature, we will consider them as belonging to one conformation, even if there is a substantial difference in their configurations. According to this view, a protein conformation is a continuous subset of the conformational space (i.e., a continuum of well-defined configurations) that is accessible to a protein confined to a certain local minimum. Although such definition is not without its own problems,* its utility becomes obvious when we consider non-native protein conformations.

1.3.4. What Are Non-native Protein Conformations? Random Coils, Molten Globules, and Folding Intermediates

In the case of unfolded proteins, which are assumed to be completely nonrigid polypeptide chains, the *random coil* (54), we must consider the ensemble of molecules displaying an impressive variety of configurations (Figure 1.6). In a truly random coil, as might be the case for a synthetic polymer with identical monomer units in a good solvent, there may well be no conformational preferences for the chain. However, proteins are decorated with side chains of different chemical nature along their length, such that in water or even a chemical denaturant one might expect there to be local preferences due to hydrophilic or hydrophobic interactions, and indeed steric effects. Thus, for a number of proteins studied in solution, some persistent local and nonlocal conformational effects have been detected, indicating that an unfolded protein is not in fact a truly random coil. On the other hand, the enthalpy of these interactions is very small in comparison to the entropy of the flexible chain, so the overall free energy of each of these conformers will be very similar. On a free energy surface, these would be represented as shallow wells in the generally flat surface of unfolded state free energy.

The relative position of a local energy minimum with respect to the native state gives rise to a further set of descriptions of intermediate states. As a protein folds it may sample stabilizing conformations that contain persistent structure constituting a local free energy minimum. At the earliest stages of folding there may only be a few interactions that may be very transient—these are termed *early intermediates*. By contrast, species may accumulate further along in the folding process that contain a large although incomplete number of native-like contacts. These are referred to as *late intermediates*, implying that they should form toward the end of the kinetic folding process. There is also the possibility that these local minima arise from stabilizing contacts that are not present in the native protein, and in fact need to be broken apart before the molecule can productively fold. These *off-pathway intermediates* may also arise from intermolecular interactions between folding chains and can lead to nonproductive aggregation that prevents further folding.

The above intermediate states form during folding in the "forward" direction from the unfolded to the native state and, since they are only partially stable, generally do not accumulate sufficiently to be detected other than transiently. It

*For example, this definition is temperature dependent. Indeed, if any two local minima are separated by a high-energy barrier ($>k_B T$), the interconversion between these two states does not occur readily at room temperature T, and these two states should be viewed as two different conformations. However, raising the temperature significantly above the room temperature T will eventually make the passage over this barrier possible, leading to a merger of the two microstates to a single conformation.

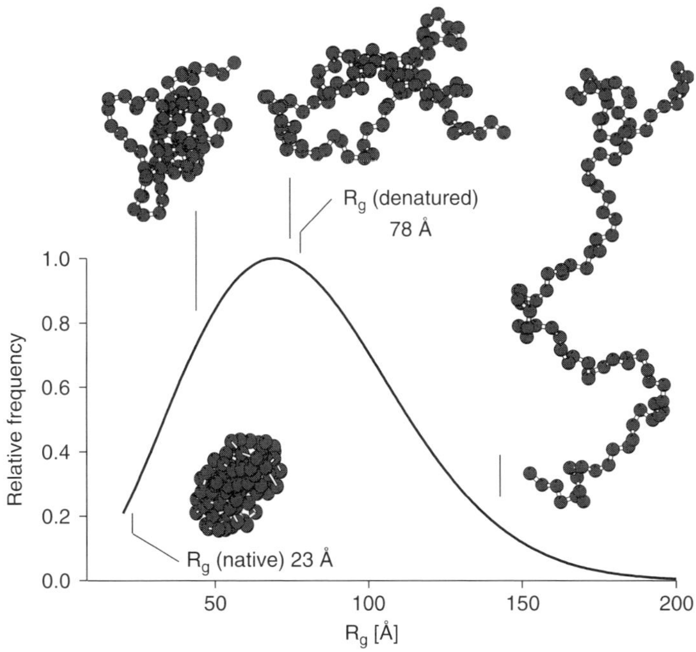

FIGURE 1.6. Representative configurations of a random coil (a freely joined chain of 100 hard spheres) and the distribution of its radius of gyration R_g. The R_g values of a model protein phosphoglycerate kinase are indicated for comparison. Adapted with permission from (54). © 1996 Elsevier.

is also possible that such intermediates may form during the reverse process (i.e. protein unfolding), allowing them to be studied by other methods. Unfortunately, the conditions for unfolding (e.g., chemical denaturant, low pH, high temperature) are generally so harsh that once the stabilizing interactions in the native state have been removed, the unfolding process occurs with high cooperativity and without accumulation of intermediates. However, under mildly denaturing conditions, partially folded states have been detected at equilibrium for a number of proteins, and these have been termed *molten globules* (55). The original definition of the molten globule state was quite specific: a structural state that has significant secondary structure but with no fixed tertiary interactions. There are various biophysical tests for this, such as the ability of the protein to bind hydrophobic dyes, consistent with a significant amount of exposed hydrophobic surface area, as would be expected for a partially folded state. The definition has become somewhat relaxed to include many other partially folded ensembles observed, kinetically or at equilibrium, which almost fit the definition. What is clear is that the molten globule itself is a much more dynamic structure than previously thought. Several new concepts have recently been introduced to reflect the structural diversity and the dynamic character of the molten globule state, such as "a precursor of the molten globule" and "a highly structured molten globule" (56).

One common question that arises is whether the equilibrium molten globule intermediate is actually the same species as that detected in the folding pathway of proteins. Thermodynamically, there is nothing to suggest they should be, since the equilibrium by definition is independent of the pathway (57, 58). However, comparisons of the characteristics of transient intermediates with the corresponding equilibrium partially folded state have concluded that the similarities are very close, at least for the proteins studied (59–62). Also, a number of states transiently populated by the native state ensemble under mildly destabilizing conditions have been shown to have similarities to folding intermediates. Thus, it seems likely that, at least in the later stages of folding, there is indeed some kind of folding/unfolding pathway with specific intermediate states visited in both the folding and unfolding directions.

1.3.5. Protein Folding Pathways

In the preceding section we began to use the term *folding pathway*, which is understood to be a series of structural changes leading from the fully denatured state of the protein to its native conformation. Introduction of the concept of a folding pathway resolves the Levinthal paradox mentioned earlier by suggesting that the folding process is a directed process involving conformational biases, rather than a merely random conformational search. Despite the vast number of degrees of freedom in macromolecules, the number of folding pathways was initially believed to be rather limited (63). A general scheme of protein folding within this paradigm is presented in terms of rapid equilibration of unfolded protein molecules between different conformations prior to complete refolding. Such equilibrium favors certain compact conformations that have lower free energies than other unfolded conformations, and some of these favored conformations are important for efficient folding. The rate-limiting step is thought to occur late in the pathway and to involve a high-energy, distorted form of the native conformation. The latter is a single transition state through which essentially all molecules refold (64).

The classic folding pathway paradigm specifically states that "proteins are not assembled via a large number of independent pathways, nor is folding initiated by a nucleation event in the unfolded protein followed by rapid growth of the folded structure" (64). Nevertheless, a large body of experimental evidence now suggests the existence of large number of folding routes. Furthermore, over the past several years it has become clear that the length of the polypeptide chain is an important factor in determining mechanistic details. Smaller proteins (<100 residues) appear to prefer a nucleation-type mechanism that does not involve any specific metastable intermediate species (65). On the other hand, for a number of larger proteins, intermediate states with specific regions of the early-formed stable structure have been established. If an intermediate is detected, then this argues strongly for a specific protein folding pathway, but in the protein lysozyme, for instance, parallel folding pathways were found, suggesting multiple possible trajectories. In smaller proteins no observable intermediates accumulate, which might indicate a less directed folding mechanism, or indeed that partially formed structures are too labile to be detected. Based on these observations, a modified view of a folding pathway invokes several common stages of folding, consistent with the progressive development of structure and stability through an ever-slowing set of reactions (66).

Three major models have arisen to attempt to explain this narrowing of the conformational search process (67). The simple *framework* model suggests that secondary structure elements would first form based on their sequence-intrinsic propensities, followed by collision of these preformed structures to form tertiary interactions. Alternatively, *nucleation* of short regions of sequence to form transient secondary structure could act as a template on which adjacent parts of the chain would condense and propagate structure. The third mechanism, *hydrophobic collapse*, calls for the hydrophobic residues to conglomerate nonspecifically to minimize solvent exposure, followed by rearrangement to the final native structure. The actual mechanism of protein folding probably involves some or all of these processes and there is evidence for each mechanism from folding studies of different proteins.

The controversy with regard to whether protein folding follows a specific pathway or whether each molecule follows a completely different trajectory to achieve its final folded state has been elegantly resolved in the *new view of protein folding* (68–70). The centerpiece of this theory is the concept of a protein energy landscape in the conformational space, which is discussed in the following section.

1.4. PROTEIN ENERGY LANDSCAPES AND THE FOLDING PROBLEM

1.4.1. Protein Conformational Ensembles and Energy Landscapes: Enthalpic and Entropic Considerations

Classically, a simple chemical reaction is considered to proceed along a reaction coordinate, in which chemical bonds are formed or broken in a well-defined manner, with a transition state at the highest point on the energy profile where bond breaking and formation are occurring, and possibly detectable metastable intermediate structures at local free energy minima. Transition state theory can be applied to relate reaction rates to the heights of the various free energy barriers along the reaction coordinate. Macromolecules are much more complex, however, and the folding of a protein involves the formation of a large number of interactions that may either be local in nature or indeed involve regions that are quite distant in the polypeptide sequence. Nevertheless, for many years the protein folding reaction was assumed to occur via a similar sequential pathway, perhaps involving a number of intermediate species along the way, but for each unfolded molecule the mechanism of folding to the native state was identical.

Significant advances in theory during the past decade have changed our understanding of the basic principles that govern the protein folding process and have offered an elegant way to resolve the Levinthal paradox (69, 71–75). In the case of small organic molecules in a reaction, there are only a small number of conformations available to the reactant species, but for an unfolded protein the conformational space sampled by the unstructured chain is vast by comparison, even for a relatively small protein. Thus, it might seem difficult to imagine the chain becoming oriented in such a way as to proceed to fold via a single pathway, and the process would surely be extremely inefficient. Realization that folding may proceed through multiple parallel pathways, rather than a single route, has led to the introduction of the concepts of protein energy landscapes (or folding funnels), a cornerstone of the "new view" of protein folding.

Protein folding can be viewed much like any other chemical reaction, which may be represented in three dimensions by a conformational energy surface: the trajectories on these surfaces lead from reactants (unstructured states) to products (the native state). Because entropy plays a much more significant role in protein folding reactions, it is necessary to consider the free energy, rather than simply potential energy (76). The enthalpic gain of forming hydrogen bonds and making favorable hydrophobic or hydrophilic contacts is compensated by a significant loss of entropy as the chain becomes more and more conformationally restricted. For simple molecules the entropic term is generally far less significant, but in the case of a folding protein the overall free energy of stabilization in the folded protein may be only a few kcal/mole, being the very small difference between large ΔH (formation of stabilizing interactions) and $T\,\Delta S$ (loss of entropy upon folding to the native state) terms. The conformational entropy loss for a protein, which continues to adopt a better defined three-dimensional structure, is often defined on the "per residue" basis. It can be estimated by using only the backbone entropy [the entropy loss due to side chain packing is significantly less (74)]:

$$\Delta s = s_{\mathrm{u}} - s_{\mathrm{f}} \approx k_{\mathrm{B}} \cdot \ln\left(\frac{\Omega_{\mathrm{u}}}{\Omega_{\mathrm{f}}}\right), \tag{1-4-1}$$

where $s_{\mathrm{u,f}}$ and $\Omega_{\mathrm{u,f}}$ represent the entropies and the numbers of microstates per residue in the unfolded and natively folded forms of the protein, respectively. Estimations of Δs for small proteins at room temperature give an entropy loss on the order of tens of J/(mol·K·residue) (74). To ensure fast folding despite the unfavorable entropy change, the corresponding free energy surface (or energy landscape) must have the form of a multidimensional "funnel" (Figure 1.7), where the vertical axis (the depth of the funnel) represents the number of native contacts made (Q) or the relative free energy of the conformational space. The horizontal axis corresponds to the conformational entropy of the system. In this representation it becomes clear that as the folding chain makes more native contacts, the chain entropy is reduced along with the overall free energy. The native state resides at the bottom of the potential well, characterized by low entropy and a global free energy minimum. A funneled energy landscape is robust to both environmental changes and sequence mutations, since most potentially competing low-energy states are similar in structure (74), as represented by the multiple local minima at the bottom of the funnel in Figure 1.7.

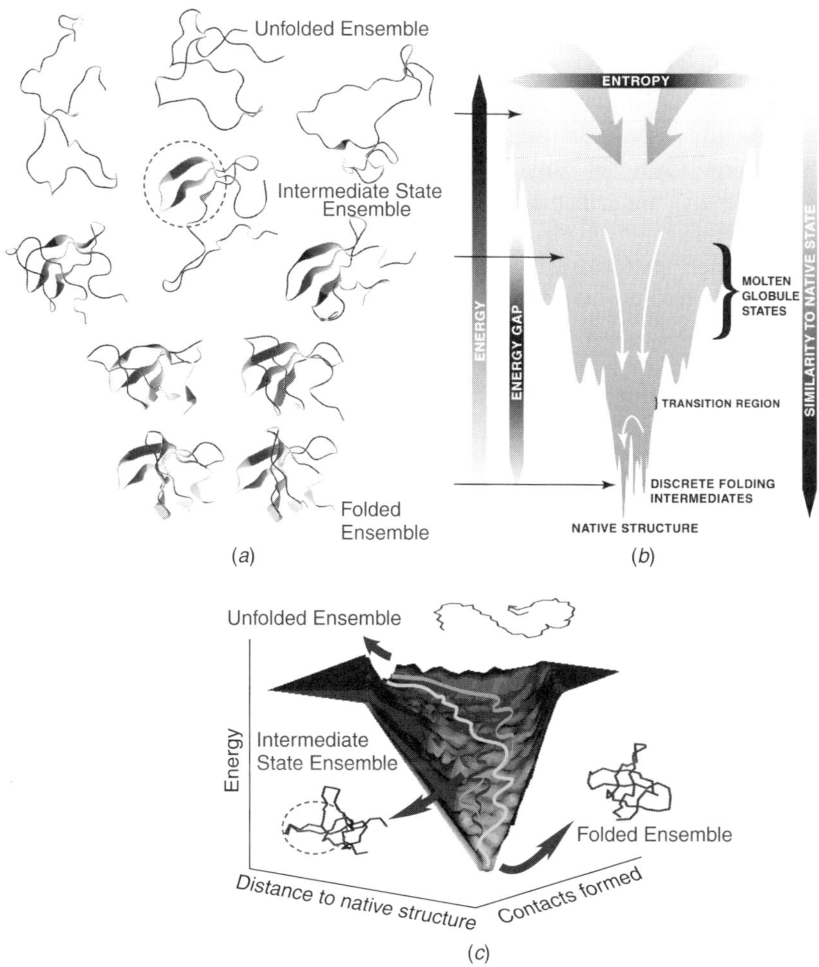

FIGURE 1.7. Schematic representation of a protein folding funnel. As the large ensemble of structures of an unfolded polypeptide compacts, forming native-like contacts through intermediate states and finally to the native state (**A**), the energy surface can be represented schematically as seen in panel (**B**). Multiplicity of folding routes is shown with different folded trajectories on the energy surface (**C**); however, asymmetry of the surface biases the trajectories toward the preferred route, which can be considered a folding pathway. Reproduced with permission from (120). © 2001 AAAS.

An important feature of the folding funnel is that its slopes are not always monotones, hence the competition between the "downhill slide" toward the native fold and the possibility of equally favorable excursions into local free energy minima, depending on the ruggedness of the energy surface. These local minima may represent transient formation of partially folded species, accumulation of intermediates, or misfolded forms, depending on the trajectory taken by the chain along the energy surface. This general model allows for multiple pathways with no specific order of structure formation, but subtle changes in the energy surface would lead to a far more directed approach by energetically favoring particular regions of conformational space. It also provides at least one possible solution for Levinthal's problem of folding time scale. If a favored conformation is visited on the energy landscape, then it would be sufficiently stabilized to reside there for a longer time and hence restrict the conformational search for the most stable native structure.

As has been stated already, one of the most important features that distinguishes these folding funnel models from more traditional reaction coordinates is the key feature of conformational entropy. Thus, we are forced to consider each stage along a folding reaction not as a single structure but instead as a conformational ensemble. The unfolded state—the nearly flat plane at the top of the funnel—demonstrates the large number of microstates in the unfolded polypeptide chain, whereas the deep energy well of the native state may be a single structure or indeed a small ensemble of similar structures closely related in energy. Along the folding trajectory as the chain becomes more ordered, these ensembles become smaller and smaller, but throughout there is always some conformational flexibility that must be considered.

It must be stressed that these folding landscapes are purely theoretical. Based on experimental data it has proved possible to style different landscapes more as a representation than a true physical picture, showing local free energy wells for intermediate species, and alternate possible routes for parallel pathways. Even with the most powerful modern computing facilities, however, it is not yet possible to predict the folding pathway(s) of a protein, although some trajectories in a computational ensemble may appear to fit well with experimental data (77). Only with a complete understanding of the energetics involved in driving a protein to fold will we be able to computationally predict how a given protein will fold.

Semiquantitative models and experiments are revealing how the folding free energy surface is sculpted by the protein sequence and its environment. Although any downhill path from the unfolded state will eventually lead to a native conformation at the bottom of the funnel, the asymmetry (energetic heterogeneity) of the surface can bias the choice of folding routes (78). The existence of such "preferred" folding routes can be observed experimentally and interpreted in terms of folding pathways (vide supra). The sometimes conflicting demands of folding, structure, and function determine which folding pathways, if any, dominate (75).

1.4.2. Equilibrium and Kinetic Intermediates on the Energy Landscape

Under native conditions, folding of small proteins usually appears to be a highly cooperative process (79, 80). Using standard biophysical techniques, only the native and unfolded states are generally sufficiently populated to be detectable. For instance, a titration of chemical denaturant studied by circular dichroism for most proteins will generate a series of spectra that can be deconvoluted to contributions solely from the two endpoints of the unfolding reaction, namely, the native and denatured states. This led to the belief that the folding of proteins was not only cooperative but also two-state, that is, without populated intermediate states.

In contrast, under certain conditions, significant accumulation of an intermediate conformational ensemble may occur if the free energy barrier is sufficiently high, referred to as an *equilibrium intermediate*. These species can be studied in great detail since the rate of conversion is low, allowing significant structural information to be obtained (81). Classic examples of these include the partially folded state of the apo-form of myoglobin (82), acid- and alcohol-induced *A*-state of ubiquitin (83), or the acid-induced molten globule of α-lactalbumin (84). As we have already mentioned in the preceding sections of this chapter, these equilibrium intermediate states in some cases may represent important conformations visited along the folding trajectory. Therefore, structural information about these states can give valuable clues as to the nature of the conformational search process. Transient formation or indeed accumulation of certain intermediates is usually induced in vitro by simply changing the protein's environment (e.g., by varying the solution pH, temperature, presence of chaotropes) (60, 85). This can result in significant alterations of the energy surface, decreasing the free energy of the intermediate states, and thus increasing their equilibrium population. One needs to be aware, however, that such equilibrium intermediates may differ significantly from the kinetically observed species. As pointed out recently by Fersht, an equilibrium intermediate by definition need not necessarily be on the preferred kinetic folding pathway since thermodynamic (equilibrium) parameters are independent of mechanism (58). The goal here though is to make these elusive states more amenable to study using a variety of biophysical techniques, in order to determine not only the conformational preferences of a partially folded protein, but also the possible conformations transiently visited by the native state that may be vital to its in vivo function.

Refolding of large proteins often does not conform to the simple two-state model discussed in the beginning of this section. Nevertheless, such proteins may be refolded spontaneously by rapid dilution from a chemically denatured state into native conditions. Kinetic studies of these processes have enabled detection of transient *kinetic intermediates*, which serve as "resting points" in a protein folding process. Since the energy difference between the global minimum and any of the surrounding local minima is usually quite high, the Boltzmann weight of the states that correspond to the local minima is very low. Under native conditions, kinetic intermediates become populated only transiently during refolding experiments. These species have been the focus of close experimental scrutiny, since their structure and behavior may reveal many intimate details of the protein folding process. In order to be detected these species must have characteristics different from either the native or unfolded states. For instance, the aromatic residues may experience an environment that makes them hyperfluorescent, or secondary structure may form in advance of tertiary interactions, making the intermediate detectable using stopped-flow optical techniques such as circular dichroism. Alternatively, the secondary structure may protect certain amide protons against exchange with bulk solvent, which can be detected by NMR or mass spectrometry, as we shall see in Chapters 2 and 5. The experimental identification and characterization of kinetic intermediates has been the focus of a great deal of research over the past decade, as researchers attempt to glean the determinants of protein folding. Due to their transient nature, however, kinetic intermediates often cannot be observed directly and their properties can only be inferred from indirect measurements.

If the energy barrier separating a kinetic intermediate from the native conformation is high, a significant accumulation of such intermediates may occur. These are referred to as *kinetically trapped* or *metastable* intermediate states. Under certain conditions, excessive accumulation of metastable kinetic intermediates in the course of the folding process may trigger nonspecific interactions among them and, in extreme cases, aggregation. Likewise, aggregation can also be caused by incorrect folding (e.g., due to a sequence mutation). Aggregation processes in vivo are prevented by chaperones, a special class of proteins that bind and sequester the misfolded and partially folded polypeptides (86–89). It is important to note, however, that the chaperones only assist, but do not direct, protein folding.

1.5. PROTEIN DYNAMICS AND FUNCTION

1.5.1. Limitations of the Structure–Function Paradigm

Proteins carry out their functions by interacting with other proteins, as well as other molecules ranging from giant biopolymers (such as DNA) to small organic molecules and monatomic ligands. In all cases, protein–ligand binding is the first stage of the interaction, which can be followed by a variety of processes ranging from sophisticated chemical transformation of the ligand (e.g., enzyme catalysis) to simple release of the ligand in the presence of other cofactors (e.g., transport proteins). In this section, we only consider the main characteristics of the protein–ligand binding process.

Binding has traditionally been considered within a framework of the structure–function paradigm, a cornerstone of molecular biology for many years. It was over a hundred years ago that Fischer coined the term *lock-and-key* to emphasize the requirement for a stereochemical fit between an enzyme and its substrate in order for binding to occur (90). The limits of this view of the binding process became obvious in the middle of the last century, when a large body of newly acquired information on enzyme kinetics appeared to be in conflict with the notion that "the enzyme was a rather rigid negative of the substrate and that the substrate had to fit into this negative to react" (91). The revision of the lock-and-key theory by Koshland led to a rise of the so-called *induced fit* theory, whose major premise was that the "reaction between the enzyme and substrate can occur only after a change in protein structure induced by the substrate itself" (91). Conformational changes occurring during enzyme catalysis are relatively small scale and affect mostly the catalytic site (92). Similarly, the conformational changes occurring in

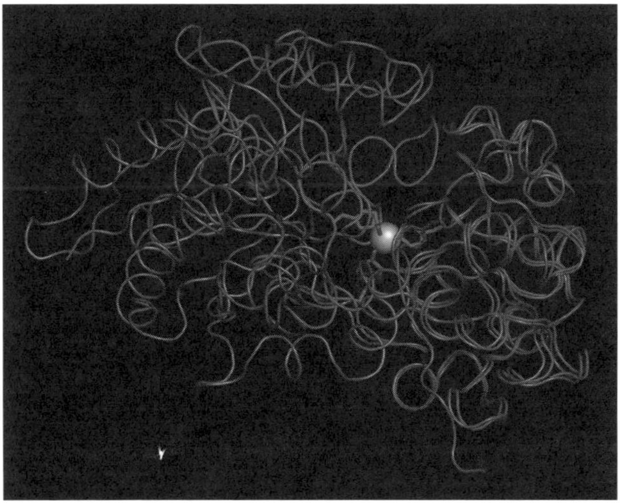

FIGURE 1.8. Superimposed crystal structures of the apo- and holo-forms of the N-lobe of human serum transferrin. Created with PyMol. © 2004 Delano Scientific LLC. See insert for color representation.

other proteins as a result of the induced-fit type binding usually affect a limited fraction of the protein structure. An example of that is ferric ion binding by the iron-transport protein transferrin, an event that results in the repositioning of two protein domains within each lobe of the protein (Figure 1.8). Although the overall effect of such repositioning is quite significant (and results in closing the cleft between the two domains), the number of affected amino acid residues does not exceed a dozen (93). More recently, numerous examples of large-scale conformational changes induced by ligand binding have been reported. The most extreme case is represented by the *intrinsically disordered* proteins, which actually lack stable structure under native conditions in the absence of the ligand (94).

The above considerations strongly suggest that structure is not the sole determinant of protein function. As elegantly put by Onuchic and Wolynes, "the twentieth century's fixation on structure catapulted folding to center stage in molecular biology. The lessons learned about folding may, in the future, increase our understanding of many functional motions and large-scale assembly processes" (78).

In the following section we will consider various aspects of protein dynamics under native conditions that may be important modulators or even determinants of function.

1.5.2. Protein Dynamics Under Native Conditions

With very few exceptions the protein structure under native conditions is not a rigid crystalline state, but undergoes local breathing motions, involving anything from side chain rotation to rearrangement of secondary structure elements relative to each other. Although the existence of such motions within the native state of the protein can be detected with a variety of experimental techniques, their exact nature remains a subject of discussion in the literature. The commonly accepted models of local dynamics within natively folded proteins invoke the notions of a *structural fluctuation* (a localized transient unfolding affecting only few atoms within the protein) (95) or a *mobile defect* (which considers not only the emergence and dissipation of local disorder, but also the possibility of its propagation through the protein structure). Although the latter model has not enjoyed as much attention as the former, thorough theoretical considerations suggest that local perturbations of the secondary structure may in fact propagate through certain elements of the secondary structure (e.g., α-helices) in the form of a soliton (96). Alternatively, the local dynamics can be described with a *solvent penetration* model (slow diffusion of the solvent molecules into and out of the protein interior) (97). Such description is actually very similar to the mobile defect model, as applied to the integral solute–solvent system, instead of the protein molecule alone.

Above and beyond local structural fluctuations, the dynamics of proteins under native conditions is exemplified by transiently sampling alternative (higher-energy or "activated") conformations. Such activated (non-native) states are often functionally important despite their low Boltzmann weight (98, 99). An example of such behavior can be seen in cellular retinoic acid binding protein I (CRABP I), which sequesters and transports insoluble all-*trans* retinoic acid (RA) in the cytosol. The structures of the *apo-* and the *holo*-forms of this protein are very similar, consistent with the lock-and-key type of binding. However, the native structure of CRABP I provides no clue as to how the ligand gets access to the internal protein cavity, which is its binding site (Figure 1.9). Obviously, in order to provide entrance into the cavity, a fraction of the native structure has to be lost transiently, an event consistent with the notion of sampling an activated protein state. Realization of the importance of transient non-native protein structures for their function has not only greatly advanced our understanding of processes as diverse as recognition, signaling, and transport, but also has had profound practical implications, particularly for the design of drugs targeting specific proteins (100).

Many proteins use dynamics as a means of communication between different domains. This process, by which a signal such as a binding event in one domain triggers a conformational change in another domain, is known as *allostery*. The paradigm for this effect is hemoglobin, a tetrameric protein mentioned earlier. Binding a molecule of oxygen at the heme site of the α-chain induces a change in the oxygen affinity of the β-chain binding site by rearrangement of the inter-domain interactions (101, 102). Another example is the chaperone protein DnaK, which assists in preventing the misfolding of nascent chains as they emerge from the ribosome. This 70 kDa protein consists of an ATPase domain joined via a short linker region to a peptide-binding domain. Binding of ATP causes a conformational change in the peptide-binding domain that increases its affinity for substrate. Subsequent hydrolysis of nucleotide in the ATPase domain signals a

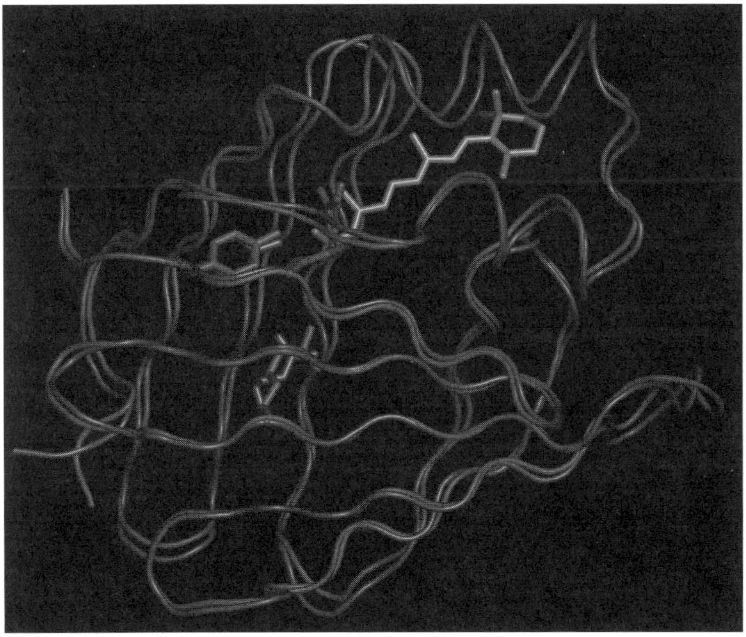

FIGURE 1.9. Crystal structures of the apo- and holo-forms of cellular retinoic acid binding protein I. Created with PyMol. © 2004 Delano Scientific LLC. See insert for color representation.

conformational change in the adjacent domain that releases the unfolded polypeptide and allows it to begin to refold. The exact mechanism by which this allosteric communication occurs is still poorly understood. It is clear, however, that it must involve dynamic events at the interdomain interface that transmits the signal between the two binding sites (103).

Another class of recently discovered proteins relies on dynamics even more heavily for function. *Intrinsically disordered* proteins are remarkable in that they appear to have very little stable folded structure in isolation, contrary to our classical view of proteins as folded species. Several hundred proteins have now been identified that contain large segments of disorder even under native conditions, and these seem to serve a wide variety of functions in vivo (104–106). A number of these are involved in activation or inhibition of transcription or translation, and while these proteins appear unstructured under native conditions, they undergo a structural transition in the presence of their cognate substrate, whether this be another protein or a recognition site on a molecule of DNA or RNA (94, 107). One important aspect of intrinsic disorder may be the necessity for this class of proteins to recognize and bind to multiple sites. Whereas a highly structured protein may only have a limited and very specific binding site available to it, one which has a highly dynamic structure should be able to adapt to a variety of different structural motifs. This intrinsic disorder phenomenon also seems contrary to the commonly accepted view that an unstructured protein should be targeted for proteolysis, or else degradation by the proteasome. It seems this class of proteins manages to avoid such scenarios either by having regions that are sterically inaccessible or else by not containing residues that are sensitive to proteases. Indeed, one other observation is that many intrinsically disordered proteins have a relatively short lifetime in the cell: they are expressed as needed in response to a signal and then rapidly removed by degradation. This would provide an efficient mechanism to switch on or off a cellular process for only a short period of time. A larger number of proteins in the eukaryotic genome have been predicted to have disordered regions compared to prokaryotes, perhaps indicating the need for higher organisms to adjust more rapidly to environmental changes.

1.5.3. Biomolecular Dynamics and Binding from the Energy Landscape Perspective

The development of the folding funnel concept has also far-reaching consequences for our understanding of how proteins interact with each other and with other ligands. A recently introduced general scheme of protein folding and binding implies that the only difference between the two processes is chain connectivity (108, 109), namely, that monomeric protein folding represents an energy funnel for a single chain, whereas protein–protein association and peptide binding is a similar landscape but with discontinuous backbone connections. In the more general case, however, the concept can be extended to encompass the chemical nature of the ligand and the energetics of the binding process, whether it be noncovalent interaction or a chemical process as in the case of enzymatic catalysis. The energy funnel concept can be applied theoretically to describe the process by which a protein recognizes and binds another molecule. Rigid proteins that bind ligands via a lock-and-key type mechanism presumably do not require significant dynamic events so will have few local minima similar in energy to the native state. One interesting exception (CRABP) will be considered later in this section. In contrast, those proteins that utilize an induced fit binding mechanism may have a rugged energy surface characterized by a number of local minimum conformational states at the bottom of the folding/binding funnel (108).

The idea of a binding funnel has also been demonstrated computationally by Zhang et al. to explain the fast protein–ligand association rates exhibited by many proteins (110). In this model, the initial collision event is accompanied by favorable interactions to form a long-lived encounter complex that significantly limits the search process to the ligand-bound conformation. The funnel energy landscape of protein binding may be a common feature in protein–protein associations (111). Knowledge of the relative energies and structural features of the local conformational minima available to proteins is clearly key to the understanding of what makes proteins efficient in binding their physiological ligands. Likewise, an extension of the protein folding problem is to understand how protein monomers assemble as functional multimers or other macromolecular complexes. In the cellular environment this must be an efficient process aided by dynamics and specific recognition events, all of which may be described in terms of the energetics of accessible conformational space.

All visibly different modes of binding (lock-and-key, induced fit, and binding of intrinsically disordered proteins) appear to have only quantitative differences within the framework provided by the binding funnel concept. We illustrate this point in Figures 1.10–1.12, which represent hypothetical folding funnels for proteins of each class in the absence and in the presence of their respective ligands. A protein whose ligand-binding behavior conforms to the lock-and-key type interaction (such as CRABP I considered earlier) is suggested to have an activated state whose structural features increase the rate of ligand entry into the binding site (Figure 1.10). The protein molecules sample this activated state relatively frequently due to its relatively low energy. Once the ligand enters the binding site, its interaction with the protein (multiple hydrophobic contacts in the case of CRABP I–RA interaction) increases the stability of the native conformation. Although the "visitations" to the activated state are still possible, they do not occur as frequently due to the increased energy difference between the two states. As a result, the protein can acquire the ligand relatively easily but does not release it unless the energy landscape is altered again, for example, by a competing receptor of the ligand.

A similar analysis can be carried out for proteins conforming to the induced fit type behavior (we will use the N-lobe of human serum transferrin as an example). Although the X-ray data suggest the existence of two distinct conformations of the protein depending on the presence of the ligand (open conformation for the apo-form and closed for the holo-form of the protein), there is experimental

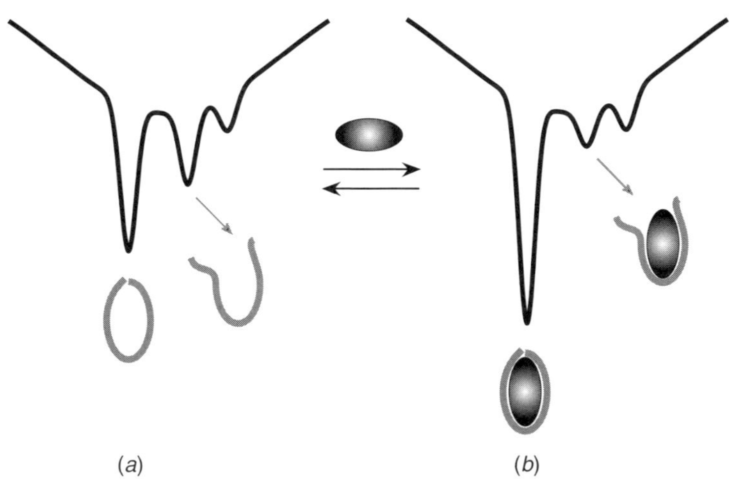

FIGURE 1.10. Schematic representations of the energy landscapes for the apo- (**A**) and holo- (**B**) forms of a protein whose ligand-binding behavior conforms to the lock-and-key type interaction.

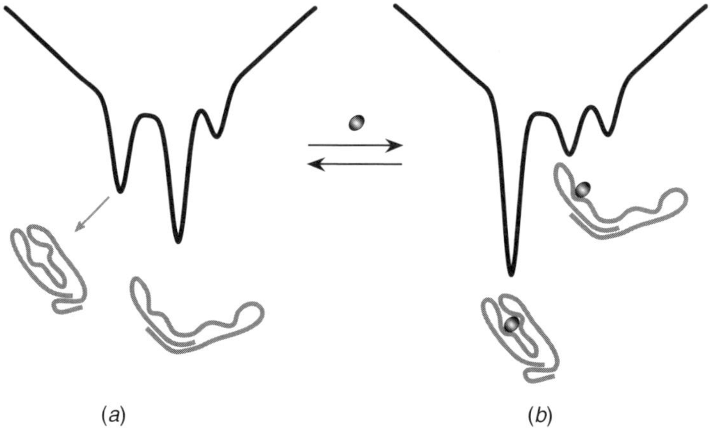

FIGURE 1.11. Schematic representations of the energy landscapes for the apo- (**A**) and holo- (**B**) forms of a protein whose ligand-binding behavior conforms to the induced fit type interaction.

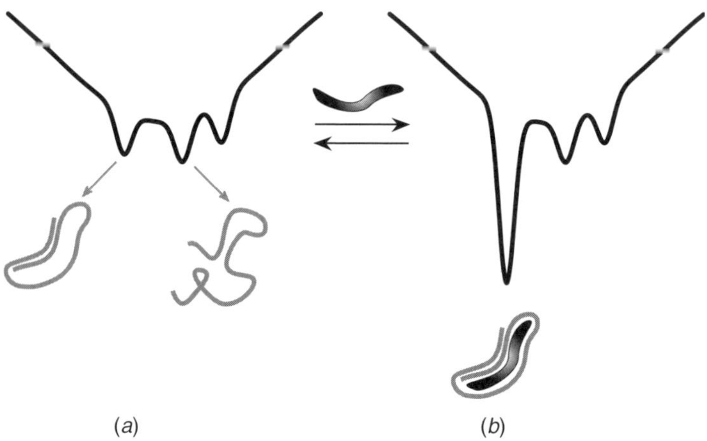

FIGURE 1.12. Schematic representations of the energy landscapes for the apo- (**A**) and holo- (**B**) forms of an intrinsically unstructured protein whose folding is induced by the ligand.

evidence suggesting that both conformations coexist in solution under equilibrium (112). The open conformation is, of course, favored in the absence of the ligand, while iron binding shifts the equilibrium toward the closed state (Figure 1.11). Such a shift is qualitatively similar to the one considered for the lock-and-key interaction. The only difference is that the protein state corresponding to the global energy minimum in the absence of the ligand becomes "downgraded" to the status of an activated state (local energy minimum) as a result of the ligand binding.

Finally, folding of an intrinsically unstructured protein in the process of ligand binding can be viewed as a preferential stabilization of one particular conformation among many available to the protein in the ligand-free form (Figure 1.12). An example of such behavior is presented by the β-chains of mammalian hemoglobins, which populate at least four different states (only one of them appears to be very close to the compact natively folded conformation) in solution in the absence of its binding partner, α-globin (113). Again, the general features of binding in this scheme appear to be very similar to those seen in the previous two examples (Figures 1.10 and 1.11), the major distinct feature being the absence of the preferred conformation in the absence of the ligand. We will revisit the physiological significance of such interactions in Chapter 11.

1.5.4. Energy Landscapes Within a Broader Context of Nonlinear Dynamics: Information Flow and Fitness Landscapes

It becomes increasingly clear that the significance of the concept of energy landscapes extends well beyond the fields of protein folding and even molecular biophysics. Recently, Huang and Ingber questioned the validity of the commonly accepted paradigm of cell regulation as a collection of pathways that link receptors with genes by asking: "Can identification of all these signaling proteins and their assignment into distinct functional pathways lead to the full understanding of developmental control and cell fate regulation?" (114). The existence of distributed information within cellular signaling, as well as the fact that a single signaling molecule may activate several genes and even produce opposite effects depending on its microenvironment led to a suggestion that the concept of *linear signaling pathways* is inappropriate. The suggestion that the signaling and regulatory pathways are not simple linear connections between receptors and genes is similar to the earlier realization of the limits of the classic concept of protein folding pathways. The paradox of signaling nonlinearity is resolved by the introduction of a concept of *cellular states*, and the switches between these states are viewed as *biological phase transitions* (114). In this new view, the cell fates are viewed as common end programs or attractors within the entire regulatory network, which can be visualized as a *potential landscape* with multiple minima. Each minimum corresponds to a certain fate of the cell (e.g., differentiation, proliferation, apoptosis). A similar idea was recently proposed as the basis of a quantitative model of carcinogenesis, in which normal cells in vivo occupy a ridge-shaped maximum in a well-defined *tissue fitness landscape*, a configuration that allows cooperative coexistence of multiple cellular populations (115).

Finally, *dynamic fitness landscapes* have proved to be a valuable concept in evolutionary biology, molecular evolution (116, 117), combinatorial optimization, and the physics of disordered systems (118). In evolutionary biology this concept is often used to visualize the relationship between genotypes (or phenotypes) and replication success, an idea initially put forward by Wright (119). An evolving population typically climbs uphill in the fitness landscape, until it reaches a local optimum (Figure 1.13), where it then remains, unless a rare mutation opens a path to a new fitness peak.

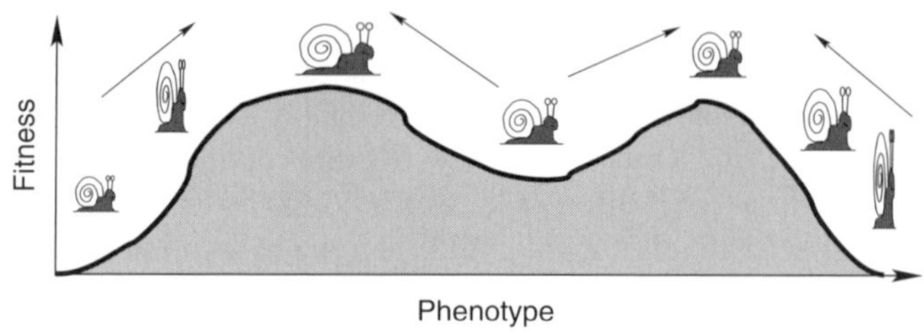

FIGURE 1.13. Schematic representation of a fitness landscape, where the fitness of a snail species depends on its shape. Mutations continuously produce variants that are selected if their fitness is larger than the fitness of the current "wild-type" snail. As a consequence, the shape of the snails changes over time until it reaches a maximum of the fitness landscape. Reproduced with permission from (118).

REFERENCES

1. Thieffry D., Sarkar S. (1998). Forty years under the central dogma. *Trends Biochem. Sci.* **23**: 312–316.
2. Nirenberg M. (2004). Historical review: deciphering the genetic code—a personal account. *Trends Biochem. Sci.* **29**: 46–54.
3. Hatfield D. L., Gladyshev V. N. (2002). How selenium has altered our understanding of the genetic code. *Mol. Cell. Biol.* **22**: 3565–3576.
4. Ibba M., Soll D. (2002). Genetic code: introducing pyrrolysine. *Curr. Biol.* **12**: R464–R466.
5. Budisa N., Minks C., Alefelder S., Wenger W., Dong F., Moroder L., Huber R. (1999). Toward the experimental codon reassignment in vivo: protein building with an expanded amino acid repertoire. *FASEB J.* **13**: 41–51.
6. Chin J. W., Cropp T. A., Anderson J. C., Mukherji M., Zhang Z., Schultz P. G. (2003). An expanded eukaryotic genetic code. *Science* **301**: 964–967.
7. Avetisov V., Goldanskii V. (1996). Mirror symmetry breaking at the molecular level. *Proc. Natl. Acad. Sci. U.S.A.* **93**: 11435–11442.
8. Podlech J. (2001). Origin of organic molecules and biomolecular homochirality. *Cell. Mol. Life Sci.* **58**: 44–60.
9. Warshel A., Papazyan A. (1998). Electrostatic effects in macromolecules: fundamental concepts and practical modeling. *Curr. Opin. Struct. Biol.* **8**: 211–217.
10. Schutz C. N., Warshel A. (2001). What are the dielectric "constants" of proteins and how to validate electrostatic models? *Proteins* **44**: 400–417.
11. Harvey S. C. (1989). Treatment of electrostatic effects in macromolecular modeling. *Proteins* **5**: 78–92.
12. Daune M. (1999). *Molecular Biophysics: Structures in Motion*. New York: Oxford University Press.
13. Mirsky A. E., Pauling L. (1936). On the structure of native, denatured, and coagulated proteins. *Proc. Natl. Acad. Sci. U.S.A.* **22**: 439–447.
14. Pauling L., Corey R. B., Branson H. R. (1951). The structure of proteins—2 hydrogen-bonded helical configurations of the polypeptide chain. *Proc. Natl. Acad. Sci. U.S.A.* **37**: 205–211.

15. Ramachandran G. N., Ramakrishnan C., Sasisekharan V. (1963). Stereochemistry of polypeptide chain configurations. *J. Mol. Biol.* **7**: 95–99.
16. Tanford C. (1978). Hydrophobic effect and organization of living matter. *Science* **200**: 1012–1018.
17. Scheraga H. A. (1998). Theory of hydrophobic interactions. *J. Biomol. Struct. Dyn.* **16**: 447–460.
18. Southall N. T., Dill K. A., Haymet A. D. J. (2002). A view of the hydrophobic effect. *J. Phys. Chem. B* **106**: 521–533.
19. Widom B., Bhimalapuram P., Koga K. (2003). The hydrophobic effect. *Phys. Chem. Chem. Phys.* **5**: 3085–3093.
20. Hartley G. S. (1936). *Aqueous Solutions of Paraffin-Chain Salts*. Paris: Hermann & Cie.
21. Tanford C. (1979). Interfacial free-energy and the hydrophobic effect. *Proc. Natl. Acad. Sci. U.S.A.* **76**: 4175–4176.
22. Hildebrand J. H. (1968). A criticism of term hydrophobic bond. *J. Phys. Chem.* **72**: 1841–1842; Nemethy G., Scheraga H. A., Kauzmann W. *Ibid.*: 1842.
23. Frank H. S., Evans M. W. (1945). Free volume and entropy in condensed systems. III. Entropy in binary liquid mixtures; partial molar entropy in dilute solutions; structure and thermodynamics in aqueous electrolytes. *J. Chem. Phys.* **13**: 507–532.
24. Tanford C. (1997). How protein chemists learned about the hydrophobic factor. *Protein Sci.* **6**: 1358–1366.
25. Chothia C., Hubbard T., Brenner S., Barns H., Murzin A. (1997). Protein folds in all-α and all-β classes. *Annu. Rev. Biophys. Biomol. Struct.* **26**: 597–627.
26. Zhang C., DeLisi C. (1998). Estimating the number of protein folds. *J. Mol. Biol.* **284**: 1301–1305.
27. Govindarajan S., Recabarren R., Goldstein R. A. (1999). Estimating the total number of protein folds. *Proteins* **35**: 408–414.
28. Hou J., Sims G. E., Zhang C., Kim S. -H. (2003). A global representation of the protein fold space. *Proc. Natl. Acad. Sci. U.S.A.* **100**: 2386–2390.
29. Caetano-Anolles G., Caetano-Anolles D. (2003). An evolutionarily structured universe of protein architecture. *Genome Res.* **13**: 1563–1571.
30. Rost B. (1997). Protein structures sustain evolutionary drift. *Fold. Des.* **2**: S19–S24.
31. Wood T. C., Pearson W. R. (1999). Evolution of protein sequences and structures. *J. Mol. Biol.* **291**: 977–995.
32. Zhang C., DeLisi C. (2001). Protein folds: molecular systematics in three dimensions. *Cell. Mol. Life Sci.* **58**: 72–79.
33. Denton M. J., Marshall C. J., Legge M. (2002). The protein folds as platonic forms: new support for the pre-Darwinian conception of evolution by natural law. *J. Theor. Biol.* **219**: 325–342.
34. Epstein C. J., Goldberger R. F., Anfinsen C. B. (1963). Genetic control of tertiary protein structure—studies with model systems. *Cold Spring Harbor Symp. Quant. Biol.* **28**: 439–449.
35. Dawkins R. (1996). *The Blind Watchmaker: Why the Evidence of Evolution Reveals a Universe Without Design*. New York: Norton.

36. Pennock R. T. (2001). *Intelligent Design Creationism and Its Critics: Philosophical, Theological, and Scientific Perspectives*. Cambridge, MA: MIT Press.
37. Levinthal C. (1968). Are there pathways for protein folding? *J. Chim. Phys. Phys. Chim. Biol.* **65**: 44–45.
38. Dill K. (1999). Polymer principles and protein folding. *Protein Sci.* **8**: 1166–1180.
39. Radford S. E., Dobson C. M. (1999). From computer simulations to human disease: emerging themes in protein folding. *Cell* **97**: 291–298.
40. Jaenicke R. (1998). Protein self-organization in vitro and in vivo: partitioning between physical biochemistry and cell biology. *Biol. Chem.* **379**: 237–243.
41. Jaenicke R. (1995). Folding and association versus misfolding and aggregation of proteins. *Philos. Trans. R. Soc. Lond. B Biol. Sci.* **348**: 97–105.
42. Rochet J. -C., Lansbury P. T. (2000). Amyloid fibrillogenesis: themes and variations. *Curr. Opin. Struct. Biol.* **10**: 60–68.
43. Uversky V. N. (2003). Protein folding revisited. A polypeptide chain at the folding–misfolding–nonfolding cross-roads: which way to go? *Cell. Mol. Life Sci.* **60**: 1852–1871.
44. Dobson C. M. (2004). Principles of protein folding, misfolding and aggregation. *Semin. Cell Dev. Biol.* **15**: 3–16.
45. Carrell R. W., Lomas D. A. (1997). Conformational disease. *Lancet* **350**: 134–138.
46. Bellotti V., Mangione P., Stoppini M. (1999). Biological activity and pathological implications of misfolded proteins. *Cell. Mol. Life Sci.* **55**: 977–991.
47. Thompson A. J., Barrow C. J. (2002). Protein conformational misfolding and amyloid formation: characteristics of a new class of disorders that include Alzheimer's and Prion diseases. *Curr. Med. Chem.* **9**: 1751–1762.
48. Bucciantini M., Giannoni E., Chiti F., Baroni F., Formigli L., Zurdo J., Taddei N., Ramponi G., Dobson C. M., Stefani M. (2002). Inherent toxicity of aggregates implies a common mechanism for protein misfolding diseases. *Nature* **416**: 507–511.
49. Caughey B., Lansbury P. T. (2003). Protofibrils, pores, fibrils, and neurodegeneration: separating the responsible protein aggregates from the innocent bystanders. *Annu. Rev. Neurosci.* **26**: 267–298.
50. Wales D. J., Scheraga H. A. (1999). Global optimization of clusters, crystals, and biomolecules. *Science* **285**: 1368–1372.
51. Schonbrun J., Wedemeyer W. J., Baker D. (2002). Protein structure prediction in 2002. *Curr. Opin. Struct. Biol.* **12**: 348–354.
52. Whisstock J. C., Lesk A. M. (2003). Prediction of protein function from protein sequence and structure. *Q. Rev. Biophys.* **36**: 307–340.
53. Price N. C. (2000). Conformational issues in the characterization of proteins. *Biotechnol. Appl. Biochem.* **31**: 29–40.
54. Smith L. J., Fiebig K. M., Schwalbe H., Dobson C. M. (1996). The concept of a random coil. Residual structure in peptides and denatured proteins. *Fold. Des.* **1**: R95–R106.
55. Ptitsyn O. B. (1995). Molten globule and protein folding. *Adv. Protein Chem.* **47**: 83–229.
56. Uversky V. N. (1998). How many molten globules states exist? *Biofizika* **43**: 416–421.

57. Clarke J., Fersht A. R. (1996). An evaluation of the use of hydrogen exchange at equilibrium to probe intermediates on the protein folding pathway. *Fold. Des.* **1**: 243–254.
58. Clarke J., Itzhaki L. S., Fersht A. R. (1997). Hydrogen exchange at equilibrium: a short cut for analysing protein-folding pathways? *Trends Biochem. Sci.* **22**: 284–287.
59. Jennings P. A., Wright P. E. (1993). Formation of a molten globule intermediate early in the kinetic folding pathway of apomyoglobin. *Science* **262**: 892–896.
60. Bai Y., Sosnick T. R., Mayne L., Englander S. W. (1995). Protein folding intermediates: native state hydrogen exchange. *Science* **269**: 192–196.
61. Bai Y. (1999). Equilibrium amide hydrogen exchange and protein folding kinetics. *J. Biomol. NMR* **15**: 65–70.
62. Parker M. J., Marqusee S. (2001). A kinetic folding intermediate probed by native state hydrogen exchange. *J. Mol. Biol.* **305**: 593–602.
63. Creighton T. E. (1984). Pathways and mechanisms of protein folding. *Adv. Biophys.* **18**: 1–20.
64. Creighton T. E. (1988). Toward a better understanding of protein folding pathways. *Proc. Natl. Acad. Sci. U.S.A.* **85**: 5082–5086.
65. Gunasekaran K., Eyles S. J., Hagler A. T., Gierasch L. M. (2001). Keeping it in the family: folding studies of related proteins. *Curr. Opin. Struct. Biol.* **11**: 83–93.
66. Matthews C. R. (1993). Pathways of protein folding. *Annu. Rev. Biochem.* **62**: 653–683.
67. Daggett V., Fersht A. R. (2003). Is there a unifying mechanism for protein folding? *Trends Biochem. Sci.* **28**: 18–25.
68. Baldwin R. L. (1995). The nature of protein folding pathways: the classical versus the new view. *J. Biomol. NMR* **5**: 103–109.
69. Dill K. A., Chan H. S. (1997). From Levinthal to pathways to funnels. *Nat. Struct. Biol.* **4**: 10–19.
70. Pande V. S., Grosberg A., Tanaka T., Rokhsar D. S. (1998). Pathways for protein folding: is a new view needed? *Curr. Opin. Struct. Biol.* **8**: 68–79.
71. Bryngelson J. D., Onuchic J. N., Socci N. D., Wolynes P. G. (1995). Funnels, pathways, and the energy landscape of protein folding: a synthesis. *Proteins* **21**: 167–195.
72. Onuchic J. N., Luthey-Schulten Z., Wolynes P. G. (1997). Theory of protein folding: the energy landscape perspective. *Annu. Rev. Phys. Chem.* **48**: 545–600.
73. Brooks C. L. III, Gruebele M., Onuchic J. N., Wolynes P. G. (1998). Chemical physics of protein folding. *Proc. Natl. Acad. Sci. U.S.A.* **95**: 11037–11038.
74. Plotkin S. S., Onuchic J. N. (2002). Understanding protein folding with energy landscape theory. Part I: Basic concepts. *Q. Rev. Biophys.* **35**: 111–167.
75. Gruebele M. (2002). Protein folding: the free energy surface. *Curr. Opin. Struct. Biol.* **12**: 161–168.
76. Dinner A. R., Sali A., Smith L. J., Dobson C. M., Karplus M. (2000). Understanding protein folding via free-energy surfaces from theory and experiment. *Trends Biochem. Sci.* **25**: 331–339.
77. Fersht A. R., Daggett V. (2002). Protein folding and unfolding at atomic resolution. *Cell* **108**: 573–582.

78. Onuchic J. N., Wolynes P. G. (2004). Theory of protein folding. *Curr. Opin. Struct. Biol.* **14**: 70–75.
79. Jackson S. E. (1998). How do small single-domain proteins fold? *Fold. Des.* **3**: R81–R91.
80. Myers J. K., Oas T. G. (2002). Mechanisms of fast protein folding. *Annu. Rev. Biochem.* **71**: 783–815.
81. Ptitsyn O. B. (1994). Kinetic and equilibrium intermediates in protein folding. *Protein Eng.* **7**: 593–596.
82. Eliezer D., Jennings P. A., Dyson H. J., Wright P. E. (1997). Populating the equilibrium molten globule state of apomyoglobin under conditions suitable for structural characterization by NMR. *FEBS Lett.* **417**: 92–96.
83. Brutscher B., Bruschweiler R., Ernst R. R. (1997). Backbone dynamics and structural characterization of the partially folded A state of ubiquitin by ^{1}H, ^{13}C, and ^{15}N nuclear magnetic resonance spectroscopy. *Biochemistry* **36**: 13043–13053.
84. Koshiba T., Yao M., Kobashigawa Y., Demura M., Nakagawa A., Tanaka I., Kuwajima K., Nitta K. (2000). Structure and thermodynamics of the extraordinarily stable molten globule state of canine milk lysozyme. *Biochemistry* **39**: 3248–3257.
85. Bai Y., Englander J. J., Mayne L., Milne J. S., Englander S. W. (1995). Thermodynamic parameters from hydrogen exchange measurements. *Methods Enzymol.* **259**: 344–356.
86. Fink A. L. (1999). Chaperone-mediated protein folding. *Physiol. Rev.* **79**: 425–449.
87. Hartl F. U., Hayer-Hartl M. (2002). Molecular chaperones in the cytosol: from nascent chain to folded protein. *Science* **295**: 1852–1858.
88. Trombetta E. S., Parodi A. J. (2003). Quality control and protein folding in the secretory pathway. *Annu. Rev. Cell Dev. Biol.* **19**: 649–676.
89. Mogk A., Bukau B. (2004). Molecular chaperones: structure of a protein disaggregase. *Curr. Biol.* **14**: R78–R80.
90. Fischer E. (1894). Einfluss der configuration auf die wirkung derenzyme. *Ber. Dtsch. Chem. Ges.* **27**: 2985–2993.
91. Koshland D. E. (1958). Application of a theory of enzyme specificity to protein synthesis. *Proc. Natl. Acad. Sci. U.S.A.* **44**: 98–104.
92. Koshland D. E. (1998). Conformational changes: how small is big enough? *Nat. Med.* **4**: 1112–1114.
93. Jeffrey P. D., Bewley M. C., MacGillivray R. T., Mason A. B., Woodworth R. C., Baker E. N. (1998). Ligand-induced conformational change in transferrins: crystal structure of the open form of the N-terminal half-molecule of human transferrin. *Biochemistry* **37**: 13978–13986.
94. Dunker A. K., Brown C. J., Lawson J. D., Iakoucheva L. M., Obradovic Z. (2002). Intrinsic disorder and protein function. *Biochemistry* **41**: 6573–6582.
95. Maity H., Lim W. K., Rumbley J. N., Englander S. W. (2003). Protein hydrogen exchange mechanism: local fluctuations. *Protein Sci* **12**: 153–160.
96. d'Ovidio F., Bohr H. G., Lindgard P. A. (2003). Solitons on H bonds in proteins. *J. Phys. Condens. Mat.* **15**: S1699–S1707.
97. Miller D. W., Dill K. A. (1995). A statistical mechanical model for hydrogen exchange in globular proteins. *Protein Sci.* **4**: 1860–1873.

98. Tsai C. D., Ma B., Kumar S., Wolfson H., Nussinov R. (2001). Protein folding: binding of conformationally fluctuating building blocks via population selection. *Crit. Rev. Biochem. Mol. Biol.* **36**: 399–433.
99. Papoian G. A., Wolynes P. G. (2003). The physics and bioinformatics of binding and folding—an energy landscape perspective. *Biopolymers* **68**: 333–349.
100. Carlson H. A. (2002). Protein flexibility and drug design: how to hit a moving target. *Curr. Opin. Chem. Biol.* **6**: 447–452.
101. Perutz M. F. (1972). Stereochemical mechanism of cooperative effects in haemoglobin. *Biochimie* **54**: 587–588.
102. Perutz M. F., Wilkinson A. J., Paoli M., Dodson G. G. (1998). The stereochemical mechanism of the cooperative effects in hemoglobin revisited. *Annu. Rev. Biophys. Biomol. Struct.* **27**: 1–34.
103. Pellecchia M., Montgomery D. L., Stevens S. Y., Vander Kooi C. W., Feng H. P., Gierasch L. M., Zuiderweg E. R. (2000). Structural insights into substrate binding by the molecular chaperone DnaK. *Nat. Struct. Biol.* **7**: 298–303.
104. Churchill M. E., Travers A. A. (1991). Protein motifs that recognize structural features of DNA. *Trends Biochem. Sci.* **16**: 92–97.
105. Sigler P. B. (1988). Transcriptional activation. Acid blobs and negative noodles. *Nature* **333**: 210–212.
106. Frankel A. D., Kim P. S. (1991). Modular structure of transcription factors: implications for gene regulation. *Cell* **65**: 717–719.
107. Iakoucheva L. M., Brown C. J., Lawson J. D., Obradovic Z., Dunker A. K. (2002). Intrinsic disorder in cell-signaling and cancer-associated proteins. *J. Mol. Biol.* **323**: 573–584.
108. Tsai C. -J., Xu D., Nussinov R. (1998). Protein folding via binding and vice versa. *Folding and Design* **3**: R71–R80.
109. Tsai C. -J., Kumar S., Ma B., Nussinov R. (1999). Folding funnels, binding funnels, and protein function. *Protein Sci.* **8**: 1181–1190.
110. Zhang C., Chen J., DeLisi C. (1999). Protein–protein recognition: exploring the energy funnels near the binding sites. *Proteins* **34**: 255–267.
111. Tovchigrechko A., Vakser I. A. (2001). How common is the funnel-like energy landscape in protein–protein interactions? *Protein Sci.* **10**: 1572–1583.
112. Baker H. M., Anderson B. F., Baker E. N. (2003). Dealing with iron: common structural principles in proteins that transport iron and heme. *Proc. Natl. Acad. Sci. U.S.A.* **100**: 3579–3583.
113. Griffith W. P., Kaltashov I. A. (2003). Highly asymmetric interactions between globin chains during hemoglobin assembly revealed by electrospray ionization mass spectrometry. *Biochemistry* **42**: 10024–10033.
114. Huang S., Ingber D. E. (2000). Shape-dependent control of cell growth, differentiation, and apoptosis: switching between attractors in cell regulatory networks. *Exp. Cell Res.* **261**: 91–103.
115. Gatenby R. A., Vincent T. L. (2003). An evolutionary model of carcinogenesis. *Cancer Res.* **63**: 6212–6220.
116. Drossel B. (2001). Biological evolution and statistical physics. *Adv. Phys.* **50**: 209–295.

117. Wilke C. O., Ronnewinkel C., Martinetz T. (2001). Dynamic fitness landscapes in molecular evolution. *Phys. Rep.* **349**: 395–446.
118. Reidys C. M., Stadler P. F. (2002). Combinatorial landscapes. *SIAM Rev.* **44**: 3–54.
119. Wright S. (1967). "Surfaces" of selective value. *Proc. Natl. Acad. Sci. U.S.A.* **58**: 165–172.
120. Brooks C. L., Onuchic J. N., Wales D. J. (2001). Statistical thermodynamics. Taking a walk on a landscape. *Science* **293**: 612–613.

2

OVERVIEW OF "TRADITIONAL" EXPERIMENTAL ARSENAL TO STUDY BIOMOLECULAR STRUCTURE AND DYNAMICS

Biophysicists traditionally use a battery of experimental techniques, which have not until relatively recently included mass spectrometry. The purpose of this chapter is to present an overview of the methodologies that have routinely been employed, explaining the details that can be gleaned from each technique. This will be used as a jump point in future chapters to demonstrate the complementary nature of mass spectrometric methodologies in the biophysical arena.

2.1. X-RAY CRYSTALLOGRAPHY

2.1.1. Fundamentals

Macromolecules are generally too small to be observed directly by optical microscopy because their size is insufficient to diffract light of wavelengths in the visible region of the electromagnetic spectrum. However, the interatomic spacing (~1.5 Å) is similar to the wavelength of X-rays. In an ordered crystal the molecules are exactly aligned in one or only few orientations in three-dimensional space in a repeating array of *unit cells*. The unit cell is defined as the smallest volume element that represents the entire crystal. Adjacent unit cells are related to one another by translation along the cell axes. Thus, the entire crystal can be constructed by laying unit cells side by side. Within a protein molecule each atom will have an exactly equivalent atom at the exact same position in a molecule in each and every other unit cell, forming crystal planes of atoms. These planes,

Mass Spectrometry in Biophysics: Conformation and Dynamics of Biomolecules
By Igor A. Kaltashov and Stephen J. Eyles
ISBN 0-471-45602-0

or more correctly the planes formed by the electron clouds between atoms, are what produce the diffraction pattern when exposed to X-ray radiation (1).

Lattice planes in crystals will diffract incident X-rays in a coherent manner depending on the angle of the incident beam, according to Bragg's law:

$$2d_{hkl} \sin\theta = n\lambda, \qquad (2\text{-}1\text{-}1)$$

which indicates that a beam of X-rays incident at an angle θ on a series of parallel planes separated by spacing d (the indices h, k, and l represent the lattice indices in three dimensions) will be reflected only if the left side of the equation is an integer multiple n of the incident wavelength λ. Other angles will, of course, cause reflections but only if the above relationship holds will the waves reflected by adjacent lattice planes produce constructive interference and emerge from the crystal in phase to produce a strong diffracted beam.

In principle, one could use Equation (2-1-1) to relate the measured diffraction angles to interatomic distances and from there calculate the molecular structure in the crystal. In practice, however, this is computationally extremely complex, especially for macromolecules containing a large number of atoms. In practice, waves may be considered as periodic functions of the form

$$f(x) = F \cos 2\pi(hx + \alpha). \qquad (2\text{-}1\text{-}2)$$

In one dimension the function describing this wave has amplitude F, frequency h, and phase α. Each diffracted X-ray beam may then be considered in terms of a Fourier series of reflections from each atom in the structure, to yield structure factors in three dimensions, F_{hkl}, of the form

$$F_{hkl} = f_a + f_b + f_c \cdots. \qquad (2\text{-}1\text{-}3)$$

The contribution from each atom ($a, b, c, \ldots$) is summed to describe the reflection. Expanding into three dimensions leads to complicated sums of complex numbers to describe the contribution of reflections from every atom in the unit cell to each observed diffraction peak, which leads to a description of the electron density at every point in the unit cell:

$$\rho(x, y, z) = \frac{1}{V} \sum_h \sum_k \sum_l |F_{hkl}| e^{-2i\pi(hx+ky+lz-\alpha'_{hkl})}. \qquad (2\text{-}1\text{-}4)$$

This equation represents the electron density map at every point (x, y, z) in three-dimensional space in the unit cell, volume V. Each F_{hkl} represents a specific reflection, giving rise to a peak in the diffraction pattern.

Clearly, solution of this equation is an extremely complex computational problem, made significantly worse by the fact that a complete solution also requires knowledge of the phase (α) of each of the waves, a parameter that cannot be obtained directly from measurement of the diffraction pattern. In fact, the so-called phase problem can be the main bottleneck in crystal structure determination.

For small molecules crystallization is often a relatively simple matter, but for proteins and other macromolecules growing crystals of sufficient quality and size to obtain high-quality diffraction patterns is a significant challenge. One of the most common approaches to crystallization of proteins is the "hanging drop" method, which involves suspending a droplet of protein solution in a closed container above a reservoir of buffer in a closed container. At equilibrium the vapor pressure inside the container is equal to that in the droplet, providing conditions under which protein crystals may be formed. In practice, a wide variety of different buffer conditions (e.g., varying pH, temperature, protein concentration, and ionic strength) are assayed to determine the optimal conditions. If only small crystals are obtained initially, these may be used as seeds to obtain large high-quality crystals for diffraction. Generally, crystals must be at least 0.5 mm in the shortest dimension to obtain high-quality diffraction data.

Once a crystal is produced it is sealed in a capillary with a small amount of mother liquor (the buffer used to produce the crystal) and mounted on a sample stage (known as a goniometer) that can be precisely rotated in three dimensions in the sample beam. The sample can then be exposed to the X-ray source and diffraction patterns recorded for different orientations of the crystal with respect to the beam. Diffraction patterns at the various angles are recorded using area detectors consisting of arrays of scintillation counters that can count X-ray photons. Crystals are often damaged by heating or free radicals produced from X-ray exposure; consequently, a number of individual crystals are generally required for a complete structure determination.

As mentioned above, the problem of determining phase is often a significant factor in analyzing the data to obtain a structure. One way around this is to also grow crystals containing heavy atoms (commonly Hg, Pb, or Au), which in many cases can be accommodated within the crystal by binding to the protein at specific sites. Provided that the protein structure and the structure of the crystals are not changed by the presence of these atoms (in this case the crystal lattice is described as isomorphic), then diffraction patterns from these *heavy atom derivatives* can be used to determine a first approximation to the phase values by a process known as *isomorphous replacement*. Subtracting out the diffraction data for the protein leaves only the reflections produced by the presence of the heavy atoms in the crystal, and since there are relatively few of these atoms it is a significantly easier task to determine their positions in the unit cell. Other approaches such as anomalous scattering and molecular replacement (using known structures of similar proteins) can also be employed to solve the phase problem and hence lead to a solution of the crystal structure of the biomolecule.

2.1.2. Crystal Structures at Atomic and Ultrahigh Resolution

Acquisition of crystallographic data at atomic resolution (1.2 Å or higher) provides several important benefits (2). Structure refinement can be done with only weak stereochemical constraints. Mismodeling can easily be corrected or avoided during the refinement, since the atom types can be distinguished at this resolution.

Finally, solvent molecules can be placed accurately within the shells (3). At a resolution of 0.8 Å or higher, hydrogen atoms become visible and the atoms can no longer be considered spherical particles due to deformation of electron clouds into bonding and nonbonding orbitals. Such "visualization" of electrons allows the charge distribution and the bond polarization and order to be determined (4), yielding a very informative picture of the protein structure.

2.1.3. Crystal Structures of Membrane Proteins

Membrane proteins* constitute a particularly "difficult" class of biopolymers because of their poor solubility characteristics. It is estimated that less than 0.2% of all membrane protein structures are known (by comparison, approximately 10% of soluble proteins are structurally covered), most of them solved by X-ray crystallography (5). Despite some spectacular successes, X-ray crystallography of membrane proteins is still considered art due to extreme difficulties associated with production of crystals suitable for X-ray diffraction analyses (5, 6). An alternative to "conventional" X-ray crystallographic analysis of a membrane protein is its reconstitution into two-dimensional membrane protein crystals in the presence of lipids (7). Unfortunately, most two-dimensional crystals are not ordered enough to provide resolution better than 4 Å (5).

2.1.4. Protein Dynamics and X-Ray Diffraction

Protein structures are represented in crystallography as a set of atoms in a three-dimensional space, which are assumed to exhibit harmonic and isotropic vibrations (quantitatively described using B-factors). It has been realized for quite some time that this assumption is an oversimplification that does not provide adequate description of protein mobility (8). Several methods have been developed to provide more assessment of protein dynamics based on observed crystal heterogeneity [reviewed in (8)]. Despite such progress, information on large-scale dynamic events is usually very difficult to obtain due to the influence of packing forces in protein crystals.

Several crystallographic techniques have been developed in recent years aiming specifically at characterization of protein dynamics by means of X-ray crystallography. It is well known that unlike crystals of small organic molecules, those of proteins contain significant amount of solvent. Not only do these solvent molecules weaken the intermolecular contacts in the crystal lattice, but they also permit in many cases diffusion of small substrates through the lattice and subsequent binding to the active sites on the enzymes (9, 10). These observations have led to a conclusion that crystals of biological macromolecules bear significant resemblance to very concentrated solutions (10). This notion, of course, downplays the fact that large-scale dynamic events are almost always suppressed by the crystal packing forces and simply cannot occur within the framework of

*We will discuss properties of membrane proteins in more detail in the final chapter of this book.

the crystal lattice. However, if the macromolecular dynamics does not involve large-scale motions, methods of X-ray crystallography can be used successfully to characterize these dynamic events. In particular, there are numerous examples of using crystallographic methods to study enzyme dynamics by specifically targeting the transient intermediate states on the enzymatic pathway. This is usually achieved by either trapping the unstable species (11) or using time-resolved methods (10). The former strategy employs adjustments of temperature, solution pH, and other experimental variables, forcing unstable reaction intermediates to accumulate and be stabilized in the protein crystals (11). The latter strategy, on the other hand, exploits both chemical and structural heterogeneity present in the sample throughout the reaction. Such (time-dependent) heterogeneity is then deconvoluted into structures of intermediates using sophisticated methods of data analysis (10).

2.2. SOLUTION SCATTERING TECHNIQUES

Unlike X-ray diffraction, the solution scattering techniques, briefly reviewed in this section, do not provide structural information on biopolymers at high (atomic) resolution. They do, however, often provide a means to observe large-scale structural heterogeneity of proteins directly in solution. As such, they are becoming indispensable tools in studies of protein dynamics in solution. In addition, they may complement the diffraction data by providing low-resolution structural information on the protein segments that fail to produce adequate diffraction patterns.

2.2.1. Static and Dynamic Light Scattering

Light scattering has been used historically to characterize synthetic polymers, although in recent years it has increasingly been applied to study biopolymers as well. In *static light scattering*, intensity of scattered light is measured as a function of the scattering vector $\vec{q}$:

$$q = |\vec{q}| = \frac{4\pi n}{\lambda} \sin\frac{\theta}{2}, \qquad (2\text{-}2\text{-}1)$$

where n is the solvent refractive index, and θ is a scattering angle (12). Intensity of the scattered light adjusted for background scattering and normalized to a reference solvent gives the Rayleigh ratio $\mathrm{R_s}(q)$, which can be expressed in dilute solutions as (12)

$$\frac{1}{\mathrm{R_s}(q)} \cdot \frac{4\pi^2 n^2 \cdot (dn/dc)^2}{N_\mathrm{A}\lambda^4} = \frac{1}{\mathrm{P}(q) \cdot \langle M \rangle} + 2Bc, \qquad (2\text{-}2\text{-}2)$$

where c is protein concentration, N_A is Avogadro's number, $\langle M \rangle$ is the weight-averaged molecular weight of protein particles in solution, B is the second virial coefficient describing interparticle interactions in solution, and $\mathrm{P}(q)$ is a particle shape factor. If a particle size is small compared to the light wavelength, then it obviously acts as a "point scatterer," and its shape is irrelevant for scattering [i.e., $P(q) = 1$ in (2-2-2)]. Most proteins fall into this category; hence, the only key parameter that can be obtained from the static light scattering methods is the molecular weight. Such sensitivity of the static light scattering measurements to the average molecular weight of the protein particles makes it very useful in the studies of protein association and aggregation. In the following chapters of this book we will refer to several examples of such measurement, namely, use of static light scattering to study association–dissociation equilibria of hemoglobin tetramers (13), as well as formation of amyloid fibrils (14). Once the particle size becomes comparable to the wavelength used in the experiments (i.e., radius of gyration $R_g \sim 1/q$), $P(q)$ can be approximated reasonably well as a quadratic function of qR_g, and the scattering profiles can be used to determine the particle gyration radius (12).

Dynamic light scattering is sensitive to the diffusion of scattering particles in solution, as it measures the intensity of light scattered at a fixed angle, which is then analyzed with an autocorrelator. The resulting correlation function has the particle diffusion coefficient as one of its arguments, which can be used to calculate the hydrodynamic radius of the particle.

2.2.2. Small-Angle X-Ray Scattering

Small-angle X-ray scattering (SAXS) is another solution scattering technique that has been (and still is) steadily gaining prominence in recent years. Although it is not suitable for structural analysis at the atomic level, SAXS often provides valuable structural information at lower resolution (overall shape of the protein, as well as its quaternary and tertiary structures). The molecular size range amenable to SAXS spans a 10–1000 Å range, thus providing a means to characterize not only single proteins, but also very large macromolecular assemblages (15). As the name infers, SAXS measures an interference pattern of elastically scattered X-rays, which is usually expressed as a function of a scattering vector length Q:

$$Q = |\vec{Q}| = \frac{4\pi}{\lambda} \sin\theta, \qquad (2\text{-}2\text{-}3)$$

where λ is a wavelength of the X-ray beam and 2θ is the scattering angle. The measured scattering profile $I(Q)$ is a contrast between the X-rays scattered from the protein molecules in solution and the equivalent volume of solvent displaced by them (15):

$$I(Q) = \left\langle \left| \int d^3\vec{r}\,(\rho(\vec{r}) - \rho_s(\vec{r})) \cdot e^{-i\vec{Q}\cdot\vec{r}} \right|^2 \right\rangle, \qquad (2\text{-}2\text{-}4)$$

where the averaging is done with respect to all orientations of vector $\vec{Q}$. The lack of translational order (lattice) within the protein solution leads to continuous distribution of $I(Q)$ (unlike distinct Bragg reflections arising due to the translationally ordered molecular structure in crystals). The contrast factor $\Delta\rho$ [difference between the electron densities of protein molecules (ρ) and solvent (ρ_s)] is typically on the order of 0.12 $e^-/Å^3$ (four times lower that ρ alone). As a result, protein scattering strength in aqueous solution is only about 10% of what it would be in a solvent-free environment (16). More rigorous treatment of the scattering profile $I(Q)$ accounts for scattering from the protein hydration shell, whose electron density (as well as other physical properties) may be different from that of the bulk solvent.

One of the most common ways to interpret the SAXS data employs the inverse Fourier transform of $I(Q)$ to yield the radial Patterson distribution:

$$P(r) = \frac{1}{2\pi^2} \int [I(Q) \cdot Qr \cdot \sin(Qr)]\, dQ. \tag{2-2-5}$$

Patterson distribution is related to the probability distribution of the distances between the scattering centers (atoms) within the scattering particle. As such, it provides information on the overall shape of the scattering particle and its dimension (15). Patterson distribution can also be used to calculate the gyration radius of the scattering particle:

$$R_g = \frac{\sqrt{\int P(r) \cdot r^2\, dr}}{\sqrt{\int P(r)\, dr}}, \tag{2-2-6}$$

a parameter that reflects both its size and shape. Recently, extensive efforts have been directed toward developing methods that restore low-resolution shape as well as internal structure of macromolecules in solution based on isotropic scattering data (17). All of the above considerations relate to a homogeneous solution of nearly identical (although randomly oriented) scattering particles. Sample heterogeneity (either presence of multiple protein species or conformational inhomogeneity) poses a very serious challenge to SAXS. In this case, a measured scattering profile would contain contributions from all components, weighted by their fractional concentrations. This difficulty can be overcome in some cases by using singular value decomposition (SVD) to characterize individual protein conformers (18) or protein aggregates (19).

2.2.3. Cryo-Electron Microscopy

Cryo-electron microscopy (cryo-EM) is an imaging technique that often allows large macromolecular complexes to be visualized at low resolution. This technique exploits the fact that images of macromolecules represent their two-dimensional projections. Three-dimensional structural data can be obtained from such projections if the data from "different views" are combined and processed using various reconstruction techniques (20). In a few cases, it actually provided structural information at a resolution high enough to derive atomic coordinates, although typically it is not possible to determine macromolecular structure at a resolution higher than 10–30 Å (21). Combination of the low-resolution cryo-EM data on macromolecular assemblies and high-resolution X-ray crystallographic structures of their constituents can provide structure of large subcellular structures at near-atomic resolution by "docking" the high-resolution structures into the low-resolution maps (22).

The energy of electrons used in the electron microscopes corresponds to a wavelength range of 0.0015–0.040 Å. The scattering power for the electrons is significantly higher than that of X-rays, hence the much smaller sample requirement. An important distinction (and a serious drawback) of electron microscopy from other scattering methods is the necessity of the high vacuum environment in the microscope. This may be achieved by rapid freezing of the sample and maintaining it at or below liquid nitrogen temperature to minimize solvent sublimation during the experiment. Freezing also reduces the radiative damage to the macromolecular structures, although even such samples typically do not tolerate more than 10–15 electrons per $Å^2$ (20). Therefore, in order to obtain cryo-EM data at a high signal-to-noise ratio, 10,000 to 100,000 individual particle images need to be aligned and averaged (20). The presence of any kind of symmetry in the sample enhances the signal-to-noise ratio in the cryo-EM images, although imaging of asymmetric randomly oriented single particles is also possible. In fact, it has been possible in some cases to obtain structural data on single molecules (provided they exceed the lower size limit of several hundred kDa) (23). Cryo-EM provides a means of capturing different conformational states of macromolecular assemblies by imaging the complexes that are "trapped" at different stages of their interactions and conformational changes (23, 24).

Methods of cryo-EM analysis can also be used to characterize biological macromolecules and their assemblies within the cells. Utilization of this technique (usually referred to as cryo-electron tomography, or cryo-ET) to study the in vivo processes is based on the precept of the cryogenic preservation of the native (fully hydrated) biological structure (25). Several recent examples of using cryo-ET to characterize structure of macromolecular complexes within their cellular context demonstrate a promise to reveal the internal structure of organelles and indeed entire cells at the molecular resolution (25–28). Resolutions of 5–8 nm have already been achieved and the prospects for further improvement are good (29). Since many intracellular activities are conducted by complexes in the MDa range with dimensions of 20–50 nm, current resolutions should suffice to identify many of them in tomograms, although the residual noise and the dense packing of cellular constituents hamper interpretation (29). Medalia et al. have recently collected cryo-ET data on vitrified eukaryotic cells (30).

In the following chapters of this book we will refer to several examples of structural studies of macromolecular complexes using cryo-EM, which include ribosomal structure at 11.5 Å resolution (31), mechanism of chaperonin–substrate interaction (32–35), structure of the nuclear pore complex (28), viruses (36), and the crowded cytosolic environment of eukaryotic cells (30). An example of a tomogram of a vitrified cell can be seen in Chapter 11 (Figure 11.24) of this book.

2.2.4. Neutron Scattering

Small-angle neutron scattering (SANS) is an emerging technology (as far as applications to biological systems). It utilizes the same principle as SAXS, although the scattering centers are predominantly nuclei (as opposed to electrons in SAXS). Analysis of the SANS profiles can reveal important information related to the mass and the overall shape of the scattering particles, as well as the distribution of interatomic vectors within a molecule (37). Unfortunately, such experiments remain a rarity, as access to the research facilities possessing neutron beam generators remains limited.

2.3. NMR SPECTROSCOPY

Nuclear magnetic resonance (NMR) spectroscopy has for many years been at the forefront of protein structure determination. Spin energy levels of nuclei, when placed in a magnetic field, become split by the Zeeman effect. Upon application of radio frequency radiation, each nucleus within a molecule can be caused to resonate at a specific frequency, which is highly dependent on its local environment. In the case of proteins, the naturally abundant proton (^{1}H) has nuclear spin $I = \frac{1}{2}$, and by isotopic enrichment methods, other spin-$\frac{1}{2}$ nuclei ^{13}C and ^{15}N can now be readily incorporated into expressed proteins.

In a magnetic field B_0 the spin states of a nucleus become split into equally spaced energy levels according to the magnetic field strength and the gyromagnetic ratio γ, a constant for any given nucleus (where h is Planck's constant):

$$E_m = -\frac{\gamma m h B_0}{2\pi}. \tag{2-3-1}$$

For a spin-$\frac{1}{2}$ nucleus there are two possible values of the magnetic quantum number $m = (-\frac{1}{2}, +\frac{1}{2})$, giving rise to two Zeeman nuclear energy levels E_m under the influence of B_0. Nuclear transitions between the two energy levels may be effected by application of electromagnetic radiation. The difference in energy levels,

$$\Delta E_m = \frac{\gamma h B_0}{2\pi}, \tag{2-3-2}$$

corresponds to radiation of frequency ν_0:

$$\nu_0 = \frac{\omega_0}{2\pi} = \frac{\gamma B_0}{2\pi}. \tag{2-3-3}$$

Thus, as in other spectroscopic methods, transitions between the nuclear energy levels may be produced by irradiation. In older NMR spectrometers the radio frequency spectrum was scanned and each individual resonance could be detected as absorbance. However, whereas in optical spectroscopy irradiation is with a constant light source and absorptive or fluorescent properties are measured, in the case of modern pulsed NMR the highest energy spin level is preferentially populated by irradiation and then the relaxation to the ground state is measured. Upon inversion of nuclear spin by application of a radio frequency (rf) electromagnetic pulse, the magnetization relaxes back to the ground state, and in doing so precesses around the B_0 magnetic vector at its characteristic Larmor precession frequency ω_0. It is this resonance frequency that is detected by the rf receiver coil of an NMR spectrometer.

The power of NMR for structural characterization comes from the fact that each nucleus within a molecule will have a slightly different Larmor frequency, depending on its local environment, that is, shielding by local magnetic fields of nearby electronic motion. Thus, an NMR spectrum will contain peaks corresponding to resonances from each nucleus in the sample, each with a frequency (known as chemical shift, δ) characteristic of the chemical environment. Rather than using absolute frequency units, by convention the resonant frequencies are denoted as parts per million differences, relative to an internal standard, of the carrier frequency (for an 11.7 tesla magnet, for ^{1}H nuclei $\nu_0 \sim 500$ MHz). The so-called chemical shift scale (δ, measured in ppm; $\delta = \omega_0/\nu_0$) is then independent of the magnetic field strength B_0, making data acquired on different instrumentation more readily comparable. For protons, the chemical shift values generally fall in the range 0–10 ppm (relative to an internal reference compound, usually trimethylsilane defined as $\delta = 0$ ppm).

In modern instrumentation, generally all protons within a molecule are simultaneously excited by a broadband rf pulse. This causes all nuclei to simultaneously resonate at their characteristic frequencies, which can be detected as a time domain signal image current by the receiver coil. This signal can then be separated into individual signals in the frequency domain by application of a Fourier transform:

$$S(\nu) = FT\{s(t)\} = \int_{-\infty}^{\infty} s(t)e^{-i2\pi\nu t}\,dt, \qquad \text{(2-3-4)}$$

where $s(t)$ and $S(\nu)$ are the time and frequency domain signals, respectively. The final NMR spectrum consists of frequency domain data with peaks spread across the chemical shift scale representing each resonance. In addition, the area under each peak is proportional to the number of nuclei resonating at that frequency.

An important factor, which makes NMR an incredibly powerful structural tool, is the "cross-talk" between nuclei within a molecule. Perturbation of local electron clouds gives rise to a scalar coupling between nuclei, which are covalently bonded, and is significant for nuclei separated by up to four bonds. The effect of coupling is to alter the energy levels of the two spin states and produce new energy levels depending on whether the spins of the two nuclei are aligned in parallel or antiparallel directions. This leads to multiplet fine structure in the observed NMR peaks. The number and nature of adjacent protons give rise to specific coupling patterns and coupling constants (differences in energy levels, denoted J and measured in hertz) which can be used to assign covalent structure. In proteins, each amino acid side chain gives a characteristic scalar coupling pattern that is invaluable for spin system assignment.

One of the most important NMR phenomena for protein structure determination is an alternative coupling scheme that operates via "through-space" interactions. Once a spin has been inverted by an rf pulse it relaxes to the ground state via various mechanisms. If other nuclei are close enough in space (<5 Å for ^1H), then a dipolar coupling known as the nuclear Overhauser effect (NOE) can also occur. The proximity of other nuclei allows alternative relaxation pathways for the spin, so this manifests itself as a change in the time taken for a nucleus to relax to its ground state. The relaxation effect falls off rapidly with internuclear distance (r) following the relationship NOE $\propto 1/r^6$; hence, the NOE intensity can be measured and used to provide distance constraints in structural calculations.

In the simplest NOE experiment, the resonance of a single nucleus is saturated by continuous radiation (this effectively equalizes the population of the two spin states), followed by broadband excitation of all other nuclei, and subsequent detection. Nuclei that are sufficiently close in space to the saturated nucleus will experience a signal enhancement. The resultant spectrum can be subtracted from a normal spectrum obtained without saturation, yielding a difference spectrum that will contain peaks only from nuclei that are spatially close to the target nucleus. The more complex two-dimensional NOE experiment correlates all nuclei that are in proximity to one another and, once Fourier transformed, produces a spectrum from which cross-peak intensities can be correlated directly to maximal distances. Using the known allowable backbone torsion angles with the network of distance constraints that may be obtained from NOE experiments allows structural models to be built and refined for proteins.

As the number of nuclei in a molecule increases, the limited chemical shift range accessible to protons (generally −2 to 12 ppm) leads to a very crowded NMR spectrum. One way to alleviate this problem is to add a variable delay, or evolution time, into the pulse sequence. In subsequent experiments, this evolution time is increased by a fixed amount each time, leading to a two-dimensional array. Fourier transformation of the array in both dimensions produces a two-dimensional spectrum in which the diagonal corresponds to a simple NMR spectrum, but now cross-peaks are observed that, depending on the chosen pulse sequence, correlate to resonances that are coupled either in a scalar or dipolar manner to one another. This greatly simplifies the process of spectral assignment, since it is now a relatively simple matter to work out which nuclei are correlated through bonds or through space. Multidimensional NMR experiments have evolved complex descriptive acronyms, such as COSY (correlation spectroscopy, to describe scalar couplings) and NOESY (nuclear Overhauser effect spectroscopy, for dipolar couplings). In the latter case, the intensity of the cross-peak is related to the relaxation effect of the neighboring nucleus and can thus be used as a distance constraint for structure determination.

2.3.1. Heteronuclear NMR

In naturally occurring proteins, the only spin-$\frac{1}{2}$ nucleus present at high abundance is ^{1}H. With modern molecular biology techniques, however, isotopically enriched media can be used quite readily to incorporate ^{15}N and ^{13}C into proteins at high levels. In the case of proteins expressed in *E. coli*, for instance, cells are grown on minimal medium containing $^{15}NH_4Cl$ and ^{13}C-glucose as the sole sources of nitrogen and carbon, respectively, which generally produces decent yields of isotopically labeled protein. Now, in the NMR spectrometer, heteronuclear as well as homonuclear coupling come into play. Protons will experience scalar coupling not only from neighboring protons but also from ^{13}C and ^{15}N nuclei. Thus, we can take advantage of the chemical shift dispersion not only of protons but also of the other nuclei, and complex pulse sequences have been developed, spanning multiple dimensions, to trace backbone resonance assignments and provide further constraints for structure refinement of biomolecules (Figure 2.1).

To date, the upper limit for complete resonance assignment of proteins remains at ~30 kDa, due to resonance overlap and unfavorable relaxation times of large biomolecules. Further technological advancements both in magnetic field strength and probe design promise to increase that limit. Recent development of the TROSY pulse sequence (38) has also allowed a major step forward by removing the line-broadening effect of slowly tumbling molecules in solution. Also, partial alignment of proteins within lipid bicelles shows much promise (39, 40).

In addition to the experiments that seek to elucidate structure, there are also NMR strategies aimed specifically at measuring protein dynamics. As mentioned above, relaxation of nuclei after an rf pulse can be affected by a number of factors, one of which is intramolecular internal motions. ^{15}N nuclei, most notably backbone amides, are particularly susceptible to dynamics-related relaxation via the nuclear Overhauser effect. Measurement of the ^{15}N NOE and relaxation times at each amide along the backbone of a protein allows extraction of order parameters, which can be used to determine the dynamic nature of individual segments within a protein. This has enabled flexible regions within proteins to be identified, with correlation times in the picosecond to microsecond time range, and can be correlated to intermolecular interactions and functional dynamics.

2.3.2. Hydrogen Exchange by NMR

As will be described further in later chapters, the rate at which labile backbone amide protons exchange with bulk solvent also serves as a measure, not only of surface accessibility but also of dynamics. Amides that are normally buried, or else hydrogen bonded, can become exposed to solvent during dynamic opening (or local unfolding) events. If the bulk solvent is heavy water ($^{2}H_2O$ or D_2O), then an exchange reaction $>NH \rightarrow >ND$ will cause the disappearance of the corresponding peak in the NMR spectrum, since deuterium is an NMR-invisible isotope (Figure 2.2). Thus, measuring intensities of the amide resonances as a function of time in deuterated solvent can be a powerful residue-specific monitor of dynamics (41, 42). Experimental conditions such as temperature and pH can

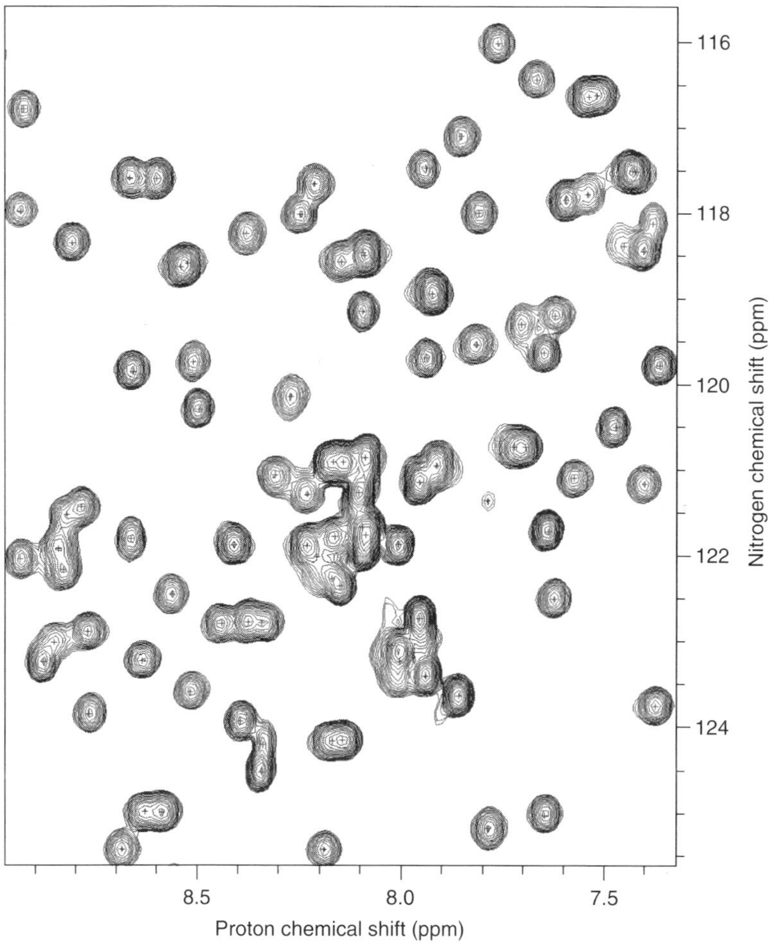

FIGURE 2.1. Amide *fingerprint* region of the ^{15}N-^{1}H HSQC spectrum of a protein. Each contoured peak represents an amide proton whose ^{1}H resonant frequency and that of the adjacent ^{15}N are correlated. In a folded protein these resonances are well resolved due to the chemical shifts induced by persistent structure. Figure courtesy of Dr. Joanna Swain (University of Massachusetts–Amherst).

be manipulated to extract thermodynamic details, and to obtain actual measures of the free energies involved in certain dynamic events (43–48).

Likewise during the folding of a protein, amides that form stable structure earliest in the folding process will become protected against exchange with solvent. In combination with a quenched-flow apparatus, pulse-labeling experiments can be performed that can be used to monitor protein folding on the millisecond time scale. Unfolded protein is allowed to refold for a variable time before labeling all exposed amides with deuterium. For the purposes of studying hydrogen exchange the COSY (correlation of amide NH and $C_{\alpha}H$ protons) and ^{15}N-HSQC

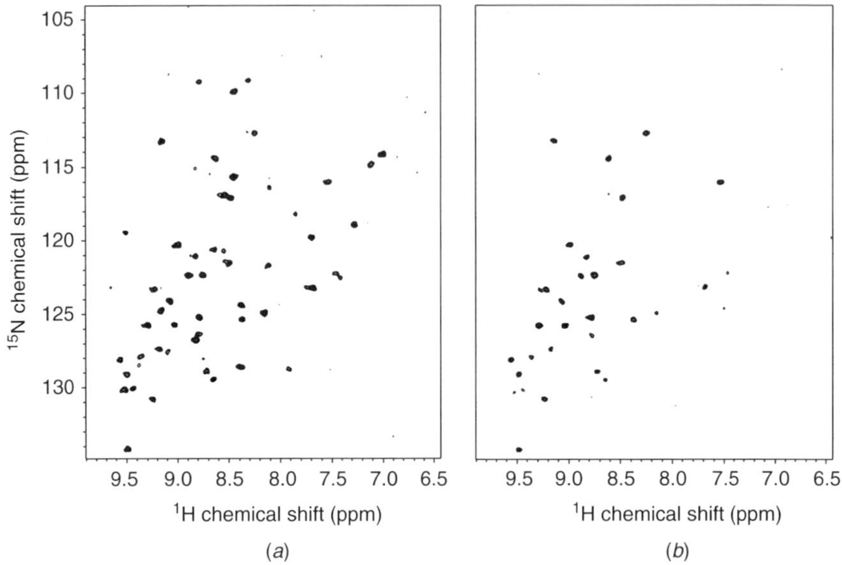

FIGURE 2.2. Dynamics of the C-terminal domain of DnaK in $^{2}H_2O$ measured by HDX NMR at (**A**) $t = 0$ and (**B**) $t = 24$ hours. NMR resonances observed at the earlier time point disappear over time as ^{1}H is replaced by ^{2}H. Figure courtesy of Dr. Joanna Swain (University of Massachusetts–Amherst).

(heteronuclear single quantum coherence, which correlates amide protons with the attached nitrogen in uniformly ^{15}N-labeled proteins) experiments are most useful. In a fully protonated sample, these correlation peaks will have maximum intensity, whereas if the proton becomes replaced by a deuteron due to an exchange event then it will no longer be visible (^{2}H nuclei are not observable in ^{1}H NMR experiments). Thus, in a refolding experiment where unprotected amides are labeled with deuterium, the only peaks observed are those that correspond to residues protected against exchange before the isotope-labeling event. Intensities of amide resonances can then be used to determine the order of formation of structural regions during folding (49–51).

A natural extension of this, which has yielded some exciting results, is the ability to use a flow system in the NMR spectrometer and effectively perform the whole experiment online. Stopped-flow NMR has been used to investigate the folding of proteins in real time (52, 53), although the time resolution is currently somewhat limited compared with stopped-flow optical methods. As instrument sensitivities are improved and more need for high throughput is required, various hybrid instruments have appeared, notably a combination LC system with an outlet to flow-through NMR and electrospray mass spectrometry.

The NMR method, while clearly extremely useful and having high resolution, is limited by the need for complete assignment of all resonances. For large proteins, resonance overlap and peak broadening become significant issues. Additionally, other factors such as paramagnetic ligands may also preclude the use of NMR methods. However, perhaps one of the more serious limitations of NMR to measure protection in pulse-labeling experiments is that it can only measure protection at amino acid residues that are sufficiently protected against exchange during the time course of sample workup and of the NMR experiment itself. Folding experiments in vitro are normally performed at low concentrations ($<10\ \mu M$) to avoid the possibility of aggregation or other nonphysiologically relevant misfolding events. However, for NMR experiments, sample concentrations in the millimolar range are required, so time-consuming concentration steps are often required that may enable further undesirable HDX processes to occur. Even with fast NMR experiments, the total time for sample preparation and measurement may be several hours. This workup time could potentially lead to loss of information in regions of the protein that are less highly protected in the native structure.

The field of NMR spectroscopy is far too wide-ranging to cover anything more than basics in this text. We refer the reader to a number of excellent books on the subject (54, 55).

2.4. OTHER SPECTROSCOPIC TECHNIQUES

2.4.1. Cumulative Measurements of Higher Order Structure: Circular Dichroism

Circular dichroism (CD) spectroscopy utilizes differential absorption of left- and right-polarized light (usually within the 170–700 nm range) by chiral molecules. Since most biological molecules possess multiple chiral chromophores, their CD spectra may provide important insights on the three-dimensional arrangement of such macromolecules with varying specificity. Specifically, CD spectroscopy is often used to measure overall secondary structural content of peptides and proteins (the far-UV region, 190–250 nm), chirality due to unique arrangement of the aromatic side chains and/or disulfide bridges within the protein (250–300 nm), as well as protein interaction with chiral ligands and cofactors (e.g., Soret band). Although the assessment of protein conformations afforded by CD spectroscopy typically yields a low-resolution picture, the experiments are relatively simple and are not very demanding in terms of sample workup and consumption. As a result, CD measurements are often used to complement measurements by other techniques.

Far-UV CD: Evaluation of Secondary Structural Content of Peptides and Proteins. The most ubiquitous chromophore in peptides and proteins is the peptide amide bond, which gives rise to two characteristic absorption bands in the far-UV region: a $\pi \rightarrow \pi^*$ transition at 190 nm and a weaker (and broader) $n \rightarrow \pi^*$ transition at 210 nm (56). Different forms of secondary structure arrange the peptide bonds in a very specific nonrandom asymmetric fashion, giving rise to characteristic and recognizable CD spectra in the far-UV region. Thus, spectra of proteins with mostly α-helical content ($\alpha\alpha$ proteins) have pronounced negative bands at 222 nm and 208 nm (Figure 2.3), a feature notably absent from the spectra of proteins rich in β-sheets. The latter ones are usually characterized by a single minimum whose position may vary within the 210–230 nm range, depending on the protein (Figure 2.3). Far-UV CD spectra of some β-rich proteins may actually resemble model spectra of unordered (random coil) polypeptides, exhibiting a minimum below 200 nm (57). Proteins that contain both α-helices and β-strands have contributions from both secondary structural elements in their CD spectra; however, it is very difficult to distinguish proteins with separate ($\alpha + \beta$) or intermixed (α/β) regions based solely on CD data. These two classes are often treated as a combined $\alpha\beta$ class for the purposes of CD analysis (58).

There are numerous algorithms that can be used to calculate the secondary structural content of a given polypeptide or a protein based on its far-UV CD spectrum. The most straightforward way to obtain a reasonable estimate of the helical content (f_h) uses mean residue ellipticity $[\theta]$ at 208 nm or 222 nm (59):

$$f_h = -\frac{[\theta]_{208} + 4000}{29000}, \tag{2-4-1}$$

$$f_h = -\frac{[\theta]_{222} + 3000}{33000}. \tag{2-4-2}$$

More sophisticated treatments of this problem take into account finite length of helical segments within the protein (both $[\theta]_{208}$ and $[\theta]_{222}$ diminish as helical segments become shorter).

The secondary structural content of the $\alpha\beta$ proteins can also be estimated using various empirical methods. An underlying assumption is that the CD spectrum can be presented as a linear combination of individual secondary structural components, including random coil conformation (more rigorous treatment also takes into account contributions from aromatic chromophores and noise). Usually CD spectra of a set of reference proteins with known structures are used to determine the component secondary structure spectra. Several algorithms have been designed for these purposes, and an interested reader is referred to several excellent recent reviews on the subject (56).

Near-UV CD: Interactions of Aromatic Side Chains and Disulfide Bonds. The near-UV absorption spectra of proteins are dominated by the bands of aromatic side chains (Phe, Tyr and Trp), and their unique arrangement within the protein structure usually gives rise to CD signal (Figures 2.4 and 2.5). The intensity of such peaks depends on a number of aromatic amino acid residues in the protein, as well as on flexibility of the protein. Environment of the aromatic groups and protonation state (for Tyr) also influence the signal. A disulfide bond is another intrinsic protein chromophore that contributes to the near-UV CD spectra (near 260 nm). However, this band is usually rather weak and broad, making it difficult to distinguish disulfide signal from that of aromatic residues. If two (or more) aromatic residues are located in close proximity, coupling of the electronic transitions usually occurs, leading to significant enhancement of the CD signal. These local interactions are more important than the sheer number of the aromatic

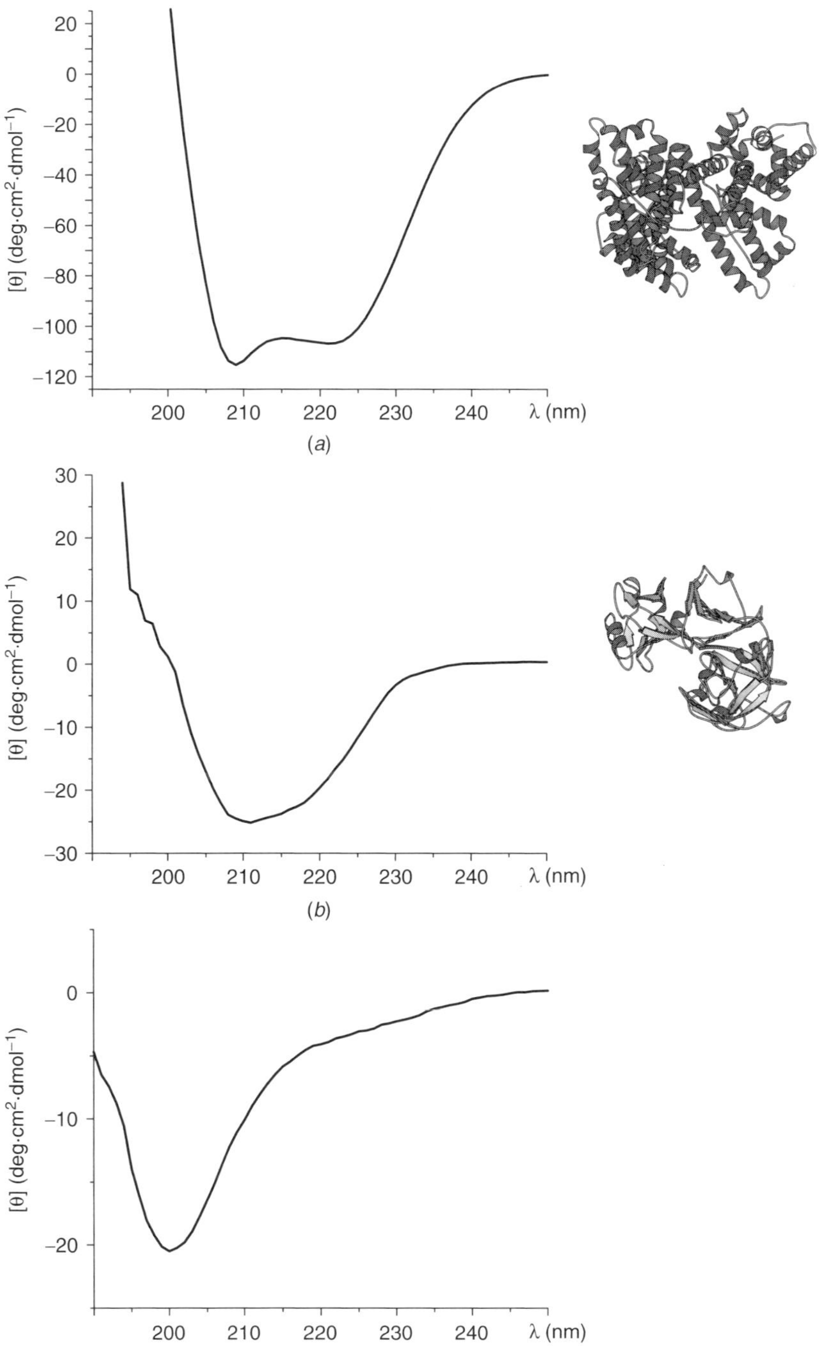

FIGURE 2.3. Far-UV CD spectra of an α-helical protein bovine serum albumin (**A**), a mostly β-sheet protein pepsin (**B**), and a mostly disordered protein β-casein (**C**).

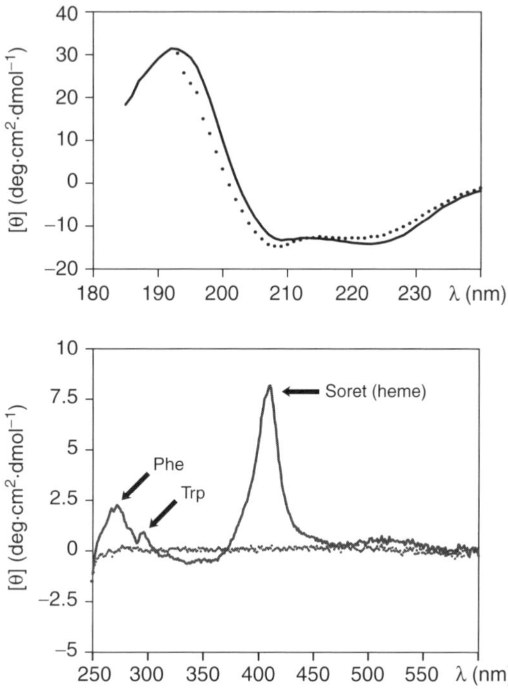

FIGURE 2.4. Acid unfolding of myoglobin monitored with CD spectroscopy in the far-UV (*top*) and near-UV/Vis (*bottom*) regions. The solid lines represent CD spectra of myoglobin acquired under the near-native conditions (10 mM ammonium acetate); the dotted lines represent the spectra acquired in water/methanol solution acidified with 1% CH_3COOH (by volume). Disappearance of the Soret band indicates heme dissociation from the protein, and the disappearance of the aromatic side chain bands (Phe, 7; Tyr, 2; Trp, 2) indicates loss of tertiary contacts, while a significant proportion of the secondary structure (α-helical) is retained.

residues in determining the appearance of the near-UV CD spectra (60). In fact, in many cases it is possible to use near-UV CD spectroscopy to detect local conformational changes that affect the arrangement (and, therefore, transition coupling) of such "proximally clustered" aromatic side chains.

Due to their complexity and the multiplicity of factors influencing the appearance of the near-UV CD spectra, the latter cannot be interpreted as easily as the far-UV CD spectra. Nonetheless, the near-UV CD measurements can also be very useful, particularly in a situation when only a qualitative assessment of a protein ternary structure (or a lack of such) is required. For example, near-UV CD spectroscopy is often used as a means to monitor progressive loss (or gain) of protein ternary structure in response to changing solvent conditions (Figure 2.4) or point mutations. Still, even such qualitative analyses need to be carried out with great care, as the appearance of the protein near-UV CD spectrum can be influenced by a variety of extrinsic factors.

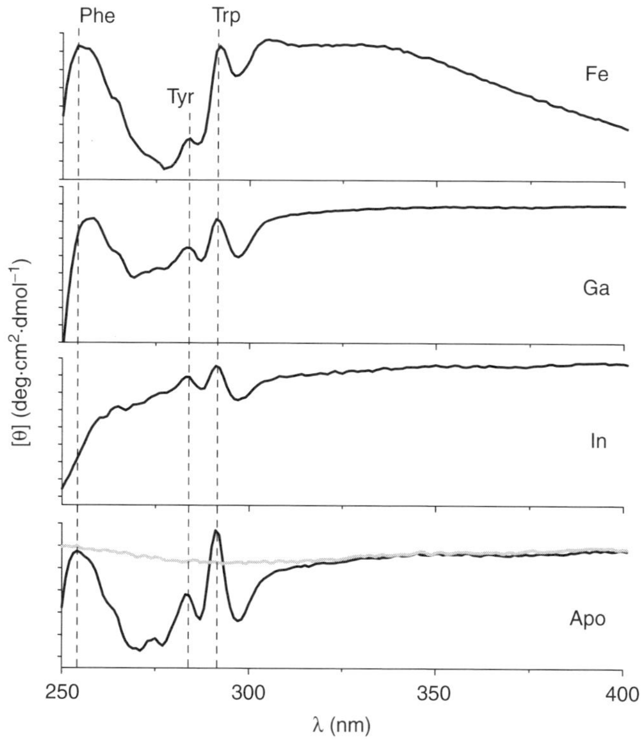

FIGURE 2.5. Near-UV CD spectra of human serum transferrin substituted with Fe^{3+} (*top trace*), Ga^{3+}, In^{3+}, and the apo-form of the protein (*bottom trace*). Spectra are courtesy of Mingxuan Zhang (Graduate Research Assistant at the University of Massachusetts–Amherst).

CD Signals Due to Extrinsic Chromophores. Proteins themselves (i.e., polypeptide chains without bound ligands) usually do not give rise to a CD signal above 300 nm. In many cases, however, this region does contain CD bands that are collectively referred to as extrinsic CD bands (56). These bands represent transitions in ligands, which may have inherent optical activity (due to chirality) or become optically active due to the coupling of electronic transitions on the protein chromophores and the extrinsic chromophore, as well as mixing of transitions of differing symmetry (56). In the latter two cases, the extrinsic chromophore does not have to be optically active to produce CD signal, which thus provides valuable information on the protein–ligand interaction.

One of the specific examples of CD spectroscopic studies of protein–ligand interaction that will be used in later chapters of this book is heme–polypeptide interaction in hemoproteins. A prosthetic heme group has an absorption band at 410 nm (Soret band), which remains "CD-silent" unless the heme is bound to a protein. Asymmetry introduced by positioning of the heme in a binding pocket within the protein leads to appearance of an abundant signal in the CD spectrum of the protein (Figure 2.4). Both position and intensity of this band are influenced by the heme environment within the binding pocket and are often used as reporters of the protein–heme interaction. Figure 2.4 illustrates how different regions of the CD spectra of myoglobin can be used to characterize higher order structure of this protein at different levels (near-UV, secondary structure; far-UV, ternary fold; and Soret band, heme–protein interaction).

It should probably be emphasized once again that significant changes in the near-UV CD spectra do not necessarily manifest large-scale protein conformational changes. Numerous extrinsic factors, as well as subtle changes in the environment of aromatic side chains, often lead to profound alterations of certain CD bands. An example of such behavior is shown in Figure 2.5, where replacement of two metal ions within a 78 kDa protein transferrin results in profound changes in the near-UV CD profiles. However, the protein tertiary structure is not affected by the Fe^{3+} to Ga^{3+} (or In^{3+}) substitution. Instead, these rather dramatic changes in the appearance of the CD spectra reflect changes in the coupling of the electronic transitions of the metal ions and those of the aromatic side chains. Another example is evolution of the Soret band in the CD spectrum of myoglobin at pH above neutral. Although the heme group remains bound to the protein within this pH range, there is a noticeable red shift and a gradual demise of the Soret band as the solution pH is increased above 8. This is caused by deprotonation of some of the residues in the heme-binding pocket of the protein, a process that does not compromise the structural integrity of the heme–protein complex. These two examples emphasize the great care that needs to be exercised when interpreting both near-UV and Vis CD data.

CD Spectroscopy of Other Biopolymers: Oligonucleotides. The nucleic acid chromophores are the bases that have rich high-intensity absorption spectra in the region below 300 nm. The bases do not have intrinsic optical activity; however, they are attached to asymmetric sugars (deoxyribose or ribose rings), which induce CD bands in the absorption region of nucleic acids. Oligonucleotide base stacking in solution forces the entire chain to assume a helical structure, resulting in "super-asymmetry." Electronic transitions in the neighboring bases interact with each other, leading to CD signal amplification. Formation of double-stranded helices by complementary oligonucleotides also leads to super-asymmetry, hence high-intensity and information-rich CD spectra. Due to its sensitivity to the base–base interactions, CD spectroscopy is often used to monitor changes in the secondary structure of oligonucleotides (61). In recent years, CD spectroscopy has also been increasingly applied to study nonclassical nucleic acid structures (e.g., triplex, quadruplex, parallel DNA). A detailed review of such applications can be found in (62).

2.4.2. Vibrational Spectroscopy

Vibrational Absorption (IR) Spectroscopy. Vibrational spectroscopy is another popular tool to study protein higher order structure and dynamics. Because of the enormous amount of normal modes of vibrations in a typical protein, the vibrational bands overlap, giving rise to very complicated IR spectra. Nevertheless, it is often possible to extract information from such spectra that provide valuable insight into protein behavior. The basic principles of protein vibrational spectroscopy can be understood by considering a simple harmonic oscillator, whose frequency is calculated as

$$\nu = \frac{\omega}{2\pi} = \frac{1}{2\pi}\sqrt{\frac{k(m_1 + m_2)}{m_1 m_2}}, \tag{2-4-3}$$

where m_1 and m_2 are the masses of the two atoms in the oscillator and k is a force constant. Any change in the force constant (determined by the interatomic bond electron density) will result in the frequency change. The intensity of the absorption band is determined by a number of factors. For an ideal harmonic oscillator, the transition can occur only to the next vibrational level. In the mid-IR region, the majority of oscillators are not excited at room temperature; hence, only transitions to the first excited state are observed. Probability of such transitions will be determined by the square of the transition dipole moment, TDM (63):

$$\text{TDM} = \left\langle \frac{\partial \mu}{\partial R(R_o)} \right\rangle \sqrt{\frac{h(m_1 + m_2)}{8\pi^2 \nu m_1 m_2}}, \tag{2-4-4}$$

where μ is a dipole moment and h is Planck's constant. The $\langle \partial\mu/\partial R(R_o)\rangle$ term represents an electronic contribution to the transition probability (by linking the latter to the change of the dipole moment at the equilibrium position R_o). Obviously, stronger absorption is favored by larger dipole changes, which are often correlated with significant bond polarities. Therefore, strong absorption bands are often observed for polar bonds (such as carbonyl), while carbon–carbon bands are typically much weaker or absent from the vibrational spectra. The following discussion is intended to provide a brief overview of some of the key features of the vibrational spectra of proteins.

Amide Group Absorption. There are several bands associated with the vibrations of the protein amide groups. The *amide I* band (near 1650 cm^{-1}) is most commonly used for secondary structure analysis, as it is affected by the protein backbone structure (63). This band is mostly due to the carbonyl stretching with minor contributions from other vibrations (out-of-phase C—N stretching, C—C—N deformation, and N—H bending). Although both theoretical considerations as well as experimental work carried out with model protein suggest that different secondary structural elements give rise to distinct amide I bands, they usually overlap in the absorption spectra. As a result, a broad amide I band is typically observed for a majority of proteins (Figure 2.6). Two different approaches are often used to analyze protein secondary structure based on the appearance of the amide I band. The first one aims at decomposing the amide I band into "component bands" corresponding to different types of secondary

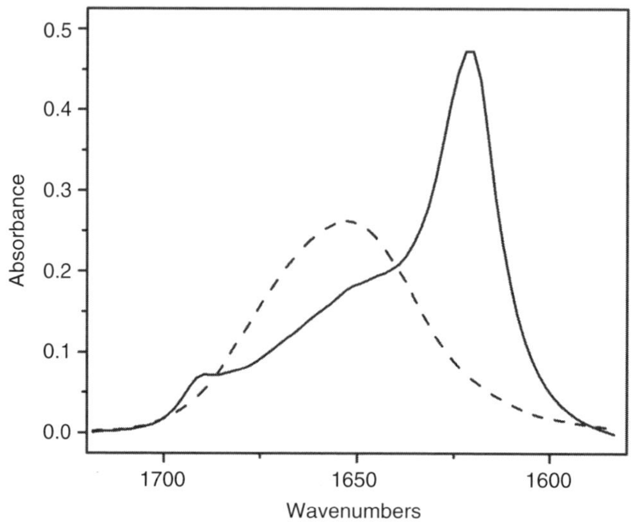

FIGURE 2.6. FTIR spectrum of a synthetic prion peptide at 25 °C (solid line) and 75 °C (dashed line). The peptide is predominantly in a β-sheet conformation at low temperatures, as evidenced by the two sharp peaks at 1621 cm^{-1} and 1691 cm^{-1} in the scan. The β sheets then melt as temperature increases, so that the spectrum at high temperature shows a single broad peak at 1653 cm^{-1} characteristic of a random coil. Figure courtesy of Dr. Sarah Petty (Mt. Holyoke College).

structure. This approach employs various mathematical procedures to resolve the component bands (e.g., by using Fourier deconvolution, taking second derivative), followed by fitting the experimental band with a sum of components (each of which is then assigned to a certain secondary structural element) (63). The second approach utilizes a "calibration set of spectra" (i.e., spectra of proteins with known structure) to perform pattern-recognition calculations. A chemometric procedure (factor analysis) is then used to reduce a large number of spectra in the calibration set to a smaller set of a few linearly independent basis spectra. This set is used to reconstruct the "unknown" spectrum.

The *amide II* (~1550 cm^{-1}) and *amide III* (1400 to 1200 cm^{-1}) bands are also dependent on the secondary structural content; however, the structure–frequency correlation is much less straightforward. As a result, these two bands are rarely used in structural studies (63).

Absorption of Amino Acid Side Chains. The absorption region of most side chains overlaps with the amide I spectral region (1610–1700 cm^{-1}, vide supra), often causing difficulties in spectral interpretation (64). Only two amino acid side chain groups are "interference-free," as they absorb in distinct spectral regions without overlapping with other groups. These are protonated carboxyl groups of Asp and Glu, 1710–1790 cm^{-1} (deprotonated Asp and Glu side chains absorb strongly in the 1550–1580 cm^{-1} region, as well as near 1400 cm^{-1}), and the sulfhydryl group of Cys (2550–2600 cm^{-1}). Assignment of bands corresponding to other side chains is not very easy and usually requires additional experiments. The most popular choice for these experiments is isotope labeling [change in the atomic mass results in a band shift according to (2-4-3)]. This may be achieved by H/D exchange of labile hydrogen atoms, although in many cases it is not very helpful for a variety of reasons (e.g., too many groups are affected by this exchange, resulting in too many changes in the spectrum). Alternatively, one can use uniform isotopic labeling of one amino acid, thus minimizing the number of band shifts.

Absorption of side chains is greatly influenced by their environment, which makes vibrational spectroscopy a sensitive tool to probe both intra- and intermolecular interactions. As such, vibrational spectroscopy often provides an opportunity to obtain information on protein properties that are difficult to discern with other biophysical methods. Only several of these features will be listed here (those that are important within the context of the discussion in the following chapters of the book). Since the bond vibration frequency depends on the electron density [in the form of k in (2-4-3)], FTIR spectroscopy may be used to monitor *redox reactions* (65). Likewise, *metal chelation* by carboxylate groups, as well as *protonation state* of amino acid side chains, often results in significant (and detectable) frequency shifts (63). Finally, bandwidth can be used as a semiquantitative measure of protein conformational heterogeneity (with flexible structures giving broader bands). Since the time scale of infrared absorption measurements is very fast, it is being increasingly used for studies that require high temporal resolution (e.g., enzyme catalysis, charge transfer in redox-active systems) (66).

FTIR spectroscopy can also be used as a means to monitor protein hydrogen/deuterium (H/D) exchange reactions. Replacing ^{1}H in an N—H group with D (^{2}H) results in detectable frequency shifts [as expected from (2-4-3)]. Unfortunately, the measurements are complicated due to significant spectral interference (from side chains, as well as $^{1}H^{2}HO$), making such studies not as conclusive as those carried out with NMR or mass spectrometry.

Finally, it should probably be mentioned again that water (being a strong IR absorber) often interferes with IR spectroscopic measurements. As a result, aqueous solutions of proteins and other macromolecules cannot easily be investigated using regular IR spectroscopic methods, necessitating the use of thin films and other solid samples.

Raman Spectroscopy. The technical difficulties mentioned in the previous section (interference of water absorption bands with those of proteins) can largely be overcome by using more sophisticated vibrational spectroscopic techniques, such as Raman spectroscopy. Although it provides information similar to that extracted form "regular" IR absorption spectrum (i.e., normal modes of vibration), fundamental mechanistic differences (Raman is an inelastic scattering, *not* an absorption technique) lead to significant variation in the quality of information obtained from these experiments. While diminished H_2O and D_2O interference in Raman spectra is certainly a big advantage, Raman spectroscopy suffers several drawbacks that often limit its application. The signal-to-noise ratio is typically inferior to that in the absorption IR spectra and quantitation is usually rather problematic (67).

While most Raman studies target amide I and amide III bands (aiming at secondary and tertiary structure elucidation), there is growing interest in evaluation of structural information that can be obtained from the side chain bands (68).

UV Resonance Raman. If the excitation wavelength lies within a region of the protein electronic absorption, the spectrum will contain only the bands from the chromophore in resonance (off-resonance Raman scattering events would have a tremendously lower probability). Thus, UV resonance Raman (UVRR) spectroscopy provides a means to obtain a spectrum that contains signatures from certain amino acid residues (e.g., Trp and Tyr at 229 nm excitation) or protein backbone (amide $\pi \rightarrow \pi^*$ transition at 206 nm). The Raman spectra obtained using the 229 nm excitation are obviously dependent on the environment of Trp and Tyr residues within the protein. As such, they would reflect tertiary structural arrangement of the protein. The 206 nm excitation is also structurally informative, since the amide bond excitation gives rise to Raman spectra dominated by the amide vibrations whose characteristics (frequency, bandwidth, etc.) depend on the protein secondary structure. Deconvolution of such spectra using a calibration set of "pure" secondary structural elements can provide quantitative information on the secondary structure of proteins being examined (69). One may be compelled to draw parallels between far- and near-UV CD and UVRR utilizing 206 and 229 nm excitation, respectively. In the later chapters of this book, we will return to some examples of using UVRR spectroscopy to monitor evolution of protein secondary structure upon denaturation.

Raman Optical Activity. Raman optical activity (ROA) spectroscopy can be viewed as a hybrid of CD and IR Raman spectroscopy. Current applications of ROA include secondary structure assignment (utilizing mostly amide I and amide III regions), evaluation of protein conformational heterogeneity (sometimes beyond that accessible by CD or NMR), as well as studies of (partially) disordered protein states (70).

2.4.3. Fluorescence: Monitoring Specific Dynamic Events

Fluorescence is another popular spectroscopic technique that is widely used to study macromolecular behavior in solution. The aromatic groups of three amino acids (tryptophan, tyrosine, and phenylalanine) offer intrinsic fluorescent probes of protein conformation, dynamics, and intermolecular interactions. Tryptophan is perhaps the most popular probe, since it occurs in one or a few residues in most proteins. The fluorescence of the indole chromophore is highly sensitive to environment, making tryptophan an ideal choice for reporting protein conformation changes and interactions with other molecules (71). Tryptophan absorbs at 275–295 nm and emits at 320–350 nm (tyrosine also fluoresces in this region, but its contribution is not as significant). In principle, if the protein structure is known, then changes in tryptophan fluorescence could be interpreted in structural terms at atomic resolution (71). However, in practice the fluorescence signature of a protein arises from a large variety of contributions that cannot be adequately modeled.

In its simplest form fluorescence can be used at equilibrium as a measure of structure due to the quenching effect of solvent exposure. In a completely denatured protein, tryptophan residues are exposed to surrounding solvent and fluorescence is quenched by collisions with solvent molecules. On the other hand, a folded protein will generally have a tightly packed hydrophobic core in which aromatic residues are buried and sequestered from solvent (Figure 2.7). This generally leads to an increased fluorescence signal from proteins in structured states, and generally a shift in the emission spectrum to higher frequency (blue shift), due to the reduced dielectric constant in the interior of a folded protein relative to solvent (72). Thus, measuring the fluorescence emission spectrum for a protein can be a useful measure of the extent of burial of aromatic residues. Since this will change during the folding of a protein, monitoring changes of fluorescence in a kinetic manner is a quite sensitive probe of folding kinetics.

Another aspect of fluorescence that has been used extensively for study of protein structure and dynamics is the principle of fluorescence resonance energy

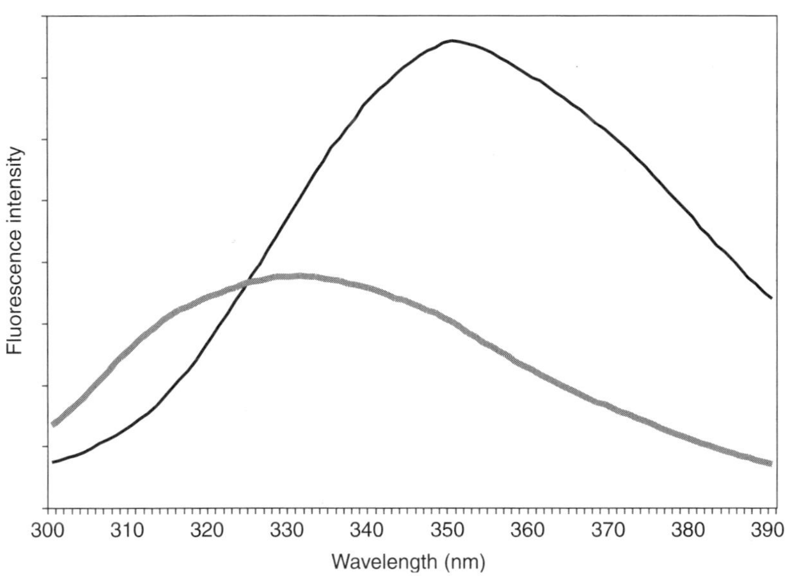

FIGURE 2.7. Tryptophan fluorescence emission spectrum of native and denatured states of cellular retinoic acid binding protein I (CRABP I), excited at 280 nm. The unfolded state shows a broad fluorescence signal centered around 350 nm, characteristic of solvent exposed tryptophan. By contrast, the folded state has a spectrum that is blue-shifted to 328 nm and significantly quenched (less intense) than that of the unfolded state. This is due to the specific quenching effect of tryptophan fluorescence by a proximal cysteine residue and is characteristic of native structure in this protein.

transfer (FRET). Where there is overlap between the emission spectra of one fluorophore (donor) and the absorption spectrum of a second (acceptor), if the two moieties are sufficiently close in space there is the possibility of energy transfer between the two upon excitation of the donor. This manifests itself as a reduction in fluorescence of the donor and enhanced fluorescence of the acceptor, the efficiency of energy transfer depending on the distance between the donor and acceptor pair. Careful choice of chromophores, each of which pair have different characteristic distance of 50% transfer efficiency, or Förster distance (73), allows these to be used as *molecular rulers* from which intermolecular distances (or simply distances between regions of a single macromolecule) can be measured. Dynamics in proteins can also be measured using FRET if the dynamic event involves changing the distance between the two fluorophores. An excellent source of information on the theory of the FRET technique may be found in (74).

Time Resolved Fluorescence. Fluorescence emission is characterized by two parameters, the quantum yield* (Φ_0) and the lifetime of the excited state τ. Both parameters (Φ_0 and τ) are modulated by a number of intrinsic and extrinsic factors, most notably by the environment of the fluorophoric groups within the macromolecule. This is often used to evaluate macromolecular structure and dynamics in solution. Instead of reemitting the photon, a number of other events can occur leading to the decay of the excited state (through a collision with another molecule or energy transfer to another group). The latter event is referred to as *quenching* and can be used to evaluate interaction of the fluorescent groups within the protein. The lifetime of the excited state of a single tryptophan varies from a few hundred picoseconds to 10 ns and can generally be expressed as (71, 75)

$$\frac{1}{\tau} = k_{\mathrm{r}} + k_{\mathrm{isc}} + k_{\mathrm{sol}} + \sum_i k_{\mathrm{q}}^i, \qquad (2\text{-}4\text{-}5)$$

where k_{r} is the radiative rate constant, k_{isc} is the rate constant for the intersystem crossing, and k_{q}^i are rate constants for quenching through different mechanisms. Solvent quenching of the tryptophan fluorescence (k_{sol}) is typically 10^7–10^8 s^{-1} for unstructured polypeptides but is significantly reduced if the protein conformation prevents solvent access to the indole group. Peptide bonds are another common quencher of the tryptophan fluorescence. The rate constants of quenching by side chains exhibit significant variation depending on the mechanism of quenching and the particular side chain involved (75).

Since the tryptophan side chain is typically involved in multiple quenching interactions within the protein, the observed fluorescence decay kinetics is usually very complex even if only one tryptophan residue is present in the protein (75). In a pulsed fluorescence experiment, the emitted fluorescence is analyzed as a function of time following a short intense light pulse (excitation). Finite duration

*Quantum yield is defined as a ratio of the numbers of emitted and absorbed photons and determines the efficiency of the fluorescence for any given molecule.

of the excitation pulse introduces another complication, as the measured total fluorescence intensity $I_F(t)$ is actually a convolution of the excitation pulse profile $g(t)$ and the intensity of radiation emitted by a single fluorescent particle $i_F(t)$ (76):

$$I_F(t) = \int_0^t g(t') \cdot i_F(t - t')\, dt'. \qquad (2\text{-}4\text{-}6)$$

Deconvolution of $i_F(t)$ based on the known shape of $g(t)$ and measured $I_F(t)$ can be carried out using an inverse Fourier transformation procedure (77). If a fluorescent molecule is excited by a polarized light, measurements of the polarization anisotropy of the emitted light can be used to study molecular tumbling (76, 78).

Because of its superior sensitivity, fluorescence can be used to study structure and behavior of macromolecules present in solutions in minute concentrations. Perhaps one of the most exciting recent developments in the field is the emergence of "single molecule spectroscopy," a method of studying individual biopolymers under physiological conditions (79). Monitoring one molecule at a time often provides unique information on distribution functions of relevant observable properties, allowing subpopulations in a heterogeneous sample to be resolved (80–82).

2.5. OTHER BIOPHYSICAL METHODS TO STUDY MACROMOLECULAR INTERACTIONS AND DYNAMICS

2.5.1. Calorimetric Methods

Calorimetry is the best overall biophysical method to extract accurate thermodynamic data about stability of proteins and binding interactions and is in fact the only method to measure the enthalpy change ΔH directly (83, 84). When proteins fold and unfold, or when a protein binds a ligand, for instance, there will be a change in the free energy associated with the event. This is a temperature-dependent phenomenon that can be described by

$$\Delta G(T) = \Delta H(T_m) + \int_{T_m}^{T} \Delta C_p\, dT - T\, \Delta S(T_m) - T \int_{T_m}^{T} \Delta C_p\, d \ln T \qquad (2\text{-}5\text{-}1)$$

Differential scanning calorimetry (DSC) measures the heat capacity change, ΔC_p, of a solution as a function of temperature (85–87). DSC instruments contain two cells, to one of which is added a solution of the biomolecule(s) of interest and to the other (reference cell) an equal volume of solvent. The system is then heated adiabatically at a constant rate and the difference in power required to maintain the two cells at the same temperature represents the excess heat capacity of the sample. To obtain the temperature dependence of the heat capacity, the mass of analyte (m_a) and the partial specific volumes of analyte and solvent

($\overline{v}_a$ and $\overline{v}_s$) must be known. Then

$$C_p(T) = C_p^{\text{solv}} \frac{\overline{v}_a}{\overline{v}_s} + \frac{\Delta C_p^{\text{sol-solv}}}{m_a} \quad \text{(2-5-2)}$$

where $\Delta C_p^{\text{sol-solv}}$ is the measured heat capacity change in the calorimeter and C_p^{solv} is the heat capacity of the solvent alone.

When a protein is thermally denatured its heat capacity will generally change linearly with temperature until it begins to unfold, after which there is a large increase in $C_p(T)$ peaking at the midpoint of unfolding, T_m, as shown in Figure 2.8. The unfolded protein will have a higher heat capacity than the native protein and also increases linearly with temperature. Deviations from the predicted C_p values for a known polypeptide chain may be indicative of residual structure in the thermally denatured state.

From integration under the curve for the DSC thermograms, the calorimetric enthalpy and entropy changes for unfolding can be measured. Using the van't Hoff equation, an alternative enthalpy can be calculated assuming two-state behavior. If the ratio of the enthalpies calculated by both methods is unity, then this is indicative of a cooperative transition between folded and unfolded states. Deviation from unity may suggest either the presence of unfolding intermediates or else an irreversible unfolding process, depending on the direction of deviation.

DSC can also be applied to study the thermodynamics of protein–ligand and protein–DNA association reactions (88, 89). If the interaction is of sufficiently high affinity, then there will be a difference between the heat capacity of the complex versus the sum of those of the isolated components. By performing calorimetric measurements on the separate species and on the complex, it is possible to obtain insight into thermodynamic forces that stabilize the interaction. However, care must be exercised since binding affinities are often very sensitive to external factors such as pH and ionic strength (and pH itself is temperature dependent). Also, deconvolution of thermograms from protein complexes can be difficult, especially if each component of the system undergoes a thermal transition (90).

To investigate interactions between molecules, a more tractable approach is to use a different calorimetric method known as isothermal titration calorimetry (ITC). In this case, the system is held at constant temperature and one binding partner is titrated into a solution containing the second binding partner in a highly controlled manner. As each aliquot of titrant is added, the heat change is measured relative to a reference cell by determining the amount of power required to keep a constant temperature difference between the two cells.

Figure 2.8A shows a typical titration profile. Initially, there is a large heat change as titrant is added, but as binding approaches saturation the heat change diminishes with further addition until finally there is no further change. From these experiments binding curves can be constructed to obtain association constants. In addition, the total heat absorbed or released by the system is the enthalpy change upon binding. Once again this calorimetric value can be compared with van't Hoff enthalpies obtained by other methods to determine whether the association is a cooperative process.

By performing ITC titrations as a function of temperature, the change in heat capacity of binding can also be determined. This represents a measure of the burial of solvent exposed surface area occurring when the two interaction partners

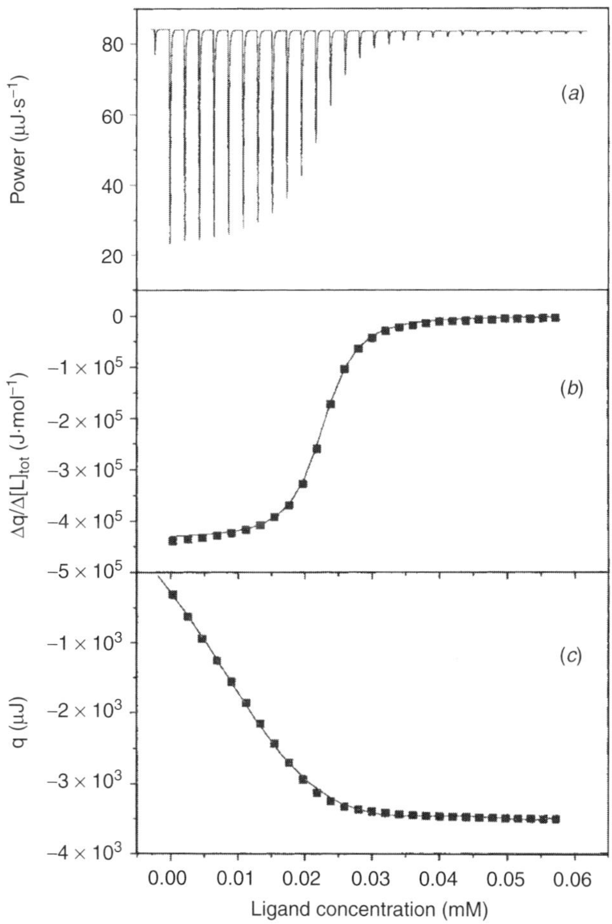

FIGURE 2.8. Calorimetric data obtained from ITC (**A–C**) or DSC (**D**). (**A**) Differential power signal upon titrating complementary strands of 16 bp DNA at 30 °C. Upon each addition heat is absorbed, leading to a spike in the power trace. Integration yields titration curves of heat change (Δq) versus ligand concentration ($[L]$), from which binding constants and enthalpies of association are determined. (**D**) Differential scanning calorimetry thermogram showing the change in excess heat capacity with temperature as a protein undergoes thermal denaturation. Various fitting procedures allow parameters such as the heat capacity change of denaturation (ΔC_p^{den}) to be extracted from the data. Adapted with permission from (83).

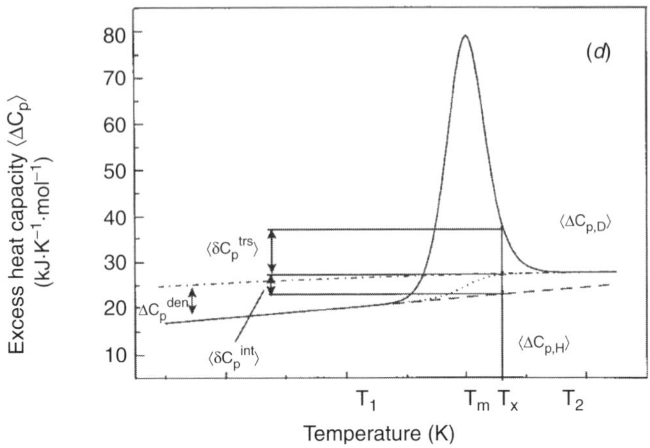

FIGURE 2.8. (*Continued*)

bind. Again, this can also be calculated empirically and deviations between theoretical and experimental values can provide useful information about the nature of the binding interactions. Only in the case of a simple lock-and-key binding mechanism would one expect the two values to coincide. With an induced-fit mechanism the rearrangements required to bind ligand would be associated with an additional change in ΔC_p. This effect is likely even further accentuated in the case of intrinsically disordered proteins, where in order to bind ligand effectively the protein folds around the template ligand, thus producing a large heat change.

2.5.2. Analytical Ultracentrifugation

The technique of analytical ultracentrifugation has been used for many years to study the molecular weights, hydrodynamic properties, and solution interactions in macromolecules (91–93). The method takes advantage of the effect of diffusion of molecules within a solution that can be balanced by the tendency of the molecules to sediment under the influence of a centrifugal force. While NMR and crystallography can yield high-resolution structural data, the ultracentrifuge is extremely versatile and can provide details on size and shape of the sedimenting species and can also be used to obtain equilibrium association constants and to measure extent of aggregation states and conformational changes.

In a spinning rotor a sedimenting force F_s exists that depends on the angular velocity, ω, of the rotor and the radial distance, r, from the axis of rotation:

$$F_s = m\omega^2 r \tag{2-5-3}$$

This force is balanced by a buoyant force F_b that is a function of the solvent density (ρ) and the partial specific volume of the solute ($\overline{v}$), and a frictional force F_f that depends on the velocity of the particle (u) and its shape and size, denoted by the frictional coefficient, f:

$$F_b = -m\overline{v}\rho\omega^2 r, \tag{2-5-4}$$

$$F_f = -fu. \tag{2-5-5}$$

These forces balance and as such these equations may be combined to produce the Svedberg equation:

$$\frac{MD(1-\overline{v}\rho)}{RT} = \frac{u}{\omega^2 r} \equiv s, \tag{2-5-6}$$

where the diffusion coefficient $D = RT/N_A f$, N_A is Avogadro's number, M is the molecular weight of the particle, and s is the sedimentation coefficient, which has the unit of svedbergs ($\mathrm{S} = 10^{-13}$ s).

The experimental setup of the analytical ultracentrifuge involves rotation at a very precisely controlled speed and temperature. In order to sediment macromolecules in a reasonable time frame, the centrifuge is normally capable of speeds up to 60,000 rpm and the rotor chamber is evacuated to reduce friction and turbulence. Additionally, and most importantly, an optical system is present that can precisely measure the radial concentration of analyte across the observation cell. Commercially available instrumentation normally measures optical absorbance or interference due to the difference in refractive index between sample and solute, although fluorometric detection systems are also available (94).

The sample is introduced into one sector of a two-sector cell while the adjacent sector contains only the reference solvent, precisely matched in concentration and ionic strength to that containing the analyte. The cell is then aligned radially in the centrifuge such that sedimentation under centrifugal force can be measured by the variation of solute concentration as a function of radial distance from the rotor axis. The optical system is precisely synchronized to the rotation of the rotor so that observation occurs at the same position in the cell with each pass of the rotor through the light path. In older systems absorbance or interference was measured using film and then measuring the intensity of the photographic image or counting interference fringes manually, but with the advent of CCD cameras and computer imaging systems this is all now computer controlled.

Two types of experiment can be performed in an analytical ultracentrifuge, the first of which is known as sedimentation velocity. This method can be used to rapidly assess the number of species present in a mixture, whether there are interacting species, and the sedimentation and diffusion coefficients. It is often used as a precursor to more detailed study by sedimentation equilibrium (see below).

An initially uniform solution of analyte is subjected to a sufficiently high angular velocity that the solute begins to sediment rapidly toward the bottom of the cell (Figure 2.9A). As solute is depleted from the meniscus region of the sample column, a sharp boundary forms that shifts toward the bottom of the cell with time and also broadens due to diffusion. The rate of sedimentation depends on the molecular weight and shape of the solute molecules. From these experiments, the rate of broadening of the boundary can be used to calculate the diffusion coefficient, and the rate of sedimentation gives the sedimentation coefficient, from which the molecular weight of the analyte can be calculated. The latter can be determined according to the equation

$$s = \frac{dr_{\mathrm{bnd}}/dt}{\omega^2 r} \tag{2-5-7}$$

$$\ln(r_{\mathrm{bnd}}/r_m) = s\omega^2 t \tag{2-5-8}$$

where r_{bnd} is the radial distance of the boundary from the axis, and r_m is the distance of the solvent meniscus. Thus, the sedimentation coefficient can be determined from a plot of $\ln(r_{\mathrm{bnd}})$ versus time. In practice, a variety of

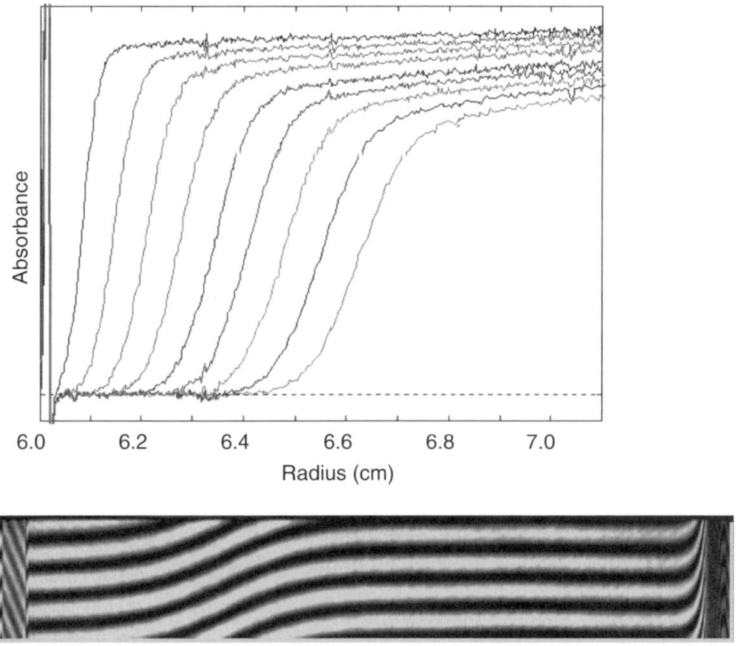

(*a*)

FIGURE 2.9. (**A**) Typical traces obtained from sedimentation velocity, and (**B**) sedimentation equilibrium experiments using analytical ultracentrifugation. In (**A**) the change in absorbance along the length of the observation cell is plotted, each curve representing a different time point. As the experiment proceeds, absorbing material begins to sediment, leading to a depletion in the absorbance in the bulk solution and the sedimenting boundary shifts along the cell. In (**B**) equilibrium has been reached where sedimentation and diffusion forces are exactly balanced, leading to a distribution of material throughout the cell whose properties can be used to determine molecular parameters such as molecular weight and dissociation constants.

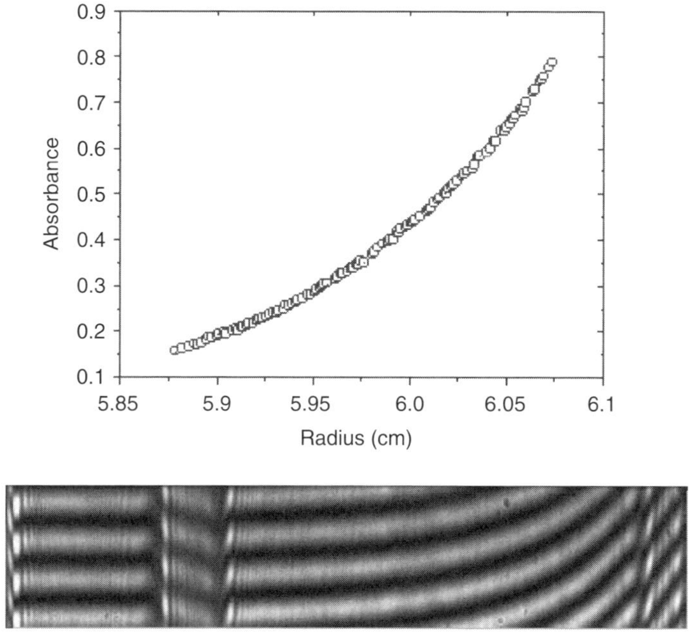

(*b*)

FIGURE 2.9. (*Continued*)

alternative methods have been developed to determine sedimentation coefficients that account for such factors as nonideality of the solution, radial dilution, self-association, concentration dependence of the sedimentation coefficient, and even sedimentation occurring during the initial rotor acceleration. Perhaps one of the most elegant of these involves computational numerical solution of the Lamm equation (92, 95, 96):

$$\frac{dc(r,t)}{dt} = \frac{1}{r}\frac{d}{dr}\left[rD\frac{dc(r,t)}{dr} - s\omega^2 r^2 c(r,t)\right], \tag{2-5-9}$$

which represents a general equation for sedimentation of a particle of sedimentation coefficient s and diffusion coefficient D. Using the wealth of time-dependent concentration data obtained from the ultracentrifuge, the above equation can rapidly be solved computationally to fit the data and obtain values for the coefficients of the sedimenting species. This method has been used successfully to measure molecular weight distributions ranging from small molecules all the way up to viral capsids (>1 MDa).

In the case of multiple populations present in the sample, the individual components will sediment at different rates according to their molecular weight and shape. If the components are sufficiently different, multiple boundaries can be observed, enabling extraction of parameters for the individual species. One should note that the sedimentation coefficients are not additive. For instance, ribosomes consist of two subunits that have molecular weights corresponding to 40S and 60S. However, the intact ribosome has a sedimentation coefficient of 80S.

The second type of experiment involves much lower rotor speeds, and instead of simply sedimenting the solute molecules, they are allowed to come to equilibrium where sedimentation forces are exactly balanced by diffusion. In this case, the boundaries between solvent and solute reach an equilibrium radial distribution; solute concentration increases exponentially with increasing radius and becomes invariant with time (Figure 2.9B). Because it is an equilibrium method, this technique can be used to obtain thermodynamic information about interacting species. In the case of a reversible association such as protein–protein or protein–nucleotide binding, one can obtain equilibrium dissociation constants. It can also be used to detect the existence of nonspecific associations and/or aggregation events. The disadvantage is that whereas a sedimentation velocity experiment can be completed in a matter of hours, it can sometimes take several days for the system to reach a satisfactory equilibrium in the ultracentrifuge. In many cases, this can be circumvented by using short solution columns, but at the potential expense of resolution from using only a small portion of the observation cell for measurements.

At equilibrium the buoyant molecular weight of a solvent may be derived from Equation (2-5-4):

$$M = \frac{2RT}{(1 - \overline{v}\rho)\omega^2} \cdot \frac{d(\ln c)}{dr^2}. \tag{2-5-10}$$

Thus, at equilibrium the molecular weight of a single species can be determined from the slope of a plot of ln c versus r^2. In practice, however, very few biomolecules exist purely as monomers and the real power of equilibrium sedimentation is the ability to probe these intermolecular interactions. As the concentration varies across the radial length of the cell, so too will the degree of association by the law of mass action. At the top of the cell the concentration of oligomer is negligible so monomer molecular weights can be extracted, whereas toward the bottom of the cell the increased molecular weight resulting from associations can be probed.

Normally, a complete data set will include experiments at a series of concentrations, rotor speeds, and even temperatures, from which a global fitting procedure can be applied to model different interaction scenarios (e.g., monomer–dimer, monomer–tetramer). Other treatments can determine whether these associations are truly reversible or are in fact aggregation events, and also electrostatic effects that can lead to solution nonideality and hence ambiguity of the results. These complicated procedures are beyond the scope of this text, but excellent review articles and introductory texts are available (91, 92, 96–99). Careful and thorough analysis, however, can be used to gain extremely valuable details about molecular weights and thermodynamics of association and has even been applied to investigate the effects of detergents on membrane proteins and the effects of macromolecular crowding, as is likely the case in the intracellular environment.

2.5.3. Surface Plasmon Resonance

This relatively recent method has become very popular for the study of protein–ligand and protein–protein interactions. The technique relies on a phenomenon where incident light passes between a high refractive index (n_1) and low refractive index (n_2, where $n_2 < n_1$) medium when there is an electrically conducting surface at the interface (100–102). Above a critical angle (θ) of incident light, the beam is reflected at the interface back into the high refractive index medium. However, in doing so it also sets up an electric field in the conducting surface that can lead to a conversion of photon energy into plasmon energy at the surface in a manner that is critically dependent on the refractive index, n_2, at the surface. If the wave vectors of the photon (k_x) and plasmon (k_{sp}) energies are equal in magnitude and direction, then a resonant effect occurs (known as surface plasmon resonance, SPR) between the two that manifests itself as a reduction in intensity of the reflected light:

$$k_x = \frac{2\pi}{\lambda} n_1 \sin\theta, \tag{2-5-11}$$

$$k_{sp} = \frac{2\pi}{\lambda} \sqrt{\frac{n_{gold}^2 n_2^2}{n_{gold}^2 + n_2}}, \tag{2-5-12}$$

where n_{gold} is a constant, the refractive index of the gold layer, and λ is the wavelength of the incident light. k_x can be tuned to match k_{sp} either by varying the wavelength or angle of the incident beam. In practice, the angle θ is varied and the angle of minimum reflectance intensity due to the SPR effect is measured. The refractive index, n_2, at the surface can then be calculated by equating (2-5-11) and (2-5-12) ($k_{sp} = k_x$). If a protein is immobilized very close to the surface, then binding of another protein or ligand will cause a change in concentration, and hence refractive index, n_2, at the surface. Thus, the SPR phenomenon can be used as a very sensitive probe of concentration changes and hence protein–ligand interactions.

A glass slide coated with a thin layer of gold can be decorated at its surface with the biomolecule of interest. A variety of chemistries have been developed to allow this to happen readily; for instance, a thin layer of carboxymethylated dextran allows attachment of proteins via a variety of reactive side chains (amine, thiol, carboxylate), while also providing a hydrophilic environment to accommodate the binding interaction. More specific coatings such as streptavidin can be used to select biotinylated proteins or nickel nitrilotriacetate will specifically bind proteins with histidine tags.

The plate is then placed into the instrument with the protein-decorated surface exposed to a liquid flow chamber. As buffer containing the interaction partner of the immobilized protein is flowed over the surface, protein–protein interactions will occur, and the binding event causes a change in the refractive index of the surface that is dependent on the concentration of bound species. This refractive index change is detected as a change in the SPR response and can thus be used quantitatively to measure binding constant. This is an extremely sensitive and versatile technique that can be used to screen interactions of proteins with ligands. Since the target protein is linked either covalently or by a high-affinity interaction with the surface, ligands can in general be washed off by adjusting buffer conditions, thus regenerating the surface so that the experiment can be repeated with different concentrations of ligand or else completely different molecules. Some information can be obtained about the kinetics of binding from the time course of development of the SPR response, and thermodynamic binding data can be extracted from the concentrations of bound ligand at steady state. However, care must be taken to ensure that the bound proteins are in the correct orientation on the surface to enable efficient binding of ligand, and also that the immobilization process does not in some way compromise the structural integrity or ligand binding efficacy of the target protein.

2.5.4. Gel Filtration

The means by which a macromolecule passes through a chromatography column packed with silica gel depends empirically on its size and shape. Larger molecules will be excluded from the pores formed within the gel bed and hence will pass through the column rapidly. On the other hand, small molecules can enter at least partially into these cavities and hence passage of these molecules will be retarded. However, gel filtration (also known as size exclusion) chromatography is not a simple method to characterize molecular weight since shape also has a significant effect. Regions of a protein with an extended conformation may become entangled in the gel pores and hence cannot pass through the column as rapidly as one would predict from their molecular weight. Similarly, molecules with unusual charge distributions or patches of hydrophobic residues may interact with the column resin in unpredictable ways.

Nevertheless, gel filtration has uses for studying protein–protein interactions. Clearly, multimeric forms should have a different mobility than the monomer. With careful calibration and taking into account the many caveats, gel filtration can be a helpful empirical tool for the investigation of protein size and intermolecular interactions.

2.5.5. Gel Electrophoresis

Electrophoresis involves applying an electrical potential across a thin (usually) polyacrylamide gel held between glass plates, to which samples of analyte biomolecules have been applied at one end. Under the influence of the electric field, the analyte will enter the gel and pass through it depending on a combination of molecular weight, molecular shape, and charge on the molecule.

The most popular method today involves denaturing the protein sample by boiling with β-mercaptoethanol and the anionic detergent sodium dodecylsulfate (SDS). The former reagent reduces all disulfide bridges and the latter causes complete denaturation and encapsulates the proteins in detergent micelles that have an approximately constant charge-to-mass ratio. Protein molecules will thus migrate through the gel matrix in a predictable manner, with smaller molecules passing through the gel more rapidly, high molecular weight ones more slowly. Once the samples have been separated but before they pass completely through the gel, the electric field is removed and the proteins may be visualized by staining with a protein binding dye. By comparison with a mixture of calibration standards, the molecular weight can be determined with some degree of precision.

Since this electrophoretic technique employs denaturing conditions, it is not useful for looking at biologically relevant macromolecular interactions. However, it is possible to use nondenaturing conditions and apply native proteins to the gel. Due to the above dependencies on size, shape, and charge, interpretation of the mobility of analyte through a native gel is not trivial. Normally, it is necessary to run a suite of gels with different degrees of cross-linking in the gel matrix to obtain useful data. Once again, this is a simple empirical technique that can be useful for initial screening of proteins and their interactions but requires very careful calibration to obtain meaningful information.

In this chapter we have briefly reviewed a number of the biophysical methods routinely used in the laboratory to study protein size, shape, structure, and dynamics. Some of these are very low resolution, others extremely high resolution, providing anything from a quick and general view of protein associations right through to a picture at the atomic level. In the following chapters we will demonstrate some of the many applications of mass spectrometry to answer biological questions. In many ways, these techniques are complementary to the methods described herein, providing confirmatory evidence and in many cases unique information. No single biophysical method is sufficient to fully describe a system, but in the next chapters we shall see the power of MS in the biophysical arena.

REFERENCES

1. Rhodes G. (2000). *Crystallography Made Crystal Clear. A Guide for Users of Macromolecular Models.* San Diego, CA: Academic Press.
2. Schmidt A., Lamzin V. S. (2002). Veni, vidi, vici—atomic resolution unravelling the mysteries of protein function. *Curr. Opin. Struct. Biol.* **12**: 698–703.
3. Longhi S., Czjzek M., Cambillau C. (1998). Messages from ultrahigh resolution crystal structures. *Curr. Opin. Struct. Biol.* **8**: 730–737.
4. Lario P. I., Vrielink A. (2003). Atomic resolution density maps reveal secondary structure dependent differences in electronic distribution. *J. Am. Chem. Soc.* **125**: 12787–12794.
5. Arora A., Tamm L. K. (2001). Biophysical approaches to membrane protein structure determination. *Curr. Opin. Struct. Biol.* **11**: 540–547.
6. Werten P. J., Remigy H. W., de Groot B. L., Fotiadis D., Philippsen A., Stahlberg H., Grubmuller H., Engel A. (2002). Progress in the analysis of membrane protein structure and function. *FEBS Lett.* **529**: 65–72.

7. Stahlberg H., Fotiadis D., Scheuring S., Remigy H., Braun T., Mitsuoka K., Fujiyoshi Y., Engel A. (2001). Two-dimensional crystals: a powerful approach to assess structure, function and dynamics of membrane proteins. *FEBS Lett.* **504**: 166–172.
8. Read R. J. (1996). As MAD as can be. *Structure* **4**: 11–14.
9. Mozzarelli A., Rossi G. L. (1996). Protein function in the crystal. *Annu. Rev. Biophys. Biomol. Struct.* **25**: 343–365.
10. Moffat K. (2001). Time-resolved biochemical crystallography: a mechanistic perspective. *Chem. Rev.* **101**: 1569–1581.
11. Petsko G. A., Ringe D. (2000). Observation of unstable species in enzyme-catalyzed transformations using protein crystallography. *Curr. Opin. Chem. Biol.* **4**: 89–94.
12. Murphy R. M. (1997). Static and dynamic light scattering of biological macromolecules: what can we learn? *Curr. Opin. Biotechnol.* **8**: 25–30.
13. Yamaguchi T., Adachi K. (2002). Hemoglobin equilibrium analysis by the multiangle laser light-scattering method. *Biochem. Biophys. Res. Commun.* **290**: 1382–1387.
14. Murphy R. M., Pallitto M. M. (2000). Probing the kinetics of beta-amyloid self-association. *J. Struct. Biol.* **130**: 109–122.
15. Wall M. E., Gallagher S. C., Trewhella J. (2000). Large-scale shape changes in proteins and macromolecular complexes. *Annu. Rev. Phys. Chem.* **51**: 355–380.
16. Doniach S. (2001). Changes in biomolecular conformation seen by small angle X-ray scattering. *Chem. Rev.* **101**: 1763–1778.
17. Svergun D. I. (1999). Restoring low resolution structure of biological macromolecules from solution scattering using simulated annealing. *Biophys. J.* **76**: 2879–2886.
18. Chen L., Hodgson K. O., Doniach S. (1996). A lysozyme folding intermediate revealed by solution X-ray scattering. *J. Mol. Biol.* **261**: 658–671.
19. Segel D. J., Eliezer D., Uversky V., Fink A. L., Hodgson K. O., Doniach S. (1999). Transient dimer in the refolding kinetics of cytochrome *c* characterized by small-angle X-ray scattering. *Biochemistry* **38**: 15352–15359.
20. Unger V. M. (2001). Electron cryomicroscopy methods. *Curr. Opin. Struct. Biol.* **11**: 548–554.
21. Chiu W., McGough A., Sherman M. B., Schmid M. F. (1999). High-resolution electron cryomicroscopy of macromolecular assemblies. *Trends Cell. Biol.* **9**: 154–159.
22. Amos L. A. (2000). Focusing-in on microtubules. *Curr. Opin. Struct. Biol.* **10**: 236–241.
23. Frank J. (2002). Single-particle imaging of macromolecules by cryo-electron microscopy. *Annu. Rev. Biophys. Biomol. Struct.* **31**: 303–319.
24. Saibil H. R. (2000). Conformational changes studied by cryo-electron microscopy. *Nat. Struct. Biol.* **7**: 711–714.
25. Steven A. C., Aebi U. (2003). The next ice age: cryo-electron tomography of intact cells. *Trends Cell Biol.* **13**: 107–110.
26. Marsh B. J., Mastronarde D. N., Buttle K. F., Howell K. E., McIntosh J. R. (2001). Inaugural article: Organellar relationships in the Golgi region of the pancreatic beta

cell line, HIT-T15, visualized by high resolution electron tomography. *Proc. Natl. Acad. Sci. U.S.A.* **98**: 2399–2406.

27. Baumeister W. (2002). Electron tomography: toward visualizing the molecular organization of the cytoplasm. *Curr. Opin. Struct. Biol.* **12**: 679–684.
28. Stoffler D., Feja B., Fahrenkrog B., Walz J., Typke D., Aebi U. (2003). Cryo-electron tomography provides novel insights into nuclear pore architecture: implications for nucleocytoplasmic transport. *J. Mol. Biol.* **328**: 119–130.
29. Grunewald K., Medalia O., Gross A., Steven A. C., Baumeister W. (2003). Prospects of electron cryotomography to visualize macromolecular complexes inside cellular compartments: implications of crowding. *Biophys. Chem.* **100**: 577–591.
30. Medalia O., Weber I., Frangakis A. S., Nicastro D., Gerisch G., Baumeister W. (2002). Macromolecular architecture in eukaryotic cells visualized by cryoelectron tomography. *Science* **298**: 1209–1213.
31. Gabashvili I. S., Agrawal R. K., Spahn C. M., Grassucci R. A., Svergun D. I., Frank J., Penczek P. (2000). Solution structure of the *E. coli* 70S ribosome at 11.5 Å resolution. *Cell* **100**: 537–549.
32. Falke S., Fisher M. T., Gogol E. P. (2001). Classification and reconstruction of a heterogeneous set of electron microscopic images: a case study of GroEL–substrate complexes. *J. Struct. Biol.* **133**: 203–213.
33. Falke S., Fisher M. T., Gogol E. P. (2001). Structural changes in GroEL effected by binding a denatured protein substrate. *J. Mol. Biol.* **308**: 569–577.
34. Ranson N. A., Farr G. W., Roseman A. M., Gowen B., Fenton W. A., Horwich A. L., Saibil H. R. (2001). ATP-bound states of GroEL captured by cryo-electron microscopy. *Cell* **107**: 869–879.
35. Carrascosa J. L., Llorca O., Valpuesta J. M. (2001). Structural comparison of prokaryotic and eukaryotic chaperonins. *Micron* **32**: 43–50.
36. Tang L., Johnson J. E. (2002). Structural biology of viruses by the combination of electron cryomicroscopy and X-ray crystallography. *Biochemistry* **41**: 11517–11524.
37. Byron O., Gilbert R. J. (2000). Neutron scattering: good news for biotechnology. *Curr. Opin. Biotechnol.* **11**: 72–80.
38. Pervushin K., Riek R., Wider G., Wuthrich K. (1997). Attenuated T-2 relaxation by mutual cancellation of dipole–dipole coupling and chemical shift anisotropy indicates an avenue to NMR structures of very large biological macromolecules in solution. *Proc. Natl. Acad. Sci. U.S.A.* **94**: 12366–12371.
39. Losonczi J. A., Prestegard J. H. (1998). Improved dilute bicelle solutions for high-resolution NMR of biological macromolecules. *J. Biomol. NMR* **12**: 447–451.
40. Tjandra N., Bax A. (1997). Direct measurement of distances and angles in biomolecules by NMR in a dilute liquid crystalline medium. *Science* **278**: 1697–1697.
41. Woodward C. (1993). Is the slow-exchange core the protein-folding core? *Trends Biochem. Sci.* **18**: 359–360.
42. Li R. H., Woodward C. (1999). The hydrogen exchange core and protein folding. *Protein Sci.* **8**: 1571–1590.
43. Parker M. J., Marqusee S. (2001). A kinetic folding intermediate probed by native state hydrogen exchange. *J. Mol. Biol.* **305**: 593–602.

44. Englander S. W. (2000). Protein folding intermediates and pathways studied by hydrogen exchange. *Annu. Rev. Biophys. Biomol. Struct.* **29**: 213–238.
45. Englander S. W., Mayne L., Bai Y., Sosnick T. R. (1997). Hydrogen exchange: the modern legacy of Linderstrom-Lang. *Protein Sci.* **6**: 1101–1109.
46. Raschke T. M., Marqusee S. (1997). The kinetic folding intermediate of ribonuclease H resembles the acid molten globule and partially unfolded molecules detected under native conditions. *Nat. Struct. Biol.* **4**: 298–304.
47. Chamberlain A. K., Handel T. M., Marqusee S. (1996). Detection of rare partially folded molecules in equilibrium with the native conformation of RNaseH. *Nat. Struct. Biol.* **3**: 782–787.
48. Bai Y. W., Sosnick T. R., Mayne L., Englander S. W. (1995). Protein-folding intermediates—native-state hydrogen-exchange. *Science* **269**: 192–197.
49. Roder H., Elove G. A., Englander S. W. (1988). Structural characterization of folding intermediates in cytochrome-*c* by H-exchange labeling and proton NMR. *Nature* **335**: 700–704.
50. Miranker A., Radford S. E., Karplus M., Dobson C. M. (1991). Demonstration by NMR of folding domains in lysozyme. *Nature* **349**: 633–636.
51. Miranker A., Robinson C. V., Radford S. E., Dobson C. M. (1996). Investigation of protein folding by mass spectrometry. *FASEB J.* **10**: 93–101.
52. Balbach J., Forge V., Lau W. S., van Nuland N. A. J., Brew K., Dobson C. M. (1996). Protein folding monitored at individual residues during a two-dimensional NMR experiment. *Science* **274**: 1161–1163.
53. Balbach J., Forge V., van Nuland N. A. J., Winder S. L., Hore P. J., Dobson C. M. (1995). Following protein-folding in real-time using NMR-spectroscopy. *Nat. Struct. Biol.* **2**: 865–870.
54. Cavanagh J., Fairbrother W. J., Palmer A. G. III, Skelton N. J. (1995). *Protein NMR Spectroscopy: Principles and Practice*. San Diego: Academic Press.
55. Levitt M. H. (2001). *Spin Dynamics: Basics of Nuclear Magnetic Resonance*. Hoboken, NJ: John Wiley & Sons.
56. Sreerama N., Woody R. W. (2000). Circular dichroism of peptides and proteins. In *Circular Dichroism: Principles and Applications.*, ed. R. W. Woody, pp. 601–620. New York: Wiley-VCH.
57. Sreerama N., Woody R. W. (2003). Structural composition of beta-I- and beta-II-proteins. *Protein Sci.* **12**: 384–388.
58. Sreerama N., Venyaminov S., Woody R. (1999). Estimation of the number of alpha-helical and beta-strand segments in proteins using circular dichroism spectroscopy. *Protein Sci.* **8**: 370–380.
59. Pelton J. T., McLean L. R. (2000). Spectroscopic methods for analysis of protein secondary structure. *Anal. Biochem.* **277**: 167–176.
60. Sreerama N., Manning M. C., Powers M. E., Zhang J. X., Goldenberg D. P., Woody R. W. (1999). Tyrosine, phenylalanine, and disulfide contributions to the circular dichroism of proteins: circular dichroism spectra of wild-type and mutant bovine pancreatic trypsin inhibitor. *Biochemistry* **38**: 10814–10822.
61. Johnson W. C. (2000). CD of nucleic acids. In *Circular Dichroism: Principles and Applications*, ed. R. W. Woody, pp. 703–718. New York: Wiley-VCH.

62. Maurizot J. C. (2000). Circular dichroism of nucleic acids: nonclassical conformations and modified oligonucleotides. In *Circular Dichroism: Principles and Applications*, ed. R. W. Woody, pp. 719–740. New York: Wiley-VCH.

63. Barth A., Zscherp C. (2002). What vibrations tell us about proteins. *Q. Rev. Biophys.* **35**: 369–430.

64. Barth A. (2000). The infrared absorption of amino acid side chains. *Prog. Biophys. Mol. Biol.* **74**: 141–173.

65. Breton J. (2001). Fourier transform infrared spectroscopy of primary electron donors in type I photosynthetic reaction centers. *Biochim. Biophys. Acta* **1507**: 180–193.

66. Zscherp C., Barth A. (2001). Reaction-induced infrared difference spectroscopy for the study of protein reaction mechanisms. *Biochemistry* **40**: 1875–1883.

67. Thomas G. J. Jr. (1999). Raman spectroscopy of protein and nucleic acid assemblies. *Annu. Rev. Biophys. Biomol. Struct.* **28**: 1–27.

68. Thomas G. J. Jr. (2002). New structural insights from Raman spectroscopy of proteins and their assemblies. *Biopolymers* **67**: 214–225.

69. Chi Z., Chen X. G., Holtz J. S., Asher S. A. (1998). UV resonance Raman-selective amide vibrational enhancement: quantitative methodology for determining protein secondary structure. *Biochemistry* **37**: 2854–2864.

70. Barron L. D., Hecht L., Blanch E. W., Bell A. F. (2000). Solution structure and dynamics of biomolecules from Raman optical activity. *Prog. Biophys. Mol. Biol.* **73**: 1–49.

71. Chen Y., Barkley M. D. (1998). Toward understanding tryptophan fluorescence in proteins. *Biochemistry* **37**: 9976–9982.

72. Lakowicz J. R. (1999). *Principles of Fluorescence Spectroscopy*. New York: Kluwer Academic/Plenum.

73. Förster T. (1948). Intermolecular energy migration and fluorescence. *Ann. Phys. (Leipzig)* **2**: 55–75.

74. van der Meer B. W., Coker G. III, Chen S.-Y. (1994). *Resonance Energy Transfer: Theory and Data*. New York: VCH.

75. Engelborghs Y. (2001). The analysis of time resolved protein fluorescence in multi-tryptophan proteins. *Spectrochim. Acta A: Mol. Biomol. Spectrosc.* **57**: 2255–2270.

76. Daune M. (1999). *Molecular Biophysics: Structures in Motion*. New York: Oxford University Press.

77. Kauppinen J., Partanen J. (2001). *Fourier Transforms in Spectroscopy*. New York: Wiley-VCH.

78. Fersht A. (1999). *Structure and Mechanism in Protein Science: A Guide to Enzyme Catalysis and Protein Folding*. New York: Freeman.

79. Kelley A. M., Michalet X., Weiss S. (2001). Chemical physics: single-molecule spectroscopy comes of age. *Science* **292**: 1671–1672.

80. Weiss S. (2000). Measuring conformational dynamics of biomolecules by single molecule fluorescence spectroscopy. *Nat. Struct. Biol.* **7**: 724–729.

81. Kohler J. (2001). Optical spectroscopy of individual objects. *Naturwissenschaften* **88**: 514–521.

82. Michalet X., Kapanidis A. N., Laurence T., Pinaud F., Doose S., Pflughoefft M., Weiss S. (2003). The power and prospects of fluorescence microscopies and spectroscopies. *Annu. Rev. Biophys. Biomol. Struct.* **32**: 161–182.

83. Jelesarov I., Bosshard H. R. (1999). Isothermal titration calorimetry and differential scanning calorimetry as complementary tools to investigate the energetics of biomolecular recognition. *J. Mol. Recogn.* **12**: 3–18.
84. Ladbury J. E., Chowdry B. Z., eds. (1998). *Biocalorimetry: Applications of Calorimetry in the Biological Sciences.* Hoboken, NJ: John Wiley & Sons.
85. Freire E. (1995). Differential scanning calorimetry. *Methods Mol. Biol.* **40**: 191–218.
86. Privalov P. L., Potehkin S. A. (1986). Scanning microcalorimetry in studying temperature–induced changes in proteins. *Methods Enzymol.* **131**: 4–51.
87. Sturtevant J. M. (1987). Biochemical applications of differential scanning calorimetry. *Proc. Natl. Acad. Sci. U.S.A.* **74**: 2236–2240.
88. Brandts J. F., Lin L. N. (1990). Study of strong to ultratight protein interactions using differential scanning calorimetry. *Biochemistry* **29**: 6927–6940.
89. Breslauer K. J., Freire E., Straume M. (1992). Calorimetry: a tool for DNA and ligand-DNA studies. *Methods Enzymol.* **211**: 533–567.
90. Freire E. (1994). Statistical thermodynamic analysis of differential scanning calorimetry data: structural deconvolution of heat capacity function of proteins. *Methods Enzymol.* **240**: 502–530.
91. Laue T. M., Stafford W. F. 3rd. (1999). Modern applications of analytical ultracentrifugation. *Annu. Rev. Biophys. Biomol. Struct.* **28**: 75–100.
92. Lebowitz J., Lewis M. S., Schuck P. (2002). Modern analytical ultracentrifugation in protein science: a tutorial review. *Protein Sci.* **11**: 2067–2079.
93. Rivas G., Stafford W., Minton A. P. (1999). Characterization of heterologous protein–protein interactions using analytical ultracentrifugation. *Methods* **19**: 194–212.
94. Laue T. M., Anderson A. L., Weber B. J. (1997). Prototype fluorimeter for the XLA/XLI analytical ultracentrifuge. *Proc. SPIE-Int. Soc. Opt. Eng.* **2985**: 196–204.
95. Schuck P. (1998). Sedimentation analysis of noninteracting and self-associating solutes using numerical solutions to the Lamm equation. *Biophys. J.* **75**: 1503–1512.
96. Schuck P. (2000). Size-distribution analysis of macromolecules by sedimentation velocity ultracentrifugation and Lamm equation modeling. *Biophys. J.* **78**: 1606–1619.
97. Laue T. (2001). Biophysical studies by ultracentrifugation. *Curr. Opin. Struct. Biol.* **11**: 579–583.
98. Ralston G. (1993). *Introduction to Analytical Ultracentrifugation.* Fullerton, CA: Beckman Instruments, Inc.
99. McRorie D. K., Voelker P. J. (1993). *Self-Associating Systems in the Analytical Ultracentrifuge.* Fullerton, CA: Beckman Instruments, Inc.
100. Welford K. (1991). Surface-plasmon polaritons and their uses. *Opt. Quantum Electron.* **23**: 1–27.
101. Markey F. (1999). What is SPR anyway? *Biacore J.* **6**: 14–17.
102. Rich R. L., Myszka D. G. (2000). Advances in surface plasmon resonance biosensor analysis. *Curr. Opin. Biotechnol.* **11**: 54–61.

3

OVERVIEW OF BIOLOGICAL MASS SPECTROMETRY

The purpose of this chapter is to provide readers who are less familiar with mass spectrometry with concise background material on modern MS instrumentation and techniques that will be referred to in the later chapters. The idea is to equip the reader with a knowledge base that could later be used to make an informed choice as to the use of a particular instrument or technique for a specific task.

3.1. BASIC PRINCIPLES OF MASS SPECTROMETRY

In almost a century that has passed since the introduction of the concept of mass spectrometry (MS) by J. J. Thomson (1, 2), this technique has become an indispensable analytical tool in many areas of science and technology. In this section we briefly review the most important concepts and definitions pertaining to MS. Emphasis is placed on those features of MS measurements that are particularly relevant for MS analysis of biological macromolecules, as outlined in the following chapters of this book.

The basic principles of mass spectral measurements utilize the ability of electric and magnetic fields to exert influence on charged particles:

$$m\ddot{\vec{\mathrm{r}}} = q(\vec{\mathrm{E}} + [\dot{\vec{\mathrm{r}}} \times \vec{\mathrm{B}}]). \tag{3-1-1}$$

Mass Spectrometry in Biophysics: Conformation and Dynamics of Biomolecules
By Igor A. Kaltashov and Stephen J. Eyles
ISBN 0-471-45602-0

(Here we assume that the interaction takes place in a vacuum: that is, the dielectric constant is equal to one and magnetic susceptibility is zero.) Since the magnitude of each of these interactions is determined by both charge q and mass m of the particle, its trajectory $\vec{r}$ is universally dependent on the mass-to-charge ratio. Only charged particles can be manipulated by the electromagnetic fields; hence, all neutral molecules must be ionized prior to analysis. Thus, a typical mass spectrometer usually carries out two distinct functions (ionization and mass analysis*). According to this view, the mass spectrometer is an analytical device that volatilizes and ionizes molecules and measures ion abundance as a function of the ionic mass-to-charge ratio. In most types of mass spectrometers, such measurements are actually carried out by first separating the ions (either spatially or temporally) according to their mass-to-charge ratios, followed by detection of each type of ion† (Figure 3.1).

Unless the sample is already in the vapor phase, it must be volatilized prior to or concurrently with the ionization. The sensitivity of the analysis obviously depends on ionization efficiency, that is, the fraction of the neutral analyte molecules that are converted to ions and transferred to the next segment of the instrument where the mass analysis is carried out. Ionization is quite often accompanied by dissociation of the analyte molecules yielding fragment ions, which in many cases can be more abundant than the (quasi) molecular ions (the latter representing the intact analyte molecules M with added charge, e.g., $M^{\bullet +}$, MH^+). Although fragmentation complicates the appearance of the mass spectrum, it often provides useful structural information, a subject to be discussed in more detail in the following sections of this chapter.

Ions produced in the ionization source are transferred to the mass analyzer, where the mass-to-charge ratios are measured. The physical principles of such measurements vary greatly among the many different types of mass analyzers and will be reviewed later in this chapter. In the remaining part of this section

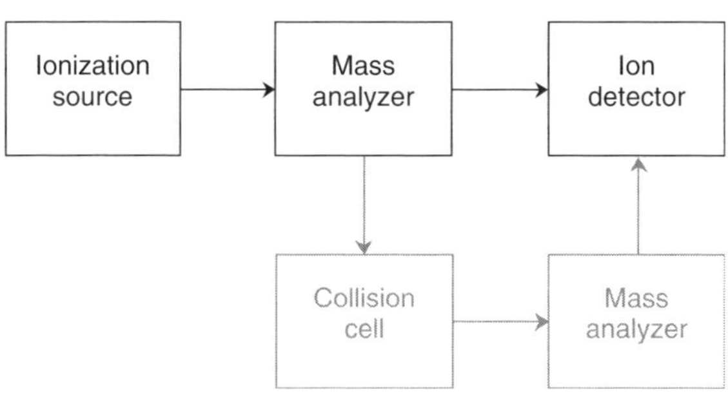

FIGURE 3.1. Schematic block diagram of a mass spectrometer. Elements shown in gray are used only for tandem MS measurements.

*It should correctly be called "mass-to-charge" measurement.

†An exception is Fourier transform ion cyclotron resonance mass spectrometry, where no physical separation of ions is required prior to detection.

we consider some common characteristics and features of mass analysis and their pertinence to mass spectrometry of biopolymers.

The total charge q on any ion is a multiple of an elementary (electron or proton) charge e (in SI, $e = 1.6022 \cdot 10^{-19}$ C; in CGS, $e = 4.8032 \cdot 10^{-10}$ esu):

$$q = \pm ze, \tag{3-1-2}$$

where z is always a positive integer. Therefore, the mass-to-charge ratio is always expressed as m/z, not m/q. The mass of an ion is generally different from that of a neutral molecule. The difference could be quite negligible (e.g., for ions produced by electron stripping or attachment,* $M^{\bullet +}$ or $M^{\bullet -}$). However, as we will see in the following sections of this chapter, charge acquisition by biopolymers is almost always accompanied by a substantial gain (or loss) of mass (e.g., $[M + nH]^{n+}$, $[M + nH + mNa]^{(n+m)+}$, $[M - nH + mNa]^{(n-m)-}$), which usually cannot be ignored. In this last statement we make an implicit assumption that a biopolymer under consideration can be assigned a mass, just like any other physical object. As it turns out, the definition of a macromolecular mass is not as trivial as it may seem. Since mass measurement is one of the main objectives of MS as an analytical technique, it is necessary to explore the concept of macromolecular mass in greater detail.

3.1.1. Stable Isotopes and Isotopic Distributions

Every molecule has a unique chemical composition; however, there is always some "physical" heterogeneity associated with the presence of stable isotopes (nuclides having the same number of protons but different number of neutrons). The existence of stable isotopes was demonstrated experimentally by Francis W. Aston (3), although their existence had been postulated at least a quarter of a century prior to this discovery (for which the Nobel Prize in Chemistry was awarded in 1922†).

Fractions of "heavy" isotopes of elements that are ubiquitous in biological molecules (C, N, H, O, P, etc.) do not usually exceed 1%‡ (see Table 3.1). Nevertheless, if a molecule contains a large number of atoms of a certain element, contributions of heavy isotopes can become very significant. We will illustrate this using carbon, an element that has two stable isotopes, ^{12}C (98.9%) and ^{13}C (1.1%), and buckminsterfullerene, a C_{60} molecule. The probability that every carbon atom in C_{60} is represented by the "light" isotope can be calculated as $P_0 = (0.989)^{60} = 0.515$, meaning that almost half of all buckminsterfullerene

*The electron mass is only $9.1095 \cdot 10^{-28}$ g, more than three orders of magnitude below that of a proton ($1.6726 \cdot 10^{-24}$ g) and neutron ($1.6749 \cdot 10^{-24}$ g), the building blocks of atomic nuclei.

†Aston's own account of his discovery can be found at `http://www.nobel.se/chemistry/laureates/1922/aston-lecture.html`.

‡Sulfur, of course, is a notable exception, with ^{33}S and ^{34}S together accounting for almost 5% of stable sulfur isotopes.

TABLE 3.1. Isotope Abundance and Accurate Masses for Selected Elements Based on Compilations[a–d]

Element	Isotope	Accurate Mass, u (STD, μu)	Natural Abundance
Hydrogen	^{1}H	1.007 825 031 90 (0.000 57)[a]	0.999 844 26(5)[b]
	^{2}H (D)	2.014 101 777 95 (0.000 62)[a]	0.000 155 74(5)[b]
Carbon[b]	^{12}C	12.000 000 000 000	0.988 944(28)
	^{13}C	13.003 354 838 3 (0.0049)[a]	0.011 056(28)[b]
Nitrogen	^{14}N	14.003 074 007 4 (0.0018)[a]	0.996 337(4)[b]
	^{15}N	15.000 108 973 (0.012)[(1)]	0.003 663(4)[b]
Oxygen	^{16}O	15.994 914 622 3 (0.0025)[a]	0.997 6206(5)[b]
	^{17}O	16.999 131 50 (0.22)[b]	0.000 3790(9)[b]
	^{18}O	17.999 160 4 (0.9)[b]	0.002 0004(5)[b]
Fluorine	^{19}F	18.998 403 2 (0.5)[c]	1*
Sodium	^{23}Na	22.989 770 (2)[c]	1*
Magnesium[b]	^{24}Mg	23.985 041 87 (0.26)	0.789 92(25)
	^{25}Mg	24.985 837 00 (0.26)	0.100 03(9)
	^{26}Mg	25.982 593 00 (0.26)	0.110 05(19)
Aluminum	^{27}Al	26.981 538 (2)[c]	1*
Silicon[b]	^{28}Si	27.976 926 49 (0.22)	0.922 223(9)
	^{29}Si	28.976 494 68 (0.22)	0.046 853(6)
	^{30}Si	29.973 770 18 (0.22)	0.030 924(7)
Phosphorus	^{31}P	30.973 761 (2)[c]	1*
Sulfur[b]	^{32}S	31.972 070 73 (0.15)	0.950 3957(90)
	^{33}S	32.971 458 54 (0.15)	0.007 4865(12)
	^{34}S	33.967 866 87 (0.14)	0.041 9719(87)
	$^{36}S^{(\beta-)}$	35.967 080 88 (0.25)	0.000 1459(21)
Chlorine[b]	^{35}Cl	34.968 852 71 (0.04)	0.757 79(46)
	^{37}Cl	36.965 902 60 (0.05)	0.242 21(46)
Potassium	^{39}K	38.963 706 9 (0.3)[a]	0.932 581 (44)[d]
	$^{40}K^{(\beta-)}$	39.963 998 67 (0.29)[a]	0.000 117 (1)[d]
	^{41}K	40.961 825 97 (0.28)[a]	0.067 302 (44)[d]
Calcium[b]	^{40}Ca	39.962 591 2 (0.3)	0.969 41(6)
	^{42}Ca	41.958 618 3 (0.4)	0.006 47(3)
	^{43}Ca	42.958 766 8 (0.5)	0.001 35(2)
	^{44}Ca	43.955 481 1 (0.9)	0.020 86(4)
	$^{46}Ca^{(\beta-)}$	45.953 692 7 (2.5)	0.000 04(1)
	$^{48}Ca^{(\beta-)}$	47.952 533 (4)	0.001 87(1)
Manganese	^{55}Mn	54.938 049 (9)[c]	1*
Iron[b]	^{54}Fe	53.939 614 7 (1.4)	0.058 45(23)
	^{56}Fe	55.934 941 8 (1.5)	0.917 54(24)
	^{57}Fe	56.935 398 3 (1.5)	0.021 191(65)
	$^{58}Fe^{(\beta-)}$	57.933 280 1 (1.5)	0.002 819(27)
Nickel	^{58}Ni	57.935 347 7 (1.6)[a]	0.680769 (89)[d]
	^{60}Ni	59.930 790 3 (1.5)[a]	0.262231 (77)[d]
	^{61}Ni	60.931 060 1 (1.5)[a]	0.011399 (6)[d]
	^{62}Ni	61.928 348 4 (1.5)[a]	0.036345 (17)[d]
	$^{64}Ni^{(\beta-)}$	63.927 969 2 (1.6)[a]	0.009256 (9)[d]

TABLE 3.1 (*Continued*)

Element	Isotope	Accurate Mass, u (STD, μu)	Natural Abundance
Copper[b]	^{63}Cu	62.929 600 7 (1.5)	0.691 74(20)
	^{65}Cu	64.927 793 8 (1.9)	0.308 26(20)
Zinc[b]	^{64}Zn	63.929 146 1 (1.8)	0.4863(20)
	^{66}Zn	65.926 036 4 (1.7)	0.2790(9)
	^{67}Zn	66.927 130 5 (1.7)	0.0410(4)
	^{68}Zn	67.924 847 3 (1.7)	0.1875(17)
	^{70}Zn$^{(\beta-)}$	69.925 325 (4)	0.0062(1)
Gallium	^{69}Ga	68.925 581 (3)[a]	0.60108 (9)[d]
	^{71}Ga	70.924 707 3 (2.0)[a]	0.39892 (9)[d]
Arsenic	^{75}As	74.921 60 (20)[c]	1*
Selenium[b]	^{74}Se	73.922 476 7 (1.6)	0.008 89(3)
	^{76}Se	75.919 214 3 (1.6)	0.093 66(18)
	^{77}Se	76.919 914 8 (1.6)	0.076 35(10)
	^{78}Se$^{(\beta-)}$	77.917 309 7 (1.6)	0.237 72(20)
	^{80}Se$^{(\beta-)}$	79.916 522 1 (2.0)	0.496 07(17)
	^{82}Se$^{(\beta-)}$	81.916 700 3 (2.2)	0.087 31(10)
Bromine	^{79}Br	78.918 337 9 (2.0)[a]	0.5069 (7)[d]
	^{81}Br	80.916 291 (3)[a]	0.4931 (7)[d]
Ruthenium	^{96}Ru	95.907 604 (9)[a]	0.0554 (14)[d]
	^{98}Ru	97.905 287 (7)[a]	0.0187 (3)[d]
	^{99}Ru	98.905 938 5 (2.2)[a]	0.1276 (14)[d]
	^{100}Ru	99.904 218 9 (2.2)[a]	0.1260 (7)[d]
	^{101}Ru	100.905 581 5 (2.2)[a]	0.1706 (2)[d]
	^{102}Ru$^{(\beta-)}$	101.904 348 8 (2.2)[a]	0.3155 (14)[d]
	^{104}Ru$^{(\beta-)}$	103.905 430 (4)[a]	0.1862 (27)[d]
Cadmium	^{106}Cd	105.906 458 (6)[a]	0.0125 (6)[d]
	^{108}Cd	107.904 183 (6)[a]	0.0089 (3)[d]
	^{110}Cd	109.903 006 (3)[a]	0.1249 (18)[d]
	^{111}Cd	110.904 182 (3)[a]	0.1280 (12)[d]
	^{112}Cd$^{(\beta-)}$	111.902 757 7 (3.0)[a]	0.2413 (21)[d]
	^{113}Cd$^{(\beta-)}$	112.904 401 4 (3.0)[a]	0.1222 (12)[d]
	^{114}Cd$^{(\beta-)}$	113.903 358 6 (3.0)[a]	0.2873 (42)[d]
	^{116}Cd$^{(\beta-)}$	115.904 756 (3)[a]	0.0749 (18)[d]
Indium	^{113}In	112.904 062 (4)[a]	0.0429 (5)[d]
	^{115}In$^{(\beta-)}$	114.903 879 (4)[a]	0.9571 (5)[d]
Iodine	^{127}I	126.904 468 (4)[a]	1*
Gadolinium	^{152}Gd	151.919 789 (3)[a]	0.0020 (1)[d]
	^{154}Gd$^{(\beta-)}$	153.920 862 (3)[a]	0.0218 (3)[d]
	^{155}Gd	154.922 619 (3)[a]	0.1480 (12)[d]
	^{156}Gd$^{(\beta-)}$	155.922 120 (3)[a]	0.2047 (9)[d]
	^{157}Gd	156.923 957 (3)[a]	0.1565 (2)[d]
	^{158}Gd$^{(\beta-)}$	157.924 101 (3)[a]	0.2484 (7)[d]
	^{160}Gd$^{(\beta-)}$	159.927 051 (3)[a]	0.2186 (19)[d]

TABLE 3.1 ***(Continued)***

Element	Isotope	Accurate Mass, u (STD, μu)	Natural Abundance
Mercury	^{196}Hg	195.965 814 (4)[a]	0.0015 (1)[d]
	^{198}Hg	197.966 752 (3)[a]	0.0997 (20)[d]
	^{199}Hg	198.968 262 (3)[a]	0.1687 (22)[d]
	^{200}Hg	199.968 309 (3)[a]	0.2310 (19)[d]
	^{201}Hg	200.970 285 (3)[a]	0.1318 (9)[d]
	^{202}Hg	201.970 625 (3)[a]	0.2986 (26)[d]
	^{204}Hg$^{(\beta-)}$	203.973 475 (3)[a]	0.0687 (15)[d]
Bismuth	^{209}Bi	208.980 38 (20)[c]	1*

[a] Audi G., Wapstra A. H. (1993). The 1993 atomic mass evaluation. 1. Atomic mass table. *Nucl. Phys. A* **565**: 1–65.

[b] Coplen T. B., Bohlke J. K., De Bievre P., Ding T., Holden N. E., Hopple J. A., Krouse H. R., Lamberty A., Peiser H. S., Revesz K., Rieder S. E., Rosman K. J. R., Roth E., Taylor P. D. P., Vocke R. D., Xiao Y. K. (2002). Isotope-abundance variations of selected elements (IUPAC technical report). *Pure Appl. Chem.* **74**: 1987–2017.

[c] Vocke R. D. (1999). Atomic weights of the elements 1997 (technical report). *Pure Appl. Chem.* **71**: 1593–1607.

[d] Rosman K. J. R., Taylor P. D. P. (1998). Isotopic compositions of the elements 1997. *Pure Appl. Chem.* **70**: 217–235.

*Single stable isotope for this element.

molecules contain at least one ^{13}C atom. The probability that k and only k out of 60 carbon atoms are ^{13}C can easily be calculated as

$$P_k = C_k^{60} \cdot (0.011)^k \cdot (0.989)^{60-k}, \tag{3-1-3}$$

where C_k^{60} is a binomial coefficient (number of possible combinations of k elements out of 60). The graph of P_k as a function of mass or, more precisely, number of ^{13}C atoms (Figure 3.2) represents an *isotopic distribution* of buckminsterfullerene. A trivial expansion of (3-1-3) gives an isotopic distribution for a hypothetical molecule built of n atoms A with two stable isotopes, ^{X}A (relative abundance p_0) and $^{X+1}$A (relative abundance $p_1 = 1 - p_0$):

$$P_k = C_k^n \cdot (p_1)^k \cdot (p_0)^{n-k}. \tag{3-1-4}$$

It is easy to show that (3-1-4) can be expanded to include more than two elements; for example, an isotopic distribution for a molecule A_nB_m can be expressed as a convolution of two binomial distributions:

$$P_k = \sum_{i,j} C_i^n \cdot C_j^m \cdot (p_1^{\mathrm{A}})^i \cdot (p_0^{\mathrm{A}})^{n-i} \cdot (p_1^{\mathrm{B}})^j \cdot (p_0^{\mathrm{B}})^{m-j} \cdot \delta_{k,i+j}, \tag{3-1-5}$$

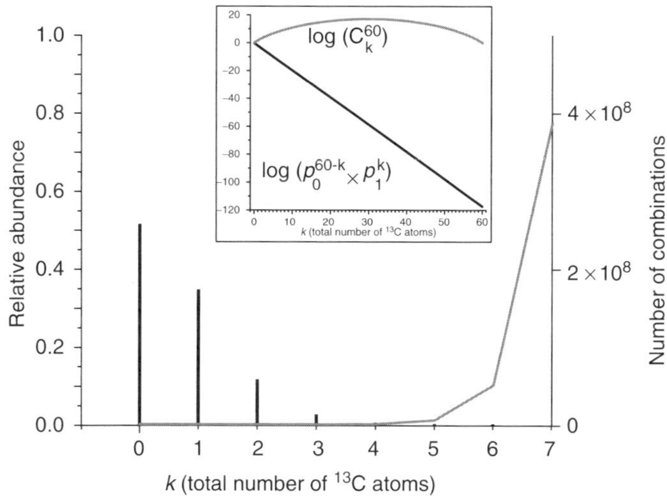

FIGURE 3.2. Isotopic distribution of a buckminsterfullerene (C_{60}) molecule (shown as a bar diagram). The inset shows logarithms of a number of combinations of k atoms out of 60 (*gray trace*) and a probability that in each such combination k atoms are ^{13}C isotopes (*black trace*).

where $p_0{}^A$, $p_1{}^A$, $p_0{}^B$, and $p_1{}^B$ represent the relative abundance of $^X A$, ^{X+1}A, $^Y B$ and ^{Y+1}B, respectively, and $\delta_{k,i+j}$ [$\delta_{p,q} = 1$ (if $p = q$) and 0 (if $p \neq q$)] is a familiar Kronecker symbol, or simply

$$P_k = \sum_{i \leq k} C_i^n \cdot C_{k-i}^m \cdot (p_1^A)^i \cdot (p_0^A)^{n-i} \cdot (p_1^B)^{k-i} \cdot (p_0^B)^{m-k+i}. \quad (3\text{-}1\text{-}6)$$

Another, more intuitive, presentation of (3-1-5) uses a product of the two binomials, $(p_0^A + p_1^A)^n \cdot (p_0^B + p_1^B)^m$, whose expansion gives the relative abundance P_k of the isotopic species whose mass exceeds that of $^X A_n {}^Y B_m$ by k mass units. Further expansion to a general case of a polyatomic molecule $A_n B_m \cdots Z_q$ is relatively straightforward:

$$(p_0^A + p_1^A + p_2^A + \cdots)^n \cdot (p_0^B + p_1^B + p_2^B + \cdots)^m \cdot \cdots \cdot (p_0^Z + p_1^Z + p_2^Z + \cdots)^q. \quad (3\text{-}1\text{-}7)$$

Despite its analytical elegance, Equation (3-1-7) is computationally demanding and is rarely used for calculating the isotopic distributions in a straightforward manner. Multiple algorithms have been developed over the last couple of decades that use various robust schemes to calculate isotopic distributions of large polyatomic molecules with high accuracy (4–9). Isotopic distributions of many elements (metals in particular) display some rather characteristic patterns, which sometimes provide unique "isotopic signatures" within small molecules (Figure 3.3). Isotopic distributions of most biopolymers are not as "telling" (as far as the elemental make-up is concerned). Instead, the large number of atoms

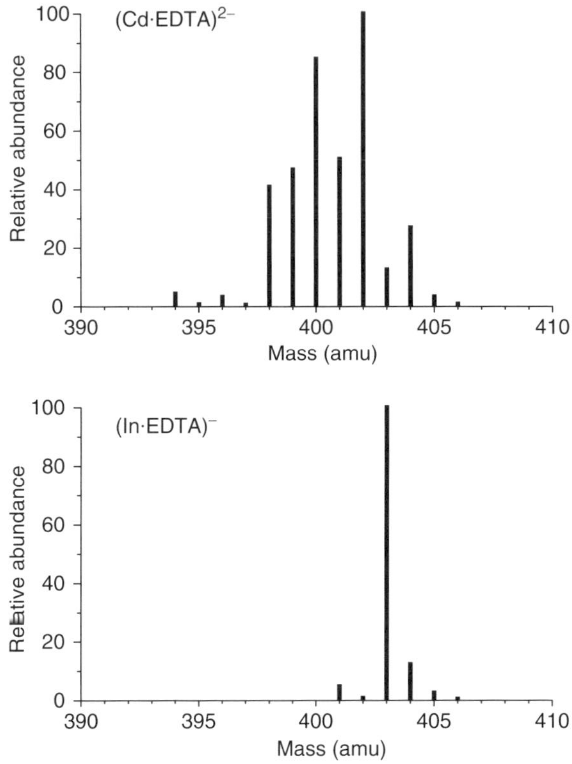

FIGURE 3.3. Isotopic signatures of metal ion–organic acid complexes: isotopic distributions of $(Cd{\cdot}EDTA)^{2-}$ and $(In{\cdot}EDTA)^{-}$.

present in a "typical" biomolecule gives rise to a convoluted isotopic distribution whose width increases as the size of the macromolecule increases (Figure 3.4).

Most commercial mass spectrometers have built-in software that calculates isotopic distributions. A large array of such programs is also available on the Internet.* For the user's convenience, data input is often available in two formats, general (using a molecular formula $A_nB_m \cdots Z_q$) or peptide- and protein-oriented (using an amino acid composition $Aaa_nBbb_m \cdots Zzz_q$ or simply an amino acid sequence). Most of the web-based programs currently calculate isotopic distributions assuming the "natural" abundance of stable isotopes for each element. Calculations of isotopic distributions for proteins expressed in a medium that is enriched with (or depleted of) a certain isotope(s) would require a more flexible algorithm.

Strictly speaking, in each of the calculated isotopic distributions shown in Figure 3.4, only the monoisotopic peak is truly a single peak, with the rest

*The ones frequented by the authors and many of their colleagues are MS-Isotope at Protein Prospector (`http://prospector.ucsf.edu/`) and Sheffield ChemPuter (`http://www.shef.ac.uk/chemistry/chemputer/isotopes.html`).

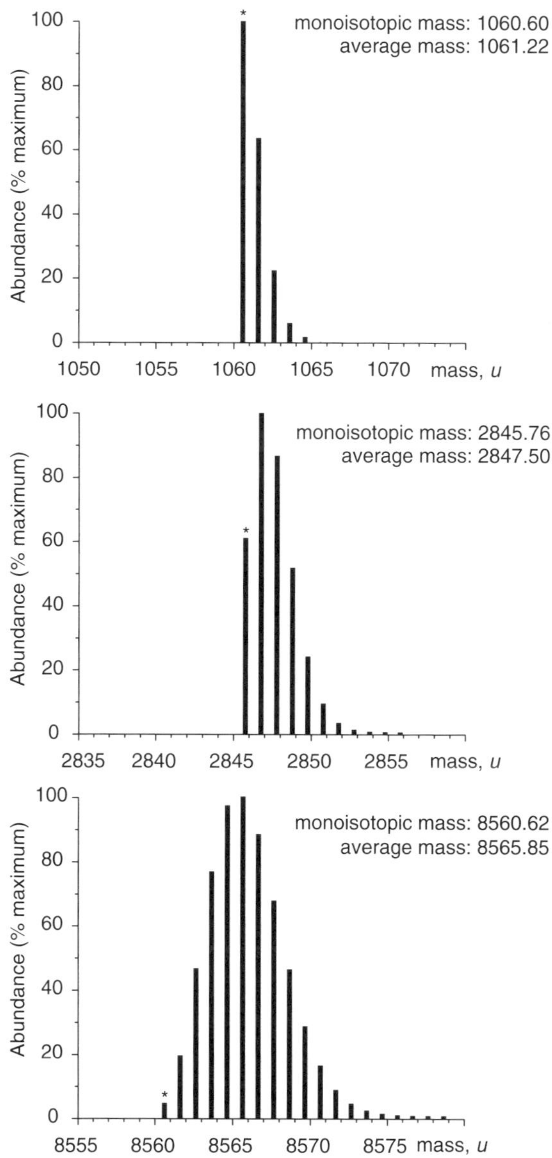

FIGURE 3.4. Calculated isotopic distributions for polypeptides bradykinin (*top*), melittin (*middle*), and ubiquitin (*bottom*). Monoisotopic peaks are marked with asterisks.

of the peaks in the cluster being composite peaks representing two or more ionic species. For example, the monoisotopic peak of ubiquitin in Figure 3.4 corresponds to an ionic species $(^{12}C_{378}{}^{1}H_{631}{}^{14}N_{105}{}^{16}O_{118}{}^{32}S)^{+}$, while the next peak represents five different species, which are usually referred to as *isobaric species*, or *isobars*. These five isobars, $(^{12}C_{377}{}^{13}C\ {}^{1}H_{631}{}^{14}N_{105}{}^{16}O_{118}{}^{32}S)^{+}$,

$(^{12}C_{378}{}^{1}H_{630}{}^{2}H^{14}N_{105}{}^{16}O_{118}{}^{32}S)^+$, $(^{12}C_{378}{}^{1}H_{631}{}^{14}N_{104}{}^{15}N^{16}O_{118}{}^{32}S)^+$, $(^{12}C_{378}$ ${}^{1}H_{631}{}^{14}N_{105}{}^{16}O_{117}{}^{17}O^{32}S)^+$, and $(^{12}C_{378}{}^{1}H_{631}{}^{14}N_{105}{}^{16}O_{118}{}^{33}S)^+$, have slightly different masses due to unequal divergence of ^{13}C, ^{2}H, ^{15}N, ^{17}O, and ^{33}S from the nearest integer number (see Table 3.1). Although the isobars of smaller peptides can be resolved in some cases (10), the present level of mass measurement technology does not generally allow such distinction to be made for larger (>1 kDa) peptides and proteins. Therefore, in the following discussion, we make no distinction between the isobars, and the "composite" nature of "isotopic peaks" will generally be ignored. (An important exception will be presented in Chapter 5, where a resolution of three isobaric peaks of ubiquitin will be demonstrated.)

3.1.2. Macromolecular Mass: Terms and Definitions

The notion of *molecular mass*, as it is used in MS, is closely related to (but not necessarily the same as) the familiar concept of *molecular weight,** a sum of the atomic weights of all atoms in the molecule. Molecular mass is measured in unified atomic mass units, defined by IUPAC as $\frac{1}{12}$ of the mass of a carbon-12 (^{12}C) atom in its ground state, $u \approx 1.660\ 5402(10) \cdot 10^{-27}$ kg. The *atomic weight* of an element is, of course, a weighted average of the atomic masses of the different isotopes; therefore, the isotopic make-up is implicitly included in the definition. Since the ionic mass measured by MS is not necessarily averaged across the entire isotopic content, several definitions of molecular mass are currently in circulation. The definitions of the molecular mass vary based on how they account for contributions from different isotopes. The *nominal mass* is calculated using a mass of lightest† isotope for each element rounded to the nearest integer. The *monoisotopic mass* is calculated in a similar fashion, but the isotopic masses are no longer rounded; that is, the nuclear mass defect is accounted for. For peptides whose molecular weight exceeds ~2 kDa, the *most abundant mass* (i.e., mass corresponding to the ionic peak of highest intensity in the isotopic cluster) no longer coincides with the monoisotopic mass (Figure 3.4). Finally, the *average mass* is calculated based on the entire isotopic distribution and is closely related to the *molecular weight* as used elsewhere in chemistry and related disciplines. The average mass is usually very close to the most abundant mass (typically within 1 u). As the number of atoms comprising the molecule increases, so does the difference between the *monoisotopic* and *average* masses. At the same time, the relative abundance of the monoisotopic peak continues to decrease (Figure 3.4), and it becomes practically undetectable even for biopolymers of a modest (>10 kDa) size.

*In fact, the terms molecular weight, molecular mass, and molar mass are all identical according to IUPAC definitions and are often used interchangeably in most areas of chemistry. They do, however, often have different meanings in MS.

†Some texts use the most abundant (instead of the lightest) isotope to define both nominal and monoisotopic masses.

3.2. METHODS OF PRODUCING BIOMOLECULAR IONS

3.2.1. Macromolecular Ion Desorption Techniques: General Considerations

Despite its early success as an analytical technique, mass spectrometry had made almost no incursions into the field of intact biopolymer analysis prior to the 1970s. The "classical" ionization techniques (such as electron impact and chemical ionization) were not suited to handle biological macromolecules. The vapor pressure of biopolymers is negligible, and their delicate nature prevented using high temperatures as a means to enhance evaporation. Although this difficulty had been circumvented in several cases by the chemical derivatization of polar biomolecules (aiming at increasing their vapor pressure), the examples of using "classical" MS to analyze even the simplest biomolecules remained very few.

Mass spectral analysis of biomolecules only became feasible with the advent of the ion desorption techniques, which initially included field desorption [reviewed in (11)] and plasma desorption (12). The introduction of fast atom bombardment (FAB) in the early 1980s (13) had perhaps even greater impact on the development of biomolecular MS. Unlike plasma desorption, FAB sources did not require special types of mass analyzers and were much easier to handle. While plasma desorption and field desorption/ionization sources are no longer used widely, FAB remains in use (14). However, its use in biophysical experiments is usually limited, and we will not consider it here in much detail. An interested reader is referred to several excellent reviews on the subject (15, 16), as well as comprehensive MS texts (17, 18). The two ionization techniques that are most relevant to our discussion are electrospray ionization mass spectrometry (ESI MS) and matrix-assisted laser desorption/ionization (MALDI).

3.2.2. Electrospray Ionization

Brief Historical Remarks. The advent of ESI MS in the mid-1980s (19) provided a means to observe spectra of intact proteins with no apparent mass limitation, a discovery honored with the Nobel Prize in Chemistry [to John Fenn in 2002 (20)]. The history of developing this technique is quite fascinating. The ESI process had been in use in various fields of science and technology (including mass spectrometry itself) for several decades prior to its direct application to biopolymer analysis. The "electrification of liquid droplets produced by spraying, bubbling, and similar methods" (21) had captured the attention of a number of researchers as early as the end of the nineteenth and beginning of the twentieth century, with the initial theoretical exploration of the ESI process carried out by Lord Rayleigh in 1882 (22). Indeed, studies of some of the phenomena related to ESI can be traced several hundred years into the past (23).

The motivation for earlier studies of ion production and fate is mostly due to the importance of such processes in atmospheric science. The realization that ESI also held great promise and a potential to become a means of producing macromolecular ions started fueling the interest of the MS community in the late 1960s (24). Malcolm Dole, a pioneer of ESI MS, wrote in 1968: "In considering how it might be possible to obtain gas phase intact ionized macromolecules for mass analysis in a mass spectrometer, the idea occurred to one of us to electrospray dilute solutions of macromolecules into air or other suitable gas at atmospheric pressure and to sample the air for macroions by means of a supersonic probe. By allowing a dilute polymer solution containing a volatile solvent to flow out of a tip of a hypodermic needle electrostatically charged . . . , a spray of finely divided and electrically charged droplets is produced. On evaporation of the solvent the charged droplets should become electrically unstable and break down into smaller droplets . . . [eventually resulting in the formation of] drops containing only one macromolecule per drop. Finally, on complete evaporation of solvent from these drops, . . . the gas phase macroions would result" (25) (Figure A.1). In fact, the first application of ESI to observe macromolecules and measure their molecular weight was reported in 1964 (26). Detection, though, was done using optical (not mass spectrometric) means. The subsequent efforts of Dole and co-workers focused on producing polymer ions in the gas phase (27, 28). Similar ideas were later used by Iribarne and Thomson to develop an ionization method capable of dealing with polar and labile analytes, which they termed atmospheric pressure ion evaporation (29, 30). The technique was shown to be quite capable of producing molecular ions for a range of small polar organic molecules, including amino acids (29, 31, 32) and adenosine triphosphate (29). Concurrently, Alexandrov and co-workers used ESI MS for peptide analysis (33, 34) and reported controlled peptide ion dissociation in the interface region yielding structurally diagnostic fragment ions, a phenomenon that would later be referred to as *cone fragmentation* or *nozzle-skimmer fragmentation* (31). Finally, an ESI source was

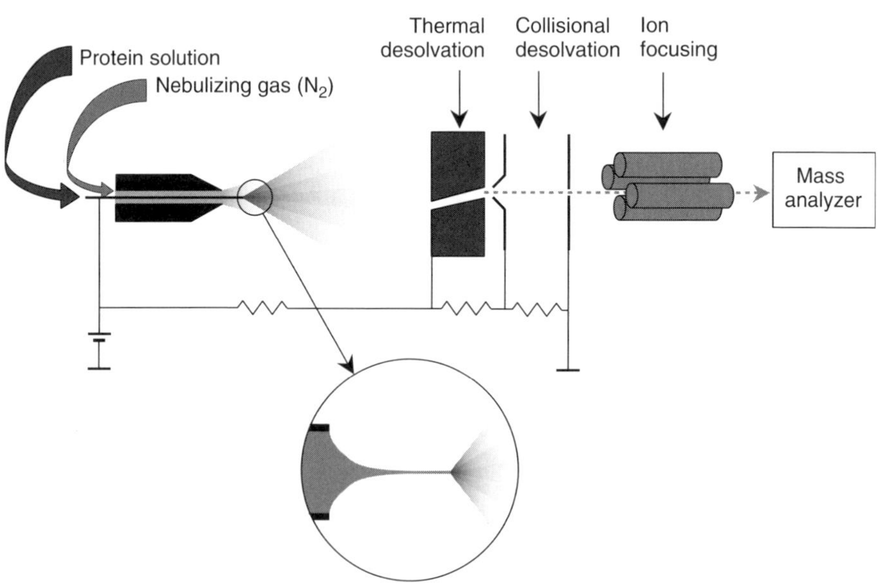

FIGURE 3.5. A schematic representation of an ESI MS interface.

interfaced with HPLC (35, 36). However, as Marvin Vestal recalls (37), most of this work was largely ignored by the majority of practitioners of biological MS.

A breakthrough occurred in 1988 when Fenn's group demonstrated ESI mass spectra of intact proteins (38) and synthetic polymers (39) and presented their findings at the 36th annual conference of the American Society for Mass Spectrometry. Finally, Dole's bold suggestion that it would be possible to "obtain gas phase intact ionized macromolecules for mass analysis in a mass spectrometer" was shown to be true. An interested reader may find more information on this subject by listening to Fenn's lecture "Electrospray Wings for Molecular Elephants," presented in December 2002 upon acceptance of the Nobel Prize in Chemistry (video stream at http://www.nobel.se/chemistry/laureates/2002/fenn-lecture.html).

Macroions in ESI: Multiple Charging. Electrospray ionization is a convoluted process that involves several steps, each having a profound effect on the outcome of the measurements. Many of these effects are particularly important in biophysical measurements and must be taken into account in order to avoid erroneous interpretation of the experimental data. A detailed discussion of various physical processes involved in ESI is presented in the Appendix.

A very distinct feature of ESI is the accumulation of multiple charges on a single analyte molecule during ionization. An important implication of the multiple charging in ESI MS is the appearance of multiple peaks corresponding to a single analyte (Figure 3.6, top trace). The position of each peak in the mass spectrum (m/z value) would be determined by the mass of a macromolecule M and the number of accommodated charges. There are several different charge carriers that are commonly associated with an ESI process, the most ubiquitous being a proton and alkali metals (particularly Na^+ and K^+). Given such heterogeneity in the charges accommodated by a macromolecular ion, its m/z value can generally be calculated as

$$\frac{m}{z} = \frac{M + \sum_i n_i m_i}{\sum_i n_i}, \qquad (3\text{-}2\text{-}1)$$

where m_i is a mass associated with a specific carrier, and n_i is a number of charges of this type associated with the macromolecule. If the charging is mostly due to protonation, then (3-2-1) is simplified to

$$\left(\frac{m}{z}\right)_{n+} = \frac{M + n}{n}. \qquad (3\text{-}2\text{-}2)$$

Mass spectra of even relatively small protein ions typically contain several peaks corresponding to different charge states. Therefore, calculating the mass M based on a series of m/z values of ion peaks corresponding to incrementally increasing (or decreasing) charge states is an overdetermined problem. Such

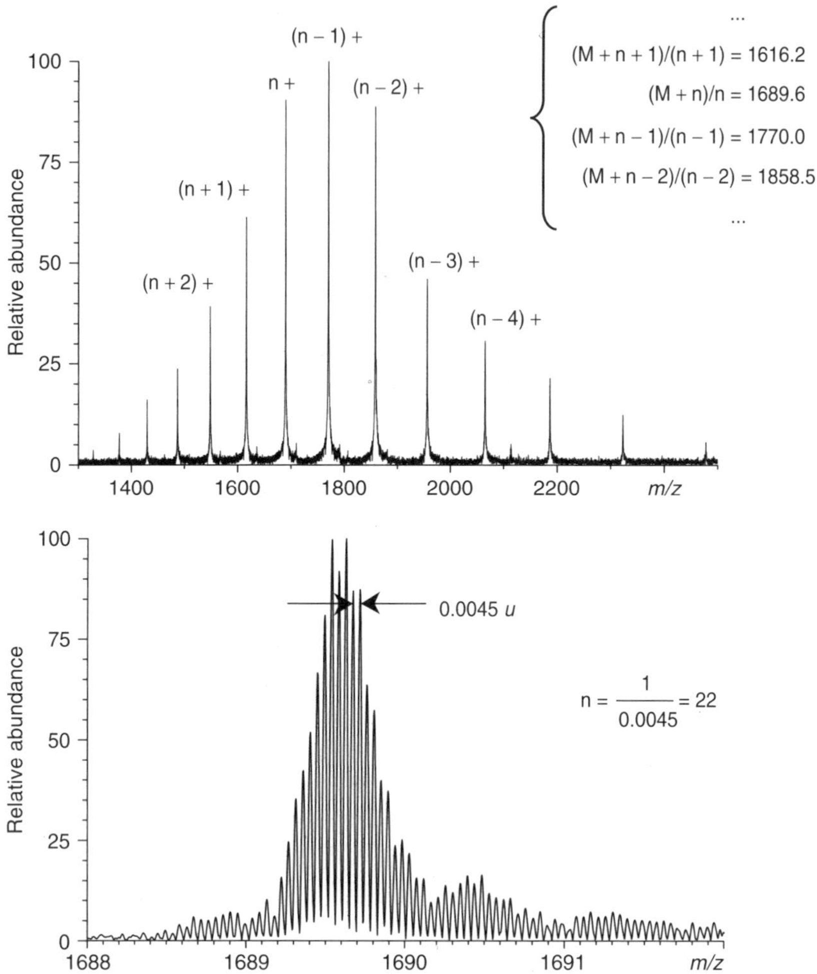

FIGURE 3.6. ESI mass spectrum of a 37 kDa protein human serum transferrin N-lobe (*h*Tf/2N) in $H_2O/CH_3OH/CH_3COOH$ (49:49:2, *v*:*v*:*v*) showing a series of multiply charged protein ion peaks (*top*) and an expanded view of the isotopic cluster of a +22 ion of *h*Tf/2N (*bottom*). Courtesy of Mingxuan Zhang (Graduate Research Assistant at the University of Massachusetts–Amherst).

calculations are easy and very straightforward (inset in Figure 3.6). Spectral interpretation becomes a bit more difficult if an ESI mass spectrum contains contributions from several different analytes. Multiple deconvolution procedures have been developed to address this problem (40–42), and most modern ESI MS data systems contain built-in deconvolution routines. High-resolution MS provides an alternative to deconvolution, as the charge state of each macromolecular ion can be deduced directly from the mass spectrum based on the distance between the adjacent isotopic peaks (Figure 3.6, bottom trace). The most sophisticated (and most accurate) methods of mass analysis of complex mixtures make use of both deconvolution procedures and high-resolution measurements (43–45).

The presence of other channels of charging a macromolecule (besides protonation) usually degrades the quality of ESI MS data, as the deconvoluted spectra in this case contain artifact peaks (e.g., corresponding to Na^+ and K^+ adducts).* Although not a significant annoyance by itself, it usually leads to a diminished signal-to-noise ratio, and in some extreme cases to an inferior accuracy of mass measurements. Other types of adducts that are encountered in ESI MS are ammonium cations and their derivatives. Adducts formed by association of ubiquitous anions (acetate, formate, etc.) with positively charged macromolecular species are also common. The extent of adduct ion formation can often be limited by using heat-induced desolvation (via elevating the temperature in the ESI interface) and/or employing carefully controlled collision-induced dissociation of the adduct ions. Each of these processes can greatly reduce the extent of biopolymer ion complexation with anions, while Na^+ and K^+ adducts remain largely unaffected. Excessive collisional activation brings about apparent reduction of the macromolecular ion charge state (the "charge stripping phenomenon") and may lead to the dissociation of covalent bonds, a process that will be considered in detail in the following sections of this chapter.

The formation of macromolecular ionic species in the negative ion mode usually proceeds via deprotonation of acidic groups. Deconvolution of negative ion ESI mass spectra is relatively straightforward, although one has to remember that acquisition of the negative charge by biopolymers most often occurs through deprotonation (mass loss), so the m/z values of protein ion peaks are

$$\left(\frac{m}{z}\right)_{n-} = \frac{M - n}{n} \qquad \text{(3-2-3)}$$

Most classes of biological polymers are amenable to ESI MS analysis, including proteins, oligonucleotides, and carbohydrates. Proteins are usually analyzed in the positive ion mode, while oligonucleotides tend to produce better signal in the negative ion mode. Alkali metal adduct ion formation often becomes a significant problem for larger oligonucleotides, although Na^+ (or K^+) complexation is often used intentionally as a means to ionize neutral carbohydrates lacking basic and acidic groups.

Applications of ESI MS for biopolymer analysis constitute a vast and still rapidly expanding field. Unfortunately, obvious space limitations do not allow us to explore this field beyond the subjects relevant to biophysics. An interested reader is referred to a couple of excellent books on the subject that contain diverse examples of ESI MS applications to various problems in the life sciences in general (46–48). We will, however, briefly consider two relatively recent

*Since the deconvolution usually reports masses of neutral (uncharged) species, the actual mass increase corresponds to a difference between a mass of Na^+ (or K^+) and that of a proton (H^+).

developments in ESI MS that are particularly important for some of the biophysical problems to be discussed in the following chapters of the book.

Nanospray Ionization. In traditional ESI MS the analyte solution is continuously supplied to the ion source at a constant flow rate. Although the flow rate can vary greatly, it rarely goes below the microliter per minute (μL/min) level, as it causes various spray instabilities when using conventional ESI sources. Wilm and Mann pointed out that achieving very low flow rates could be beneficial in many ways (49). Since the diameter of the emitted droplets is determined by the liquid flow rate through the capillary (A-3), only very small droplets will be formed when the flow rate is dramatically cut. Large surface-to-volume ratio would result in facile droplet evaporation and significant improvement of the efficiency of ion formation. Of course, another benefit would be a dramatic extension of the analysis time, given the sample volume remains constant. Realization of these ideas led to development of the so-called nano-electrospray ionization (nano-ESI) (50). The very low flow rate is realized by loading a small volume (typically 1 μL) of sample solution into a narrow bore (orifice 1–2 μm) metal-coated capillary. The liquid flow is actually induced by applying high voltage to the capillary tip; that is, the solution is drawn from the capillary electrodynamically without the use of a conventional syringe pump. The resulting flow rates are typically on the order of 20–40 nL/min and are quite stable (the small orifice prevents the formation of multiple Taylor cones at the tip of the capillary). In addition to increased sensitivity as compared to conventional ESI, nano-ESI has another important advantage. Several studies have indicated that it has much higher salt tolerance, at least an order of magnitude exceeding that of conventional ESI (51, 52). This is explained in terms of the lower size and higher charge density of droplets emitted in nano-ESI, which result in early fission events without extensive solvent evaporation (which would otherwise lead to significant increase in salt ion concentration prior to fission).

Cold Spray. The "cold" spray ionization (CSI) is another variation of ESI technique that has been introduced very recently (53, 54). A major point of distinction between ESI and CSI is that the former usually makes use of elevated temperature in the interface region to assist desolvation, while the latter operates at very low temperatures (just as the name implies). Typically, both the source block and the desolvation chamber in CSI MS are kept at temperatures below 15 °C (and can be as low as −50 °C). This often allows the labile organic species to be observed as intact ionic species that would normally be destroyed during the desolvation stage of common ESI.

3.2.3. Matrix Assisted Laser Desorption/Ionization

ESI is an example of the ionization methods that produce macromolecular ions directly from the solution bulk (other "bulk ionization" methods have met with less success when applied to biopolymers). Desorption from solid surfaces provides another opportunity to transfer analyte molecules to the gas phase. Perhaps the simplest method—heating to vaporize the sample—has become the de facto standard for analysis of small organic molecules, where ionization is achieved by passing the gaseous analyte through an electron beam. The harsh conditions of heating and electron impact, however, generally lead to extensive fragmentation, and this method is only practical for low molecular weight (<800 Da) and relatively nonpolar analytes. A reasonable alternative to the indiscriminate heating of the sample is the utilization of high-energy projectiles that serve as "external agents" providing very localized (both spatially and temporally) heating to the sample. Such rapid and local heating results in the near-adiabatic expansion of the microscopically small portions of the sample, leading to their expulsion from the surface to the gas phase. This principle is utilized by fast atom bombardment (FAB), a desorption/ionization technique that uses a beam of high-energy (tens of keV) atoms or ions* to desorb macromolecular ions from the surface of a liquid matrix (13). The matrix is a viscous nonvolatile solvent that "protects" the dissolved analyte molecules from the damage that may be inflicted upon the direct impact of the high-energy "projectile." The liquid matrix also enables the analyte to circulate freely and hence constantly brings new material to the surface. FAB has proved very successful for nonvolatile materials up to several kDa, but again fragmentation often reduces the signal intensity of the (generally protonated or alkali metal ion-cationized) pseudo-molecular ion, giving this only limited usage in the biological arena. Limitations are also created by the requirement of analyte solubility in the nonvolatile liquid matrix (glycerol or 3-nitrobenzyl alcohol are commonly used). Perhaps the most serious limitation of FAB is its inability to produce larger (>10 kDa) macromolecular ions. Nevertheless, FAB MS has been the technique of choice for many years for the analysis of peptides and other small bio- and synthetic polymers.

Rapid and highly localized heating of the sample can also be achieved by using photons instead of high-velocity atoms. The invention of the laser enabled light energy of high intensity to be focused on a small area, facilitating desorption and ionization of species from either a solid or liquid sample. The rate of energy transfer, dependent on the laser fluence, determines whether vaporization is favored over decomposition of the analyte. If sample heating upon irradiation is sufficiently rapid, then it becomes possible, in many cases, to desorb intact species before decomposition can occur. This technique, known as laser desorption (or laser ablation) mass spectrometry (55), generally required post-ionization of the analyte molecules in the gas phase (usually accomplished by either electron impact or multiphoton ionization). A major limitation of laser desorption was the inadequate efficiency of energy transfer to the irradiated sample, without which a facile desorption of intact macromolecules cannot be achieved. This difficulty has been solved by the utilization of chromophores as matrices that facilitate the energy transfer and effectively shield the analyte molecules

*This ionization technique is called liquid secondary ion mass spectrometry (LSI MS) when the charged projectiles (usually Cs^+ ions) are used for the analyte ion desorption from the surface of a liquid matrix.

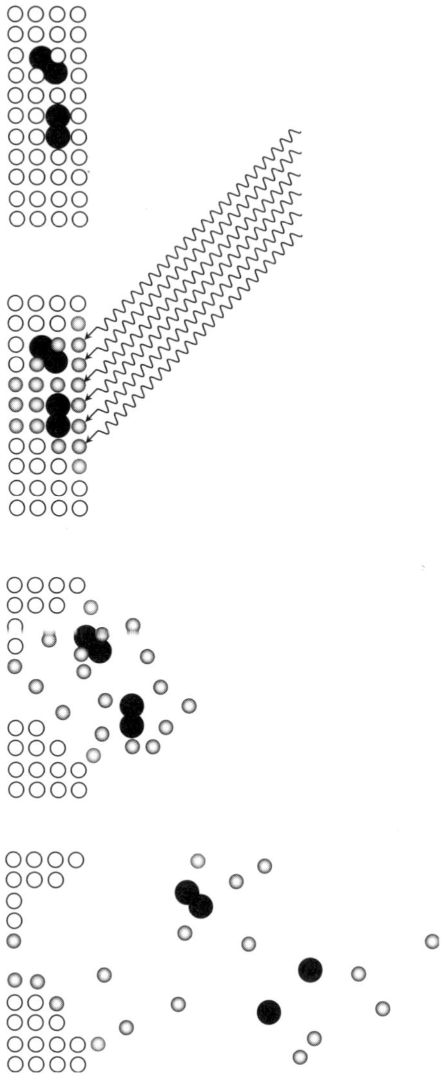

FIGURE 3.7. A schematic representation of the MALDI process. White circles represent matrix molecules packed in a crystal; embedded analyte molecules are shown in black. Grey circles represent photoexcited matrix molecules.

from radiation damage (Figure 3.7). This technique, presently known as *matrix-assisted laser desorption ionization* (MALDI), was developed simultaneously by Koichi Tanaka (56) and Michael Karas and Franz Hillenkamp (57), an invention for which the Nobel Prize in Chemistry was awarded in 2003 (58).

Most current MALDI schemes use aryl-based acids (e.g., nicotinic acid) as UV-absorbing matrices, an approach pioneered by Karas and Hillenkamp (57).

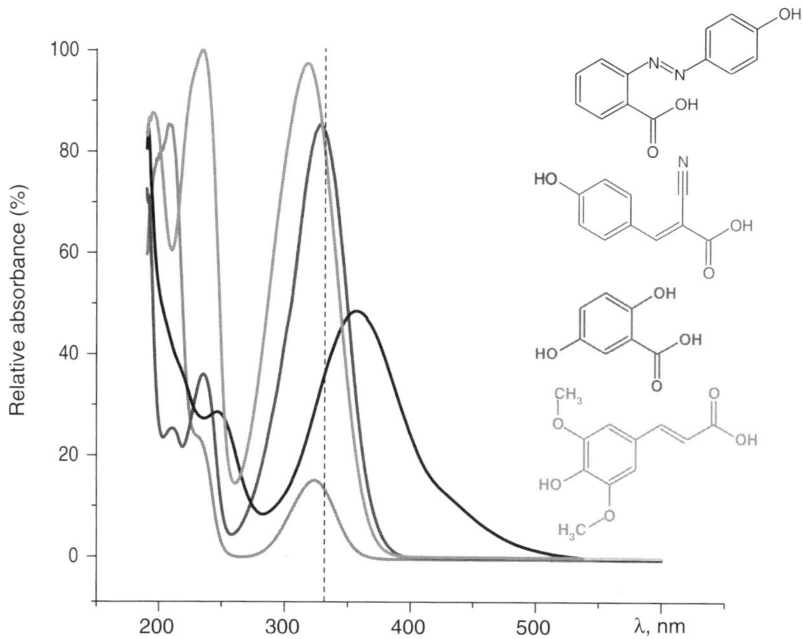

FIGURE 3.8. UV-Vis absorption spectra and chemical structures of popular MALDI matrices (from top to bottom): 2-(4-hydroxyphenylazo)-benzoic acid (HABA), α-cyano-4-hydroxy cinnamic acid (αCHCA), 2,5-dihydroxybenzoic acid (DHBA), and 3,5-dimethoxy-4-hydroxy cinnamic acid (sinapinic acid or SA). The dotted line shows the emission wavelength of a nitrogen laser. Courtesy of Wendell P. Griffith (Graduate Research Assistant at the University of Massachusetts–Amherst).

Many organic compounds with conjugated double bonds absorb UV light in the 250–370 nm range (Figure 3.8), hence the popularity of relatively inexpensive nitrogen lasers (emitting wavelength 337 nm) in MALDI mass spectrometry [another popular choice is the Nd:YAG laser, emitting wavelength 266 nm (the fourth harmonic)]. Soon after the introduction of UV-MALDI mass spectrometry, several groups started experimenting with IR lasers as a means of macromolecule desorption from the solid surface (59, 60). A particularly intriguing aspect of IR-MALDI is the possibility of utilizing frozen water as a matrix (water has a strong absorption band near 3 μm due to the O—H stretching mode). (Other IR-MALDI matrices are succinic acid, glycerol, and urea.) Using water as a MALDI matrix may open up several interesting opportunities. For example, it may allow the biological macromolecules to be kept in their native environments prior to mass analysis (whereas most UV matrices denature proteins; several important exceptions will be discussed in Chapter 4). Another interesting opportunity is the direct "imaging" of frozen biological tissues using IR-MALDI MS. Although UV irradiation remains the most popular choice for ionization in MALDI mass spectrometry, in recent years there has been steady growth in the number of

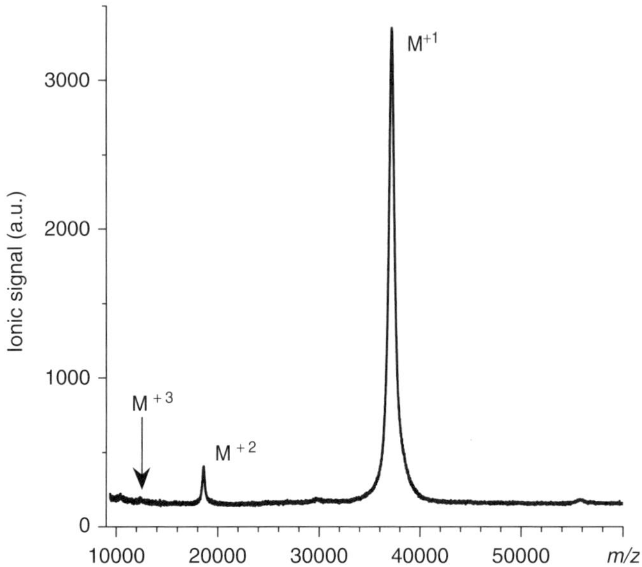

FIGURE 3.9. A conventional (UV) MALDI TOF mass spectrum of human serum transferrin N-lobe. Compare the extent of multiple charging with that displayed in the ESI mass spectrum of the same protein (Figure 3.6). Courtesy of Rachael Leverence (Graduate Research Assistant at the University of Massachusetts–Amherst).

studies that utilize IR-MALDI to achieve various analytical goals in biopolymer analysis (61–66).

Macromolecular ions produced by MALDI can also carry multiple charges; however, the extent of protonation is significantly below that achieved with ESI (Figure 3.9). This requires the utilization of mass analyzers with an extended m/z range (such as time-of-flight, to be discussed in the following sections of this chapter), although analyzers with limited mass range can still be used for detection of smaller peptides produced by MALDI. Generally, MALDI surpasses "conventional" ESI in terms of sensitivity [detection levels in the low attomole (10^{-18} moles) range have been reported] and is more tolerant to salts. Superior sensitivity, relative simplicity of operation, and ease of automation made it a top choice as an analytical technique for a variety of proteomics-related (high-throughput) applications. On the other hand, MALDI mass spectra generally are not as reproducible as those obtained with ESI. Sample preparation is clearly the major critical factor in obtaining usable and reproducible data and this depends strongly on a number of components, including analyte type and choice of matrix, solvent, and added salts.

Like ESI, MALDI is a "soft ionization" method that enables intact macromolecular ions to be transported into the gas phase for analysis by mass spectrometry. However, increased laser fluence often results in deposition of "excessive" energy, leading to analyte ion fragmentation. The extent of the fragmentation can often be controlled by modulating the power of the laser beam, allowing this phenomenon to be used analytically as a means of producing structurally diagnostic fragment ions (a discussion of various ion fragmentation processes will be presented in the following sections of this chapter). It appears that collisional cooling in the plume region is the major suppressor of the metastable ion dissociation. Collisional cooling can greatly be enhanced by elevating the background pressure in the MALDI source region to intermediate ($\sim$1 torr) or high (1 atm) levels. MALDI at atmospheric pressure [AP-MALDI (67)] offers an additional advantage of allowing liquid matrices to be used (68). A recent study by Doroshenko and co-workers demonstrated that the AP-IR-MALDI can be used to analyze peptides using *liquid water* as a matrix (69). One important issue that still has to be resolved is the mass range of biopolymers amenable to AP-IR-MALDI analysis. The study clearly demonstrates that ESI no longer monopolizes the field of biopolymer analysis directly in aqueous solutions.

Despite the spectacular success of MALDI MS and its enormous popularity as a potent bioanalytical tool, many mechanistic aspects of the MALDI processes (both physical and chemical) remain poorly understood as yet (70). An interested reader is referred to a series of review articles focusing on both desorption (71) and ionization (72, 73) aspects of MALDI that were published recently in a thematic issue of *Chemical Reviews*.

3.3. MASS ANALYSIS

3.3.1. General Considerations: m/z Range and Mass Discrimination, Mass Resolution, Duty Cycle, Data Acquisition Rate

Once the macromolecular ions have been produced and transferred to the vacuum, they can be selectively manipulated and detected using the vast arsenal of MS. In this respect, macromolecular ions can be dealt with using the principles developed originally for handling small organic and inorganic ions. An important difference, however, is the large size of biopolymers, which makes certain aspects of ion manipulation and detection much more challenging.

Generally, several issues have to be considered to ensure high-quality mass measurement for macromolecular ions produced by ESI or MALDI. Perhaps the most important one is the requisite m/z range of a mass analyzer. For example, polymer (both bio- and synthetic) ions generated by MALDI typically have few charges, hence the requirement that the mass analysis be carried out using an analyzer with an extended m/z range. This can best be accomplished by the time-of-flight (TOF) mass analyzers, which will briefly be reviewed at the end of this chapter. ESI, on the other hand, produces polymer ions with significantly higher number of charges, as compared to MALDI. As can be seen from Figures 3.6 and 3.9, the most abundant ionic species of a 37 kDa protein *h*Tf/2N in the ESI mass spectrum carries 22 charges, while the most abundant ionic species of the same protein in the MALDI spectrum carries only one positive charge. This, of course, relaxes the requirements vis-à-vis the "working" m/z range of the analyzer employed for the detection of the ESI-generated biopolymer ions. In many cases, relatively inexpensive analyzers with modest m/z range (m/z values of protein ions generated by ESI under denaturing conditions usually fall within the 500–2500 range) would suffice (e.g., low-end quadrupole filters and ion traps). An important exception is a situation when ESI MS is used to detect "native" proteins. As we will see in Chapter 4 of this book, the number of charges accumulated by such protein ions generated by ESI under "near-native" conditions in solution is usually relatively low. As a result, m/z values of such "low charge density" ions will be very high, warranting the use of analyzers with extended m/z range. In the extreme cases of large macromolecular complexes, an m/z range of up to 20,000 may be required. (Several examples of such large complexes, including intact ribosomes and viruses, will be considered in Chapter 11.)

The second important characteristic of the mass analysis process is the resolution (or mass resolving power) attained. Mass resolution defines the minimal difference between the m/z values of the two ionic species that still allows a clear distinction to be made during the mass analysis. Historically, the two peaks of equal intensity were considered to be resolved if the valley between them was 10% of the maximum intensity. Hence, the definition of the resolution was

$$R = \frac{M}{\Delta M}, \tag{3-3-1}$$

where M is the m/z value of a particular ion peak, and ΔM is the width of this peak at 5% of its height. More recently, a more liberal (and much more forgiving) definition of the mass resolution was adopted, the one using ΔM at 50% of the peak height. Regardless of the definition used, resolution is a function of m/z, rather than a "fixed" characteristic of a given instrument.

The observation of distinct isotopic species of peptides and proteins of progressively increasing size can be achieved by increasing the mass resolving power of the analyzer (Figure 3.10). Ultrahigh resolution may allow in some cases the isobaric species to be separated and distinctly detected (10). Resolution is very important for accurate mass measurements, as it allows in many cases the monoisotopic mass to be accurately measured. However, the very low abundance of monoisotopic peaks of large proteins makes their detection all but impossible. In such cases high-resolution mass measurements do not offer any significant advantages over "conventional" MS with modest resolution, unless the abundance of the monoisotopic peaks is increased [e.g., by using isotope depletion during protein expression (74)].

Finally, issues related to the duty cycle of a mass analyzer and its data acquisition rate need to be taken into consideration when selecting the mass analyzer appropriate for a specific task at hand. MALDI, by definition, is a pulsed process and is best interfaced with "fast cycling" analyzers that can be synchronized with a laser (e.g., TOF, ion trap). Interfacing MALDI with "scanning" devices, such as a magnetic sector MS, results in significant losses in sensitivity and requires

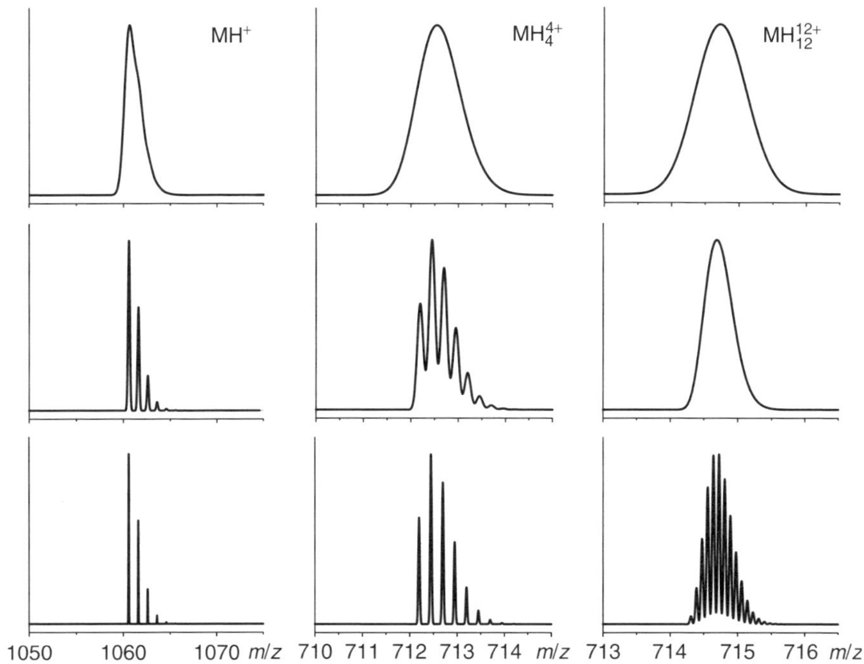

FIGURE 3.10. Simulated isotopic patterns for three peptide ions: bradykinin (*left*), melittin (*middle*), and ubiquitin (*right*) at a resolution of 1000 (*top*), 5000 (*center*), and 20,000 (*bottom*).

extended acquisition times (75). On the other hand, ESI is a "continuous" ionization process; hence there is a wide variety of mass analyzers to which it can be interfaced. Still, the accumulation of ESI-generated ions in an external "storage" device followed by a pulsed introduction into a TOF or ion trap mass analyzer greatly increases the duty cycle (and, therefore, sensitivity) of the analysis compared to a conventional scheme when the ESI source is interfaced directly to a slow scanning analyzer.

A very important advantage offered by MS as an analytical technique is the possibility to implement various "hyphenated" techniques, ranging from direct (on-line) coupling of MS with separation techniques for increased selectivity and sensitivity of the analysis and implementation of tandem MS (MS/MS and MS^n) strategies for structural analysis.

3.3.2. Mass Spectrometry Combined with Separation Methods

The combination of separation techniques with mass spectrometric detection is a mature field, which has been reviewed recently in several excellent papers (76–78). In this section we only recount certain features of this combined technique that are particularly relevant to the biophysical experiments discussed in the following chapters. ESI is a particularly attractive interface between liquid chromatography (LC) and MS due to the "continuous" nature of the ionization process, which allows the analysis of the eluate content to be carried out in real time (the so-called "on-line" LC-MS). Interestingly, the very first "on-line" LC-ESI MS analyses of peptides were carried out in the early 1980s (35), prior to the wide acceptance of electrospray as a reliable analytical tool in biomolecular analysis. In the twenty years that have passed since this first report, HPLC-ESI MS has become a mature technology, with many commercial instruments available as "integrated" LC-MS systems. Current development of this technology follows two major routes, namely, reduction of the scale ("smaller columns" for ever improving detection limits) and reduction of the analysis time ("faster" separations to satisfy the ever growing demands of high-throughput analyses) (76).

"Miniaturization" of the separation step allows the amount of sample needed for analysis to be reduced very dramatically. Even for "conventional" ESI sources the typical flow rate is usually on the order of several μL/min. Such flow rate requirements are optimally matched by micro-HPLC systems, while utilization of a "regular" HPLC column* requires either postcolumn eluate flow split prior to its introduction to the ESI source or utilization of "ionspray" sources capable of handling high eluate flow rates. Besides matching the flow rates of the HPLC and ESI MS, one needs to be aware of another important technical issue related to solvent compatibility. Reversed-phase LC separation of peptides and proteins is usually carried out using trifluoroacetic acid (TFA) as an eluent modifier. TFA is a very strong electrolyte ($pK_a < 0.5$), and its presence in the solvent usually decreases the quality of the ESI MS data. If necessary, the detrimental effect of TFA on the ESI MS measurements can be reduced/eliminated using two different approaches. In the first approach, TFA concentration in the eluate is reduced by mixing it with a flow of liquid containing volatile ion-pairing agents, a step that can be carried out either "postcolumn" or in the ESI source. More sophisticated schemes utilize a second chromatographic step, whose mobile phase is modified with volatile reagents (79). An alternative approach utilizes weaker acids (such as acetic and formic acids instead of TFA) during the separation step. An unavoidable loss of chromatographic fidelity if compensated by MS detection, which allows the poorly resolved analytes to be easily distinguished, based on the differences in their masses.

Traditionally, liquid chromatography was not viewed as a "fast" method of chemical analysis and was poorly suited for high-throughput analyses. The time-consuming nature of chromatography has stimulated design of methods that improve the speed of separations (80). As we will see in Chapters 4 and 5, fast HPLC separation of peptic fragments is crucial for high-quality analysis of protein dynamics by monitoring hydrogen–deuterium exchange (HDX) reactions in a site-specific fashion (81, 82). Significant improvements in the speed of separation can be achieved by using smaller and nonporous or superficially porous particles as a column packing material. The potential for fast separations can be

*Optimal separation of a peptide mixture with a "typical" analytical column (4.6 mm) requires a flow rate of eluent on the order of 1 mL/min.

further increased by using capillary columns, monolithic columns, open tubular columns, and small-diameter packed capillaries (80). Elevated temperatures can also be effective for decreasing analysis times. However, the HDX HPLC MS measurements (the most demanding fast-LC applications in biophysical studies so far) are usually carried out at low temperatures to avoid excessive exchange of deuterium atoms between the peptides and the mobile phase during the separation step. Increasing the separation speed is usually achieved at the expense of the separation quality. Again, this is usually compensated by the high detection specificity of MS. A great deal of useful practical information on various HPLC MS techniques can also be found in a collection compiled and edited by Niessen (83).

Capillary electrophoresis (CE) is another separation technique that can be interfaced directly with ESI MS. However, the design of the CE–ESI MS interfaces is technically more difficult due to the problems associated with the high voltage utilized by CE for analyte separation. There are several experimental schemes that can be used to circumvent this problem, a detailed overview of which can be found in a recent review article by Tomer (76). Although the number of applications of the "on-line" CE–ESI MS systems is still lagging behind that of HPLC–ESI MS, the former technique is steadily gaining popularity due to its superior sensitivity and resolution. Several excellent reviews on the subject detail recent developments in this field (84–87).

The interfacing of ESI MS with other types of separations, such as immobilized metal-ion affinity chromatography [IMAC (88)] or bioaffinity chromatography (89), is also possible, although the coupling cannot be direct and involves an "intermediary" HPLC step. A concise overview of these multidimensional chromatographic techniques can be found in Tomer's review (76). Our attention has been focused in this section on ESI MS as a "detector device." Utilization of liquid matrices in AP-MALDI (briefly discussed in the previous section of this chapter) opens up a possibility for "on-line" LC-MALDI MS. Direct coupling of LC and MALDI MS can also be accomplished by using continuous flow devices (90, 91) or aerosol MALDI MS (92). Unfortunately, even a brief discussion of these devices is beyond the scope of the book; however, an interested reader is referred to several recent reviews on the subject (93–95).

3.4. TANDEM MASS SPECTROMETRY

One of the most attractive features of both ESI and MALDI for biomolecular analysis is their ability to generate intact macromolecular ions in the gas phase. While molecular weight information is very useful, it is not sufficient in most instances for unequivocal identification of even small peptides. Furthermore, while high-resolution mass measurements can sometimes reveal the molecular formula of a small peptide, they do not provide any information on its covalent structure. The latter can be obtained by inducing dissociation of the molecular ion and measuring the masses of the resulting fragment ions. Since most proteins and

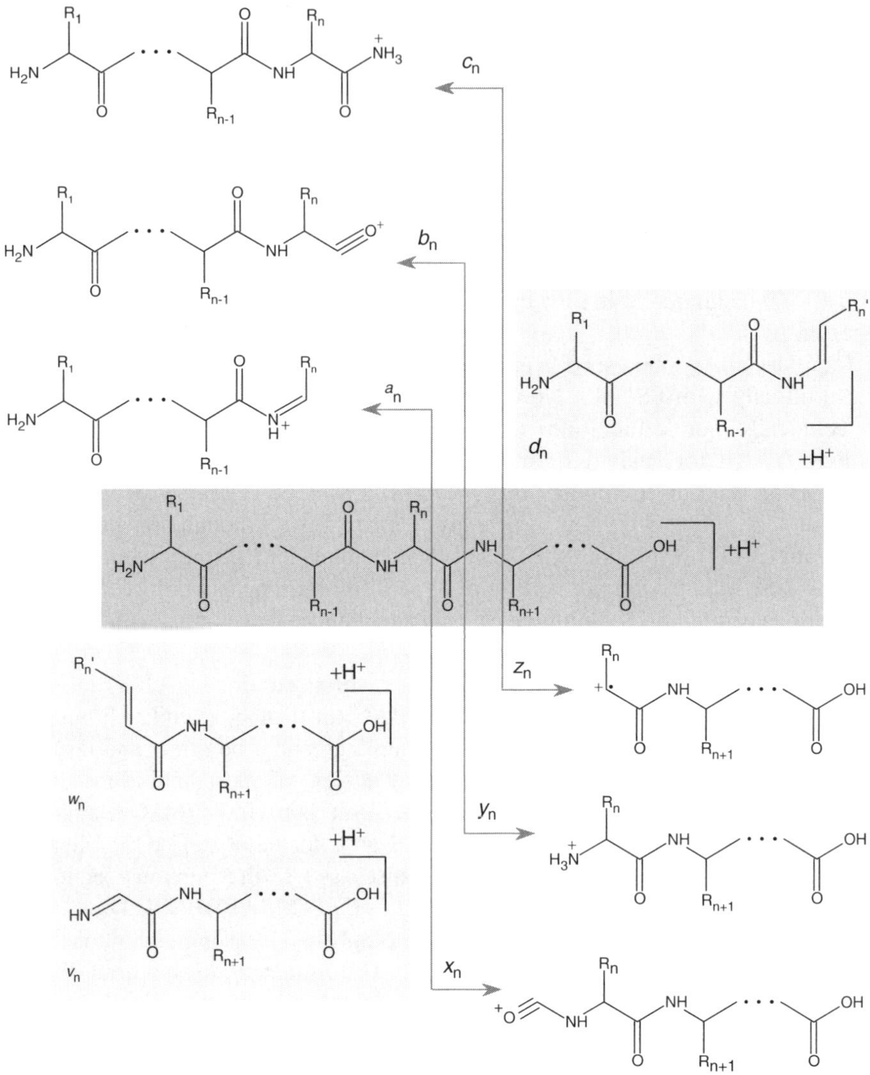

FIGURE 3.11. Biemann's nomenclature of peptide ion fragments (96). Fragment ions shown in gray boxes correspond to either complete or partial loss of the side chains and are usually observed only in high-energy CID.

peptides are linear polymers, cleavage of a single covalent bond along the backbone generates a fragment ion (or two complementary fragment ions) that is classified as an *a*-, *b*-, *c*- or *x*-, *y*-, *z*-type (96, 97), depending on (i) the type of bond cleaved and (ii) whether the fragment ion contains an N- or C-terminal portion of the peptide (Figure 3.11). Cyclic and disulfide-linked polypeptides are a special case, since a single bond cleavage does not necessarily produce distinct (physically separated) fragments. Nomenclature for cyclic peptide fragmentation can be found in (98).

3.4.1. Basic Principles of Tandem Mass Spectrometry

Ion dissociation can often be carried out in the ionization source, for example, by increasing the desolvation potential in the ESI interface ["nozzle-skimmer dissociation" first observed by Alexandrov and co-workers (34)] or by using high laser power in MALDI. If the sample is homogeneous, such in-source fragmentation can be used for biopolymer sequencing (Figures 3.12 and 3.13). In most cases, however, the sample to be analyzed is a rather heterogeneous mixture, and the "in-source" fragmentation spectrum would be very difficult to interpret due to the presence of fragment ions derived from different "precursor" ions. One way to resolve this problem is to use an on-line separation of analytes prior to their introduction to the ionization source (e.g., an HPLC–ESI MS interface discussed in the previous section). However, a much more flexible and powerful solution to this problem utilizes a mass spectrometer itself as an ion separation device, which allows the fragmentation of mass-selected precursor ions to be carried out in a variety of ways. This approach to structural characterization of molecular

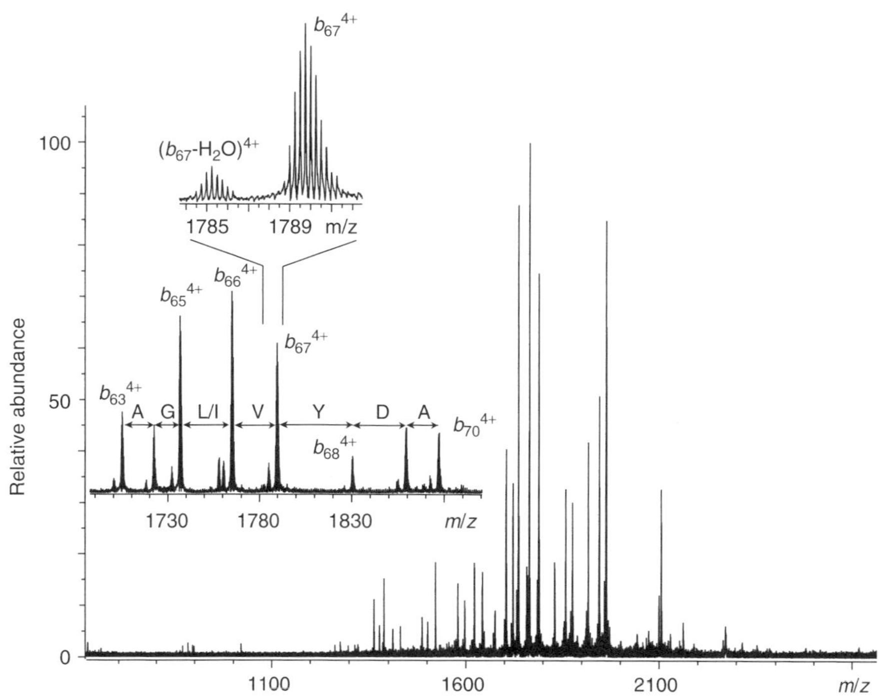

FIGURE 3.12. Prompt ion fragmentation in ESI MS: mass spectrum of human serum transferrin N-lobe (*h*Tf) acquired under elevated skimmer potential. A mass spectrum of this protein acquired under gentle conditions in the ESI interface is shown in Figure 3.6.

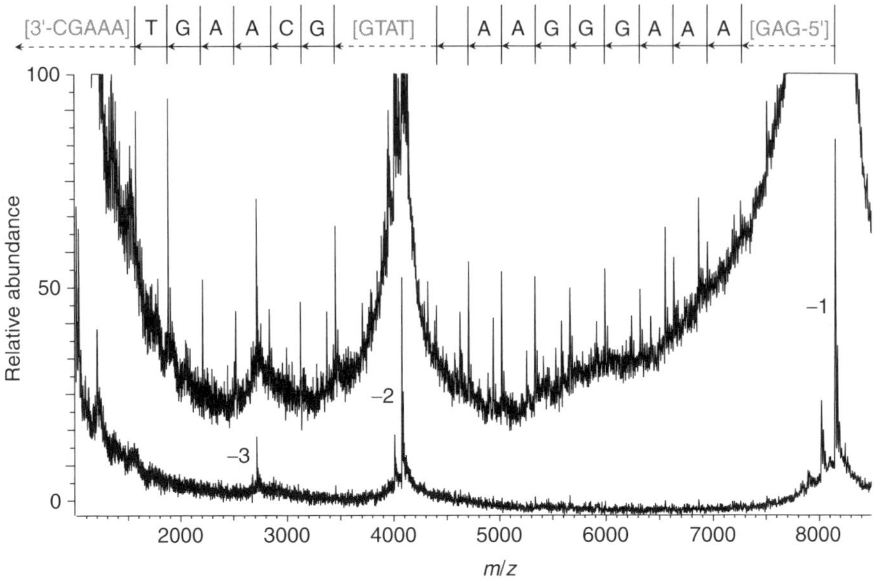

FIGURE 3.13. Prompt fragmentation in MALDI MS: UV-MALDI spectra of an oligonucleotide strand acquired at increased (*top trace*) and moderate laser power.

ions was pioneered by McLafferty and is known as *tandem mass spectrometry* or *MS/MS* (99). Tandem spectra of peptide ions are easily interpretable (provided the fragmentation yield is sufficiently high), as they only contain contributions from a single precursor ion (Figure 3.14). Although the process of mass selection greatly reduces the overall ionic signal intensity, the signal-to-noise ratio achieved in MS/MS experiments is often better than in the MS1 experiment due to elimination of chemical noise (Figure 3.14).

The term "tandem" implies physical separation of the precursor ion prior to fragmentation (MS1 selects the precursor ion and MS2 records a spectrum of the fragment ions derived from it); nonetheless, the two stages (MS1 and MS2) can be combined in one step. As we will see in the following sections of this chapter, tandem MS experiments can be carried out without physical separation of the chosen precursor ion from other ions generated in the source. A generalized definition of tandem MS extends to any experiment that defines an m/z relationship between a precursor ion and a fragment ion (100).

3.4.2. Collision-Induced Dissociation: Collision Energy, Ion Activation Rate, Dissociation of Large Biomolecular Ions

The majority of tandem MS experiments employ various means of increasing internal energy of the precursor ions to induce the dissociation of covalent bonds in the gas phase. Collisional activation (conversion of a fraction of the ion's kinetic energy to vibrational excitation upon its collision with a neutral

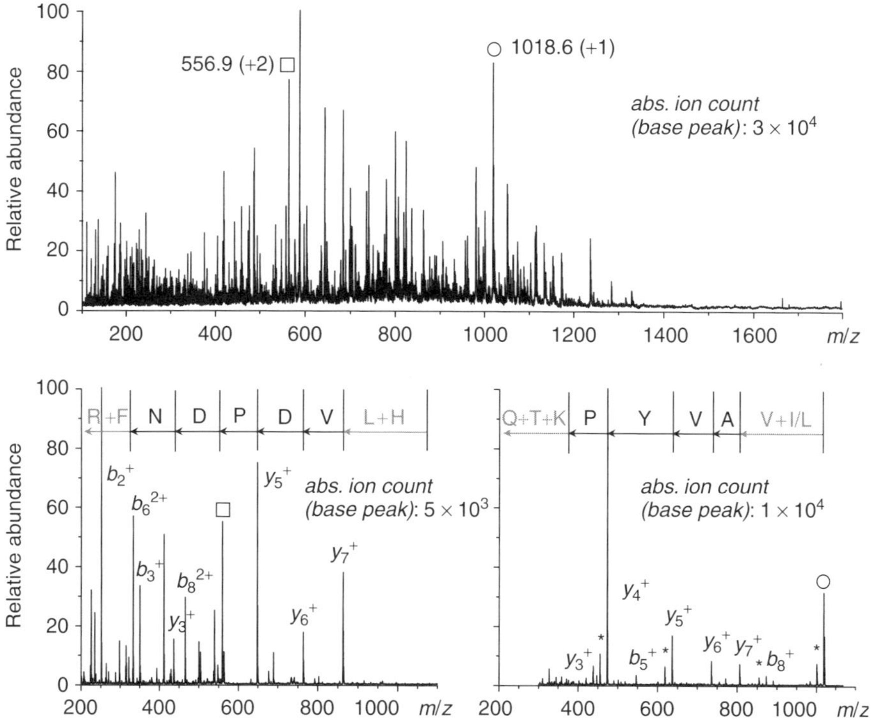

FIGURE 3.14. Tandem mass spectrometry (CID) of tryptic peptides from a digest of *Gadus morhua* hemoglobin. Two peptide ions marked with a square (doubly charged β_{97-105}, LHVDPDNFR, m/z 557) and a circle (singly charged α_{33-41}, LVAVYPQTK, m/z 1019) in a spectrum of the mixture (*top trace*) were mass selected and fragmented in a quadrupole trap. In each case, a series of abundant y-ions provides almost complete sequence coverage of the peptide ion (*bottom traces*). Courtesy of Wendell P. Griffith (Graduate Research Assistant at the University of Massachusetts–Amherst).

molecule) is perhaps the most widely used method of elevating an ion internal energy. This technique is usually referred to as collision-induced dissociation, CID* (101). Two distinct regimes of collisional activation are usually recognized, high- and low-energy. The low-energy collisional activation refers to a broad range of ion kinetic energy prior to collision (usually in the sub-eV range) and typically requires multiple collisions in order to accumulate enough internal energy to afford dissociation of a covalent bond.† Therefore, low-energy CID is a very slow process, with the activation time typically exceeding 10^{-2} s (103).

*A slightly different term (collision-activated dissociation, CAD) is also in frequent use.

†It must be noted, however, that *low-energy* and *slow-heating fragmentation* are not necessarily synonymous; a good example is the so-called SORI CID (which will be discussed in the following sections of this chapter). While the energy of each collision in SORI may be relatively high, the frequency of such collisions is typically very low. As a result, the internal energy accumulation is slow due to the efficient radiative cooling of the ions between consecutive collisions (102).

The low-energy CID spectra of peptide ions are usually dominated by fragment ion peaks corresponding to *b*- and *y*-ions (Figure 3.15A, B). One of the consequences of multiple collisions is the formation of internal fragment ions, whose presence in the CID spectra makes their interpretation more difficult, and the greater propensity for rearrangement prior to fragmentation.

The high-energy CID is usually carried out by accelerating ions to several keV prior to colliding them with neutral targets (typically atoms of inert gases). For smaller ions, a single high-energy collision is often sufficient to cause dissociation of a covalent bond. However, as the ion size increases, so does the number of vibrational degrees of freedom among which the excitation is distributed. Furthermore, collisional energy in the center-of-mass frame decreases with increasing mass of the projectile ion when the kinetic energy of the projectile and the mass of the neutral target are fixed. Nonetheless, in many cases it is possible to obtain nearly complete sequence coverage even for relatively large (up to 5 kDa) polypeptide ions (104, 105). In sharp contrast to the fragmentation patterns observed with multiple low-energy collisions, fragment ions produced by high-energy CID are amply formed by extensive cleavages of the backbone (not limited to the amide bond dissociation, producing *b*- and *y*-type fragment ions), as well as the side chains (Figure 3.15C). Dissociation of the covalent bonds along the amino acid side chains leading to elimination of either the entire side chain (*v*-ions) or a portion of it (*d*- and *w*-ions) is unique to high-energy CID and can provide information on the identity of isomeric side chains, for example, by distinguishing between leucine and isoleucine (106).

3.4.3. Other Fragmentation Techniques: Electron-Capture Dissociation, Photoradiation-Induced Dissociation, Surface-Induced Dissociation

We have already mentioned that one of the major factors limiting the fragmentation yield of CID is the low efficiency of energy conversion (from translational to internal degrees of freedom). Collisional activation of large biomolecular ions is particularly problematic due to the large disparity between the masses of the target (neutral atom or molecule of the collision gas) and the projectile (ion).* Such mass disparity results in relatively modest collisional energy in the center-of-mass frame even if the energy of the projectile in the laboratory frame is very high. One way to circumvent this problem is to employ a collision target with large (ideally, infinite) mass. Practical implementation of this approach led to the development of surface-induced dissociation (SID), a fragmentation technique that utilizes ion–surface collisions as a means to convert ion kinetic energy to internal excitation (107, 108). However, the fraction of energy deposited into the ion upon such collision approaches unity very seldom, as it is largely determined by the mass of the chemical moiety representing an immediate collision partner for the ion impacting the surface (109). One practical aspect of SID that makes

*Even the mass of the heaviest target (Xe) is several orders of magnitude below that of a relatively modest biomolecule.

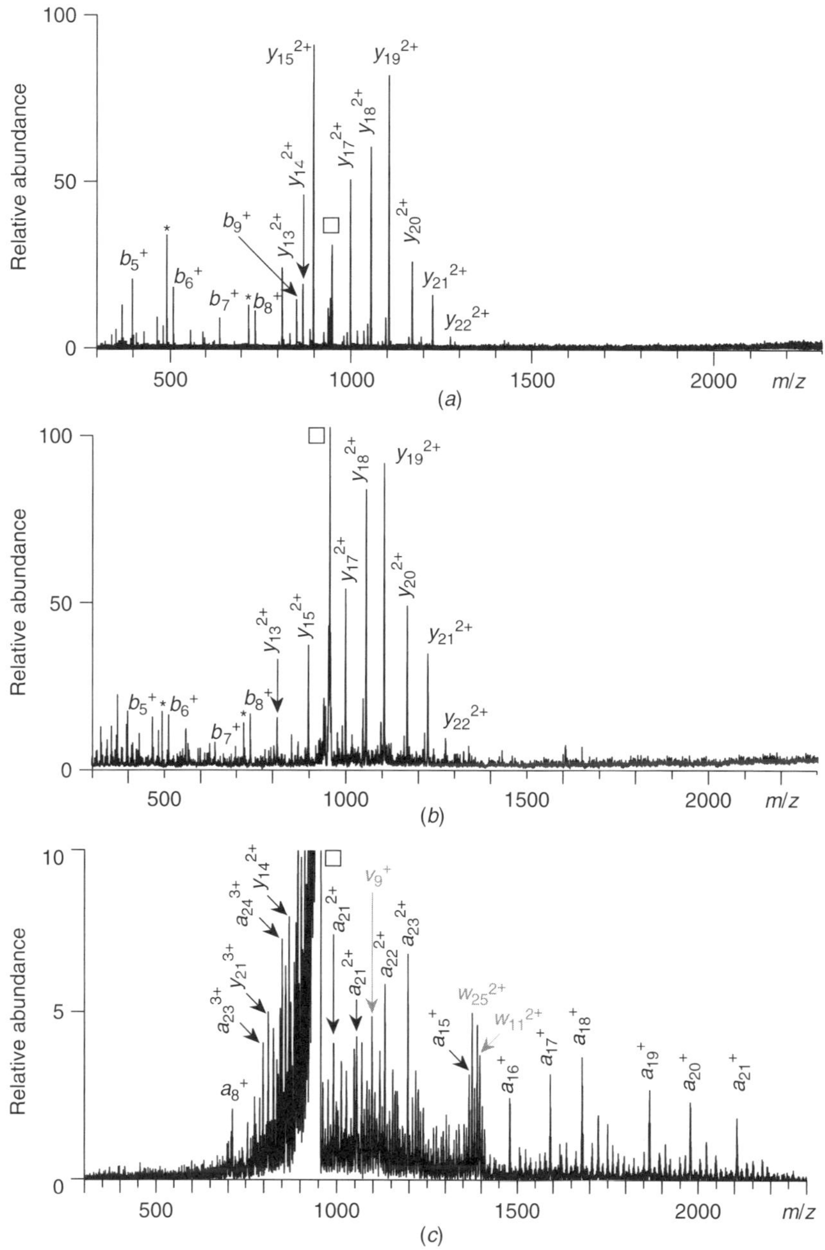

FIGURE 3.15. Low- versus high-energy peptide ion fragmentation. Low-energy CID spectra of melittin (+3 charge state) were acquired with an FT ICR MS (SORI CID, *top trace*) and a quadrupole ion trap (*middle trace*). High-energy CID spectrum was acquired with a magnetic sector MS (B/E linked scan mode, *lower trace*). Only the most abundant fragment ions are labeled in the spectra. Courtesy of Anirban Mohimen and Joshua Hoerner (Graduate Research Assistants at the University of Massachusetts–Amherst).

its applications rather limited is the difficulty of collecting the fragment ions off the surface.

Ion-neutral collision is not the only process that can be used to increase ion internal energy. In principle, any exothermic process can be employed for ion activation purposes. For example, photoexcitation of ions in the gas phase often leads to their dissociation. While numerous studies of small ions utilize photodissociation induced by UV light, dissociation of large macromolecular ions is most effective when IR photons [typically, 10.6 μm (CO_2 laser)] are used for the excitation [a technique known as *infrared multiphoton dissociation*, IRMPD (110)].

Finally, a recently introduced technique of *electron capture dissociation* (ECD) (111) holds great promise due to its unique specificity, which in many cases allows "targeted" fragmentation to be carried out (with no or little energy partitioning). One particularly attractive feature of ECD is its ability to cleave disulfide bonds in the gas phase (112), a very challenging task when other methods of ion activation are employed.

3.4.4. Ion–Molecule Reactions in the Gas Phase: Internal Rearrangement, Charge Transfer

Ion activation times can range from less than 10^{-15} s to over 10^3 s, depending on the particular activation technique used to induce fragmentation, physical size of the ion (number of internal degrees of freedom), as well as the time frame of the experiment (103). In the case of a slow activation process, covalent bond cleavage may be preceded by an internal rearrangement. One particular type of such rearrangement is hydrogen scrambling, which will be discussed in some detail in Chapter 5.

Charge transfer is another gas phase process that frequently occurs in the ESI interface region. Although charge transfer reactions do not usually result in ion fragmentation, they obviously affect the appearance of charge state distributions in ESI mass spectra. As we will see in Chapters 4 and 5, charge state distributions of protein ions are often used to assess "compactness" of protein structures in solution. Therefore, close attention needs to be paid to gas phase processes in order to avoid misinterpretation of the ESI MS data. In some instances it is possible to trap ions of different polarities simultaneously and confine them to a small volume (113). Electrostatic attraction between the ions of opposite charges may lead to a variety of ion–ion reactions, some of which will be considered in Chapter 9.

3.5. BRIEF OVERVIEW OF COMMON MASS ANALYZERS

The mass analyzer is the part of a mass spectrometer where the ions are separated according to their m/z values. As outlined in the preceding section, combination of two (or even more) mass analyzers often allows spontaneous or induced fragmentation of mass-selected ions to be studied using methods of tandem MS. Certain types of mass analyzers allow tandem experiments to be carried out without utilization of a second analyzer (the "tandem-in-time" as opposed to "tandem-in-space" MS). There is a wide range of mass analyzers differing in their compatibility with various ion sources; ability to handle ions of certain types; analytical figures of merit; user-friendliness; and, of course, price tags. In this section we attempt to provide a brief review of mass analyzers that are most popular in the bio-MS community. A brief discussion of each analyzer will include principles of its operation, compatibility with ESI and MALDI sources, m/z limitations and commonly attained resolution, and tandem capabilities. A much more comprehensive discussion of mass analyzers (which also includes devices and designs not covered in this section) can be found in an excellent recent review by McLuckey and Wells (100). A more "in-depth" discussion of physical aspects of mass analysis of ions can be found in other review articles (114, 115).

3.5.1. Mass Analyzer as an Ion Dispersion Device: Magnetic Sector MS

The idea to use a combination of electrostatic and magnetic fields as a means of ion separation in space (dispersion) was introduced by J. J. Thomson in his parabola mass spectrograph (1). A simpler and more efficient method of ion separation in homogeneous magnetic fields was introduced several years later by Dempster (116). This instrument became a prototype of the highly successful *magnetic sector* mass analyzer, which is widely used in mass spectrometry to this day. According to (3-1-1), an ion introduced to the magnetic field orthogonally will follow a circular trajectory whose radius will be determined by the ion's m/z ratio, its velocity v, and the magnetic field strength B [in other words, the magnet will act as a momentum separator or, more precisely, a momentum-to-charge (mv/z) separator]. If all ions of the same charge have been accelerated to the same kinetic energy prior to their introduction into the magnetic field (i.e., $\mathrm{KE} = \frac{1}{2}mv^2 = zeV$), then the radius of circular trajectory for each ion will be uniquely defined by its m/z. Later modifications of the magnetic sector mass analyzers led to significant improvements in resolution and other performance characteristics.

Perhaps the most important among such modifications was the implementation of a "double-focusing" scheme (117), which adds another analyzer (a sector with radial electrostatic field, often termed electrostatic analyzer, ESA) acting as a "kinetic energy separator." A mass spectrum is usually obtained by scanning the magnetic field strength over a desired range, while the electrostatic field of ESA remains constant (linked to the acceleration voltage V_0 to allow passage of ions whose kinetic energy-to-charge ratio is equal to eV_0). In addition to a dramatic increase in mass resolution, the presence of a second analyzer allows tandem MS experiments to be carried out. For example, a fragment ion ${m_f}^{zf+}$ produced upon dissociation of a metastable ion ${m_0}^{z0+}$ immediately following their acceleration (in the so-called first field-free region, 1 FFR, Figure 3.16) will have the same velocity as the precursor ion itself; however, the momenta

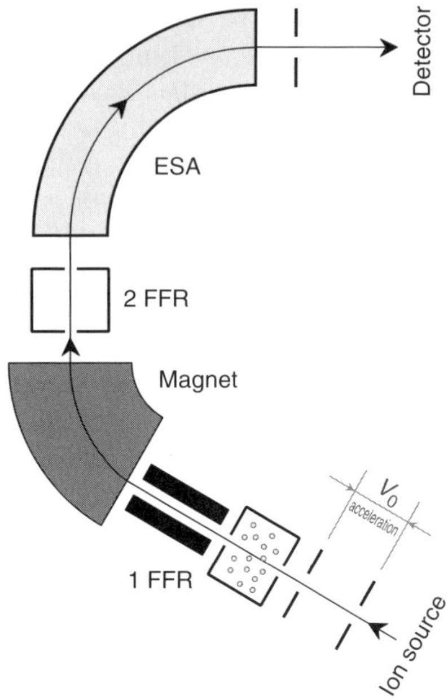

FIGURE 3.16. Schematic representation of a magnetic sector MS (BE geometry). Collision gas in the first field-free region (as shown in the diagram) is used only for acquisition of CID spectra (B/E, B^2/E, and CNL scans).

and the kinetic energies of these two ions will be different (both the momentum and the kinetic energy of the fragment ion will relate to those characteristics of the precursor ion as m_f/m_0). Therefore, in order to guarantee the passage of the fragment ion through *both* magnetic sector and ESA, both fields have to be reduced by a factor of m_f/m_0 compared to those needed for the passage of the precursor ion. If both fields are scanned, while their ratio remains constant, a full range of fragment ions originating from the same precursor will be detected, an experiment commonly known as a *B/E scan* or *linked scan*). The representation of the B/E scan line on the (B, E) plane connects the $(B_0, E_0)^*$ point with the coordinate origin (Figure 3.17). Obviously, if $z_0 > 1$, the B/E scan should start at the point $(z_0 B_0, z_0 E_0)$, otherwise none of the fragment ions whose charge is less than that of the precursor ion would be detected. This is exactly how a fragmentation spectrum of melittin (Figure 3.15C) has been acquired.

The B/E scan is an example of a "pseudo-tandem" MS experiment (the precursor ion is not mass-selected prior to its dissociation). Other examples of "pseudo-tandem" MS are a *B^2/E scan* (providing a means to detect all ions giving rise to a certain fragment) and a *constant neutral loss* (CNL) *scan*. A

* B_0 is the magnetic field strength needed for passage of the intact precursor ion through the magnetic sector and E_0 is the electrostatic field strength required for passage of any "fully accelerated" ion ($KE/z = eV$) through ESA.

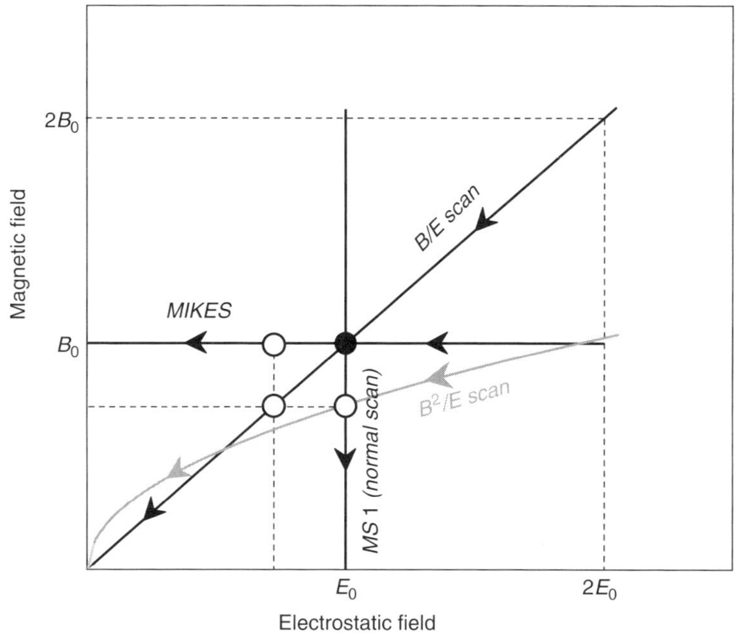

FIGURE 3.17. Graphic representation of various scans used in sector mass spectrometers similar to one whose schematics is shown in Figure 3.16. The solid circle indicates a position of a precursor ion and the open circles indicate positions of one particular fragment ion as detected in the course of various experiments: magnetic field scan (prompt fragment in MS1), as well as a linked (B/E) scan and an electrostatic field scan (MIKES). The gray line represents the precursor ion scan (B^2/E).

physical selection of the precursor ion prior to its dissociation can be achieved by placing a collisional cell after the magnetic sector (second field-free region, 2 FFR, Figure 3.16). Since the kinetic energy of each fragment ion will be a m_f/m_0 fraction of the precursor kinetic energy, scanning the electrostatic field of the ESA will allow a mass spectrum of all fragment ions to be acquired. Although this technique [termed *mass-analyzed ion kinetic energy spectrometry*, MIKES] has rather limited analytical use (due to very low mass resolution), it can provide information on macromolecular ion geometry, a topic we will discuss in some detail in Chapter 9. A combination of at least three sectors (e.g., *BEB**) is required in order to obtain high-resolution tandem MS (MS/MS) data.

The magnetic sector MS is a very flexible analytical device that allows a variety of experiments to be carried out without any hardware modification. Sectors allow high-resolution measurements to be carried out within a wide m/z range. Ion fragmentation experiments (CID) are carried out in a high collisional energy regime, enabling observation of fragments that are not usually detected when most other analyzers are used (e.g., *d*-, *v*- and *w*-ions in peptide fragmentation

*Magnetic sector 1—ESA—magnetic sector 2.

spectra). However, the magnetic sector instruments have significant disadvantages as well. Since the data acquisition rate is usually limited by the magnet scan rate, sector MS cannot be interfaced readily with "pulsed" ionization sources (such as MALDI). Even when interfaced with a continuous ionization source (such as ESI), the sensitivity of the analysis is often limited by the unfavorable duty cycle (particularly when the data acquisition is carried out over a wide m/z range).

3.5.2. Temporal Ion Dispersion: Time-of-Flight MS

The concept of *time-of-flight* (TOF) mass analyzers was first introduced over half a century ago under the names "time dispersion mass spectrometer" (118) and "ion velocitron" (119). The basic principle of TOF MS is very simple (Figure 3.18): ions of different masses are accelerated to the same kinetic energy (by traversing a potential difference V) within a very short period of time and introduced into a field-free "drift region" (or "flight tube"). If the initial velocities of all ions are neglected, then the final velocity of each ion in the drift region will be uniquely determined by its mass-to-charge ratio and the acceleration potential U_0:

$$v = \sqrt{\frac{2zeU_0}{m}} \tag{3-5-1}$$

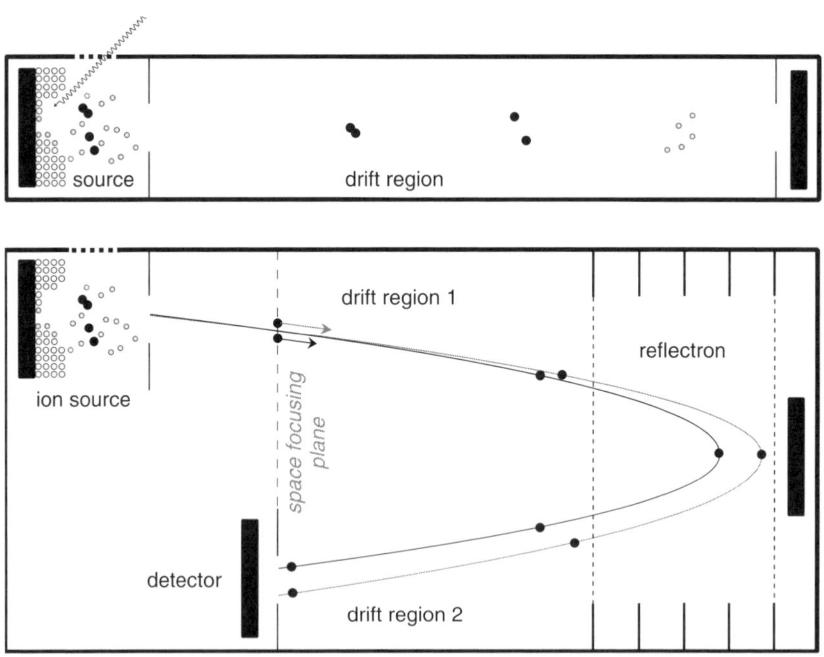

FIGURE 3.18. Schematic diagrams of linear (*top*) and single-stage reflectron (*bottom*) time-of-flight mass spectrometers.

If the duration of the pulse during which ions were introduced into the drift region is very short, then ions arriving at the detector plane will be grouped according to their m/z values:

$$t = \frac{D}{v} = \sqrt{\frac{m}{2zeU_0}} \cdot D, \tag{3-5-2}$$

where D is the length of the drift region and t is the time required to traverse this region. Therefore, measuring the ionic signal intensity as a function of the "arrival time" would allow the mass spectrum to be recorded. In reality, this simplistic approach would lead to very poor mass resolution, mostly due to significant spread in the initial kinetic energies of ions. A "correction" for the initial energy distribution can be done using an "ion mirror" or reflectron (120). The principle of the reflectron operation is illustrated in Figure 3.18: if the two ions have identical mass and charge, but different velocities, the faster ion will penetrate deeper into the decelerating region of the reflectron. As a result, its trajectory path will be longer. After its reemergence from the reflectron, this ion would still have a higher velocity, but it will be "lagging" behind the slower ion due to the extra time spent in the decelerating region. It is easy to show that the total "travel time" (from the source to the detector) of an ion having initial kinetic energy eV can be calculated as

$$t = \sqrt{\frac{m}{2ze(V + U_0)}} \cdot (D_1 + D_2 + 4d), \tag{3-5-3}$$

where D_1 and D_2 are the lengths of the "upstream" and "downstream" drift regions and d is the reflectron penetration depth (a function of eV). It is possible to adjust the reflectron parameters ($D_1 + D_2$ and U_0) in such a way that the flight times become independent (within a narrow range) of the initial kinetic energy eV. The highest mass resolving power is attained when $D_1 + D_2 = 4d$; that is, the ion spends equal time in the reflectron and the field-free drift regions (121). Although such a "single-stage" reflectron can only perform first-order velocity focusing of ions initially located at the space focal plane (Figure 3.18), the second-order focusing can be achieved using a double-stage ion mirror. Theoretically, it is possible to achieve an ideal focusing (totally independent) of the ion kinetic energy by using a continuous quadratic field (without the field-free drift region). The quadratic ion mirror is impractical for a variety of reasons; however, its modification known as a "curved-field" reflectron (122) has become very popular. While the detailed discussion of ion focusing techniques in TOF MS is beyond the scope of this book, an interested reader is referred to an excellent book by Cotter (121), as well as several recent tutorials on the subject (123).

Tandem Experiments with TOF MS. Metastable ions dissociating in the field-free region of a TOF mass analyzer would give rise to fragment ions having the same velocity as their precursor. [Ion dissociation in the drift region is usually referred to as a postsource decay (PSD).] As a result, it would be impossible to distinguish such fragment ions from the intact precursor ions in a linear TOF mass spectrum. The situation will be very different if a reflectron–TOF MS is used for mass analysis. Although the velocity of the fragment ions produced in the first drift region would still be the same as that of the intact precursor ions in both field-free regions, they will be "turned around" in the decelerating field of the reflectron much faster. The resultant time-of-flight of the fragment ion m_f will be (121)

$$t = \sqrt{\frac{m}{2zeU_0}} \cdot \left(D_1 + D_2 + 4\frac{m_f}{m}d\right). \qquad (3\text{-}5\text{-}4)$$

Mass selection of the precursor ions is usually accomplished in the first drift region by simple electrostatic gating. To achieve adequate and proper focusing of fragment ions of different masses, a series of spectra have to be collected at different decelerating potentials. Alternatively, a curved-field reflectron (vide supra) can be used to obtain a single PSD spectrum with product ions from a wide m/z range focused simultaneously (124).

Overall, fragmentation spectra provided by PSD are imperfect, since the activation of the ions takes place in the ion source and dissociation of the metastable ions can occur during acceleration, which has a detrimental effect on the resolution. Furthermore, precursor ion selection cannot be accomplished with high precision, since it is carried out in the region where ions are generally out of focus (121). This problem can be circumvented by using tandem (TOF/TOF) mass spectrometers, where the drift regions are separated by a collision cell, in which fragmentation of the mass-selected ions is carried out using CID (125, 126) or by other means of ion activation (127). Finally, interfacing TOF with other mass analyzers provides an opportunity to carry out tandem experiments with a "hybrid" mass spectrometer.

The advantages offered by TOF mass analyzers (ideal compatibility with "pulsed" ionization sources, duty cycle close to 100%, high ion transmission efficiency, virtually unlimited m/z range,* ability to carry out very fast analyses with repetition rates up to 500 kHz) make them extremely popular mass analyzers, which are suited for a variety of applications. In Chapter 11 of this book we will discuss several studies that specifically require TOF MS as a means of ion detection and cannot be carried out with any other type of mass analyzers.

3.5.3. Mass Analyzer as an Ion Filter

The ion separating properties of the "mass filter" devices result from their ability to selectively maintain stable trajectories for ions of certain m/z ratios, while the others become unstable. The idea of using quadrupolar electrical fields as a means of "filtering" ions according to their m/z ratios was introduced and

*Although the m/z range of TOF analyzers is theoretically unlimited, there is a fundamental limitation imposed by the ion detector: very large ions would have low velocities and fail to initiate a cascade of electrons at the multiplier.

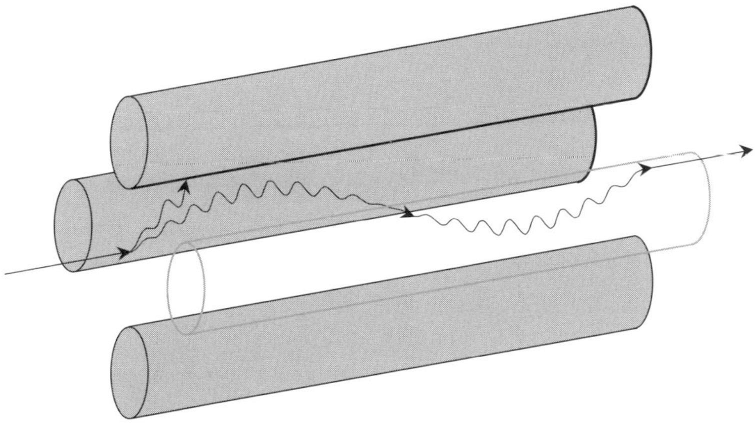

FIGURE 3.19. Schematic representation of a quadrupole mass filter with examples of stable and unstable ion trajectories.

implemented in the 1950s (128–130). A quadrupole mass spectrometer acts as a "tunable" mass filter that transmits ions within a narrow m/z range (Figure 3.19). A quadrupolar electrical field is usually configured using four parallel electrodes [metal rods of circular (or, ideally, hyperbolic) cross section] that are connected diagonally. A periodic potential of the form

$$\Phi_0 = \pm(U - V\cos\omega t) \tag{3-5-5}$$

is applied to each pair of electrodes, resulting in a periodic hyperbolic field configuration in the (x, y) plane:

$$\Phi(x, y) = (U - V\cos\omega t)\cdot\frac{x^2 - y^2}{2r_0^2}, \tag{3-5-6}$$

where r_0 is the distance from the central axis of the filter (z axis) to the surfaces of the electrodes. Combining (3-1-1) and (3-5-6) gives complicated equations from which ion trajectory within the filter can be calculated. Generally, this is a very involved mathematical procedure that is usually carried out by introducing a new variable u, which can represent either x or y:

$$\frac{d^2u}{d^2\xi} = \pm(a + 2q\cdot\cos(2\xi))\cdot u, \tag{3-5-7}$$

where $\xi = t\omega/2$ and

$$a = a_x = -a_y = \frac{4zeU}{m\omega^2 r_0^2} \quad \text{and} \quad q = q_x = -q_y = \frac{2zeV}{m\omega^2 r_0^2} \tag{3-5-8}$$

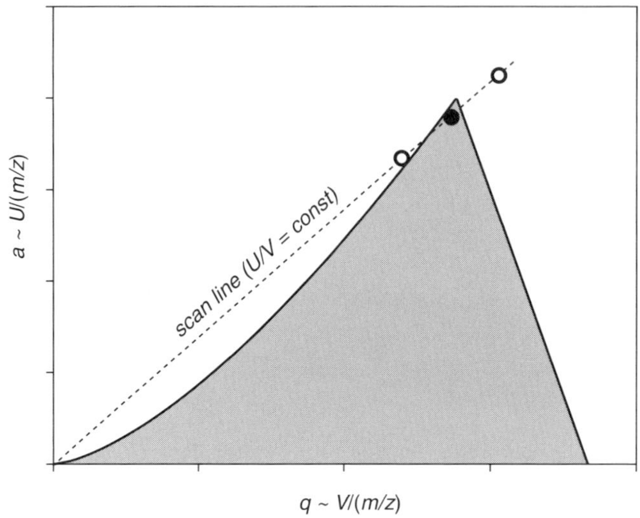

FIGURE 3.20. Schematic representation of the stability diagram for the quadrupole mass filter depicted in Figure 3.19 (see explanation in the text).

Equation (3-5-7) is known as Mathieu's differential equation. Stable solutions of Mathieu's equation are usually presented as "stability diagrams" in the (a, q) coordinates (Figure 3.20). The highest mass-resolving capability will be achieved if the filter is tuned in such a way that the "point" corresponding to the ions of interest is placed close to the apex of the stability region. In this case, a slight increase or decrease of ion m/z value will result in an unstable trajectory. To acquire a mass spectrum, both DC and RF potentials (U and V) have to be varied, while maintaining their ratio constant. This would bring previously "unstable" ions into the apex region of the stability region, thus allowing their passage through the filter. Another important conclusion from the consideration of the stability diagram is that the quadrupole will become an "ion guide," rather than an ion filter, if the DC component of the electric field (3-5-5) is equal to zero (this regime is commonly referred to as an "RF-only" mode of operation). A very intuitive treatment of the ion motions in quadrupole filters can be found in recent tutorials (131, 132).

The m/z range of a typical quadrupole MS is limited to 4000. The resolution of a quadrupole mass spectrometer can be adjusted by changing the U/V ratio, the slope of the "scan line" (see Figure 3.20). Mass resolution is not constant across the m/z axis, since the width of the transmission window is "fixed" once the "working" U/V ratio is selected. Typically, mass resolution cannot be increased significantly above the level of several thousand. The scan rate of a typical quadrupole MS is high enough to allow direct coupling to HPLC. MS/MS experiments can be carried out if three quadrupoles are arranged in tandem (a configuration referred to as QqQ). The first quadrupole is set to transmit ions of certain m/z value (precursor ions), while the second is used as a collision cell. It is operated in the RF-only mode to allow indiscriminate transmission of all ions (precursor and fragments alike) into the third quadrupole, which is scanned to obtained a fragment ion spectrum. Alternatively, the third quadrupole can be set to allow the transmission of certain fragment ions, while the first one is scanned. Mass spectra acquired in this mode contain peaks of all ions whose fragmentation gives rise to a selected fragment (so-called precursor ion scans). Finally, both first and second quadrupoles can be scanned in concert (maintaining the constant difference), yielding spectra of "constant neutral loss." Quadrupoles are often interfaced with other types of mass analyzers to produce "hybrid" tandem mass spectrometers.

3.5.4. Mass Analyzer as an Ion Storing Device: Quadrupole (Paul) Ion Trap

The idea of using a quadrupolar field to store ions is a logical extension of the concept of a quadrupolar mass filter and was introduced by the inventor of the latter (133) (for which the Nobel Prize in Physics was awarded to Wolfgang Paul in 1989). The quadrupole ion trap can be viewed as a linear quadrupole filter that was "collapsed," so that the electrical field becomes quadrupolar in all dimensions, not just an (x, y) plane. The confining capacity of a quadrupole ion trap device is due to the formation of a "trapping" potential well when appropriate potentials are applied to three electrodes of hyperbolic cross sections (two end caps and one ring electrode, see Figure 3.21). An oscillating (RF) potential on the ring electrode $\Phi_R = V \cos \omega t$ creates a dynamic parabolic (or, more correctly, saddle) field inside the trapping volume, which focuses ions to its center. A potential applied to the end caps $\Phi_R = U$ is constant (DC), and the field at any

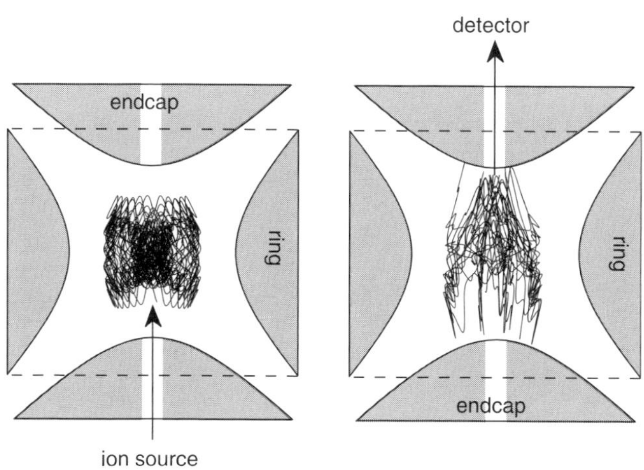

FIGURE 3.21. Schematic representation of a quadrupolar ion trap. Simulated trajectories of trapped ions (*left diagram*) and ions undergoing resonant excitation (*right diagram*) are shown. Courtesy of Prof. Richard W. Vachet (University of Massachusetts–Amherst).

point inside the trapping volume is (134, 135)

$$\Phi(r, z) = (U - V \cos \omega t) \cdot \frac{r^2 - 2z^2}{2r_0^2} + \frac{(U - V \cos \omega t)}{2}, \tag{3-5-9}$$

where r and z are cylindrical coordinates ($r^2 = x^2 + y^2$), and r_0 is the shortest distance from the center of the trap to the surface of the ring electrode. (In practice, the DC potential is applied by providing DC offset to the RF potential, which is applied to the ring electrode.) Ion trajectories in such a field will be determined by solutions of Mathieu's equation similar to (3-5-7). Solutions that correspond to trajectories confined to the trapping volume will form a "stability" region in the (a_z, q_z) plane (Figure 3.22). Trapping of the externally generated ions that are injected into the trap is facilitated by collisional cooling (using He gas in the trap at a pressure of $\sim$ 1 mtorr). Consideration of the stability diagram in Figure 3.22 suggests that in order to maximize the "stable" m/z range, the ion trap has to be operated at $a_z = 0$. This corresponds to no DC potential applied to the end caps (the "mass-selective instability mode"). Under such conditions, ions will become ejected from the trap only if their q_z value exceeds 0.908 (Figure 3.22). This is utilized for the purposes of ion detection in a mass-specific (or, more correctly, m/z-specific) fashion. Since $q_z = 4zeV/(mr_0{}^2\omega^2)$, gradual

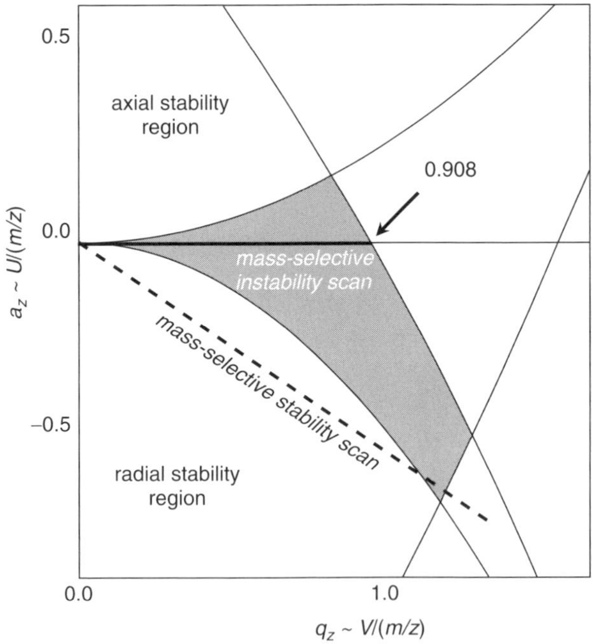

FIGURE 3.22. Schematic representation of the stability diagram for the quadrupole ion trap depicted in Figure 3.21 (see explanation in the text).

increase of V will result in an increase of q_z and will lead to the ejection of ions of progressively increasing m/z values from the trap, followed by their detection.

Trapped ("stable") ions of a given m/z oscillate at a frequency (known as the secular frequency) proportional to ω. If a harmonic potential is applied to the end caps, resonance conditions will be achieved for those ions whose secular frequency matches that of the applied potential. Resonant absorption of energy by such ions will progressively increase the amplitudes of their oscillations until they become unstable and are ejected from the trapping volume (Figure 3.21, right). Resonant excitation can also be used for mass-selective ejection/detection by creating a "hole" in the stability diagram at relatively low q_z values. A gradual increase of V will bring ions of progressively increasing m/z values to this hole, making their trajectories unstable and eventually forcing the ions out of the trap.

An ion of interest can also be isolated in the trap using a variety of methods. For example, the ion can be "brought" to the apex of the stability diagram, which would make all other ions unstable (Figure 3.22). Alternatively, the ion could be "left" on the q_z axis and the RF amplitude V is then scanned (increased) to eject all ions with lower m/z values. Next, a "resonant hole" is created at higher m/z and V is scanned to force all high-m/z ions to exit the trap through this hole. Once the ion of interest has been isolated, resonant excitation can be induced by applying a harmonic potential to the end caps (vide supra). The amplitude of the resonant signal can be adjusted such that the ions are not ejected from the trap but rather undergo a series of collisions with the molecules of the damping gas. If the energy of such collisions is high enough, the ion internal energy will continuously increase and eventually cause ion fragmentation. A scan of RF amplitude V after a period of such collisional activation will allow a mass spectrum of fragment ions to be acquired. Precursor ion selection (isolation), collisional activation and fragmentation, as well as mass analysis of the fragment ions all occur sequentially in the same location, hence the term *tandem-in-time* (as opposed to *tandem-in-space*) mass spectrometry.

Any one of the fragment ions, produced in the course of the MS/MS experiments just described, can be isolated in the trap, activated (the frequency of the resonant potential will have to be adjusted to a new m/z value), and fragmented, followed by the acquisition of a mass spectrum of the second generation fragment ions. This process can be repeated any number of times, as long as the number of ions remaining in the trap is high enough to provide a decent signal-to-noise ratio. Such experiments are referred to as *multistage tandem MS*, or simply MS^n. Fragmentation efficiency in ion trap MS often approaches 100% for smaller (<1 kDa) peptide ions.

Due to the significant improvements in the performance of ion traps, ease of operation, and relatively low cost, these analyzers have recently become very popular, both as standalone mass spectrometers and part of hybrid instruments. Fast scan time allows ion traps to be coupled directly to HPLC (using ESI as an interface). Ion traps are also ideally suited for pulsed ionization sources, such as MALDI, although the m/z range of most commercial ion trap instruments is limited to 6000 (with the optimal performance below 2500), which limits the scope of biomolecules amenable to analysis. Utilization of ESI can extend the range of biopolymers amenable to the analysis (because of the multiple charging, ESI-generated protein ions usually fall within the "allowed" m/z range). However, one needs to be aware of another potential problem, which is related to the space-charge effect. Since the trapping volume is usually very small, accumulation of a large number of charges* may give rise to strong electrical fields inside the trap. Space-charging phenomena have a detrimental effect on sensitivity and resolution and often result in mass spectral artifacts, such as ghost peaks (136, 137). More information on the general principles of quadrupole ion traps can be found in several recent review and tutorial articles (134, 135, 138, 139).

3.5.5. Mass Analyzer as an Ion Storing Device: FT ICR MS

Ion confinement to a limited volume can also be achieved by using a combination of electrostatic and magnetic fields. The simplest configuration of such a trapping device is depicted in Figure 3.23. A DC potential applied to the front and back plates will restrict the ion motion in the z-axial direction, while a constant magnetic field in the same direction will induce a circular (cyclotron) motion in the (x, y) plane. In accordance with (3-1-1), the centripetal force will be equal to the Lorentz force exerted on the ion by the magnetic field, and so the cyclotron

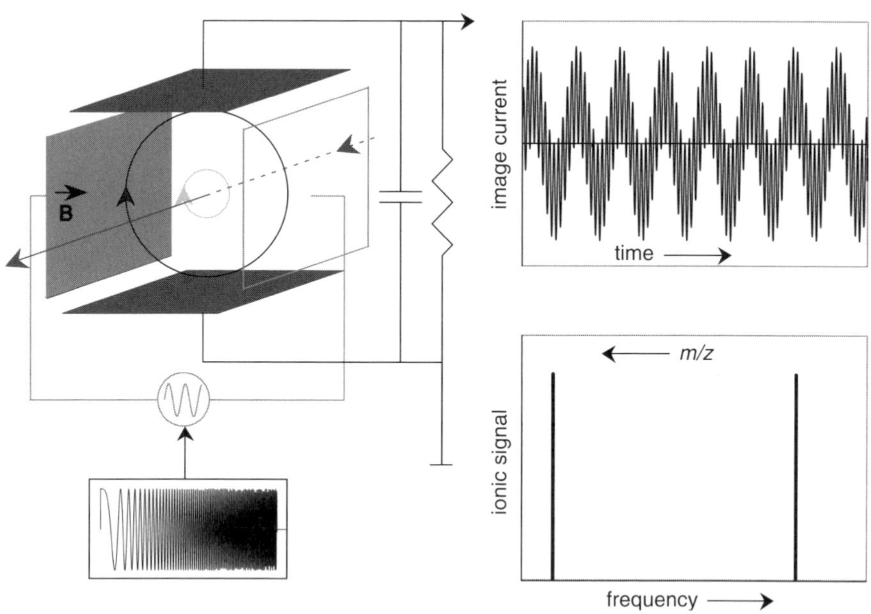

FIGURE 3.23. Principle of ion trapping, broadband excitation, and detection in FT ICR MS.

*Number of charges can exceed number of trapped proteins by at least an order of magnitude (due to the multiple charging).

motion frequency will be uniquely determined by the magnetic field strength B and the m/z ratio of the ion:

$$\omega_c = \frac{zeB}{m} \tag{3-5-10}$$

Therefore, measuring the cyclotron motion frequency can be used for mass analysis. Various modifications of this concept were brought forward in the late 1940s and early 1950s under the names *magnetic time-of-flight* mass spectrometer* (140, 141), *magnetic period* mass spectrometer or *mass synchrometer* (142, 143), *omegatron* (144, 145), and *magnetic resonance* mass spectrometer (146). Since frequency is a physical parameter that can be measured very accurately, mass spectrometers based on the principle of cyclotron motion can provide the highest accuracy m/z measurements. However, it was not until the mid-1970s that ion cyclotron resonance (ICR) mass spectrometry (147) finally became a powerful analytical technique as we know it now due to the introduction of digital Fourier transformation (FT) as a means of producing the mass spectrum (148). A historical account of the development of ICR MS can be found in a recent review by Marshall (149).

Ion detection in FT ICR MS is done by measuring the magnitude of the "image current" induced on the plates by the ion orbiting between them (Figure 3.23). Unsynchronized motion of a large number of ions will result in no net current due to the random distribution of phases among the ion population. Therefore, ion detection must be preceded by ion excitation (e.g., by applying a uniform harmonic electric field in the direction orthogonal to the magnetic field). If the field frequency is the same as the cyclotron frequency of the orbiting ions, they will be synchronized (brought in phase with the field). Furthermore, such resonant conditions will elevate ion kinetic energy, increasing the radii of their orbits (and, therefore, the magnitude of the image current induced by each ion). A homogeneous population of ions whose orbiting motions are in phase will induce an image current on the "receiver" plates (the top and bottom plates in Figure 3.24) of the form $I = I_0 \sin(\omega t + \alpha)$, whose angular frequency ω will be equal to the cyclotron frequency ω_c of the orbiting ions. The amplitude of the current I_0 will be proportional to the number of ions in the population and essentially independent of the cyclotron frequency ω_c under typical ICR conditions (150). If several ion populations (of different m/z ratios) are present in the cell and are excited to higher orbits, the resulting image current will be a superposition of several sinusoidal signals, each having a characteristic angular frequency uniquely determined by the magnetic field strength and the m/z ratio of the corresponding ion population (3-5-10). [The actual cyclotron frequency in a real ICR cell will be lower than ω_c due to the presence of an electrostatic potential, which traps the ions in the axial (z) direction.] Fourier transformation of such a spectrum (from the time domain to the frequency domain) will allow the cyclotron frequency of each ion population to be determined and m/z values calculated.

*The time-of-flight in this case is actually the period of the cyclotron motion $T = 2\pi/\omega$.

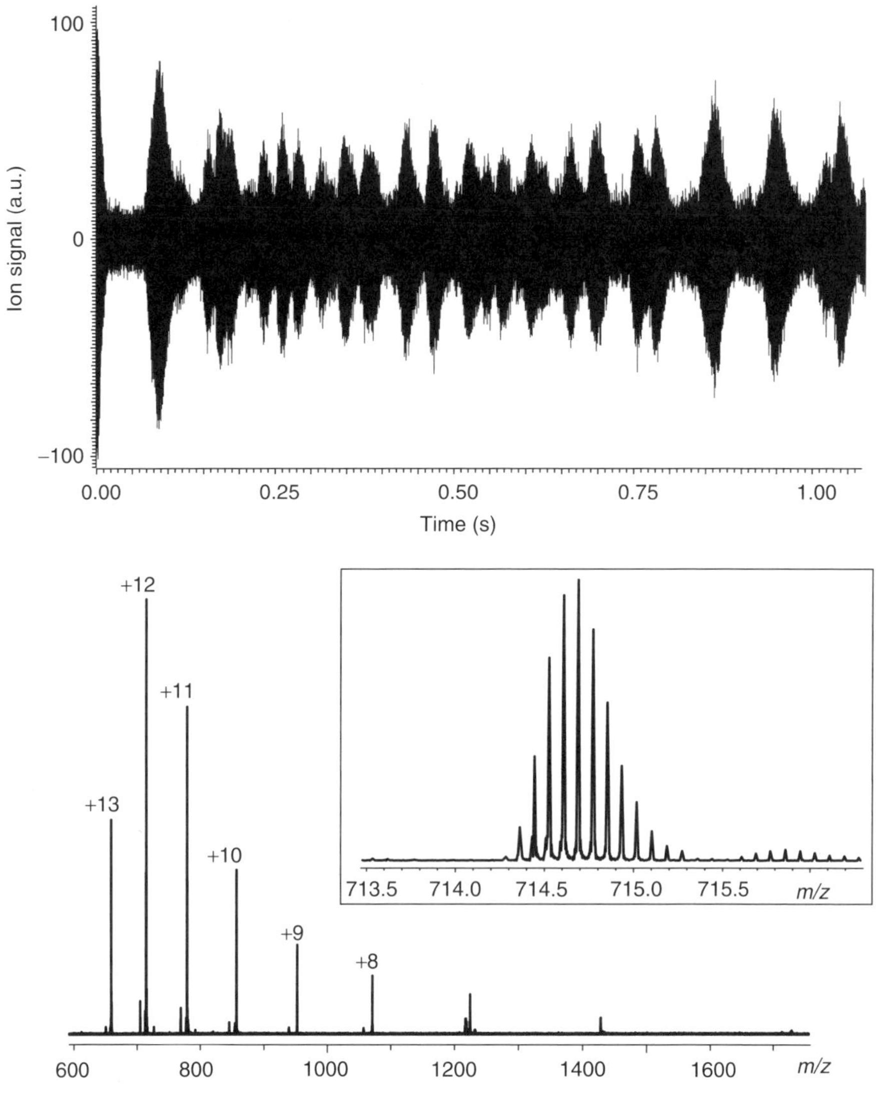

FIGURE 3.24. FT ICR mass spectra of ubiquitin: raw data in the time domain (*top trace*) and the Fourier-transformed spectrum (frequency domain).

FT ICR MS offers an advantage of being able to detect all ions simultaneously across a wide m/z range within a very short period of time. Increasing the measurement time in an FT ICR MS experiment enhances not only the signal-to-noise ratio but also the resolution. The latter is usually limited by "collisional damping" of the cyclotron motion (in contrast to quadrupole ion traps, ion-neutral collisions in the ICR cell have detrimental effects on the quality of mass spectra). Such collisions result in broadening and shape distortion of the ion peaks (without altering the m/z values), hence the very high vacuum requirements for high-resolution measurements (more demanding compared to most other mass analyzers). Due to the limited volume of the ICR cell, space-charge effects also negatively impact the quality of the FT ICR MS data (causing systematic shifts in the observed cyclotron frequencies of ions and, therefore, limiting the accuracy of mass measurements). Since all ions experience the same space-charge induced frequency shift, accurate mass measurements can be carried out using internal calibration (151). Commercial FT ICR instruments offer resolution exceeding 100,000 in the broadband mode, far outperforming any other types of mass analyzer. Image current detection is generally less sensitive compared to the "classical" ion counting techniques (utilized by other types of mass analyzers). However, the FT ICR MS detection is nondestructive and the data acquisition can be carried out with the same ion population over an extended period of time using multiple remeasurements (152, 153). In fact, it is possible to use FT ICR MS to trap and detect individual multiply charged macromolecular ions (154).

Inverse Fourier transformation (from the frequency to the time domain) allows "custom" excitation spectra to be designed. Such *stored waveform inverse Fourier transform* (SWIFT) excitation is used for highly selective ion excitation and/or ejection from the trap. SWIFT is an extremely effective method for ion isolation in the ICR cell. Fragmentation of the isolated ions can be induced by a variety of means, including collisional activation (e.g., *sustained off-resonance irradiation*, SORI CAD), photoactivation [e.g., *infrared multiphoton dissociation*, IRMPD (110)], and electron capture [ECD (111)]. Most of these methods (with the exception of ECD) are slow-heating methods according to McLuckey's classification (103). Higher-energy CAD can be implemented by using collisional activation of ions in the external source (e.g., nozzle-skimmer CAD in the ESI interface). Trapping the activated ions [either in the ICR cell or in the external reservoir, such as a hexapole ion guide (155)] allows the yield of the dissociation to be increased dramatically compared to other mass analyzers. Since FT ICR MS is a trapping device, it allows multiple stages of ion fragmentation to be carried out in sequence. Combination of several ion fragmentation techniques in one experiment often provides significant improvement of the sequence coverage of macromolecular ions (156). More information on general principles of FT ICR MS and its applications to biomolecular analyses can be found in recent reviews (157, 158).

3.5.6. Hybrid Mass Spectrometers

In addition to the mass analyzers described in the preceding sections, there are a large number of *hybrid* mass spectrometers that combine more than one type of mass analyzer in one instrument. Hybrid mass spectrometers often feature enhanced performance as a result of utilizing the advantages provided by each component. While the earlier versions of hybrid MS systems almost always utilized a sector instrument as a first stage, more recently the focus has been shifting toward TOF MS and quadrupole ion trap MS as the central components. Combination of a quadrupole mass filter and a reflectron–TOF mass analyzer has been particularly popular (159). It offers tandem capabilities, high resolution, an extended m/z range, and fast analysis time. Another hybrid instrument that has steadily been gaining popularity in recent years combines a quadrupole ion trap and a TOF mass analyzer (160). Finally, two recent high-end commercial products combine FT ICR MS with a quadrupole filter and with an ion trap.

REFERENCES

1. Thomson J. J. (1913). *Rays of Positive Electricity and Their Application to Chemical Analyses*. London: Longmans Green and Co.
2. Griffiths I. W. (1997). J. J. Thomson—the centenary of his discovery of the electron and of his invention of mass spectrometry. *Rapid Commun. Mass Spectrom.* **11**: 3–16.
3. Aston F. W. (1920). Isotopes and atomic weights. *Nature* **105**: 617–619.
4. Yergey J. A. (1983). A general approach to calculating isotopic distributions for mass spectrometry. *Int. J. Mass Spectrom. Ion Proc.* **52**: 337–349.
5. Hsu C. S. (1984). Diophantine approach to isotopic abundance calculations. *Anal. Chem.* **56**: 1356–1361.
6. Yergey J., Heller D., Hansen G., Cotter R. J., Fenselau C. (1983). Isotopic distributions in mass spectra of large molecules. *Anal. Chem.* **55**: 353–356.
7. Kubinyi H. (1991). Calculation of isotope distributions in mass spectrometry—a trivial solution for a nontrivial problem. *Anal. Chim. Acta* **247**: 107–119.
8. Rockwood A. L., VanOrden S. L., Smith R. D. (1995). Rapid calculation of isotope distributions. *Anal. Chem.* **67**: 2699–2704.
9. Rockwood A. L., VanOrden S. L. (1996). Ultrahigh-speed calculation of isotope distributions. *Anal. Chem.* **68**: 2027–2030.
10. He F., Hendrickson C. L., Marshall A. G. (2001). Baseline mass resolution of peptide isobars: a record for molecular mass resolution. *Anal. Chem.* **73**: 647–650.
11. Prâokai L. (1990). *Field Desorption Mass Spectrometry*. New York: Marcel Dekker.
12. Macfarlane R. D., Torgerson D. F. (1976). Californium-252 plasma desorption mass spectroscopy. *Science* **191**: 920–925.
13. Barber M., Bordoli R. S., Sedgwick R. D., Tyler A. N. (1981). Fast atom bombardment of solids (FAB)—a new ion source for mass spectrometry. *J. Chem. Soc. Chem. Commun.* **1981**: 325–327.
14. Henry C. (1997). FAB MS: Still fabulous? *Anal. Chem.* **69**: 625A–627A.
15. Seifert W. E. Jr., Caprioli R. M. (1996). Fast atom bombardment mass spectrometry. *Methods Enzymol.* **270**: 453–486.
16. Das P. R., Pramanik B. N. (1998). Fast atom bombardment mass spectrometric characterization of peptides. *Mol. Biotechnol.* **9**: 141–154.
17. Chapman J. R. (1993). *Practical Organic Mass Spectrometry: A Guide for Chemical and Biochemical Analysis*. Chichester: John Wiley & Sons.
18. Watson J. T. (1997). *Introduction to Mass Spectrometry*. Philadelphia: Lippincott-Raven.
19. Fenn J. B., Mann M., Meng C. K., Wong S. F., Whitehouse C. M. (1989). Electrospray ionization for mass spectrometry of large biomolecules. *Science* **246**: 64–71.

20. Fenn J. B. (2003). Electrospray wings for molecular elephants (Nobel Lecture). *Angew. Chem. Int. Ed. Engl.* **42**: 3871–3894.
21. Chapman S. (1937). Carrier mobility spectra of spray electrified liquids. *Phys. Rev.* **52**: 184–190.
22. Rayleigh J. W. S. (1882). On the equilibrium of liquid conducting masses charged with electricity. *Philos. Mag.* **14**: 184–186.
23. Gilbert W. (1600). *De Magnete, Magneticisqve Corporibvs, et de Magno Magnete Tellure; Physiologia Noua, Plurimis & Argumentis, & Experimentis Demonstrata.* Londini: excvdebat P. Short.
24. Dole M., Cox H. L. (1973). Electrospray mass spectroscopy. *Adv. Chem. Ser.* **125**: 73–84.
25. Dole M., Hines R. L., Mack L. L., Mobley R. C., Ferguson L. D., Alice M. B. (1968). Gas phase macroions. *Macromolecules* **1**: 96–97.
26. Richardson M. J. (1964). The direct observation of polymer molecules and determination of their molecular weight. *Proc. R. Soc. London A* **279**: 50–61.
27. Dole M., Mack L. L., Hines R. L. (1968). Molecular beams of macroions. *J. Chem. Phys.* **49**: 2240–2249.
28. Mack L. L., Kralik P., Rheude A., Dole M. (1970). Molecular beams of macroions. 2. *J. Chem. Phys.* **52**: 4977–4986.
29. Thomson B. A., Iribarne J. V., Dzledzic P. J. (1982). Liquid ion evaporation/mass spectrometry/mass spectrometry for the detection of polar and labile molecules. *Anal. Chem.* **54**: 2219–2224.
30. Iribarne J. V., Dzledzic P. J., Thomson B. A. (1983). Atmospheric pressure ion evaporation mass spectrometry. *Int. J. Mass Spectrom. Ion Proc.* **50**: 331–347.
31. Alexandrov M. L., Gall L. N., Krasnov N. V., Nikolaev V. I., Pavlenko V. A., Shkurov V. A. (1984). Ion extraction from solutions at atmospheric pressure—a method of mass spectrometric analysis of bioorganic substances. *Dokl. Acad. Nauk SSSR* **277**: 379–383.
32. Yamashita M., Fenn J. B. (1984). Negative ion production with the electrospray ion source. *J. Phys. Chem.* **88**: 4671–4675.
33. Alexandrov M. L., Baram G. I., Gall L. N., Krasnov N. V., Kusner Y. S., Mirgorodskaya O. A., Nikolaev V. I., Shkurov V. A. (1985). Formation of beams of quasimolecular ions of peptides from solutions. *Bioorg. Khim.* **11**: 700–704.
34. Alexandrov M. L., Baram G. I., Gall L. N., Grachev M. A., Knorre V. D., Krasnov N. V., Kusner Y. S., Mirgorodskaya O. A., Nikolaev V. I., Shkurov V. A. (1985). Application of a novel mass spectrometric method to sequencing of peptides. *Bioorg. Khim.* **11**: 705–708.
35. Alexandrov M. L., Gall L. N., Krasnov N. V., Nikolaev V. I., Pavlenko V. A., Shkurov V. A., Baram G. I., Grachev M. A., Knorre V. D., Kusner Y. S. (1984). Direct coupling of a microcolumn liquid chromatograph and a mass-spectrometer. *Bioorg. Khim.* **10**: 710–712.
36. Whitehouse C. W., Dreyer R. N., Yamashita M., Fenn J. B. (1985). Electrospray interface for liquid chromatographs and mass spectrometers. *Anal. Chem.* **57**: 675–679.
37. Vestal M. L. (2001). Methods of ion generation. *Chem. Rev.* **101**: 361–375.

38. Meng C. K., Mann M., Fenn J. B. (1988). Of protons or proteins. *Z. Phys. D* **10**: 361–368.
39. Wong S. F., Meng C. K., Fenn J. B. (1988). Multiple charging in electrospray ionization of polyethylene glycols. *J. Phys. Chem.* **92**: 546–550.
40. Mann M., Meng C. K., Fenn J. B. (1989). Interpreting mass-spectra of multiply charged ions. *Anal. Chem.* **61**: 1702–1708.
41. Ferrige A. G., Seddon M. J., Jarvis S. (1991). Maximum entropy deconvolution in electrospray mass spectrometry. *Rapid Commun. Mass Spectrom.* **5**: 374–377.
42. Reinhold B. B., Reinhold V. N. (1992). Electrospray ionization mass spectrometry—deconvolution by an entropy based algorithm. *J. Am. Soc. Mass Spectrom.* **3**: 207–215.
43. Zhang Z. Q., Guan S. H., Marshall A. G. (1997). Enhancement of the effective resolution of mass spectra of high-mass biomolecules by maximum entropy-based deconvolution to eliminate the isotopic natural abundance distribution. *J. Am. Soc. Mass Spectrom.* **8**: 659–670.
44. Zhang Z. Q., Marshall A. G. (1998). A universal algorithm for fast and automated charge state deconvolution of electrospray mass-to-charge ratio spectra. *J. Am. Soc. Mass Spectrom.* **9**: 225–233.
45. Horn D. M., Zubarev R. A., McLafferty F. W. (2000). Automated reduction and interpretation of high resolution electrospray mass spectra of large molecules. *J. Am. Soc. Mass Spectrom.* **11**: 320–332.
46. Snyder A. P., American Chemical Society. Division of Analytical Chemistry, American Chemical Society. Meeting. (1995). *Biochemical and Biotechnological Applications of Electrospray Ionization Mass Spectrometry*. Washington, DC: American Chemical Society.
47. Cole R. B. (1997). *Electrospray Ionization Mass Spectrometry: Fundamentals, Instrumentation, and Applications*. Hoboken, NJ: John Wiley & Sons.
48. Pramanik B. N., Ganguly A. K., Gross M. L. (2002). *Applied Electrospray Mass Spectrometry*. New York: Marcel Dekker.
49. Wilm M. S., Mann M. (1994). Electrospray and Taylor-cone theory, Dole's beam of macromolecules at last? *Int. J. Mass Spectrom. Ion Proc.* **136**: 167–180.
50. Wilm M., Mann M. (1996). Analytical properties of the nanoelectrospray ion source. *Anal. Chem.* **68**: 1–8.
51. Juraschek R., Dulcks T., Karas M. (1999). Nanoelectrospray—more than just a minimized-flow electrospray ionization source. *J. Am. Soc. Mass Spectrom.* **10**: 300–308.
52. Karas M., Bahr U., Dulcks T. (2000). Nano-electrospray ionization mass spectrometry: addressing analytical problems beyond routine. *Fresenius' J. Anal. Chem.* **366**: 669–676.
53. Sakamoto S., Fujita M., Kim K., Yamaguchi K. (2000). Characterization of self-assembling nano-sized structures by means of coldspray ionization mass spectrometry. *Tetrahedron* **56**: 955–964.
54. Yamaguchi K. (2003). Cold-spray ionization mass spectrometry: principle and applications. *J. Mass Spectrom.* **38**: 473–490.
55. Conzemius R. J., Capellen J. M. (1980). A review of the applications to solids of the laser ion-source in mass-spectrometry. *Int. J. Mass Spectrom. Ion Proc.* **34**: 197–271.

56. Tanaka K., Ido Y., Akita S., Yoshida Y., Yoshida T. 1987. Detection of high mass molecules by laser desorption time-of-flight mass spectrometry. In: *2-d Japan–China Joint Symposium on Mass Spectrometry*, pp. 185–188. Osaka, Japan: Bando Press.
57. Karas M., Hillenkamp F. (1988). Laser desorption ionization of proteins with molecular masses exceeding 10,000 daltons. *Anal. Chem.* **60**: 2299–2301.
58. Tanaka K. (2003). The origin of macromolecule ionization by laser irradiation (Nobel Lecture). *Angew. Chem. Int. Ed.* **42**: 3860–3870.
59. Nelson R. W., Rainbow M. J., Lohr D. E., Williams P. (1989). Volatilization of high molecular weight DNA by pulsed laser ablation of frozen aqueous solutions. *Science* **246**: 1585–1587.
60. Overberg A., Karas M., Bahr U., Kaufmann R., Hillenkamp F. (1990). Matrix-assisted infrared laser (2.94 μm) desorption ionization mass-spectrometry of large biomolecules. *Rapid Commun. Mass Spectrom.* **4**: 293–296.
61. Schieltz D. M., Chou C. W., Luo C. W., Thomas R. M., Williams P. (1992). Mass spectrometry of DNA mixtures by laser ablation from frozen aqueous solution. *Rapid Commun. Mass Spectrom.* **6**: 631–636.
62. Berkenkamp S., Karas M., Hillenkamp F. (1996). Ice as a matrix for IR-matrix-assisted laser desorption/ionization: mass spectra from a protein single crystal. *Proc. Natl. Acad. Sci. U.S.A.* **93**: 7003–7007.
63. Hunter J. M., Lin H. A., Becker C. H. (1997). Cryogenic frozen solution matrixes for analysis of DNA by time-of-flight mass spectrometry. *Anal. Chem.* **69**: 3608–3612.
64. Kraft P., Alimpiev S., Dratz E., Sunner J. (1998). Infrared, surface-assisted laser desorption ionization mass spectrometry on frozen aqueous solutions of proteins and peptides using suspensions of organic solids. *J. Am. Soc. Mass Spectrom.* **9**: 912–924.
65. Baltz-Knorr M. L., Schriver K. E., Haglund R. F. (2002). Infrared laser ablation and ionization of water clusters and biomolecules from ice. *Appl. Surf. Sci.* **197**: 11–16.
66. Berry J. I., Sun S., Dou Y., Wucher A., Winograd N. (2003). Laser desorption and imaging of proteins from ice via UV femtosecond laser pulses. *Anal. Chem.* **75**: 5146–5151.
67. Moyer S. C., Cotter R. J. (2002). Atmospheric pressure MALDI. *Anal. Chem.* **74**: 468A–476A.
68. Von Seggern C. E., Moyer S. C., Cotter R. J. (2003). Liquid infrared atmospheric pressure matrix-assisted laser desorption/ionization ion trap mass spectrometry of sialylated carbohydrates. *Anal. Chem.* **75**: 3212–3218.
69. Laiko V. V., Taranenko N. I., Berkout V. D., Yakshin M. A., Prasad C. R., Lee H. S., Doroshenko V. M. (2002). Desorption/ionization of biomolecules from aqueous solutions at atmospheric pressure using an infrared laser at 3 microns. *J. Am. Soc. Mass Spectrom.* **13**: 354–361.
70. Georgiou S., Hillenkamp F. (2003). Introduction: laser ablation of molecular substrates. *Chem. Rev.* **103**: 317–319.
71. Dreisewerd K. (2003). The desorption process in MALDI. *Chem. Rev.* **103**: 395–425.
72. Karas M., Kruger R. (2003). Ion formation in MALDI: the cluster ionization mechanism. *Chem. Rev.* **103**: 427–439.

73. Knochenmuss R., Zenobi R. (2003). MALDI ionization: the role of in-plume processes. *Chem. Rev.* **103**: 441–452.
74. Marshall A. G., Senko M. W., Li W., Li M., Dillon S., Guan S., Logan T. M. (1997). Protein molecular mass to 1 Da by ^{13}C, ^{15}N double-depletion and FT ICR mass spectrometry. *J. Am. Chem. Soc.* **119**: 433–434.
75. Kolli V. S., Orlando R. (1997). A new strategy for MALDI on magnetic sector mass spectrometers with point detectors. *Anal. Chem.* **69**: 327–332.
76. Tomer K. B. (2001). Separations combined with mass spectrometry. *Chem. Rev.* **101**: 297–328.
77. Gelpi E. (2002). Interfaces for coupled liquid phase separation/mass spectrometry techniques. An update on recent developments. *J. Mass Spectrom.* **37**: 241–253.
78. Romijn E. P., Krijgsveld J., Heck A. J. R. (2003). Recent liquid chromatographic-(tandem) mass spectrometric applications in proteomics. *J. Chromatogr. A* **1000**: 589–608.
79. Tomlinson A. J., Chicz R. M. (2003). Microcapillary liquid chromatography/tandem mass spectrometry using alkaline pH mobile phases and positive ion detection. *Rapid Commun. Mass Spectrom.* **17**: 909–916.
80. Kennedy R. T., German I., Thompson J. E., Witowski S. R. (1999). Fast analytical-scale separations by capillary electrophoresis and liquid chromatography. *Chem. Rev.* **99**: 3081–3132.
81. Wang L., Pan H., Smith D. L. (2002). Hydrogen exchange—mass spectrometry: optimization of digestion conditions. *Mol. Cell. Proteomics* **1**: 132–138.
82. Wang L., Smith D. L. (2003). Downsizing improves sensitivity 100-fold for hydrogen exchange—mass spectrometry. *Anal. Biochem.* **314**: 46–53.
83. Niessen W. M. A. (1999). *Liquid Chromatography–Mass Spectrometry*. New York: Marcel Dekker.
84. Cai J., Henion J. (1995). Capillary electrophoresis–mass spectrometry. *J. Chromatogr. A* **703**: 667–692.
85. von Brocke A., Nicholson G., Bayer E. (2001). Recent advances in capillary electrophoresis/electrospray mass spectrometry. *Electrophoresis* **22**: 1251–1266.
86. Moini M. (2002). Capillary electrophoresis mass spectrometry and its application to the analysis of biological mixtures. *Anal. Bioanal. Chem.* **373**: 466–480.
87. Schmitt-Kopplin P., Frommberger M. (2003). Capillary electrophoresis–mass spectrometry: 15 years of developments and applications. *Electrophoresis* **24**: 3837–3867.
88. Gaberc-Porekar V., Menart V. (2001). Perspectives of immobilized-metal affinity chromatography. *J. Biochem. Biophys. Methods* **49**: 335–360.
89. Lee W.-C., Lee K. H. (2004). Applications of affinity chromatography in proteomics. *Anal. Biochem.* **324**: 1–10.
90. Preisler J., Foret F., Karger B. L. (1998). On-line MALDI-TOF MS using a continuous vacuum deposition interface. *Anal. Chem.* **70**: 5278–5287.
91. Preisler J., Hu P., Rejtar T., Moskovets E., Karger B. L. (2002). Capillary array electrophoresis–MALDI mass spectrometry using a vacuum deposition interface. *Anal. Chem.* **74**: 17–25.
92. Murray K. K., Lewis T. M., Beeson M. D., Russell D. H. (1994). Aerosol matrix-assisted laser-desorption ionization for liquid-chromatography time-of-flight mass-spectrometry. *Anal. Chem.* **66**: 1601–1609.

93. Murray K. K. (1997). Coupling matrix-assisted laser desorption/ionization to liquid separations. *Mass Spectrom. Rev.* **16**: 283–299.

94. Gusev A. I. (2000). Interfacing matrix-assisted laser desorption/ionization mass spectrometry with column and planar separations. *Fresenius' J. Anal. Chem.* **366**: 691–700.

95. Orsnes H., Zenobi R. (2001). Interfaces for on-line liquid sample delivery for matrix-assisted laser desorption ionisation mass spectrometry. *Chem. Soc. Rev.* **30**: 104–112.

96. Biemann K. (1988). Contributions of mass spectrometry to peptide and protein structure. *Biomed. Environ. Mass Spectrom.* **16**: 99–111.

97. Biemann K. (1990). Appendix 5. Nomenclature for peptide fragment ions (positive ions). *Methods Enzymol.* **193**: 886–887.

98. Ngoka L. C., Gross M. L. (1999). A nomenclature system for labeling cyclic peptide fragments. *J. Am. Soc. Mass Spectrom.* **10**: 360–363.

99. McLafferty F. W. (1980). Tandem mass spectrometry (MS-MS)—promising new analytical technique for specific component determination in complex mixtures. *Acc. Chem. Res.* **13**: 33–39.

100. McLuckey S. A., Wells J. M. (2001). Mass analysis at the advent of the 21st century. *Chem. Rev.* **101**: 571–606.

101. Jennings K. R. (2000). The changing impact of the collision-induced decomposition of ions on mass spectrometry. *Int. J. Mass Spectrom.* **200**: 479–493.

102. Dunbar R. C. (1992). Infrared radiative cooling of gas-phase ions. *Mass Spectrom. Rev.* **11**: 309–339.

103. McLuckey S. A., Goeringer D. E. (1997). Slow heating methods in tandem mass spectrometry. *J. Mass Spectrom.* **32**: 461–474.

104. Fabris D., Kelly M., Murphy C., Wu Z., Fenselau C. (1993). High-energy collision-induced dissociation of multiply charged polypeptides produced by electrospray. *J. Am. Soc. Mass Spectrom.* **4**: 652–661.

105. Kolli V. S. K., Orlando R. (1995). Complete sequence confirmation of large peptides by high energy collisional activation of multiply protonated ions. *J. Am. Soc. Mass Spectrom.* **6**: 234–241.

106. Johnson R. S., Martin S. A., Biemann K. (1988). Collision-induced fragmentation of $(M + H)^+$ ions of peptides. Side chain specific sequence ions. *Int. J. Mass Spectrom. Ion Proc.* **86**: 137–154.

107. Dongre A. R., Somogyi A., Wysocki V. H. (1996). Surface-induced dissociation: an effective tool to probe structure, energetics and fragmentation mechanisms of protonated peptides. *J. Mass Spectrom.* **31**: 339–350.

108. Laskin J., Futrell J. H. (2003). Surface-induced dissociation of peptide ions: kinetics and dynamics. *J. Am. Soc. Mass Spectrom.* **14**: 1340–1347.

109. Laskin J., Futrell J. H. (2003). Energy transfer in collisions of peptide ions with surfaces. *J. Chem. Phys.* **119**: 3413–3420.

110. Little D. P., Speir J. P., Senko M. W., O'Connor P. B., McLafferty F. W. (1994). Infrared multiphoton dissociation of large multiply charged ions for biomolecule sequencing. *Anal. Chem.* **66**: 2809–2815.

111. Zubarev R. A., Kelleher N. L., McLafferty F. W. (1998). Electron capture dissociation of multiply charged protein cations. *J. Am. Chem. Soc.* **120**: 3265–3266.

112. Zubarev R. A. (2003). Reactions of polypeptide ions with electrons in the gas phase. *Mass Spectrom. Rev.* **22**: 57–77.

113. Wells J. M., Chrisman P. A., McLuckey S. A. (2002). "Dueling" ESI: instrumentation to study ion/ion reactions of electrospray-generated cations and anions. *J. Am. Soc. Mass Spectrom.* **13**: 614–622.

114. Tarantin N. I. (1999). Methods of measuring masses in nuclear physics. The basis of mass analysis is the dispersion of ions or charged particles. *Phys. Part. Nucl.* **30**: 167–194.

115. Wollnik H. (1999). Ion optics in mass spectrometers. *J. Mass Spectrom.* **34**: 991–1006.

116. Dempster A. J. (1918). A new method of positive ray analysis. *Phys. Rev.* **11**: 316–325.

117. Mattauch J. (1936). A double-focusing mass spectrograph and the masses of N15 and O18. *Phys. Rev.* **50**: 617–623.

118. Stephens W. E. (1946). A pulsed mass spectrometer with time dispersion. *Phys. Rev.* **69**: 691–691.

119. Cameron A. E., Eggers D. F. (1948). An ion velocitron. *Rev. Sci. Instrum.* **19**: 605–607.

120. Mamyrin B. A. (2001). Time-of-flight mass spectrometry (concepts, achievements, and prospects). *Int. J. Mass Spectrom.* **206**: 251–266.

121. Cotter R. J. (1997). *Time-of-Flight Mass Spectrometry: Instrumentation and Applications in Biological Research.* Washington, DC: American Chemical Society.

122. Cordero M. M., Cornish T. J., Cotter R. J., Lys I. A. (1995). Sequencing peptides without scanning the reflectron—post-source decay with a curved-field reflectron time-of-flight mass-spectrometer. *Rapid Commun. Mass Spectrom.* **9**: 1356–1361.

123. Uphoff A., Grotemeyer J. (2003). The secrets of time-of-flight mass spectrometry revealed. *Eur. J. Mass Spectrom.* **9**: 151–164.

124. Cordero M. M., Cornish T. J., Cotter R. J., Lys I. A. (1995). Sequencing peptides without scanning the reflectron—post-source decay with a curved-field reflectron time-of-flight mass spectrometer. *Rapid Commun. Mass Spectrom.* **9**: 1356–1361.

125. Cornish T. J., Cotter R. J. (1993). Tandem time-of-flight mass spectrometer. *Anal. Chem.* **65**: 1043–1047.

126. Medzihradszky K. F., Campbell J. M., Baldwin M. A., Falick A. M., Juhasz P., Vestal M. L., Burlingame A. L. (2000). The characteristics of peptide collision-induced dissociation using a high-performance MALDI-TOF/TOF tandem mass spectrometer. *Anal. Chem.* **72**: 552–558.

127. Beussman D. J., Vlasak P. R., Mclane R. D., Seeterlin M. A., Enke C. G. (1995). Tandem reflectron time-of-flight mass spectrometer utilizing photodissociation. *Anal. Chem.* **67**: 3952–3957.

128. Paul W., Steinwedel H. (1953). Ein neues massenspektrometer ohne magnetfeld. *Z. Naturforsch. A* **8**: 448–450.

129. Paul W., Raether M. (1955). Das elektrische massenfilter. *Z. Phys.* **140**: 262–273.

130. Paul W., Reinhard H. P., Vonzahn U. (1958). Das elektrische massenfilter als massenspektrometer und isotopentrenner. *Z. Phys.* **152**: 143–182.

131. Leary J. J., Schmidt R. L. (1996). Quadrupole mass spectrometers: an intuitive look at the math. *J. Chem. Educ.* **73**: 1142–1145.

132. Steel C., Henchman M. (1998). Understanding the quadrupole mass filter through computer simulation. *J. Chem. Educ.* **75**: 1049–1054.

133. Paul W. (1990). Electromagnetic traps for charged and neutral particles. *Rev. Mod. Phys.* **62**: 531–540.

134. March R. E. (1997). An introduction to quadrupole ion trap mass spectrometry. *J. Mass Spectrom.* **32**: 351–369.

135. Jonscher K. R., Yates J. R. III. (1997). The quadrupole ion trap mass spectrometer—a small solution to a big challenge. *Anal. Biochem.* **244**: 1–15.

136. Rocher F., Favre A., Gonnet F., Tabet J. C. (1998). Study of ghost peaks resulting from space charge and non-linear fields in an ion trap mass spectrometer. *J. Mass Spectrom.* **33**: 921–935.

137. Favre A., Gonnet F., Tabet J. C. (2001). Perturbation of ion trajectories by resonant excitation leads to occurrence of ghost peaks. *Rapid Commun. Mass Spectrom.* **15**: 446–450.

138. March R. E. (1998). Quadrupole ion trap mass spectrometry: theory, simulation, recent developments and applications. *Rapid Commun. Mass Spectrom.* **12**: 1543–1554.

139. March R. E. (2000). Quadrupole ion trap mass spectrometry: a view at the turn of the century. *Int. J. Mass Spectrom.* **200**: 285–312.

140. Goudsmit S. A. (1948). A time-of-flight mass spectrometer. *Phys. Rev.* **74**: 622–623.

141. Hays E. E., Richards P. I., Goudsmit S. A. (1951). Mass measurements with a magnetic time-of-flight mass spectrometer. *Phys. Rev.* **84**: 824–829.

142. Smith L. G. (1951). A new magnetic period mass spectrometer. *Rev. Sci. Instrum.* **22**: 115–116.

143. Smith L. G. (1952). Recent developments with the mass synchrometer. *Phys. Rev.* **85**: 767–767.

144. Sommer H., Thomas H. A. (1950). Detection of magnetic resonance by ion resonance absorption. *Phys. Rev.* **78**: 806–806.

145. Sommer H., Thomas H. A., Hipple J. A. (1951). The measurement of *e/m* by cyclotron resonance. *Phys. Rev.* **82**: 697–702.

146. Ionov N. I., Mamyrin B. A., Fiks V. B. (1953). Resonance magnetic mass spectrometer of high resolving power. *Z. Tekhn. Fiz.* **23**: 2104–2106.

147. Wobschall D. (1965). Ion cyclotron resonance spectrometer. *Rev. Sci. Instrum.* **36**: 466–475.

148. Comisarow M. B., Marshall A. G. (1974). Fourier transform ion cyclotron resonance spectroscopy. *Chem. Phys. Lett.* **25**: 282–283.

149. Marshall A. G. (2000). Milestones in Fourier transform ion cyclotron resonance mass spectrometry technique development. *Int. J. Mass Spectrom.* **200**: 331–356.

150. Marshall A. G., Hendrickson C. L. (2002). Fourier transform ion cyclotron resonance detection: principles and experimental configurations. *Int. J. Mass Spectrom.* **215**: 59–75.

151. Taylor P. K., Amster I. J. (2003). Space charge effects on mass accuracy for multiply charged ions in ESI FTICR. *Int. J. Mass Spectrom.* **222**: 351–361.

152. Williams E. R., Henry K. D., McLafferty F. W. (1990). Multiple remeasurement of ions in Fourier transform mass spectrometry. *J. Am. Chem. Soc.* **112**: 6157–6162.

153. Speir J. P., Gorman G. S., Pitsenberger C. C., Turner C. A., Wang P. P., Amster I. J. (1993). Remeasurement of ions using quadrupolar excitation Fourier transform ion cyclotron resonance spectrometry. *Anal. Chem.* **65**: 1746–1752.

154. Bruce J. E., Cheng X., Bakhtiar R., Wu Q., Hofstadler S. A., Anderson G. A., Smith R. D. (1994). Trapping, detection, and mass measurement of individual ions in a Fourier transform ion cyclotron resonance mass spectrometer. *J. Am. Chem. Soc.* **116**: 7839–7847.

155. Sannes-Lowery K. A., Hofstadler S. A. (2000). Characterization of multipole storage assisted dissociation: implications for electrospray ionization mass spectrometry characterization of biomolecules. *J. Am. Soc. Mass Spectrom.* **11**: 1–9.

156. Horn D. M., Ge Y., McLafferty F. W. (2000). Activated ion electron capture dissociation for mass spectral sequencing of larger (42 kDa) proteins. *Anal. Chem.* **72**: 4778–4784.

157. Amster I. J. (1996). Fourier transform mass spectrometry. *J. Mass Spectrom.* **31**: 1325–1337.

158. Marshall A. G., Hendrickson C. L., Jackson G. S. (1998). Fourier transform ion cyclotron resonance mass spectrometry: a primer. *Mass Spectrom. Rev.* **17**: 1–35.

159. Chernushevich I. V., Ens W., Standing K. G. (1999). Orthogonal injection TOF MS for analyzing biomolecules. *Anal. Chem.* **71**: 452A–461A.

160. Doroshenko V. M., Cotter R. J. (1998). A quadrupole ion trap time-of-flight mass spectrometer with a parabolic reflectron. *J. Mass Spectrom.* **33**: 305–318.

4

MASS SPECTROMETRY-BASED APPROACHES TO STUDY BIOMOLECULAR HIGHER-ORDER STRUCTURE

Obtaining higher-order structures of biopolymers is usually a first step in analyzing their behavior and understanding function. While X-ray crystallography and high-field NMR undoubtedly provide the highest quality information, limitations of these techniques (as discussed in Chapter 2) often make such high-resolution structures unavailable for a variety of important proteins and their assemblies. This chapter discusses various MS-based approaches to evaluate higher-order structure at various levels of spatial resolution when the crystallographic and NMR data are unavailable or insufficient. The chapter begins with a discussion of methods used to probe biomolecular topology and topography. These methods generally utilize chemical cross-linking to characterize tertiary and quaternary contacts by generating "proximity maps." We then proceed to a discussion of various methods of mapping solvent accessibility of protein segments. In addition to covalent modification of protein surfaces in both specific and nonspecific fashion, we briefly discuss application of hydrogen–deuterium exchange to mapping protein–protein interfaces. The chapter concludes with a brief overview of two emerging low-resolution methods that aim at evaluating the overall geometry of biopolymers and their assemblies. The first one uses controlled fragmentation of protein complex ions in the gas phase to identify neighboring subunits and hierarchical organization of protein quaternary structure. The second emerging method provides an estimate of the solvent-accessible surface area in proteins and protein complexes (based on the number of charges accommodated by the protein).

Mass Spectrometry in Biophysics: Conformation and Dynamics of Biomolecules
By Igor A. Kaltashov and Stephen J. Eyles
ISBN 0-471-45602-0

4.1. BIOMOLECULAR TOPOGRAPHY: CONTACT AND PROXIMITY MAPS VIA CHEMICAL CROSS-LINKING

Chemical cross-linking is perhaps one of the oldest biochemical techniques used to characterize protein conformation by mapping contact topology within proteins and their complexes. As early as the mid-1950s, Zahn and Meienhofer used a bifunctional reagent 1,5-difluoro-2,4-dinitrobenzol to verify a dimeric structure of insulin in solution (1, 2), which was believed at the time to be a dimer formed by two stable helices spanning the entire lengths of the peptide chains. Such a view was based on hydrogen–deuterium exchange measurements carried out earlier by Hvidt and Linderstrøm-Lang, who reported very high protection of the backbone amides (3). However, the interchain contacts detected in the cross-linking experiments did not fully support this model, leading to a suggestion that the C-terminal part of insulin is flexible in solution. As it turned out, Hvidt and Linderstrøm-Lang had actually overestimated the protection of insulin due to the significant back-exchange in their initial measurements, an error that was soon discovered and corrected (4). This early triumph of chemical cross-linking highlighted its utility as a means of producing *proximity maps*, which can be very useful in reconstructing protein tertiary and quaternary structures at low resolution.

In the decades that followed the initial work of Zahn and Meienhofer, the arsenal of cross-linking reagents expanded steadily to the dozens of various bifunctional agents that are now commercially available. Cross-linking reagents are generally classified based on their chemical specificity and the length of the "spacer arm" (cross-bridge formed between the two cross-linked sites when the reaction is complete). The chemical specificity of a cross-linker determines the overall pool of reactive groups within the polypeptide that may participate in the cross-linking reaction. Only eight of the twenty amino acid side chains are chemically active: Arg (guanidinyl), Lys (ε-amine), Asp and Glu (β- and γ-carboxylates), Cys (sulfhydryl), His (imidazole), Met (thioether), Trp (indoyl), and Tyr (phenolic hydroxylate) (5). It should probably be mentioned that with very few exceptions, no reagent is absolutely group-specific.* However, many reagents show a reasonable degree of selectivity (Table 4.1).

Monofunctional cross-linkers induce direct coupling of two functional groups of the protein without incorporating any extraneous material into the protein (Figure 4.1A). Obviously, this becomes possible only if the two functional groups are in very close proximity (direct contact).† Bifunctional cross-linkers, on the other hand, contain two reagents linked through a spacer, thus allowing the coupling of functional groups whose separation does not exceed the spacer's

*In most cases, relative chemical reactivity of a side chain is determined largely by its nucleophilicity (reactive groups of most cross-linkers interact with the side chains via nucleophilic substitution reactions). In addition to the general rules that determine nucleophilicities, the latter are affected by the state of protonation (and, therefore, solution pH and microenvironment of a particular side chain).

†Such cross-linkers operate as condensing agents, resulting in the cross-linked residues becoming directly interjoined.

TABLE 4.1. Commonly Used Group-Specific Reagents

Class of Reagents	Example	Side Chain Specificity
Dicarbonyls	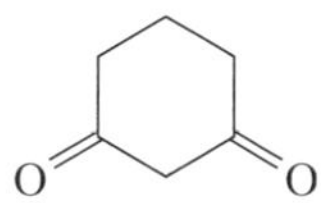	Arg
Imidoesters	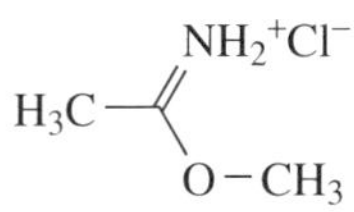	Lys
Isocyanates	HNCO	Lys
Isothiocyanates	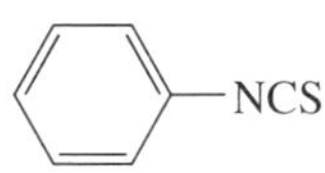	Lys
Sulfonyl halides	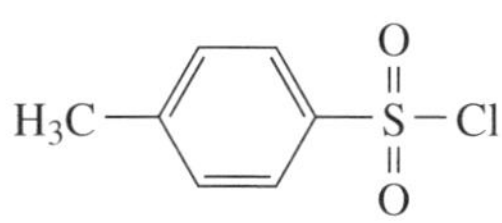	Lys
Diazonium salts	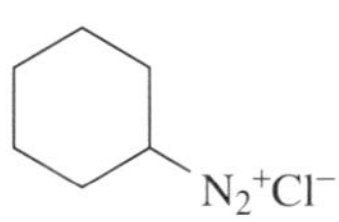	His, Tyr
Acid anhydrides	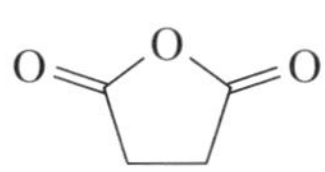	Lys, Tyr
Aryl halides	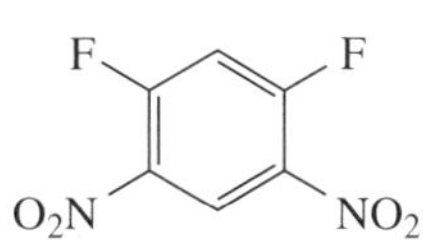	Cys, Lys, His, Tyr
α-Haloacetyls	ICH_2CO_2H	Cys, Lys, His, Tyr, Met
N-Maleimides	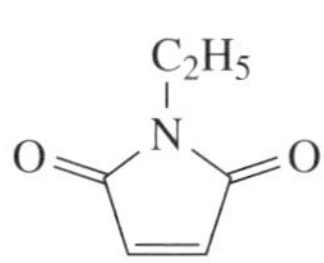	Cys, Lys
Mercurials	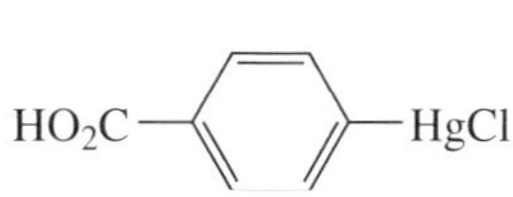	Cys

TABLE 4.1 ***(Continued)***

Class of Reagents	Example	Side Chain Specificity
Disulfides	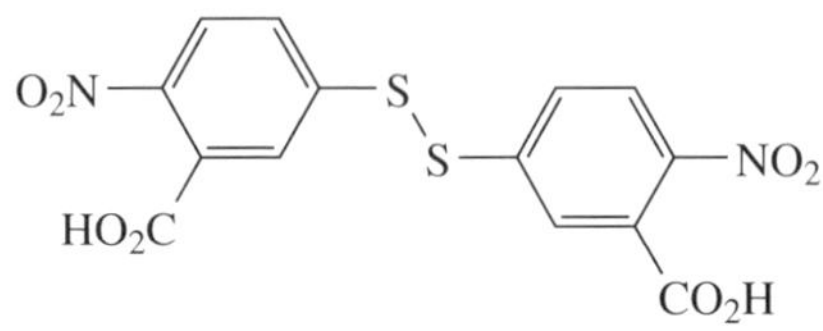	Cys
Diazoacetates	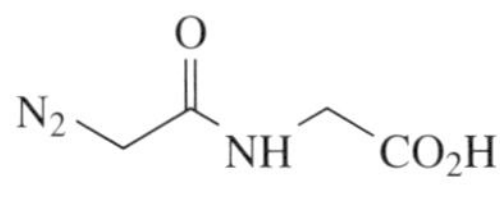	Asp, Glu, Cys

Adapted from Wong (5).

length (Figure 4.1B,C). Initially most bifunctional reagents in use were *homobifunctional* (i.e., both cross-linking groups within the reagent targeting the same reactive groups on the protein). More recently, *heterobifunctional* cross-linkers (coupling different functional groups on the protein) have seen a dramatic surge in popularity due to the wealth of structural information that they provide. Some heterobifunctional cross-linkers may incorporate a *photosensitive* (nonspecific) *reagent* in addition to a conventional (group-specific) functionality. Such photosensitive groups react indiscriminately upon activation by irradiation. Once the specific end of such a cross-linker is anchored to an amino acid residue, the photoreactive end can be used to probe the surroundings of this amino acid (Figure 4.1D). A variety of photoreactive reagents can be activated at wavelengths above 300 nm, so that there is no damage to the biological macromolecule (protein or nucleic acid) due to the photoirradiation itself. Utilizing two photoreactive reagents within a cross-linker will make it totally nondiscriminatory. Trifunctional cross-linkers have also been introduced, although their utilization in biophysical studies remains limited as yet. Finally, a very interesting approach to carrying out cross-linking experiments was introduced recently by Craik and co-workers, who suggested the use of an *intrinsic reagent* that can be expressed as a part of a fusion protein (6, 7). This strategy makes use of a Ni(II) complex of a tripeptide GGH, which mediates efficient protein–protein cross-linking in the presence of oxidants such as oxone and monoperoxyphthalic acid. When fused to the amino terminus of a protein, GGH can still support cross-linking. Several popular cross-linkers are listed in Table 4.2. More information can be found in several excellent reviews on the subject (8, 9) and an outstanding book by Wong (5). Finally, a variety of cross-linking reagents provide a means to study higher-order structure of oligonucleotides (both DNA and RNA) (10–12) or oligonucleotide–protein complexes (13, 14), an issue that will be discussed in more detail in Chapter 9.

In structural studies, cross-linking reagents are often used as "molecular rulers," as the lengths of their spacer arms are obviously related to the distances between

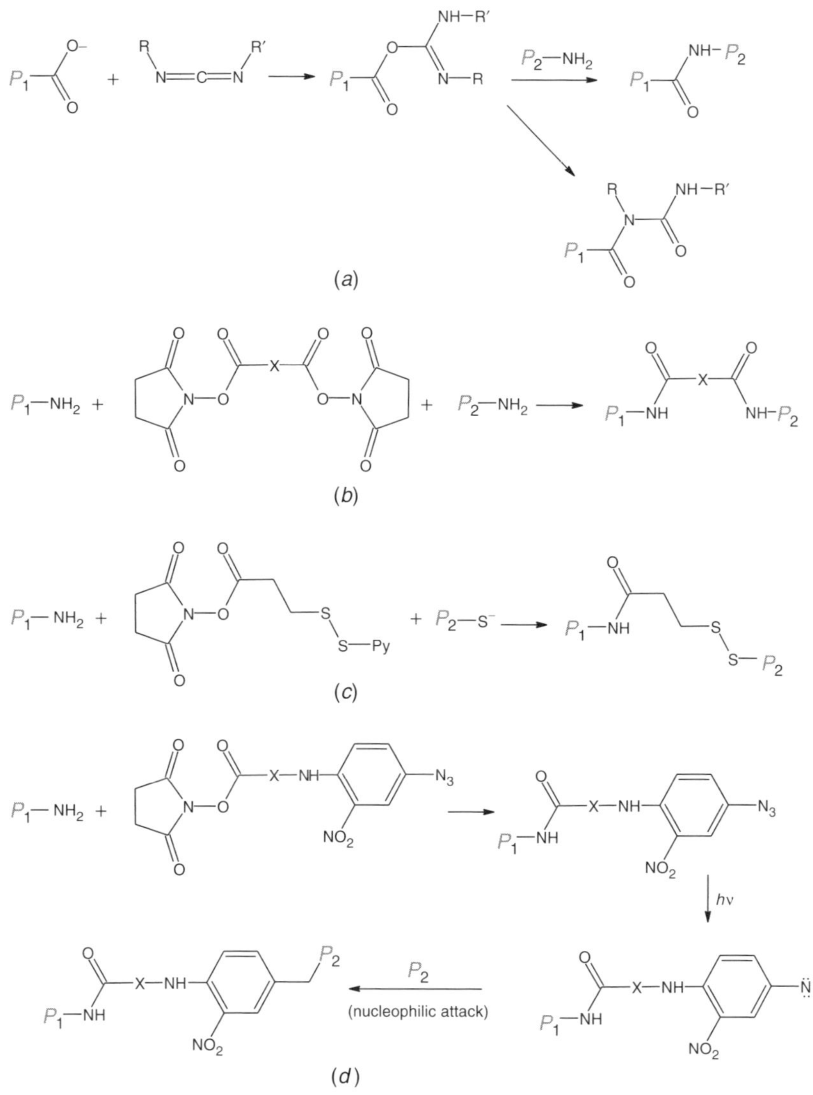

FIGURE 4.1. Schematic diagrams of cross-linking reactions utilizing monofunctional (**A**), homobifunctional (**B**), heterobifunctional (**C**), and photoreactive heterobifunctional (**D**) reagents.

the cross-linked functional groups within the protein (or protein complex). If cross-linking is carried out under nondenaturing conditions, such distance constraints will provide important information regarding the tertiary and/or quaternary structural organization. The great variety of cross-linkers available provides a wide spectrum of lengths ranging from a "zero-length" reagent EDC (1-ethyl-3-[3-dimethylaminopropyl]carbodiimide hydrochloride) to several cross-linkers whose space arms approach 20 Å (Table 4.2). The most reliable information is provided by the zero-length reagents, as they induce formation of a "direct" covalent bond between the two cross-linked sites. The spacer arms of most bifunctional reagents, on the other hand, are at least somewhat flexible (the lengths presented in Table 4.2 are mean values). As a result, assigning a single number to a distance between the cross-linked sites may lead to misinterpretation of the protein structure and its dynamic character (15). In a recent study, Houk and co-workers evaluated cross-linking "spans" using stochastic dynamics simulations for a range of homobifunctional reagents that target either amines or sulfhydryls (15). Most distributions were found to be near-Gaussian with a standard deviation ranging from 0.5 Å (for relatively "stiff" spacer arms) to 2 Å and above (for more flexible ones). However, the most striking conclusion of this work is that the distances cited for a number of popular cross-linkers* are highly improbable (see Table 4.2).

The low-resolution cross-linking studies of protein assemblies seek only to identify the neighboring units within large protein complexes, a task that can usually be accomplished by using gel electrophoresis. Detection of the cross-linked units is often aided by using coupling reagents that incorporate radioactive labels (usually done by incorporating ^{125}I into a cross-linking reagent) or other reporter groups (commonly used reporters are UV-Vis chromophores, as well as spin and fluorescent labels). Mass spectrometry, on the other hand, allows quick and direct identification of the cross-linked subunits without the need to use radioactive labels or other reporter groups. Mann and co-workers used a proteomic approach to obtain a nearest-neighbor relationship for a large protein complex (16). In-gel tryptic digest was followed by database searches to identify cross-linked proteins based on their partial sequence.

Higher resolution cross-linking studies aim at identification of the pairs of cross-linked residues within the protein or protein complex. Such information, together with the knowledge of the spacer arm length of the cross-linking reagent, provides "through space" distance constraints that are extremely valuable for defining both tertiary (intrasubunit cross-links) and quaternary (intersubunit cross-links) organization of the protein when no other structural information is available. However, confident assignment of the pairs of coupled residues within the cross-linked protein(s) is a rather challenging experimental task. A combination of proteolysis, separation methods (e.g., liquid chromatography), and mass spectrometry (and, particularly, tandem mass spectrometry or MS/MS) provides perhaps the most elegant and efficient way of solving this problem (17). Figure 4.2 show the workflow of a typical cross-linking experiment. Separation of proteolytic fragments prior to MS (or MS/MS) analysis usually results in significant improvements in sensitivity by eliminating possible signal suppression effects that may otherwise result in discrimination against larger (cross-linked)

*Typically, such distances are obtained by measuring the separation between the two reactive groups in the most extended conformation of the cross-linker and ignoring all other possible (and often preferred!) conformations.

TABLE 4.2. Examples of Commonly Used Cross-Linking Reagents

Commercial Name and (Full Name)	Chemical Structure	Targeted Groups 1	Targeted Groups 2	Spacer Arm Length and Average Distance[a]	H_2O Soluble	Comments
Zero-Length						
EDC, 1-ethyl-3-(3-dimethyl-aminopropyl) carbodiimide	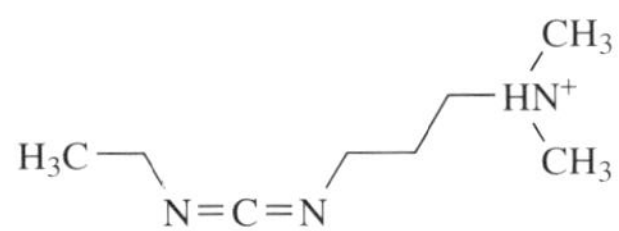	$—CO_2^-$	$—NH_2$	0 Å	Yes	
Homobifunctional						
DFDNB, 1,5-difluoro-2,4-dinitrobenzene	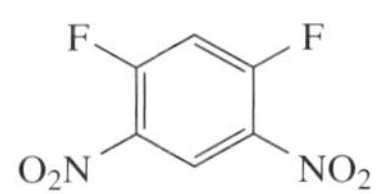	$—NH_2$	$—NH_2$	3/4.91 ± 0.07 Å	No	
Sulfo-DST, Sulfodisuccinimidyl tartrate	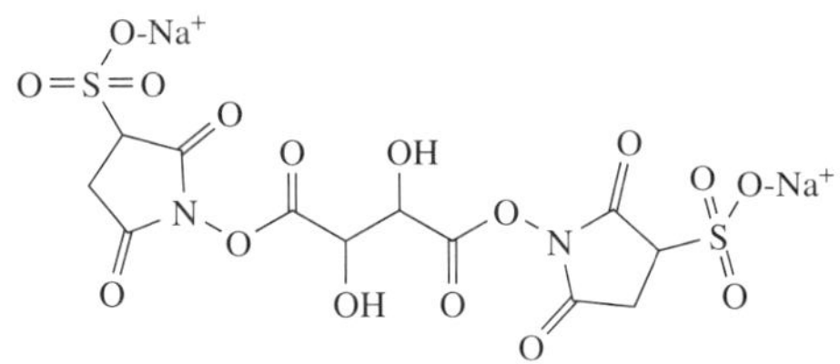	$—NH_2$	$—NH_2$	6.4 Å	Yes	
BS[3], bis(sulfo-succinimidyl) suberate	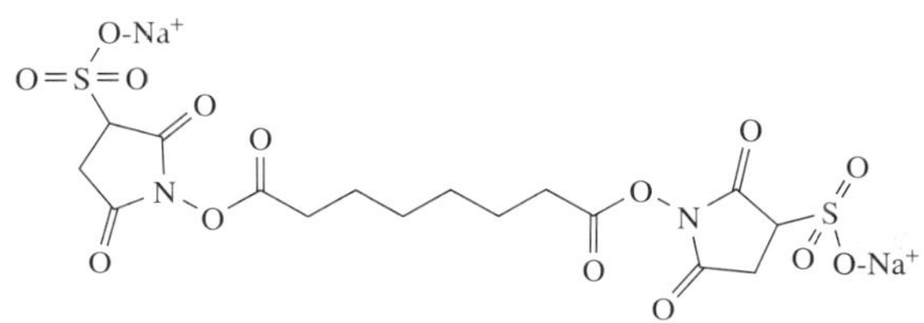	$—NH_2$	$—NH_2$	11.4 Å	Yes	

TABLE 4.2 (*Continued*)

Commercial Name and (Full Name)	Chemical Structure	Targeted Groups 1	Targeted Groups 2	Spacer Arm Length and Average Distance[a]	H_2O Soluble	Comments
Sulfo-EGS	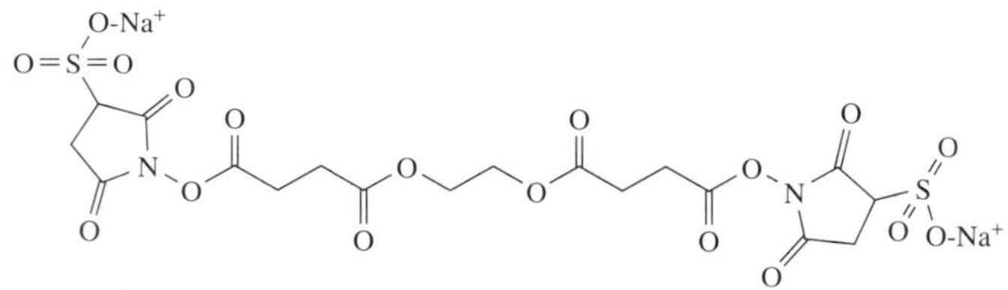	—NH_2	—NH_2	16.1 Å	Yes	
$BM[PEO]_3$, 1,8-bis-maleimido-triethylene glycol	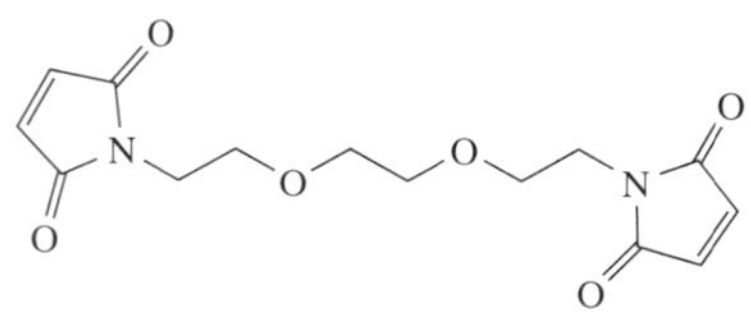	—SH	—SH	14.7 Å		
Heterobifunctional						
AMAS, *N*-(α-maleimido-acetoxy)-succinimide ester	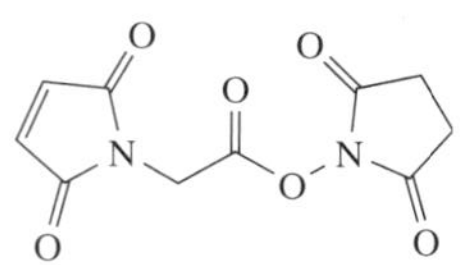	—SH	—NH_2	4.4 Å		
AEDP, 3-([2-aminoethyl] dithio)propionic acid	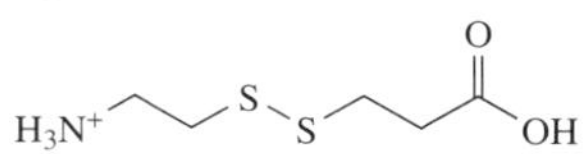	—NH_2	—CO_2^-	9.5 Å		
SAED, sulfos-uccinimidyl 2-[7-azido-4-methycoumarin-3-acetamido]-ethyl-1,3′-dithiopropionate	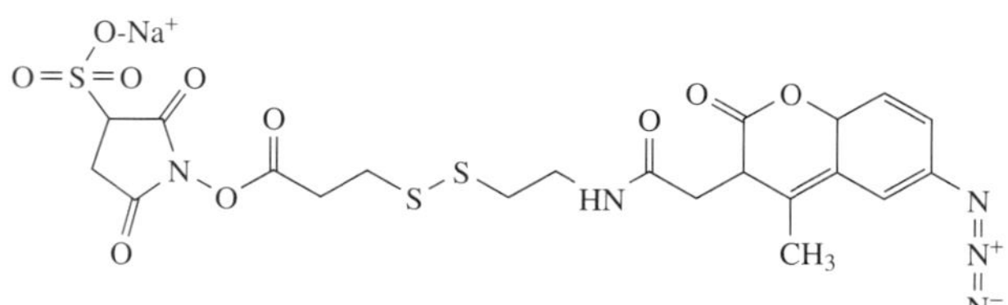	—NH_2	Any (photo-activation)	23.6 Å		

APG, *p*-azidophenyl glyoxal	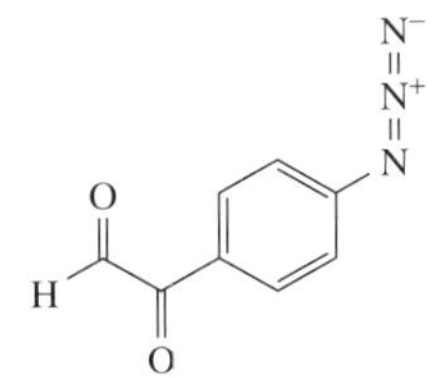	Guanidino (Arg)*	Any (photo-activation)	9.3 Å	No	Reacts selectively with Arg at pH 7–8
ASBA, 4-(*p*-azidosalicylamido)-butylamide	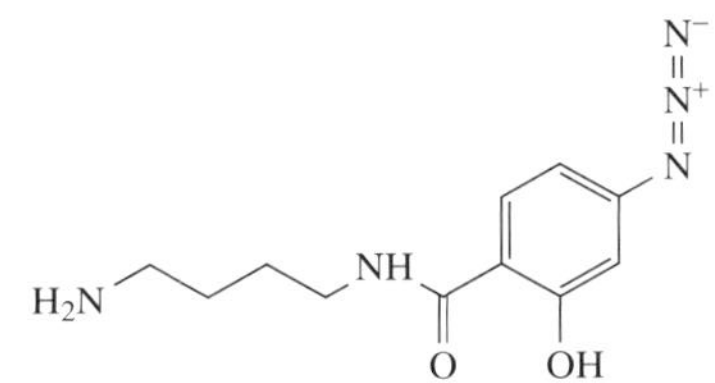	$—CO_2^-$	Any (photo-activation)	16.3 Å		
BASED, bis(*β*-[4-azidosalicylamido]ethyl) disulfide	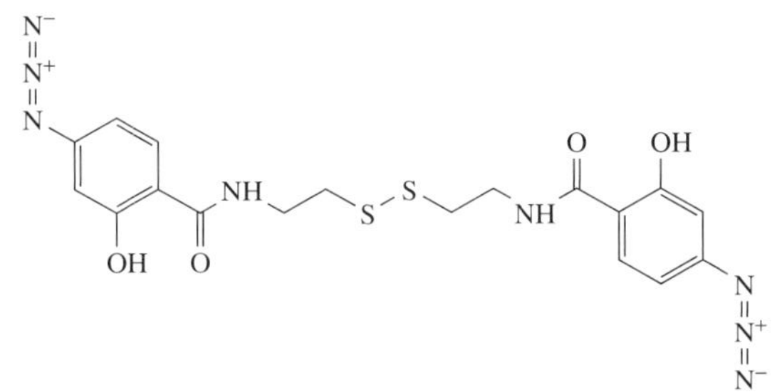	Any (photo-activation)	Any (photo-activation)	21.3 Å		

[a] From Ref. (15).

fragments (17). Although gel electrophoresis, HPLC, or gel permeation chromatography remain the most popular choices for "sorting out" products of the cross-linking reactions prior to proteolysis (Figure 4.2), utilization of capillary zone electrophoresis may result in significantly better resolution of different cross-linked species (18).

Peptide mapping alone can sometimes lead to the confident identification of the cross-linked residues (19–21). For example, using trypsin as a proteolytic agent allows the quick identification of Lys residues cross-linked by homobifunctional

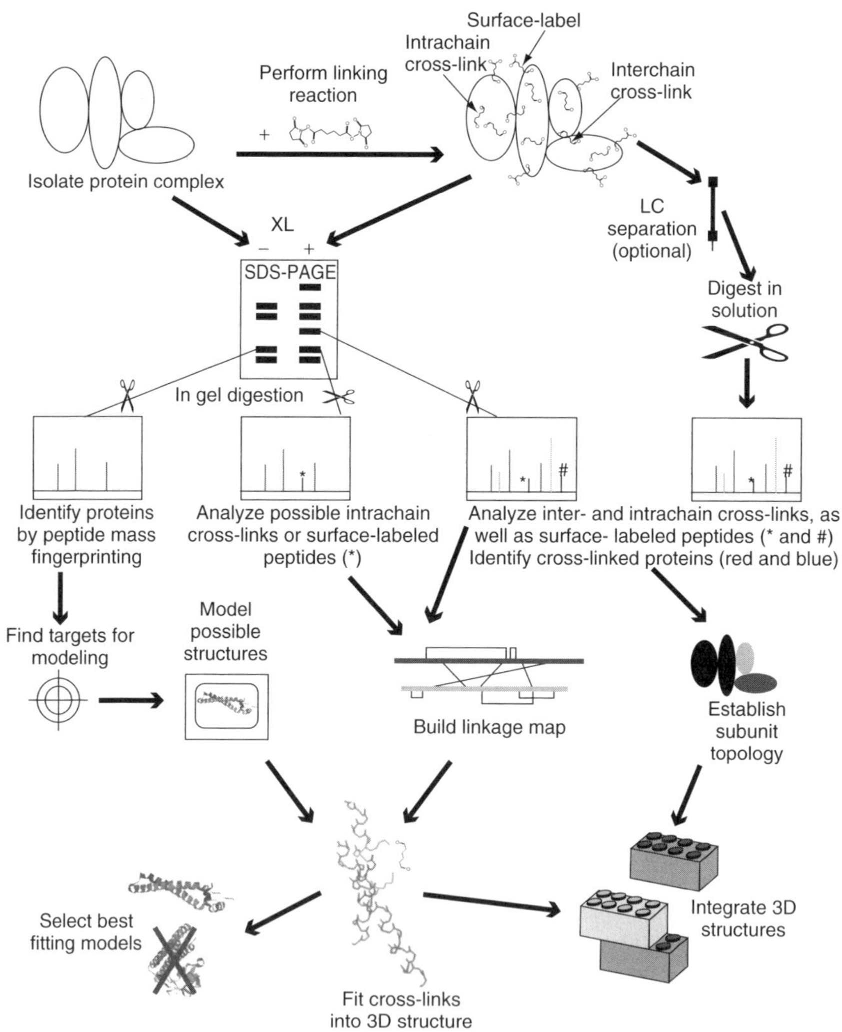

FIGURE 4.2. Schematic diagram of workflow of cross-linking a multiprotein complex and integrating the levels of information into a three-dimensional model of the structure. Reprinted with permission from (17). © 2003 Elsevier.

reagents (19). Roepstorff and co-workers developed an analytical strategy that utilizes thiol-cleavable cross-linking reagents (20). Cross-linked proteins are proteolytically digested and peptide maps are produced by MALDI MS with and without prior reduction of the thiol linkers. Peptide ion peaks disappearing from the spectra after the reduction are assigned as putative cross-linked peptides. The assignment is confirmed if the ion peak corresponding to one (or both) of the counterparts appears in the spectrum following reduction.

In many cases, however, the unambiguous assignment requires that MS/MS sequencing of the proteolytic fragments be carried out (22, 23). Recently, Gibson and co-workers analyzed MS/MS spectra of cross-linked peptide ions derived from numerous cross-linking experiments aiming at interpreting the fragmentation mechanisms. As a result, a nomenclature to be used for MS/MS analysis of cross-linked peptides has been introduced that categorizes both different types of cross-linked peptides (products of chemical cross-linking followed by proteolysis) and gas-phase fragmentation patterns (24). Peptides are classified according to the outcome of the cross-linking reaction, assigning type 0 for the "dead-end" products, type 1 for internal cross-links, and type 2 for interpeptide cross-links (Figure 4.3). Classification of fragment ions utilizes standard Biemann's nomenclature (25) for linear peptides and Gross' nomenclature for cyclic peptides (26).

Identification of the cross-linked peptides in the mass spectra (and, more importantly, cross-link containing fragment ions in the MS/MS spectra) can be greatly facilitated by utilizing stable isotope-tagging techniques (27). In many cases, isotope tagging not only simplifies the assignment of the cross-linked residues in the MS/MS spectra (27) but also improves the detection limits and even allows parallel use of two cross-linkers of similar reactivity but different spacer length (28). Another approach to developing cross-linking reagents that are particularly suited for MS and MS/MS makes use of "signature ions." A newly synthesized cross-linker, *N*-benzyliminodiacetoyloxysuccinimide (BID), yields stable benzyl cation marker ions upon low-energy CID of cross-linked peptide ions, which eliminates the necessity of purifying cross-linked peptides prior to analysis (29).

As the amount of information deduced from cross-linking experiments increases, so too does the complexity of data interpretation. This task can be greatly assisted by using a variety of automated algorithms that use MS or MS/MS data as input (17). We have already mentioned a "database mining" approach to identification of cross-linked peptides (16). This approach can be used even in a situation when the protein complex composition is not known a priori (30), one of the tasks that is actively pursued by interactomics, a branch of proteomics (31). Another approach to this problem (Automated Spectrum Assignment Program, or ASAP*) utilizes known sequence(s) of cross-linked protein(s) and searches for peaks in the mass spectra that match the masses of the expected proteolytic fragments corrected for the mass shifts caused by the chemical modification (23). Similar approaches are used by other groups of

*A Web-based version of ASAP is currently available for beta-testing at `http://roswell.ca.sandia.gov/~mmyoung`.

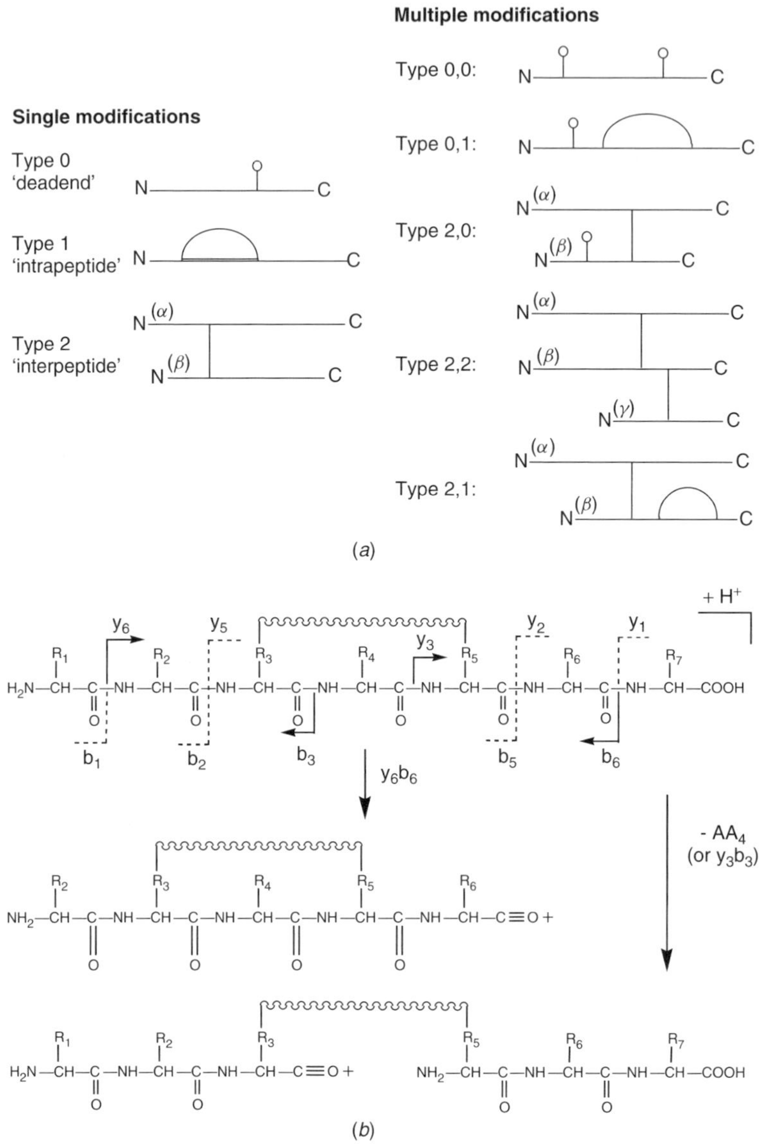

FIGURE 4.3. Classification of cross-linked peptides and nomenclature of fragment ions derived from the cross-linked peptides. (**A**) Classification of cross-linked peptides into *type 0, type 1*, and *type 2* outcomes and an extension of this nomenclature to combinations of two outcomes for cases of multiple cross-linking and/or modification. Chain length or mass ($\alpha > \beta$) and sequence position (N- to C-terminus) determine the order of the two numbers that designate the type of a cross-link. (**B**) Examples of fragmentation of the *type 1* peptide involving a *y*, *b*-cleavage (e.g., y_6b_6) or alternatively, the loss of an amino acid from the cyclic portion of the cross-linked peptide (e.g., $-AA_4$ or y_3b_3 cleavage reaction). (**C**) Examples of fragmentation of the *type 2* peptides. Adapted with permission from (24). © 2003 Elsevier.

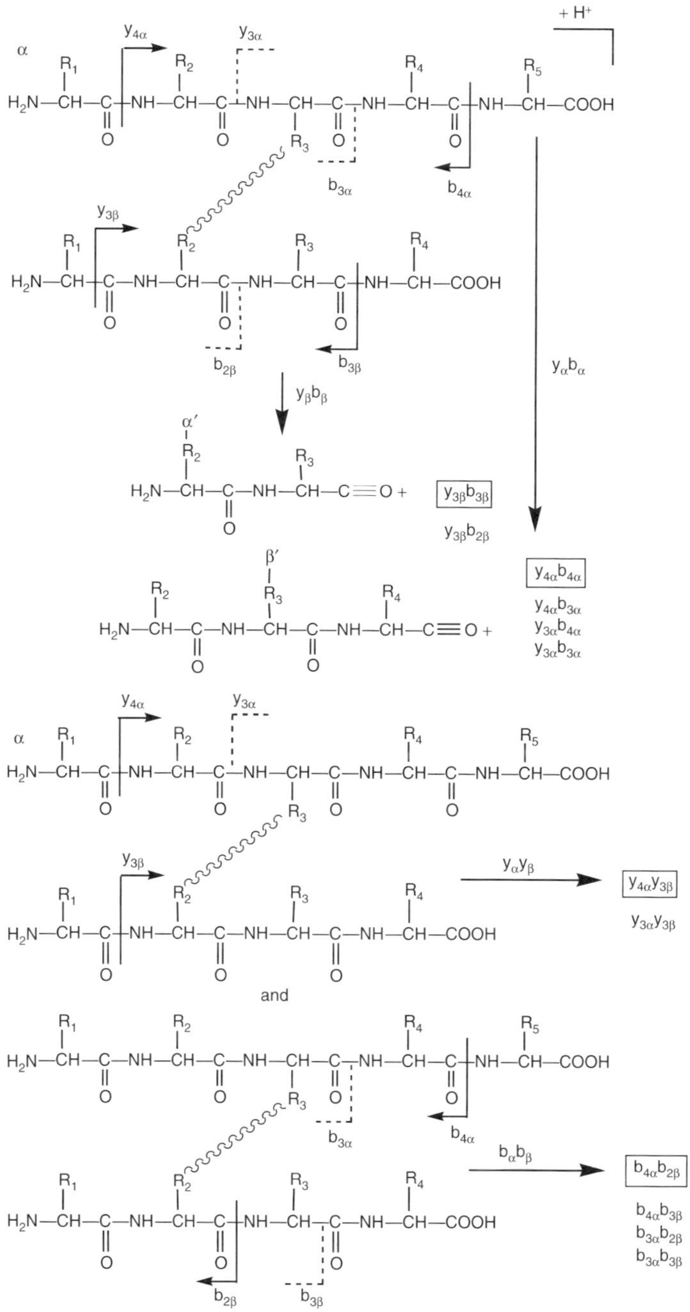

FIGURE 4.3. *(Continued)*

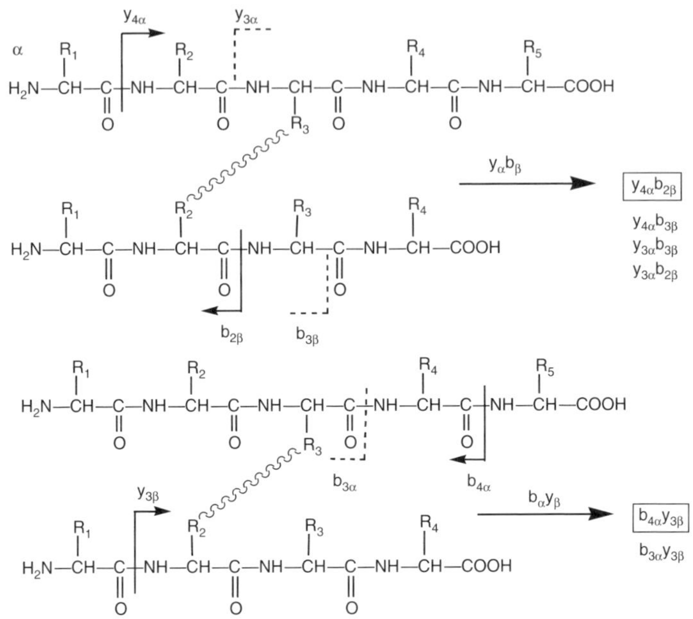

(*c*)

FIGURE 4.3. (*Continued*)

authors to assign the cross-linked peptides based on knowledge of the protein sequence and nature of the reagent (21, 32). The MS2Assign Program (24) uses the cross-linked peptide(s) amino acid sequence, mass shifts caused by cross-linking, nature of the cross-linking sites, as well as a list of all observed fragment ions to generate a library of all possible fragments and to assign the observed ones.* An algorithm developed by Chen and co-workers (33) seeks to identify the cross-linked sites using a two-step optimization procedure. First, all possible candidates are found for each product of the cross-linking reaction (based on the measured mass of the modified peptide and the known sequence of the protein). Second, a pair of the cross-linked amino acids is identified as the one that is optimally correlated to the MS/MS spectrum.

The interpretation of MS and MS/MS data on cross-linked peptides is greatly simplified as the resolution and mass accuracy of MS measurements increase (34). Identification of the cross-linked sites can also be facilitated by using multiple protease digestions in combination with MS and MS/MS analysis of the proteolytic fragments (35). A very different approach to identification of the pairs of coupled residues within the cross-linked peptides utilizes a "top-down" approach to protein sequencing in the gas phase discussed in Chapter 3. A crude cross-linked protein mixture is injected into an ESI source of an FT ICR mass

*A Web-based version of MS2Assign is currently available for beta-testing at http://roswell.ca.sandia.gov/~mmyoung.

spectrometer, followed by multistage fragmentation of cross-linked protein ions in the gas phase (36). An important advantage of this approach is complete elimination of the proteolytic and separation steps. It remains to be seen, however, if this technique will become practical when applied to large protein complexes (since the fragmentation efficiency becomes very low as protein ion size increases).

Information obtained in the course of cross-linking MS or MS/MS experiments is currently used mainly in conjunction with other experimental techniques (a number of cross-links in a typical experiment is not high enough to provide a sufficient number of constraints that would define a unique three-dimensional structure). However, as the field of sequence-based structure prediction matures (37–40), it may be possible in the very near future to use cross-linking experiments as an efficient and reliable tool for filtering out candidate structures of proteins or protein complexes that conflict with the *through-space distance constraints* derived from the cross-linking experiments.

Finally, it should be mentioned that although the majority of structural studies utilizing chemical cross-linking aim at the characterization of stable macromolecular structures, in some cases cross-linking can also be used to detect the loss of order within certain protein segments. A very interesting example of such a study was recently reported by Reisler and co-workers (41), who used cross-linking and MALDI MS to study destabilization of two key regions of scallop myosin subfragment S1 induced by bound nucleotides. When bifunctional reagents of various lengths were used, no chemical cross-linking was evident on a short time scale in the absence of nucleotides. However, addition of MgADP to the system resulted in significant enhancement of the cross-linking reaction yields for all reagents used in the study, indicating significant structural disorder induced by the nucleotide. Mapping the cross-linked sites with MALDI MS allowed the basis of structural destabilization to be elucidated (41).

4.2. MAPPING SOLVENT-EXPOSED REGIONS: FOOTPRINTING METHODS

4.2.1. Selective Chemical Labeling

Selective chemical modification (42) is another technique that has enjoyed increasing popularity in a variety of biophysical studies. Although initially developed primarily as a tool to modulate enzymatic activity (43–45), the technique has proved in recent decades to be an efficient probe of macromolecular structure at low resolution (46). The extent of chemical modification of a certain amino acid residue is used to determine its solvent accessibility. When collected in a residue-specific fashion, such information can be used to map solvent-exposed areas of the protein (Figure 4.4). The most popular amino acid-specific modifiers are acetic anhydride (ε-amine groups of lysine residues), tetranitromethane (phenyl ring of tyrosine), and diethyl pyrocarbonate (imidazole ring of His). More information on the chemical modifiers is presented in Table 4.3. The amount of information

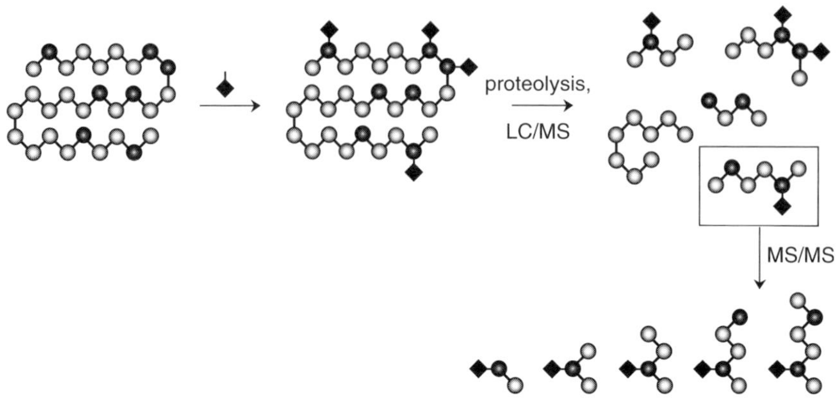

FIGURE 4.4. Mapping solvent-accessible areas of protein using selective chemical modification.

provided by selective chemical modification can be increased if such experiments are carried out in parallel with chemical cross-linking (47). A bifunctional cross-linking agent can also serve as a "labeling" reagent with type 1 and type 2 (see Figure 4.3 for explanation of nomenclature) cross-links providing information on the tertiary and quaternary organization of the protein and type 0 (dead-ended) modifications serving as gauges of solvent accessibility (48).

Although the interpretation of the results of chemical modification experiments is often based on the assumption that reactivity of side chains is correlated to their location within the protein, this is not universally true. Soon after the first high-resolution structures of several proteins became available, it became clear that assignment of "buried" and "exposed" groups within a native protein, based solely on chemical modification data, may be incorrect in any given instance (49). Obviously, functional groups can remain sterically hindered even at the surface of the protein. Other factors may play an important (and sometimes even decisive) role as well. One particularly interesting example is modification of arginine side chains with 1,2-cyclohexanedione, CHD (50, 51), with the differential reactivity of Arg residues within a protein often interpreted as resulting from varying solvent accessibility [e.g., see (52)]. However, Przybylski and co-workers used a model protein lysozyme (whose structure has been extensively characterized) to demonstrate that the reactivity of arginine residues toward CHD did not correlate with their solvent accessibility (46).* This intriguing observation has been explained in terms of intramolecular catalysis (by a nearby proton acceptor group) as being a prerequisite for a "successful" modification of the guanidino group. Such apparent importance of the chemical microenvironment on the efficiency of modification reactions highlights the difficulties associated with interpretation of

*In fact, an inverse correlation was obtained in these experiments, with the two most reactive arginine residues having the lowest solvent accessibility factors. Interestingly, acetylation of lysine residues carried out on the same protein yielded "correct" reactivity order (i.e., in perfect agreement with the solvent accessibility factors calculated based on the known crystal structure of lysozyme).

TABLE 4.3. Examples of Commonly Used Reagents for Selective Chemical Modification of Amino Acid Side Chains in Proteins

Amino Acid	Reactive Group	Reagent	Side Chain Derivative	Comments
Lys	R—NH_2	Acetic anhydride	R—NH—CO—CH_3	Large excess of the reagent is added stepwise; optimal solution conditions pH 7–9.5, 0 °C
		Cyanate	R—NH—CO—NH_2	Optimal reaction conditions pH 8, 37 °C (12–24 h); side reactions involve reversible modification of —SH and—C_6H_5OH groups
Arg	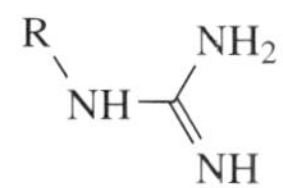	HPG (*p*-hydroxyphenyl glyoxal)	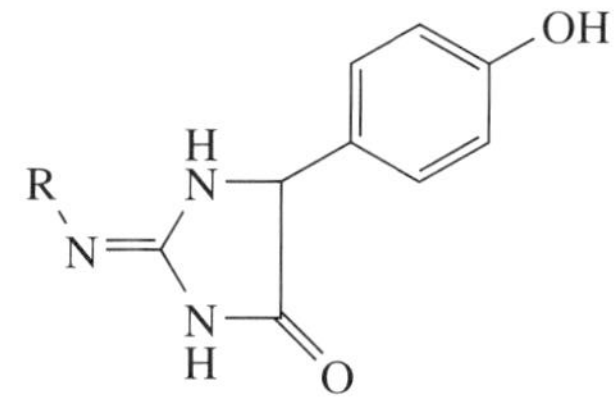	Reacts specifically under mild conditions (pH 7–9, 25 °C)
Tyr	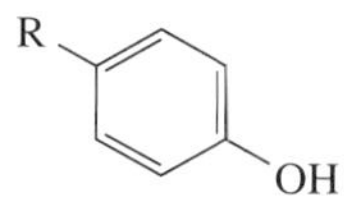	Acetyl imidazole	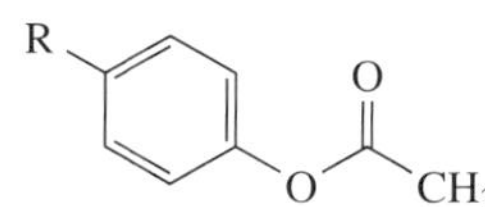	Optimal pH 7.5–8.0; side reactions involve acetylation of —NH_2 and—SH groups (acetylation of serine residues is also possible)
		TNM (tetranitromethane)	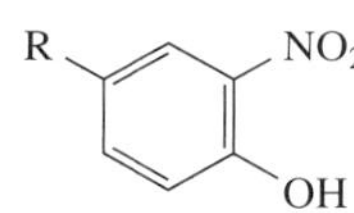	Optimal pH 8.0; side reactions involve modification of Met and Trp; cross-linking and polymerization are also possible. See comments in the text.
Asp, Glu	R—CO_2H	Diazoacetamide	R—CO—O—CH_2CO—NH_2	Requires Cu^{2+} as a catalyst; optimal conditions pH 5, 15 °C for 1–2 h
Cys	R—SH	Iodoacetic acid	R—S—CH_2—CO_2H	Modification of Met, Lys and His may also occur

the experimental results. In many other cases, reasons why a certain functional group fails to react with its specific reagents are more obscure. Acetylation of tyrosine hydroxyl groups [with *N*-acetyl imidazole (53, 54)] seems to be a particularly tricky case with the solvent-exposed residues often showing little or no reactivity [a review of such cases is presented in (49)]. Finally, a surprisingly high reactivity is sometimes displayed by "internal" (or "buried") side chains. Such an anomaly can be caused by a variety of factors, including significant alteration of the side chain's pK_a in the protein interior, catalytic effect of neighboring groups (as discussed above), and efficient trapping (binding) of the reagent inside the protein in close proximity to the modification site (e.g., due to the hydrophobic character of the reagent and/or steric compatibility). Awareness of the problems that may arise in the course of selective chemical modification studies is essential. However, the usefulness of such studies in assessing protein higher-order structure is unquestionable.

Any chemical modification of an amino acid side chain alters the protein mass, hence the appeal of mass spectrometry as a readout tool for the outcome of such experiments. Interpretation of the MS and MS/MS data on chemically modified proteins is usually relatively straightforward (as compared to the analysis of cross-linked proteins) and greatly benefits from a vast arsenal of experimental tools and techniques developed over recent years to analyze post-translational modifications of proteins (55). Modified proteins are typically processed with a suitable proteolytic enzyme, followed by mass mapping of the fragment peptides. The position(s) of the modified residue(s) within each proteolytic fragment can be established reliably using tandem mass spectrometry (56–59), as the presence of chemical modification manifests itself as a break in the ladder of the expected fragment ions (60). In a typical experiment, binding topology is determined by comparing modification patterns of the protein obtained in the presence and in the absence of its binding partner (56), although the two experiments can be combined if the labeling agent contains a stable isotope tag (57). An added benefit of using isotope tags is the easy recognition and quantitation of label-containing peptides and their fragments in MS and MS/MS spectra. Finally, incorporation of a chromophore into a reagent provides a parallel route of detection of the chemically modified species, which can be used to monitor the progress of selective chemical modification, as well as to quickly identify "labeled" fragments following proteolytic digestion of the modified protein (48).

Although generally regarded as a low-resolution technique, selective chemical modification can sometimes provide structural information at surprisingly high resolution. One particularly intriguing example was reported recently by Tomer, who used a combination of selective chemical modification, proteolysis, liquid chromatography, and tandem MS to study interaction between a full-length glycosylated HIV gp120 and a soluble segment of human CD4 (56), known to function as a primary receptor for HIV, and CD4–gp120 complex formation represents the first step of virus entry into various circulating T cells. Of the two adjacent Arg residues located at the CD4–gp120 interface (Arg^{58} and Arg^{59} according to the available crystal structure), only the latter became shielded from the solvent in the presence of gp120, while both residues reacted with hydroxyphenylglyoxal (HPG, a reactant targeting arginine side chains) in the absence of the viral protein (Figure 4.5). Therefore, only Arg^{59} (but not its nearest neighbor) was considered to be a part of the CD4–gp120 interaction site.

An even more intriguing suggestion regarding further improvements in spatial resolution in chemical modification experiments was made by Pucci and co-workers (61). They argued that modification of tyrosine residues with reagents targeting different functional groups [i.e., O-acetylation of the hydroxyl group with *N*-acetyl imidazole (62) and modification of the aromatic ring with tetranitromethane (53, 54)] may provide information on microenvironments of these functional groups, thereby extending the spatial resolution beyond the level of a single amino acid. When the two reagents were used to probe solvent accessibility of six Tyr residues of Minibody (a small de novo designed β-protein), only three were modified by *N*-acetyl imidazole, while five were reactive toward tetranitromethane (61). These observations led to a conclusion that the two "extra" Tyr residues (the ones that can be modified only by tetranitromethane) are only partially accessible to the solvent, with their phenolic hydroxyl groups being involved in hydrogen bonds* (61).

4.2.2. Nonspecific Chemical Labeling

Although the majority of chemical modification studies employ selective (amino acid-specific) reagents, nonspecific modifiers, such as methyl bromide (63), have also been used in select studies. Nonspecific photochemical labeling agents, such as diazirine,† have also seen a surge in popularity in recent years (64, 65). Another relatively nonspecific chemical modifier is the hydroxyl radical, which can be generated either chemically (59, 66), electrochemically (67) or radiolytically (68–70). Chemical generation of $OH^{\bullet}$ radicals can be carried out using hydrogen peroxide (H_2O_2) in the presence of transition metal ion complexes, such as iron–EDTA (71). Electrochemical generation of hydroxyls and/or other oxygen-containing radical species can be accomplished within an ESI ion source using oxygen as a reactive nebulizer gas at elevated needle potential, with the protein oxidation occurring at the electrospray needle tip (67). A radiolytic generation of $OH^{\bullet}$ radicals can be carried out by using either radioactive material, such as ^{137}Cs (68), or a high-intensity radiation source, such as a synchrotron (69). Oxidative side chain modifications occur significantly faster as compared to backbone cleavage, with the following reactivity order in the solvent-acccssible environment: Cys, Met $\gg$ Phe, Tyr $\gg$ Trp $>$ Pro $>$ His, Leu (72). Both types of reactions can be used to map the solvent accessibility of various protein segments.

*It should be kept in mind, however, that Tyr acetylation experiments are known to be rather uninformative as far as position of the phenolic hydroxyl groups (see earlier discussion), and any conclusion based solely on the outcome of chemical modification of the protein with *N*-acetyl imidazole should be met with certain skepticism.

†Diazirine N_2CH_2 generates methylene carbene upon photolysis, which reacts nonselectively with its molecular cage, inserting even into C—H bonds.

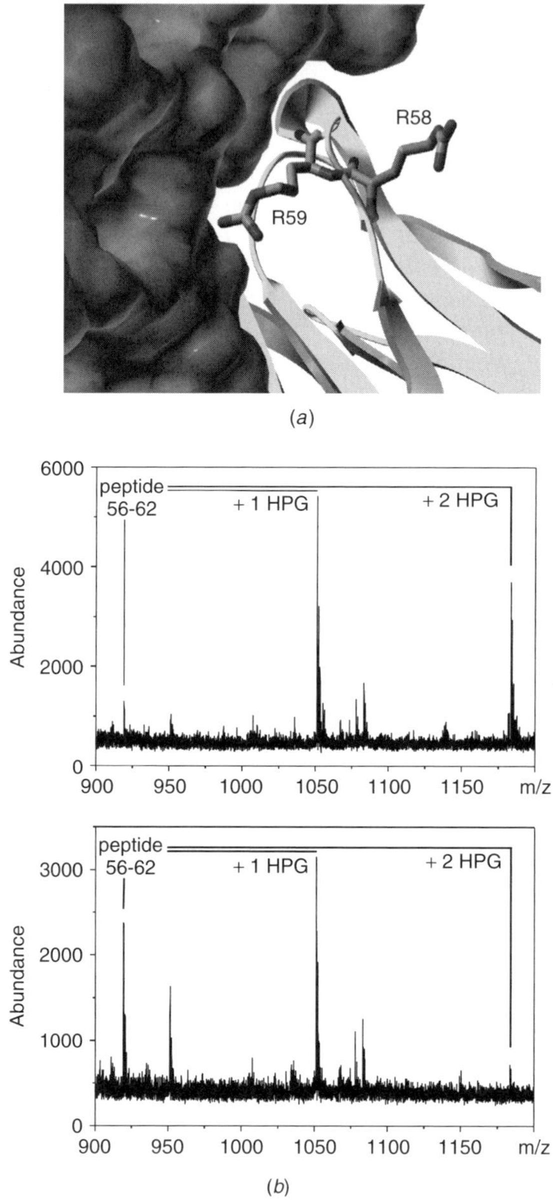

FIGURE 4.5. (**A**) Interaction site of CD4 (ribbon) with a truncated HIV envelope protein gp120 (surface) derived from the crystal structure. Arg^{59} of CD4 interacts with gp120, while Arg^{58} does not (both arginine residues depicted as sticks). (**B**) MALDI mass spectra of proteolytic peptides (AspN) derived from CD4 modified with hydroxyphenyl glyoxal (HPG) in the absence (*top*) and in the presence of gp120 (*bottom*). (**C**) Localization of the modified arginine residue within the proteolytic peptide [56–62] using tandem mass spectrometry. Reproduced with permission from (56). © 2002 American Chemical Society.

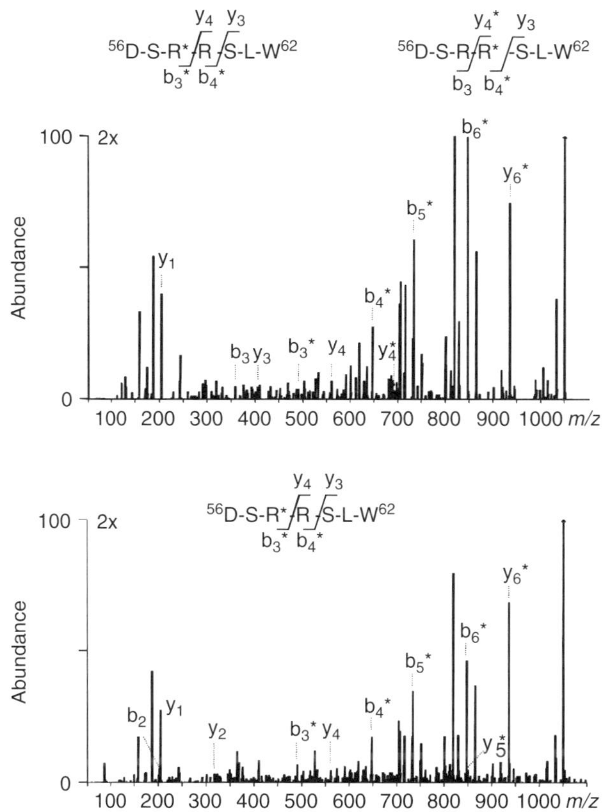

(*c*)

FIGURE 4.5. (*Continued*)

A very elegant approach to protein modification with $O_2^{\bullet -}$ and/or $OH^{\bullet}$ radicals utilizes "intrinsic" oxidation catalysts, such as transition metal ions bound to the protein (66). Combined with proteolytic degradation of the modified protein and MS/MS analysis, this scheme allows sensitive and reliable identification of amino acid side chains positioned in the metal binding site.

4.2.3. Hydrogen/Deuterium Exchange

Hydrogen/deuterium exchange (HDX) is a general experimental technique that detects the presence or absence of hydrogen bonding within the protein in solution (73). The analytical value of HDX as a tool for probing macromolecular structure was recognized almost immediately after the discovery of deuterium (74) and subsequent development of the methods of production of heavy water (75). Initial studies of the exchange reactions between organic molecules and ^2H_2O carried out by Bonhoeffer and colleagues indicated that the exchange rate is very high for heteroatoms (e.g., —OH groups), while the hydrogen atoms attached to carbon atoms (e.g., —CH_3 groups) do not undergo exchange (76). Hvidt and Linderstrøm-Lang later used HDX exchange to measure solvent accessibility of labile hydrogen atoms as a probe of polypeptide structure (3, 4). While this study was the first attempt to use HDX to probe protein structure in solution, the early studies of hydrogen exchange reactions between water and biopolymers preceded this work by more than fifteen years (77). Burley and co-workers suggested that the extent of ^{2}H incorporation into a protein molecule can be measured by monitoring its mass increase. Mass measurements of the dried sample before and after the exchange were carried out with a balance (quartz spring) (78). However, it was not until several decades later that MS was used to determine the number of labile hydrogen atoms within a polypeptide (79). The advent of ESI and MALDI MS dramatically expanded the range of biopolymers for which the extent of ^{2}H incorporation could be measured directly under a variety of conditions (80). As a result, HDX MS has now become a powerful experimental tool for probing protein higher-order structure.

Each hydrogen exchange reaction requires the formation of a hydrogen bond within the donor–acceptor pair (e.g., between the labile amide proton on the protein and the solvent catalyst, usually H_2O or D_2O). Therefore, HDX may be viewed as a selective chemical modification technique targeting labile hydrogen atoms (backbone amide, as well as all nonaliphatic side chain hydrogen atoms), whose major difference from those discussed in the preceding sections of this chapter is the reversible character of modification. Protein HDX involves two different types of reactions: (i) reversible protein unfolding that disrupts the H-bonding network, and (ii) isotope exchange at individual unprotected amides. As a result, the overall HDX kinetics may be very complicated and will be discussed in detail in the following chapter. An important feature of HDX that is particularly relevant within the context of this chapter is that the exchange rate of labile protons that are inaccessible to solvent is usually very slow. This feature has been used widely in recent years to probe the topology of protein–protein interfaces (81–90). The basic premise of the analysis is that HDX rate reduction for a given protein segment within the complex indicates that this segment is located at the protein–protein binding interface (87). Analogously, HDX can be used to identify "conformational switches" within monomeric proteins, such as changes in protein tertiary structure induced by small ligand binding (91).

Replacement of each proton with a deuteron (or vice versa) results in a protein mass change of about 1 Da, which makes mass spectrometry a very sensitive and reliable detector of the progress of protein HDX reactions. Site-specific assignment of ^{2}H incorporation is a much more challenging task due to the reversible character of HDX. Still, there are conditions (pH 2.3–3, $T = 0\,^{\circ}C$) under which the exchange of the backbone amide hydrogen atoms is relatively slow (92). (As we will see in the following chapter, the intrinsic exchange rate of the unprotected amide hydrogen atoms can be as low as 0.1 min^{-1} at room temperature and will obviously decrease further once the temperature is lowered to 0 °C.) If proteolytic degradation is performed under such *slow exchange conditions*, local details of $^{1}H-^{2}H$ distribution along the protein backbone can be maintained (93). This, of course, limits the choice of enzymes, as most of them are inactive under the slow exchange conditions. The only proteolytic agent that is active under these conditions is pepsin, although the search for suitable alternatives continues.* Back-exchange occurring during proteolysis and separation of the fragments prior to mass analysis can often be accounted for by introducing a *back-exchange* correction factor (94). Spatial resolution offered by this technique is usually limited only by the extent of proteolysis. In general, a large number of fragments, particularly overlapping ones, would lead to greater spatial resolution, and hence more precise localization of the structural regions that have undergone exchange.

Initially, site-specific characterization of protein–protein interaction via HDX MS techniques utilized electrospray ionization mass spectrometry. A typical experimental scheme would involve complete exchange of both binding partners as a first step, followed by their binding and back-exchange. Proteolytic digestion (with pepsin) and fast chromatographic separation† of the proteolytic fragments are carried out under slow exchange conditions, followed by on-line identification of the peptic fragments and measurement of their ^{2}H content (81). Noticeable back-exchange occurs even under such conditions, hence the requirement for a fast HPLC step [which appears to be the major contributor to the back-exchange (86)]. Inadequate separation of the peptic fragments by fast chromatography can be compensated for by utilizing high-resolution MS for peptide identification (86).

ESI MS as a method of detection in these experiments can be replaced with MALDI MS. Short exchange time allows preferential labeling of rapidly exchanging surface amides, so that primarily solvent accessibility changes (not conformational changes) are detected (82). Identification of the peptic fragments and measurement of their ^{2}H contents can be done without any separation step (i.e., within a single MALDI MS measurement). Again, the interprotein interface(s) can be reliably recognized by identifying peptides retaining more deuterons in the complex (82).

There are several reports on characterization of the inter-protein interfaces when the proteolytic step was replaced with fragmentation of the entire protein in the gas phase to obtain site-specific information (83, 85). It remains to be seen, however, if such experiments truly reflect solution-phase interactions and are not

*Forest and co-workers recently reported (51st ASMS Conference on Mass Spectrometry and Allied Topics, Montreal, Quebec, June 2003, ThPU 376) that combination of pepsin with proteases from *Aspergillus satoi* (type XIII) and *Rhizopus* sp. (type XVIII) had increased sequence coverage of a 77 kDa protein under "slow exchange" conditions (0 °C, pH 2.5).

†This step also provides desalting prior to ESI MS analysis, thus easing constraints on using non-volatile buffers in the course of HDX reactions and proteolytic degradation.

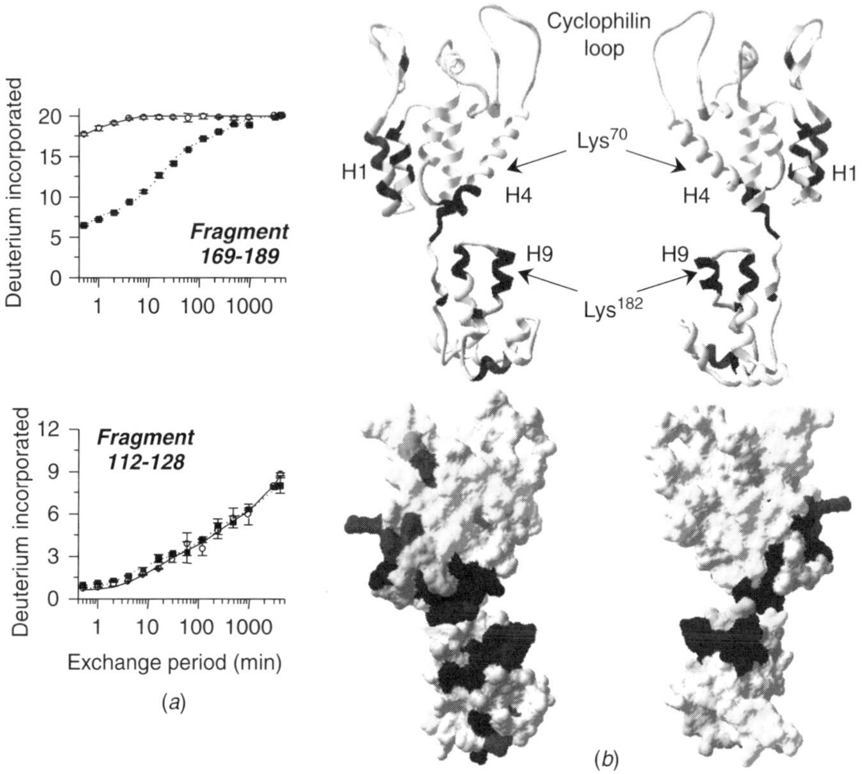

FIGURE 4.6. (**A**) Time profiles of deuterium incorporation into assembled (*open circles*) and monomeric (*shaded squares*) capsid protein for peptic fragments that span residues 169–189 (H8-H9 helices) and 112–128 (H6 helix). (**B**) Changes in amide hydrogen exchange rates between monomer capsid and assembled capsid mapped onto the monomeric capsid protein as varying shades of gray. Adapted with permission from (87). © 2003 Elsevier.

affected by gas phase processes, such as hydrogen scrambling (95). A thorough discussion of this problem will follow in Chapter 5.

Identification of the interface regions in protein complexes can sometimes provide information necessary for identifying the requisite quaternary structure of such assemblies. A very interesting example was presented recently by Prevelige and co-workers, who used HDX ESI MS in combination with chemical cross-linking to identify the previously unknown intersubunit interface regions in HIV-1 capsid protein assembly (87). Mapping such interfaces allowed the isolated domains (of known structure) to be "packed" into supramolecular structures (Figure 4.6). Another interesting approach to building quaternary structures of protein assemblies was introduced by Komives and co-workers, who used the results of HDX MALDI MS measurements to "filter" a vast array of suboptimal candidate structures* produced by computational docking of the subunits, whose monomeric structures were known beforehand (89). The majority of the "filtered" solutions formed a single cluster of the complex, whose structure was consistent with available biochemical data and from which new predictions could be made.

4.3. EMERGING LOW-RESOLUTION METHODS: ZERO-INTERFERENCE APPROACHES

4.3.1. Stoichiometry of Protein Assemblies and Topology of the Interface Regions: Controlled Dissociation of Noncovalent Complexes

Utilizing ion chemistry in the gas phase as a probe of protein higher-order structure in solution is sometimes met with a certain skepticism in the biophysical community. Indeed, water is essential to biological processes in general and to protein structure and function in particular (96). Significant alterations of the protein structure upon dramatic changes in its environment (i.e., complete removal of solvent) are expected and in some cases have been scrupulously documented (97, 98). Protein behavior in a solvent-free environment is now being actively looked at, as such studies hold the promise of providing a distinction between the "intrinsic" and "externally imposed" properties of biological macromolecules (99). We will briefly discuss such studies and their ramifications within the broader context of biophysics in Chapter 10. One general conclusion that is particularly important within the context of this chapter relates to the fate of protein assemblies upon desorption from solution to the gas phase. The existing body of knowledge suggests that many types of intra- and intermolecular interactions in proteins (with the notable exception of hydrophobic interactions) are preserved in vacuo (100, 101). Therefore, noncovalent macromolecular complexes can "survive" the transition from solution to the gas phase when using "mild" conditions in the ESI MS interface. The two parameters that are usually most critical for the "survival" of the noncovalent complex upon transition from solution to the gas phase are interface temperature and the electrical field in the ion desolvation region. Often, this allows the composition of macromolecular assemblies to be determined reliably and with minimal sample consumption (Figure 4.7). When executed carefully, such experiments almost always produce correct (i.e., confirmed by other methods) information on the stoichiometry of multiprotein complexes (102, 103).

The determination of the number of polypeptide subunits in each complex usually does not require very high accuracy in mass measurements. However, a serious problem may arise if the complex also contains small molecular weight components (metal ions, organic cofactors, etc.), which need to be accounted for. Preservation of the noncovalent interactions is often done at the expense of mass

*A total of 100,000 structures were filtered based on how well they agreed with the experimentally obtained interface protection data.

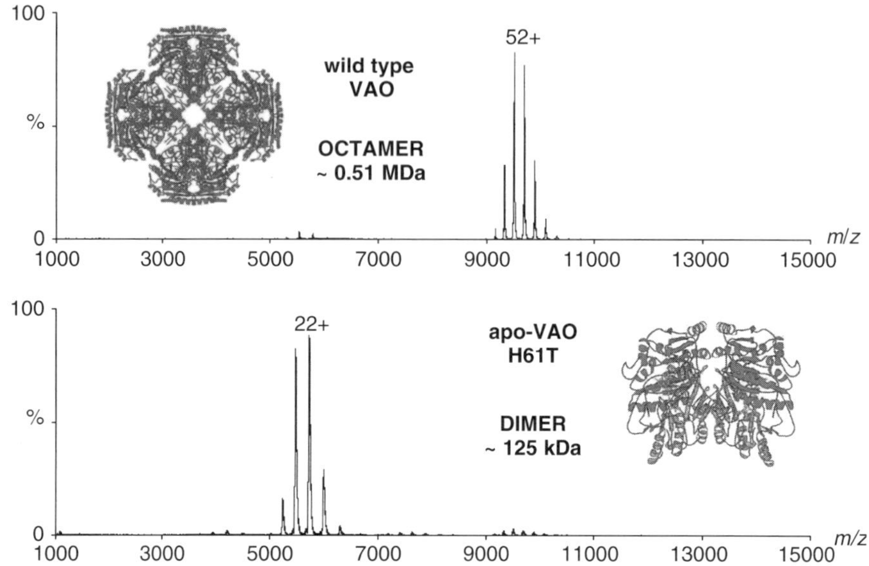

FIGURE 4.7. Nano-ESI mass spectra of a wild-type flavoenzyme vanillyl-alcohol oxidase (VAO) from *Penicillium simplicissimum* (*top*) and its mutant H61T (*bottom*). Both spectra were acquired under identical experimental conditions (50 mM ammonium acetate, pH 6.8). Under these conditions, the wild-type VAO is almost exclusively an octamer (in agreement with the crystal structure, see inset). Only very small ion signals originating from dimeric species are observed in this spectrum. In sharp contrast to the wild-type VAO, the mutant species exists almost exclusively in a dimeric form under the conditions used in this study. Reproduced with permission from (125).

accuracy,* which is critical for correct assignment of the small molecular weight components of a high molecular weight complex. To circumvent this problem, Amster suggested supplementing "mild" ESI MS measurements with those carried out under "harsher" conditions (104). In addition to efficient protein ion declustering and dissociation of the complex ion into constituent subunits, stepwise increase of the electrostatic field in the interface region (which increases average kinetic energy attained by the ions in the time period between two consecutive collisions with neutral molecules) eventually results in dissociation of cofactors from the subunits (Figure 4.8), thus allowing a reliable identification and an exact count of the low molecular weight species present in each subunit (104).

A very important step toward more detailed characterization of protein quaternary structure has been made recently by Robinson and co-workers, who utilized stepwise dissociation of macromolecular complexes in the gas phase (105–107). These experiments provide an elegant way to characterize protein quaternary

*Selecting "mild" desorption conditions typically leads to poor desolvation of protein ions in the ESI MS interface.

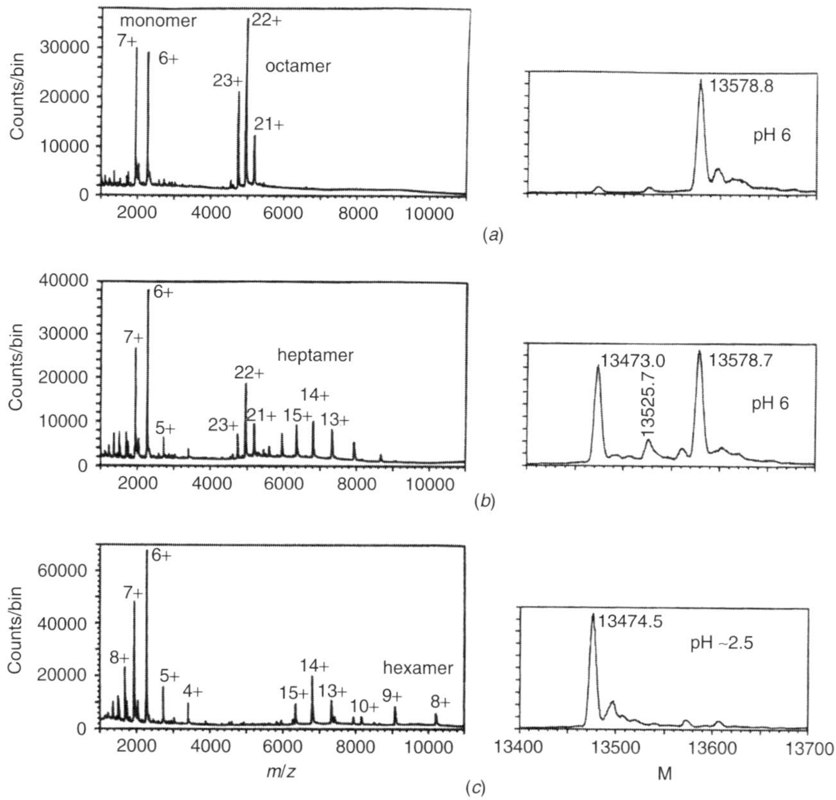

FIGURE 4.8. *Left*: ESI mass spectra of recombinant *Phascolopsis gouldii* hemerythrin (*Pg*-Hr), obtained under nondenaturing conditions and at increasing [from (**a**) to (**c**)] electrostatic fields (declustering potential) in the ion desolvation region. Peaks corresponding to the holoprotein octamer (m/z 4000–6000) decrease in intensity, while peaks corresponding to monomer (m/z 1000–3000), heptamer (m/z 6000–8000), and hexamer (m/z 8000–10000) ions increase in intensity as the ion collisional energy is increased. *Right*: Deconvoluted spectra of the monomer region in the mass spectrum of *Pg*-Hr ($z = 5$–8), showing peaks whose masses correspond to a subunit with 0 to 2 iron atoms attached to it. Under "mild" conditions (low declustering potential), the only observed peak is that corresponding to the apo-subunit plus two iron atoms (**a**). Once the declustering potential is elevated, two other peaks become prominent in the spectrum, corresponding to the apo-subunit with one and no iron atoms attached (**b**). Only the latter ionic species (apo-monomer) is observed under denaturing conditions (**c**). Adapted with permission from (104). © 2001 American Chemical Society.

structure at low resolution by revealing information on subunit interactions and topology (similar to contact maps produced in cross-linking experiments discussed earlier). Even in a situation when all subunits are identical, controlled dissociation experiments can be quite useful, as they can reveal the details of the "superquaternary" structure (108, 109). Application of this approach to characterize supramolecular structure of *Yersinia pestis* capsular F1 antigen (108) reveals a multilayered hierarchical organization of the protein complex, as demonstrated on Figure 4.9.

Until 1996, electrospray ionization remained the only technique that was widely used for the detection and characterization of noncovalent complexes (110) due to its unique ability to carry out measurements under "near-native" conditions [this is typically achieved by using aqueous solvent whose pH and ionic strength are adjusted to near-physiological levels using volatile salts and buffers (e.g., ammonium acetate, ammonium bicarbonate)]. Contrary to that, MALDI MS almost always relied on utilization of acidic matrices and organic solvents in sample preparation, a combination that is hardly suited for preserving noncovalent associations. In 1996, Przybylski and co-workers suggested a sample preparation method that utilized a basic matrix [6-aza-2-thiothymine (ATT)] without any

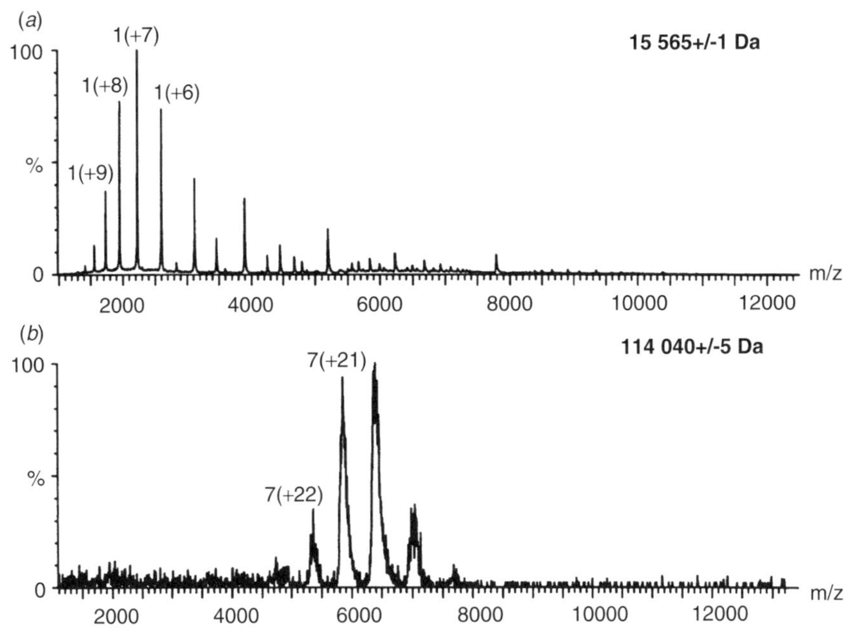

FIGURE 4.9. Nano-ESI mass spectra of the *Yersinia pestis* F1 antigen acquired under harsh (**A**) and mild (**B, C**) conditions in the ESI interface. The spectrum (**C**) was acquired under conditions of continuous collisional cooling in the ion guide prior to mass analysis. The ion peaks are labeled with the number of protein subunits followed by the charge (in parentheses). Bottom diagram shows a proposed schematic representation of the dissociation of the capsular F1 antigen based on the mass spectral data. A portion of the high molecular weight structure is shown composed of seven subunits in a helical arrangement. On dissociation with increasing collision energy, this structure gives rise to 14-mers, 7-mers, and 1-mers observed in the mass spectra. The overall average charge recorded for each species is shown with the average charge per subunit shown in parentheses. Adapted with permission from (108). © 2001 Cold Spring Harbor Laboratory Press.

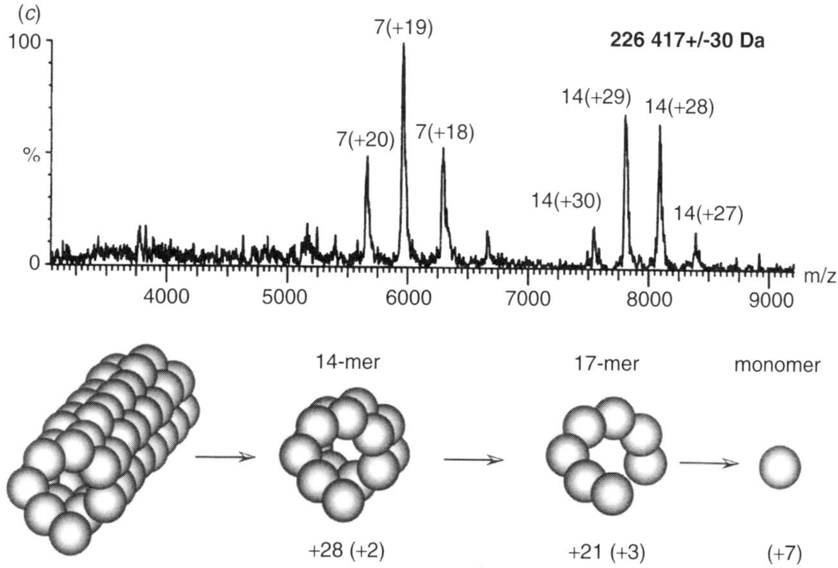

FIGURE 4.9. *(Continued)*

addition of organic cosolvents, which allowed intact noncovalent protein complexes to be observed by MALDI MS (111). In the following years, several groups have reported utilization of MALDI MS to study protein–protein and protein–oligonucleotide noncovalent interactions (112–118). Nevertheless, such applications of MALDI MS are far from being routine and some principal problems and questions still exist. An interested reader is referred to a recent article by Zehl and Allmaier, who systematically investigated several main problems, such as the effect of sample preparation, instrument-related effects on the stability of noncovalent complexes, as well as formation of nonspecific cluster ions (119).

4.3.2. Evaluation of Total Solvent-Accessible Area: Extent of Charging of Protein Molecules

The stoichiometric and topological determinations described in the previous section rely on measuring the masses of the protein complex and its constituents. As we have seen in Chapter 3, protein ions have another important characteristic, namely, a charge. The average charge of a protein ion tends to increase as the protein size (mass) increases. This is particularly obvious under denaturing conditions, as the m/z values of most proteins tend to fall within the same range (800–2500 u) irrespective of their masses. The average charge of protein ions generated under near-native conditions also increase as the protein mass increases, although not as fast, giving rise to a progressive increase in the m/z values (Figure 4.10). It is also interesting to note that unlike the mass, the charge of the protein complex ions does not follow the additivity principle; that is, the

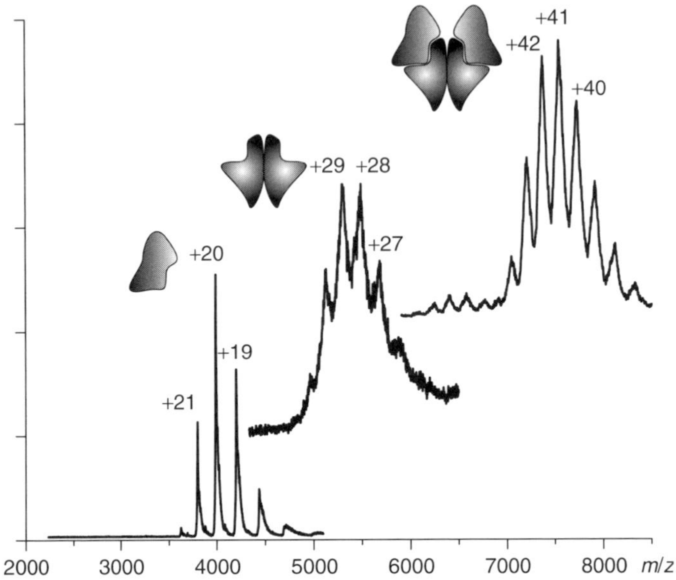

FIGURE 4.10. ESI spectra of a diferric form of human serum transferrin, Fe_2Tf (*bottom trace*), transferrin receptor dimer, TfR (*middle trace*), and a complex formed by these proteins, $(Fe_2Tf)_2TfR$ (*top trace*), acquired under near-native conditions in the authors' laboratory.

average charge of the protein complex ion is significantly less than the sum of its constituents (Figure 4.10).

A link between the protein ion charge and its size (geometry) can be expressed in terms of Rayleigh instability criteria [see Appendix (A-7)]. It could be argued that the surface of a "fission-incompetent" droplet enveloping a protein molecule is related to the solvent-accessible surface S of the latter:

$$S = 4\pi R^2 \cdot \tilde{a}, \tag{4-3-1}$$

where R is the droplet radius (or, more correctly, its characteristic linear dimension in case of nonspherical droplets) and $\tilde{a}$ is a protein-specific (or, more precisely, conformer-specific) "nonideality" parameter. This parameter should not deviate from 1 very significantly, unless the protein molecule contains extensive cavities, or else its shape departs dramatically from a spheroid (globular) form. If the entire "residual charge" Ne is accommodated by the protein upon loss of residual solvent from the droplet, one should expect to observe a power function-type dependence of the protein ion charge on the solvent-exposed surface of a corresponding conformer in solution [combining (A-7) and (4-3-1)]:

$$N = \frac{4}{e} \cdot (4\pi\gamma^2\varepsilon_0^2)^{1/4} \cdot \left(\frac{S}{\tilde{a}}\right)^{3/4} \tag{4-3-2}$$

De la Mora analyzed the maximum charge values for a variety (a data compilation from literature sources was used) of multiply charged ions of globular proteins generated by ESI under near-native conditions in comparison with the Rayleigh charge (estimated as the critical charge of a water sphere whose mass is equal to that of the protein ion) (120). For most proteins, the reported maximum charge fell within the range of 65% to 110% of the Rayleigh charge. To eliminate the possibility of protein ion charge variation caused by varying experimental parameters, we recently carried out a systematic study of the charge–geometry relationship using a set of 23 proteins, ranging from 5 kDa insulin to 0.5 MDa ferritin (Figure 4.11). The protein surface areas were calculated based on the structures available in the Protein Data Bank (121, 122), while the average charge

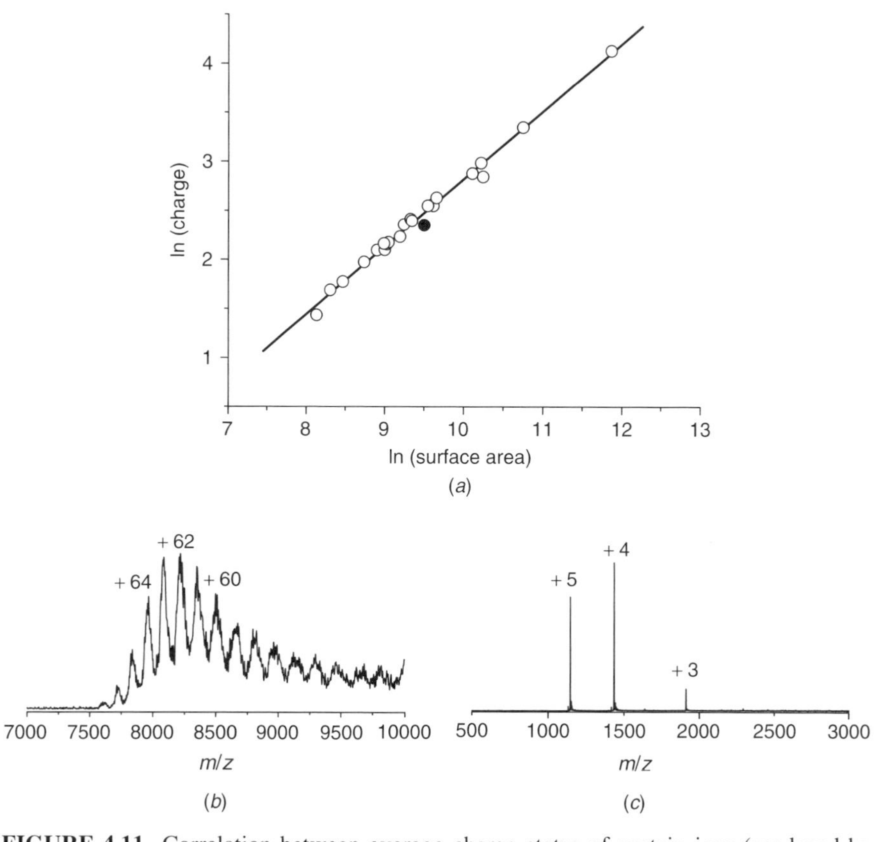

FIGURE 4.11. Correlation between average charge states of protein ions (produced by ESI under "near-native" conditions) and their surface areas calculated based on the available crystal structures (**A**). The shaded circle indicates a data point representing pepsin, an acidic protein possessing only four basic residues. The mass spectra on the right show charge state distributions of the two end-points of the "charge-surface" curve, ferritin (**B**) and insulin (**C**). All mass spectra have been acquired under identical experimental conditions.

values N were deduced from the ESI spectra acquired under identical solution (10 mM $CH_3CO_2NH_4$) and ESI interface conditions. As expected from (4-3-2), the $\ln(N)$ versus $\ln(S)$ plot appears to be linear (correlation coefficient 0.996). It is rather tempting to view this plot as a "calibration curve" that can be used to estimate interface areas in protein complexes (which become shielded from solvent and, therefore, cannot accommodate charges upon desorption from solution to the gas phase). Although the initial results of using this approach are quite encouraging,* there are reports that contradict our assertion that the protein size is a major determinant of its "charging capacity" in ESI MS (123, 124). Obviously, more experimental and theoretical work needs to be done prior to this technique becoming a routine biophysical tool.

Although the focus of this chapter is the characterization of native protein structures, the experimental methods presented here can be used to probe dynamic features of biomolecules as well. In the following two chapters we will discuss applications of both chemical cross-linking and HDX to characterize the structure of various non-native states of proteins that become populated under denaturing conditions (Chapter 5). We will also consider various uses of HDX MS to probe the transient protein states that play important roles in folding processes (Chapter 6).

REFERENCES

1. Zahn H., Meienhofer J. (1958). Reaktionen von 1,5-difluor-2,4-dinitrobenzol mit insulin. 1. Synthese von modellverbindungen. *Makromol. Chem.* **26**: 126–152.
2. Zahn H., Meienhofer J. (1958). Reaktionen von 1,5-difluor-2,4-dinitrobenzol mit insulin. 2. Mitt. versuche mit insulin. *Makromol. Chem.* **26**: 153–166.
3. Hvidt A., Linderstrøm-Lang K. (1954). Exchange of hydrogen atoms in insulin with deuterium atoms in aqueous solutions. *Biochim. Biophys. Acta* **14**: 574–575.
4. Hvidt A., Linderstrøm-Lang K. (1955). The kinetics of deuterium exchange of insulin with D_2O. An amendment. *Biochim. Biophys. Acta* **16**: 168–169.
5. Wong S. S. (1991). *Chemistry of Protein Conjugation and Cross-linking*. Boca Raton: CRC Press.
6. Brown K. C., Yang S. H., Kodadek T. (1995). Highly specific oxidative cross-linking of proteins mediated by a nickel–peptide complex. *Biochemistry* **34**: 4733–4739.
7. Brown K. C., Yu Z., Burlingame A. L., Craik C. S. (1998). Determining protein–protein interactions by oxidative cross-linking of a glycine–glycine–histidine fusion protein. *Biochemistry* **37**: 4397–4406.
8. Han K.-K., Richard C., Delacourte A. (1984). Chemical cross-links of proteins by using bifunctional reagents. *Int. J. Biochem.* **16**: 129–145.
9. Mattson G., Conklin E., Desai S., Nielander G., Savage M. D., Morgensen S. (1993). A practical approach to crosslinking. *Mol. Biol. Rep.* **17**: 167–183.

*For example, we have been able to obtain a very good estimate of the solvent-shielded surface area in a 24-meric protein ferritin.

10. Giovannangeli C., Diviacco S., Labrousse V., Gryaznov S., Charneau P., Helene C. (1997). Accessibility of nuclear DNA to triplex-forming oligonucleotides: the integrated HIV-1 provirus as a target. *Proc. Natl. Acad. Sci. U.S.A.* **94**: 79–84.

11. Garver K., Guo P. (2000). Mapping the inter-RNA interaction of bacterial virus phi29 packaging RNA by site-specific photoaffinity cross-linking. *J. Biol. Chem.* **275**: 2817–2824.

12. Mat-Arip Y., Garver K., Chen C., Sheng S., Shao Z., Guo P. (2001). Three-dimensional interaction of Phi29 pRNA dimer probed by chemical modification interference, cryo-AFM, and cross-linking. *J. Biol. Chem.* **276**: 32575–32584.

13. Gao K., Butler S. L., Bushman F. (2001). Human immunodeficiency virus type 1 integrase: arrangement of protein domains in active cDNA complexes. *EMBO J.* **20**: 3565–3576.

14. Verdine G. L., Norman D. P. (2003). Covalent trapping of protein–DNA complexes. *Annu. Rev. Biochem.* **72**: 337–366.

15. Green N. S., Reisler E., Houk K. N. (2001). Quantitative evaluation of the lengths of homobifunctional protein cross-linking reagents used as molecular rulers. *Protein Sci.* **10**: 1293–1304.

16. Rappsilber J., Siniossoglou S., Hurt E. C., Mann M. (2000). A generic strategy to analyze the spatial organization of multi-protein complexes by cross-linking and mass spectrometry. *Anal. Chem.* **72**: 267–275.

17. Back J. W., de Jong L., Muijsers A. O., de Koster C. G. (2003). Chemical cross-linking and mass spectrometry for protein structural modeling. *J. Mol. Biol.* **331**: 303–313.

18. Bossi A., Patel M. J., Webb E. J., Baldwin M. A., Jacob R. J., Burlingame A. L., Righetti P. G. (1999). Analysis of cross-linked human hemoglobin by conventional isoelectric focusing, immobilized pH gradients, capillary electrophoresis, and mass spectrometry. *Electrophoresis* **20**: 2810–2817.

19. Yang T., Horejsh D. R., Mahan K. J., Zaluzec E. J., Watson T. J., Gage D. A. (1996). Mapping cross-linking sites in modified proteins with mass spectrometry: an application to cross-linked hemoglobins. *Anal. Biochem.* **242**: 55–63.

20. Bennett K. L., Kussmann M., Bjork P., Godzwon M., Mikkelsen M., Sorensen P., Roepstorff P. (2000). Chemical cross-linking with thiol-cleavable reagents combined with differential mass spectrometric peptide mapping—a novel approach to assess intermolecular protein contacts. *Protein Sci.* **9**: 1503–1518.

21. Sinz A., Wang K. (2001). Mapping protein interfaces with a fluorogenic cross-linker and mass spectrometry: application to nebulin–calmodulin complexes. *Biochemistry* **40**: 7903–7913.

22. Yu Z., Friso G., Miranda J. J., Patel M. J., Lo-Tseng T., Moore E. G., Burlingame A. L. (1997). Structural characterization of human hemoglobin crosslinked by bis(3,5-dibromosalicyl) fumarate using mass spectrometric techniques. *Protein Sci.* **6**: 2568–2577.

23. Young M. M., Tang N., Hempel J. C., Oshiro C. M., Taylor E. W., Kuntz I. D., Gibson B. W., Dollinger G. (2000). High throughput protein fold identification by using experimental constraints derived from intramolecular cross-links and mass spectrometry. *Proc. Natl. Acad. Sci. U.S.A.* **97**: 5802–5806.

24. Schilling B., Row R. H., Gibson B. W., Guo X., Young M. M. (2003). MS2Assign, automated assignment and nomenclature of tandem mass spectra of chemically crosslinked peptides. *J. Am. Soc. Mass Spectrom.* **14**: 834–850.
25. Biemann K. (1990). Appendix 5. Nomenclature for peptide fragment ions (positive ions). *Methods Enzymol.* **193**: 886–887.
26. Ngoka L. C., Gross M. L. (1999). A nomenclature system for labeling cyclic peptide fragments. *J. Am. Soc. Mass Spectrom.* **10**: 360–363.
27. Chen X., Chen Y. H., Anderson V. E. (1999). Protein cross-links: universal isolation and characterization by isotopic derivatization and electrospray ionization mass spectrometry. *Anal. Biochem.* **273**: 192–203.
28. Müller D. R., Schindler P., Towbin H., Wirth U., Voshol H., Hoving S., Steinmetz M. O. (2001). Isotope-tagged cross-linking reagents. A new tool in mass spectrometric protein interaction analysis. *Anal. Chem.* **73**: 1927–1934.
29. Back J. W., Hartog A. F., Dekker H. L., Muijsers A. O., de Koning L. J., de Jong L. (2001). A new crosslinker for mass spectrometric analysis of the quaternary structure of protein complexes. *J. Am. Soc. Mass Spectrom.* **12**: 222–227.
30. Winters M. S., Day R. A. (2003). Detecting protein–protein interactions in the intact cell of *Bacillus subtilis* (ATCC 6633). *J. Bacteriol.* **185**: 4268–4275.
31. Govorun V. M., Archakov A. I. (2002). Proteomic technologies in modern biomedical science. *Biochemistry (Moscow)* **67**: 1109–1123.
32. Trester-Zedlitz M., Kamada K., Burley S. K., Fenyo D., Chait B. T., Muir T. W. (2003). A modular cross-linking approach for exploring protein interactions. *J. Am. Chem. Soc.* **125**: 2416–2425.
33. Chen T., Jaffe J. D., Church G. M. (2001). Algorithms for identifying protein cross-links via tandem mass spectrometry. *J. Comput. Biol.* **8**: 571–583.
34. Dihazi G. H., Sinz A. (2003). Mapping low-resolution three-dimensional protein structures using chemical cross-linking and Fourier transform ion-cyclotron resonance mass spectrometry. *Rapid Commun. Mass Spectrom.* **17**: 2005–2014.
35. Person M. D., Brown K. C., Mahrus S., Craik C. S., Burlingame A. L. (2001). Novel inter-protein cross-link identified in the GGH-ecotin D137Y dimer. *Protein Sci.* **10**: 1549–1562.
36. Kruppa G. H., Schoeniger J., Young M. M. (2003). A top down approach to protein structural studies using chemical cross-linking and Fourier transform mass spectrometry. *Rapid Commun. Mass Spectrom.* **17**: 155–162.
37. Schonbrun J., Wedemeyer W. J., Baker D. (2002). Protein structure prediction in 2002. *Curr. Opin. Struct. Biol.* **12**: 348–354.
38. Burley S. K., Bonanno J. B. (2002). Structuring the universe of proteins. *Annu. Rev. Genom. Human Genet.* **3**: 243–262.
39. Chance M. R., Bresnick A. R., Burley S. K., Jiang J.-S., Lima C. D., Sali A., Almo S. C., Bonanno J. B., Buglino J. A., Boulton S., Chen H., Eswar N., He G., Huang R., Ilyin V., McMahan L., Pieper U., Ray S., Vidal M., Wang L. K. (2002). Structural genomics: a pipeline for providing structures for the biologist. *Protein Sci.* **11**: 723–738.
40. Goldsmith-Fischman S., Honig B. (2003). Structural genomics: computational methods for structure analysis. *Protein Sci.* **12**: 1813–1821.

41. Nitao L. K., Loo R. R., O'Neall-Hennessey E., Loo J. A., Szent-Gyorgyi A. G., Reisler E. (2003). Conformation and dynamics of the SH1-SH2 helix in scallop myosin. *Biochemistry* **42**: 7663–7674.

42. Glazer A. N. (1970). Specific chemical modification of proteins. *Annu. Rev. Biochem.* **39**: 101–130.

43. Gundlach H. G., Stein W. H., Moore S. (1959). The nature of the amino acid residues involved in the inactivation of ribonuclease by iodoacetate. *J. Biol. Chem.* **234**: 1754–1760.

44. Shaw D. C., Stein W. H., Moore S. (1964). Inactivation of chymotrypsin by cyanate. *J. Biol. Chem.* **239**: 671–673.

45. Rajagopalan T. G., Stein W. H., Moore S. (1966). The inactivation of pepsin by diazoacetylnorleucine methyl ester. *J. Biol. Chem.* **241**: 4295–4297.

46. Suckau D., Mak M., Przybylski M. (1992). Protein surface topology-probing by selective chemical modification and mass spectrometric peptide mapping. *Proc. Natl. Acad. Sci. U.S.A.* **89**: 5630–5634.

47. D'Ambrosio C., Talamo F., Vitale R. M., Amodeo P., Tell G., Ferry L., Scaloni A. (2003). Probing the dimeric structure of porcine aminoacylase 1 by mass spectrometric and modeling procedures. *Biochemistry* **42**: 4430–4443.

48. Happersberger H. P., Przybylski M., Glocker M. O. (1998). Selective bridging of bis-cysteinyl residues by arsonous acid derivatives as an approach to the characterization of protein tertiary structures and folding pathways by mass spectrometry. *Anal. Biochem.* **264**: 237–250.

49. Glazer A. N. (1976). The chemical modification of proteins by group-specific and site-specific reagents. In: *The Proteins*, ed. C. L. Boeder, pp. 1–103. New York: Academic Press.

50. Toi K., Bynum E., Norris E., Itano H. A. (1967). Studies on the chemical modification of arginine. I. The reaction of 1,2-cyclohexanedione with arginine and arginyl residues of proteins. *J. Biol. Chem.* **242**: 1036–1043.

51. Toi K., Bynum E., Norris E., Itano H. A. (1965). Chemical modification of arginine with 1,2-cyclohexanedione. *J. Biol. Chem.* **240**: PC3455–3457.

52. Calvete J. J., Campanero-Rhodes M. A., Raida M., Sanz L. (1999). Characterisation of the conformational and quaternary structure-dependent heparin-binding region of bovine seminal plasma protein PDC-109. *FEBS Lett.* **444**: 260–264.

53. Riordan J. F., Sokolovsky M., Vallee B. L. (1967). Environmentally sensitive tyrosyl residues. Nitration with tetranitromethane. *Biochemistry* **6**: 358–361.

54. Sokolovsky M., Riordan J. F., Vallee B. L. (1966). Tetranitromethane. A reagent for the nitration of tyrosyl residues in proteins. *Biochemistry* **5**: 3582–3589.

55. Mann M., Jensen O. N. (2003). Proteomic analysis of post-translational modifications. *Nat. Biotechnol.* **21**: 255–261.

56. Hager-Braun C., Tomer K. B. (2002). Characterization of the tertiary structure of soluble CD4 bound to glycosylated full-length HIVgp120 by chemical modification of arginine residues and mass spectrometric analysis. *Biochemistry* **41**: 1759–1766.

57. Hochleitner E. O., Borchers C., Parker C., Bienstock R. J., Tomer K. B. (2000). Characterization of a discontinuous epitope of the human immunodeficiency virus (HIV) core protein p24 by epitope excision and differential chemical modification followed by mass spectrometric peptide mapping analysis. *Protein Sci.* **9**: 487–496.

58. Leite J. F., Cascio M. (2002). Probing the topology of the glycine receptor by chemical modification coupled to mass spectrometry. *Biochemistry* **41**: 6140–6148.
59. Sharp J. S., Becker J. M., Hettich R. L. (2003). Protein surface mapping by chemical oxidation: structural analysis by mass spectrometry. *Anal. Biochem.* **313**: 216–225.
60. Petersson A. S., Steen H., Kalume D. E., Caidahl K., Roepstorff P. (2001). Investigation of tyrosine nitration in proteins by mass spectrometry. *J. Mass Spectrom.* **36**: 616–625.
61. Zappacosta F., Ingallinella P., Scaloni A., Pessi A., Bianchi E., Sollazzo M., Tramontano A., Marino G., Pucci P. (1997). Surface topology of Minibody by selective chemical modifications and mass spectrometry. *Protein Sci.* **6**: 1901–1909.
62. Avaeva S. M., Krasnova V. I. (1975). Diethylpyrocarbonate reaction with imidazole and histidine derivatives. *Bioorg. Khim.* **1**: 1600–1605.
63. Scaloni A., Ferranti P., De Simone G., Mamone G., Sannolo N., Malorni A. (1999). Probing the reactivity of nucleophile residues in human 2,3-diphosphoglycerate/deoxy-hemoglobin complex by a specific chemical modifications. *FEBS Lett.* **452**: 190–194.
64. Richards F., Lamed R., Wynn R., Patel D., Olack G. (2000). Methylene as a possible universal footprinting reagent that will include hydrophobic surface areas: overview and feasibility: properties of diazirine as a precursor. *Protein Sci.* **9**: 2506–2517.
65. Craig P. O., Ureta D. B., Delfino J. M. (2002). Probing protein conformation with a minimal photochemical reagent. *Protein Sci.* **11**: 1353–1366.
66. Lim J., Vachet R. W. (2003). Development of a methodology based on metal-catalyzed oxidation reactions and mass spectrometry to determine the metal binding sites in copper metalloproteins. *Anal. Chem.* **75**: 1164–1172.
67. Maleknia S. D., Chance M. R., Downard K. M. (1999). Electrospray-assisted modification of proteins: a radical probe of protein structure. *Rapid Commun. Mass Spectrom.* **13**: 2352–2358.
68. Goshe M. B., Chen Y. H., Anderson V. E. (2000). Identification of the sites of hydroxyl radical reaction with peptides by hydrogen/deuterium exchange: prevalence of reactions with the side chains. *Biochemistry* **39**: 1761–1770.
69. Kiselar J. G., Maleknia S. D., Sullivan M., Downard K. M., Chance M. R. (2002). Hydroxyl radical probe of protein surfaces using synchrotron X-ray radiolysis and mass spectrometry. *Int. J. Radiat. Biol.* **78**: 101–114.
70. Xu G., Takamoto K., Chance M. R. (2003). Radiolytic modification of basic amino acid residues in peptides: probes for examining protein–protein interactions. *Anal. Chem.* **75**: 6995–7007.
71. Heyduk T., Baichoo N., Heyduk E. (2001). Hydroxyl radical footprinting of proteins using metal ion complexes. *Met. Ions Biol. Syst.* **38**: 255–287.
72. Maleknia S. D., Brenowitz M., Chance M. R. (1999). Millisecond radiolytic modification of peptides by synchrotron X-rays identified by mass spectrometry. *Anal. Chem.* **71**: 3965–3973.
73. Englander S. W., Mayne L., Bai Y., Sosnick T. R. (1997). Hydrogen exchange: the modern legacy of Linderstrøm-Lang. *Protein Sci.* **6**: 1101–1109.
74. Urey H. C., Brickedde F. G., Murphy G. M. (1932). A hydrogen isotope of mass 2. *Phys. Rev.* **39**: 164–165.

75. Lewis G. N., Macdonald R. T. (1933). Some properties of pure H^2H^2O. *J. Am. Chem. Soc.* **55**: 3057–3059.

76. Bonhoeffer K. F., Klar R. (1934). Uber den austausch von schweren wasserstoffatomen zwischen wasser und organischen verbindungen. *Naturwissenschaften* **22**: 45.

77. Champetier G., Viallard R. (1938). Reaction d'echange de la cellulose et de l'eau lourde. Hydratation de la cellulose. *Bull. Soc. Chim.* **33**: 1042–1048.

78. Burley R. W., Nicholls C. H., Speakman J. B. (1955). The crystalline/amorphous ratio of keratin fibres. Part II. The hydrogen–deuterium exchange reaction. *J. Text. Inst.* **46**: T427–T432.

79. Sethi S. K., Smith D. L., McCloskey J. A. (1983). Determination of active hydrogen content by fast atom bombardment mass spectrometry following hydrogen–deuterium exchange. *Biochem. Biophys. Res. Commun.* **112**: 126–131.

80. Katta V., Chait B. T. (1991). Conformational changes in proteins probed by hydrogen-exchange electrospray-ionization mass spectrometry. *Rapid Commun. Mass Spectrom.* **5**: 214–217.

81. Ehring H. (1999). Hydrogen exchange/electrospray ionization mass spectrometry studies of structural features of proteins and protein/protein interactions. *Anal. Biochem.* **267**: 252–259.

82. Mandell J. G., Falick A. M., Komives E. A. (1998). Identification of protein–protein interfaces by decreased amide proton solvent accessibility. *Proc. Natl. Acad. Sci. U.S.A.* **95**: 14705–14710.

83. Akashi S., Takio K. (2000). Characterization of the interface structure of enzyme–inhibitor complex by using hydrogen–deuterium exchange and electrospray ionization Fourier transform ion cyclotron resonance mass spectrometry. *Protein Sci.* **9**: 2497–2505.

84. Mandell J. G., Baerga-Ortiz A., Akashi S., Takio K., Komives E. A. (2001). Solvent accessibility of the thrombin–thrombomodulin interface. *J. Mol. Biol.* **306**: 575–589.

85. Yamada N., Suzuki E., Hirayama K. (2002). Identification of the interface of a large protein–protein complex using H/D exchange and Fourier transform ion cyclotron resonance mass spectrometry. *Rapid Commun. Mass Spectrom.* **16**: 293–299.

86. Lam T. T., Lanman J. K., Emmett M. R., Hendrickson C. L., Marshall A. G., Prevelige P. E. (2002). Mapping of protein–protein contact surfaces by hydrogen/deuterium exchange, followed by on-line high-performance liquid chromatography–electrospray ionization Fourier transform ion cyclotron resonance mass analysis. *J. Chromatogr. A* **982**: 85–95.

87. Lanman J., Lam T. T., Barnes S., Sakalian M., Emmett M. R., Marshall A. G., Prevelige P. E. (2003). Identification of novel interactions in HIV-1 capsid protein assembly by high-resolution mass spectrometry. *J. Mol. Biol.* **325**: 759–772.

88. Engen J. R. (2003). Analysis of protein complexes with hydrogen exchange and mass spectrometry. *Analyst* **128**: 623–628.

89. Anand G. S., Law D., Mandell J. G., Snead A. N., Tsigelny I., Taylor S. S., Eyck L. F. T., Komives E. A. (2003). Identification of the protein kinase A regulatory RI α-catalytic subunit interface by amide $H/^2H$ exchange and protein docking. *Proc. Natl. Acad. Sci. U.S.A.* **100**: 13264–13269.

90. Nazabal A., Laguerre M., Schmitter J.-M., Vaillier J., Chaignepain S. T., Velours J. (2003). Hydrogen/deuterium exchange on yeast ATPase supramolecular protein

complex analyzed at high sensitivity by MALDI mass spectrometry. *J. Am. Soc. Mass Spectrom.* **14**: 471–481.

91. Nemirovskiy O., Giblin D. E., Gross M. L. (1999). Electrospray ionization mass spectrometry and hydrogen/deuterium exchange for probing the interaction of calmodulin with calcium. *J. Am. Soc. Mass Spectrom.* **10**: 711–718.
92. Dempsey C. E. (2001). Hydrogen exchange in peptides and proteins using NMR-spectroscopy. *Progr. Nucl. Magn. Res. Spectrosc.* **39**: 135–170.
93. Engen J. R., Smith D. L. (2001). Investigating protein structure and dynamics by hydrogen exchange MS. *Anal. Chem.* **73**: 256A–265A.
94. Smith D. L., Deng Y., Zhang Z. (1997). Probing noncovalent structure of proteins by amide hydrogen exchange and mass spectrometry. *J. Mass Spectrom.* **32**: 135–146.
95. Kaltashov I. A., Eyles S. J. (2002). Crossing the phase boundary to study protein dynamics and function: combination of amide hydrogen exchange in solution and ion fragmentation in the gas phase. *J. Mass Spectrom.* **37**: 557–565.
96. Mattos C. (2002). Protein–water interactions in a dynamic world. *Trends Biochem. Sci.* **27**: 203–208.
97. Breuker K., Oh H. B., Horn D. M., Cerda B. A., McLafferty F. W. (2002). Detailed unfolding and folding of gaseous ubiquitin ions characterized by electron capture dissociation. *J. Am. Chem. Soc.* **124**: 6407–6420.
98. Oh H., Breuker K., Sze S. K., Ge Y., Carpenter B. K., McLafferty F. W. (2002). Secondary and tertiary structures of gaseous protein ions characterized by electron capture dissociation mass spectrometry and photofragment spectroscopy. *Proc. Natl. Acad. Sci. U.S.A.* **99**: 15863–15868.
99. Weinkauf R., Schermann J. P., de Vries M. S., Kleinermanns K. (2002). Molecular physics of building blocks of life under isolated or defined conditions. *Eur. Phys. J. D* **20**: 309–316.
100. Hoaglund-Hyzer C. S., Counterman A. E., Clemmer D. E. (1999). Anhydrous protein ions. *Chem. Rev.* **99**: 3037–3080.
101. Jarrold M. F. (2000). Peptides and proteins in the vapor phase. *Annu. Rev. Phys. Chem.* **51**: 179–207.
102. Veenstra T. D. (1999). Electrospray ionization mass spectrometry in the study of biomolecular noncovalent interactions. *Biophys. Chem.* **79**: 63–79.
103. Loo J. A. (2000). Electrospray ionization mass spectrometry: a technology for studying noncovalent macromolecular complexes. *Int. J. Mass Spectrom.* **200**: 175–186.
104. Lei Q. P., Cui X., Kurtz D. M. Jr., Amster I. J., Chernushevich I. V., Standing K. G. (1998). Electrospray mass spectrometry studies of non-heme iron-containing proteins. *Anal. Chem.* **70**: 1838–1846.
105. Rostom A. A., Robinson C. V. (1999). Disassembly of intact multiprotein complexes in the gas phase. *Curr. Opin. Struct. Biol.* **9**: 135–141.
106. Rostom A. A., Fucini P., Benjamin D. R., Juenemann R., Nierhaus K. H., Hartl F. U., Dobson C. M., Robinson C. V. (2000). Detection and selective dissociation of intact ribosomes in a mass spectrometer. *Proc. Natl. Acad. Sci. U.S.A.* **97**: 5185–5190.

107. Tito M. A., Miller J., Walker N., Griffin K. F., Williamson E. D., Despeyroux-Hill D., Titball R. W., Robinson C. V. (2001). Probing molecular interactions in intact antibody: antigen complexes, an electrospray time-of-flight mass spectrometry approach. *Biophys. J.* **81**: 3503–3509.

108. Tito M. A., Miller J., Griffin K. F., Williamson E. D., Titball R. W., Robinson C. V. (2001). Macromolecular organization of the *Yersinia pestis* capsular F1 antigen: insights from time-of-flight mass spectrometry. *Protein Sci.* **10**: 2408–2413.

109. Pinkse M. W., Maier C. S., Kim J. I., Oh B. H., Heck A. J. (2003). Macromolecular assembly of *Helicobacter pylori* urease investigated by mass spectrometry. *J. Mass Spectrom.* **38**: 315–320.

110. Loo J. A. (1997). Studying noncovalent protein complexes by electrospray ionization mass spectrometry. *Mass Spectrom. Rev.* **16**: 1–23.

111. Glocker M. O., Bauer S. H. J., Kast J., Volz J., Przybylski M. (1996). Characterization of specific noncovalent protein complexes by UV matrix-assisted laser desorption ionization mass spectrometry. *J. Mass Spectrom.* **31**: 1221–1227.

112. Moniatte M., van der Goot F. G., Buckley J. T., Pattus F., van Dorsselaer A. (1996). Characterisation of the heptameric pore-forming complex of the Aeromonas toxin aerolysin using MALDI-TOF mass spectrometry. *FEBS Lett.* **384**: 269–272.

113. Moniatte M., Lesieur C., Vecsey-Semjen B., Buckley J. T., Pattus F., van der Goot F. G., Van Dorsselaer A. (1997). Matrix-assisted laser desorption-ionization time-of-flight mass spectrometry in the subunit stoichiometry study of high-mass non-covalent complexes. *Int. J. Mass Spectrom. Ion Proc.* **169–170**: 179–199.

114. Cohen L. R. H., Strupat K., Hillenkamp F. (1997). Analysis of quaternary protein ensembles by matrix assisted laser desorption/ionization mass spectrometry. *J. Am. Soc. Mass Spectrom.* **8**: 1046–1052.

115. Lin S. H., Cotter R. J., Woods A. S. (1998). Detection of non-covalent interaction of single and double stranded DNA with peptides by MALDI-TOF. *Proteins* **S2**: 12–21.

116. Farmer T. B., Caprioli R. M. (1998). Determination of protein–protein interactions by matrix-assisted laser desorption/ionization mass spectrometry. *J. Mass Spectrom.* **33**: 697–704.

117. Kiselar J. G., Downard K. M. (2000). Preservation and detection of specific antibody–peptide complexes by matrix-assisted laser desorption ionization mass spectrometry. *J. Am. Soc. Mass Spectrom.* **11**: 746–750.

118. Woods A. S., Huestis M. A. (2001). A study of peptide–peptide interaction by matrix-assisted laser desorption/ionization. *J. Am. Soc. Mass Spectrom.* **12**: 88–96.

119. Zehl M., Allmaier G. (2003). Investigation of sample preparation and instrumental parameters in the matrix-assisted laser desorption/ionization time-of-flight mass spectrometry of noncovalent peptide/peptide complexes. *Rapid Commun. Mass Spectrom.* **17**: 1931–1940.

120. de la Mora J. F. (2000). Electrospray ionization of large multiply charged species proceeds via Dole's charged residue mechanism. *Analyt. Chim. Acta* **406**: 93–104.

121. Berman H. M., Battistuz T., Bhat T. N., Bluhm W. F., Bourne P. E., Burkhardt K., Feng Z., Gilliland G. L., Iype L., Jain S., Fagan P., Marvin J., Padilla D., Ravichandran V., Schneider B., Thanki N., Weissig H., Westbrook J. D., Zardecki C. (2002). The protein data bank. *Acta Crystallogr. D Biol. Crystallogr.* **58**: 899–907.

122. Berman H. M., Westbrook J., Feng Z., Gilliland G., Bhat T. N., Weissig H., Shindyalov I. N., Bourne P. E. (2000). The Protein Data Bank. *Nucleic Acids Res.* **28**: 235–242.

123. Grandori R. (2003). Origin of the conformation dependence of protein charge-state distributions in electrospray ionization mass spectrometry. *J. Mass Spectrom.* **38**: 11–15.

124. Samalikova M., Grandori R. (2003). Role of opposite charges in protein electrospray ionization mass spectrometry. *J. Mass Spectrom.* **38**: 941–947.

125. Tahallah N., Van Den Heuvel R. H., Van Den Berg W. A., Maier C. S., Van Berkel W. J., Heck A. J. (2002). Cofactor-dependent assembly of the flavoenzyme vanillyl-alcohol oxidase. *J. Biol. Chem.* **277**: 36425–36432.

5

MASS SPECTROMETRY-BASED APPROACHES TO STUDY BIOMOLECULAR DYNAMICS: EQUILIBRIUM INTERMEDIATES

In the preceding chapter, we surveyed various MS-based approaches to study higher-order structure of proteins under native conditions. For many decades, such well-defined and highly organized structures were thought of as the most important (if not the only) determinants of protein function. Protein folding was often considered a linear process leading from fully unstructured (and, therefore, dysfunctional) states to the highly organized native (function-competent) state. The advent of NMR has changed our perception of what "functional" protein states are, with the realization that native proteins are very dynamic species. Perhaps the most illustrious examples of the intimate link between protein dynamics and function were found in enzyme catalysis, where the chemical conversion of substrate to product is often driven by relatively small-scale dynamic events within (and often beyond) the active site. It became clear in recent years that large-scale macromolecular dynamics may also be an important determinant of protein function. A growing number of proteins are found to be either partially or fully unstructured under native conditions, and such flexibility (intrinsic disorder) appears to be vital for their function. Proteins that do have native folds under physiological conditions can also exhibit dynamic behavior via local structural fluctuations or by sampling alternative (higher-energy or "activated") conformations transiently. In many cases, such activated (non-native) states are functionally important despite their low Boltzmann weight. Realization of the importance of transient non-native protein structures for their function not only greatly advanced our understanding of processes as diverse as recognition, signaling, and transport

Mass Spectrometry in Biophysics: Conformation and Dynamics of Biomolecules
By Igor A. Kaltashov and Stephen J. Eyles
ISBN 0-471-45602-0

but has also had profound practical implications, particularly for the design of drugs targeting specific proteins. Because of their transient nature, these non-native states present a great challenge vis-à-vis detection and characterization. This chapter presents a concise introduction to an array of techniques that are used to study structure and behavior of the so-called "equilibrium intermediate states." We begin our discussion by considering protein ion charge state distributions in ESI mass spectra as indicators of protein unfolding. We then proceed to various "trapping" techniques that exploit protein reactivity to reveal structural details of various non-native states. We conclude the chapter with a detailed discussion of hydrogen exchange, arguably one of the most widely used methods to probe the structure and dynamics of non-native (partially unstructured) protein states.

5.1. MONITORING EQUILIBRIUM INTERMEDIATES: PROTEIN ION CHARGE STATE DISTRIBUTIONS (ESI MS)

It was not long after the introduction of ESI MS that the potential of this technique to probe higher-order structure of proteins was realized. Chait and co-workers (1) and later Loo and co-workers (2) observed dramatic changes in the charge state distributions of protein ions resulting from the changes in solvent composition that were known to induce protein denaturation. It has been noted that "a protein in a tightly folded conformation will have fewer basic sites available for protonation compared to the same protein in an unfolded conformation" (1), hence the decreased number of charges on ions representing "native" proteins (Figure 5.1). A more general explanation of this phenomenon was later provided within a framework of Fenn's modification of the ion evaporation model (IEM) of electrospray ionization (3), although the charged residue model (CRM) can be invoked as well (refer to the Appendix for a detailed discussion of these models of ion formation in ESI). In general, two macromolecular conformations whose overall shapes [or, more precisely, characteristic dimensions understood as an average cross section (IEM) or either surface area or gyration radius (CRM)] are very different from each other will produce ions with very significant differences in charge density (number of elementary charges per protein molecule). If both states are populated in solution under certain conditions (so-called equilibrium states), the resultant ESI spectrum is expected to exhibit a bimodal character.

Soon after these initial reports (1, 2), Fenselau and co-workers demonstrated that protein ion charge state distributions also provide information on protein conformational changes in solution induced by cofactors (such as metal ions) under near-native conditions (4). In these experiments, addition of divalent metals (Zn^{2+} or Cd^{2+}) to a solution of a small protein apo-metallothionein not only resulted in increases of the measured masses of the protein ions [consistent with binding of a requisite number (up to seven) of metal cations to the protein] but also induced noticeable shifts of their charge state distributions. The latter were interpreted as a manifestation of formation of a compact protein structure cemented by coordination of the metal ions by distant cysteine residues (4). The ability of

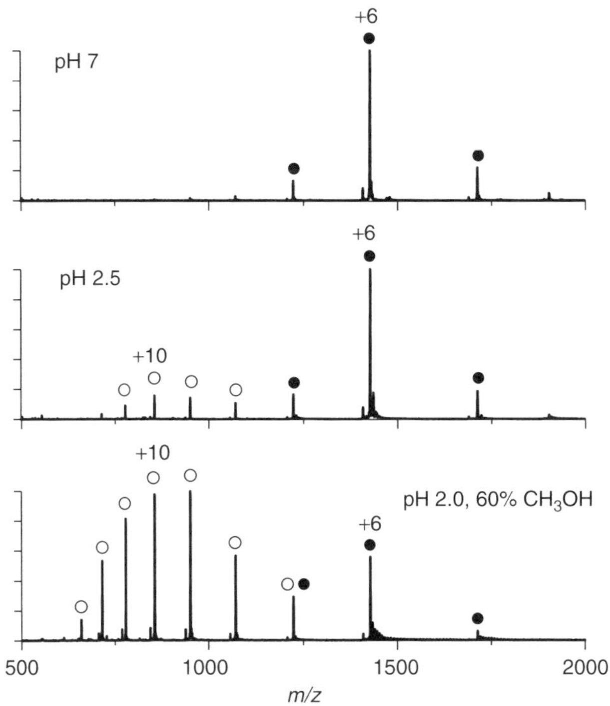

FIGURE 5.1. ESI mass spectra of ubiquitin acquired under near-native (*top trace*, 10 mM $CH_3CO_2NH_4$, pH adjusted to 7) and denaturing conditions (*middle trace*, 10 mM $CH_3CO_2NH_4$, pH adjusted to 2.5 with CH_3CO_2H; *bottom trace*, low ionic strength, pH 2.0, 60% methanol). Protein ions corresponding to the "native" and "non-native" (denatured) states of the protein are labeled with solid and open circles, respectively.

such relatively simple measurements to probe conformational changes in proteins has been utilized extensively in subsequent years to study protein unfolding and conformational heterogeneity, as well as to monitor large-scale conformational changes within proteins in solution. Tang and Wang employed this technique to monitor conformational stability of the apo- and holo-forms of myoglobin as a function of solution pH, ionic strength, and alcohol content (5). Konermann and Douglas used a similar approach to study the equilibrium alcohol- and acid-induced unfolding of several other small proteins (6–8). Further development of this technique by Konermann and co-workers utilizes time-resolved measurements, which allow kinetic studies of protein refolding/unfolding reactions to be carried out (9–12). (Such experiments will be discussed in more detail in Chapter 6.) Several groups have utilized this approach to study the effect of metal ion binding in solution on protein conformational stability (13–16). Likewise, protein ion charge state distributions provide an effective means of monitoring protein deactivation induced by a variety of extraneous factors (17, 18), as well as the influence of mutations on protein stability (19). The influence of conformational heterogeneity on protein–protein (20) and protein–DNA (21) binding can also be evaluated by monitoring protein ion charge state distributions. Changes in the protein ion charge state distributions can also be used to detect, with high sensitivity, the loosening of the protein tertiary structure in solution following reduction of disulfide bonds (22).

Most proteins are known to have multiple equilibrium states, not just folded and unfolded conformations. Although in some cases the existence of at least one intermediate* state can be inferred from the charge state distributions [e.g., see (21)], it is possible to have multiple intermediate states coexisting under equilibrium, some of which may have only minor structural differences. In most cases, such subtle conformational changes do not lead to significant variations in the overall shape of the protein and, therefore, in the average number of charges accommodated by the protein ion upon its desorption from solution. As a result, the charge state distributions corresponding to various protein conformational isomers may be unresolved or poorly resolved (i.e., two or more different conformers may give rise to ions carrying the same number of charges). An important exception is the native conformation, whose ionic signal in most cases can be distinguished with relative ease from that of non-native states. Therefore, changes in the protein ion charge state distributions have traditionally been regarded as qualitative indicators of re- or denaturation that cannot provide much information beyond loss or gain of the "native" (i.e., compact) fold.

Recently, we introduced a procedure that utilizes chemometric tools to extract semiquantitative data on multiple protein conformational isomers coexisting in solution under equilibrium (23, 24). Experiments are carried out by acquiring an array of spectra over a range of both near-native and denaturing conditions to ensure adequate sampling of various protein states and significant variation of their populations within the range of experimental conditions. The total number of protein conformers sampled in the course of the experiment can be determined by subjecting the set of collected spectra to singular value decomposition (SVD). The ionic contributions of each conformer to the total signal can then be determined by using a supervised minimization routine (Figure 5.2). Application of this method to several small model proteins yielded a picture of protein behavior consistent with that based on the results of earlier studies that utilized a variety of biophysical techniques (24). One needs to be aware, however, that the protein shape (compactness) is not the only factor that determines the appearance of charge state distributions in the ESI mass spectra. Any factor that affects either the desorption process or the gas phase ion chemistry may have a profound effect on the appearance of the spectra. Solvent composition is perhaps the most important parameter that in many cases can affect protein charge state distributions by either altering the conditions of ion desorption (e.g., by altering solvent dielectric constant or surface tension) (25, 26) or supplying proton-transfer reagents to the gas phase (25, 27). Changes in the solvent composition may also indirectly affect charge state distributions, for example, via formation of protein–anion adducts

*Here we use the term "intermediate state" as a general descriptor of any non-native partially structured protein conformation.

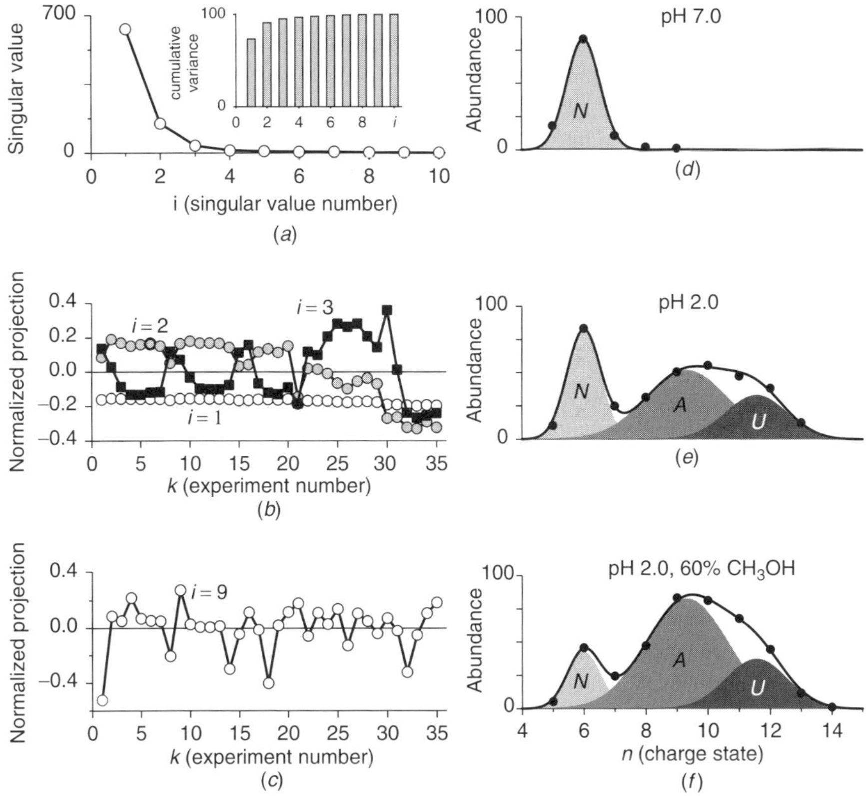

FIGURE 5.2. Factor analysis of an array of ESI mass spectra recorded at various stages of acid- and alcohol-induced unfolding of ubiquitin. The total number of protein states is determined by a singular value decomposition of the entire data matrix of dimension $N \times K$ (charge state $\times$ experiment number). While SVD produces a total of ten singular values, only the first three are required to account for 95% of variation exhibited in the spectra within the entire range of conditions (**A**). "Abstract" solutions corresponding to these three "significant" singular values show consistent behavior throughout the entire range of conditions (**B**), while the "insignificant" singular values randomly oscillate around zero (an example is shown on panel **C**). Contributions of the three protein states (assigned as ***N***ative, ***A***-state and fully ***U***nfolded) to the total ion current are determined using supervised optimization (three examples are shown on panels **D**–**F**). Adapted with permission from (24). © 2003 American Chemical Society.

in solution that dissociate in the gas phase without charge partitioning, resulting in apparent charge state reduction (28). Therefore, care needs to be exercised in order to avoid overinterpretation of any experimental data based on the measurements of the protein ion charge state distributions. In many cases, it is possible to select a range of solution conditions that do not introduce any variation in the protein ion charge state distributions other than those related to shifts of equilibria among various protein conformers.

Several groups have used ionic signals corresponding to various protein states as a means to track the progress of unfolding/refolding processes in a quantitative way (14, 23, 24, 29, 30). It is intuitively clear that an increase of the ionic contribution of a certain conformer to the total protein ion current should be indicative of an increase of the relative population of this state in solution. However, it would probably be a mistake to simply equate the conformer's relative ionic signal to its fractional concentration in solution. Two general models have been developed in recent years by Kebarle and Ho (31) and Enke and co-workers (32) in an attempt to predict ESI response for smaller analytes based on their physical characteristics, such as surface activity and solvation energy (31), as well as hydrophobicity and charge density (32). (A more detailed discussion of these models is presented in the Appendix.)

More recently, McLuckey demonstrated that ESI responses were similar within a set of structurally unrelated proteins after charge normalization in the low-concentration limit ($\leq$10 μM) (33). Only water/alcohol mixtures at low pH were used as cosolvents to avoid potential solubility problems. No significant surface activity effect was noted in this work. In light of these results, it would be reasonable to expect that the difference in ESI response between two conformers of the same protein will also be correlated primarily with the average charge carried by the corresponding ionic species in the gas phase. If this is correct, fractional concentrations of all equilibrium states populated under certain conditions can be found simply by charge normalization of their respective ionic signals. However, one should be aware of potential complications associated with such straightforward calculation of fractional concentrations of various conformers. It seems likely that in many cases there will be discrimination against the less compact species (represented by ions at higher charge states). Indeed, since the proton affinity of multiply charged peptide and protein ions reduces dramatically as the charge state increases (27, 34, 35), the higher charged ions will be more prone to proton-transfer reactions in the gas phase.

In the previous chapter we discussed the "charge–surface" relationship for protein ions representing natively folded proteins in ESI mass spectra. It remains to be seen if this simple relationship can be used to provide quantitative information on the solvent-exposed surface areas of (partially) unstructured states of proteins in solution as well. As we have already discussed, processing protein ion charge state distributions with chemometric tools can yield "pure" ionic signals of various equilibrium states of the protein. This would allow an average charge to be assigned to each observed conformer. This average charge could then be used to estimate the solvent-exposed surface area for each conformer. Once again, the utility of this approach may be limited in many cases, for example, due to an inadequate number of protonation sites within the polypeptide chain [e.g., insufficient number of basic residues when charging proceeds via protonation (36)]. Likewise, low "charge retention efficiency" of highly charged protein ions [e.g., due to highly effective proton-transfer reactions (27) or dissociation of the protein–small organic cation complex in the gas phase]. Still, in those cases when the conformer's average charge does reflect its solvent-exposed surface area, such information would be very valuable. As we have seen in Chapter 2, there are very few experimental techniques capable of providing reliable information on the shapes of non-native protein states. X-ray crystallography is intrinsically limited in its ability to deal with non-native conformations. Although several methods have been developed to provide some assessment of protein dynamics based on observed crystal heterogeneity (37), information on large-scale dynamic events is usually very difficult to obtain due to the influence of packing forces in protein crystals. So far, solution scattering techniques remain the only means to discern the shapes of such species in solution (38–41); however, they require prior knowledge of the number of different species present in the system. Furthermore, characterization of individual species in a multicomponent system becomes increasingly difficult as their number grows. Therefore, scattering techniques have had only limited success when applied to study dynamics of multistate proteins, let alone multiprotein assemblies.

A very important advantage offered by using protein ion charge state distributions as a means to monitor protein conformational changes is the ability of this technique to deal with complex multicomponent systems. Such systems present a serious problem, vis-à-vis analysis of multiple protein conformations in protein mixtures by other spectroscopic methods, since it is generally very difficult to distinguish signals corresponding to different protein species present in solution. Such signal interference does not present a challenge to the approach based on the analysis of protein ion charge state distributions in ESI mass spectra. Indeed, the ion peaks corresponding to different protein components of the mixture will generally appear at different m/z values and will not interfere with each other (separation by mass). Therefore, conformational dynamics of each protein species can be studied in a conformer-specific fashion. A utility of this approach to monitor protein dynamics in multicomponent systems was recently demonstrated using the assembly of hemoglobin tetramers as an example (20). Seven different protein species become populated in solution in the course of the assembly process, yet the dynamics of each species can be studied in a conformer-specific fashion, revealing intimate details of the assembly process (Figure 5.3).

As a final remark in our discussion of protein ion charge state distribution in ESI mass spectra, we should note that most (if not all) measurements reported so far are based solely on monitoring charge state distributions of *positive* ions. A comparison of evolutions of positive and negative ion charge state distributions in the ESI spectra of several small polypeptides has been done by Konermann and Douglas, who concluded that only positive ion spectra provide information on the protein shapes in solution (8). The apparent discrepancy between the information derived from the positive and negative ion ESI spectra most likely relates to the differences in the ionization processes. While there is compelling experimental evidence suggesting that desorption of positive protein ions follows Dole's charged residue model (42, 43), the mechanism responsible for the formation of negative protein ions has not yet been explored in detail. Resolving this apparent controversy will clearly require a better understanding of the mechanisms of ESI of macromolecules, which is currently a focus of extensive research efforts (32, 44, 45).

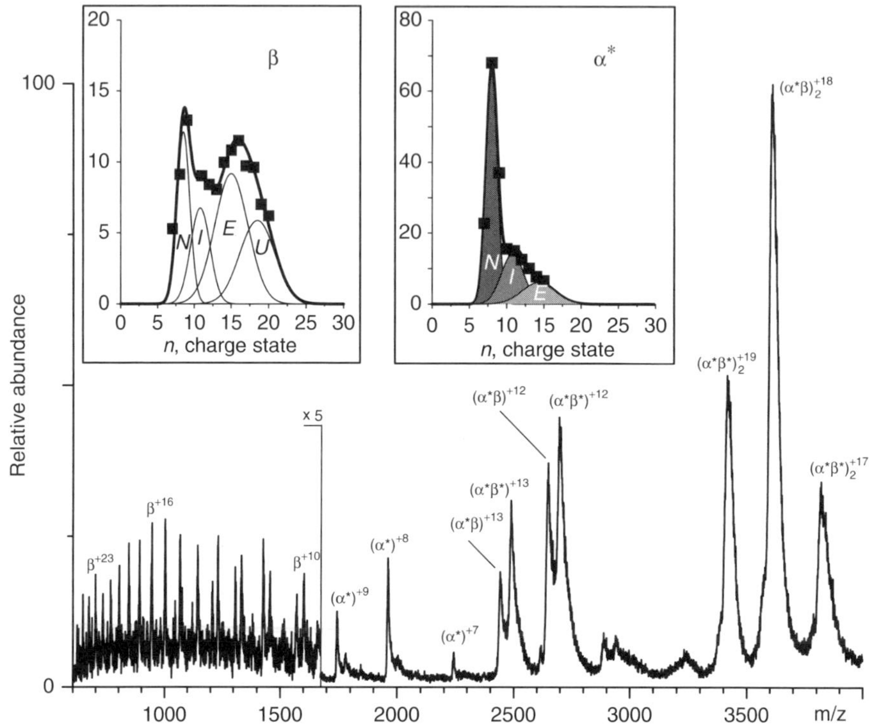

FIGURE 5.3. ESI mass spectrum of dilute (10 μM) bovine hemoglobin acquired at neutral pH (10 mM $CH_3CO_2NH_4$). Two monomeric (heme-bound α-chain, and heme-deficient β-chain) and three multimeric species (tetramer, dimer, and heme-deficient dimer or "semi-hemoglobin") are identified based on mass measurements (presence of heme is indicated with an asterisk). The insets show chemometric analysis of the charge state distributions of the two monomeric species (the rest of the spectra, acquired under mildly denaturing conditions and used for deconvolution, are not shown). The α^*-chain shows minimal conformational heterogeneity unlike the β-chain, which exhibits significant flexibility. The "rigid" α^*-chain serves as a "structural template" to which a flexible β-chain can easily adapt, forming a tightly folded heme-deficient dimer, $\alpha^*\beta$. "Locking" otherwise flexible β-chain in a native conformation enables it to bind a heme group, leading to formation of a hemoglobin dimer $\alpha^*\beta^*$ and, eventually, a tetramer $(\alpha^*\beta^*)_2$. Adapted with permission from (20). © 2003 American Chemical Society.

5.2. CHEMICAL LABELING AND TRAPPING EQUILIBRIUM STATES IN UNFOLDING EXPERIMENTS

5.2.1. Characterization of the Solvent-Exposed Surfaces with Chemical Labeling

Although chemical cross-linking and selective chemical modifications are typically used to assess native protein structures, the same approaches can be used in principle to characterize nonhomogeneous protein structures (e.g., equilibrium intermediate states). There are, however, very few examples of using either chemical cross-linking or labeling in conjunction with mass spectrometry to characterize protein conformational heterogeneity under equilibrium. One possible reason is that the structural information obtained in such experiments is ensemble-averaged, which often makes data interpretation rather difficult. This was exemplified in recent work by Craig and co-workers, who used a photochemical reagent to characterize the degree of unfolding of a small protein α-lactalbumin (46). The unfolded state of the protein (unfolding was induced by 8 M urea) was labeled 25% to 30% more than the native state due to apparent increase of the solvent-accessible surface area. However, such an increase is still below the increment expected from the theoretical estimates of the solvent-accessible surface area of the random coil (a value of 100% was predicted based on theoretical calculations). This discrepancy was attributed to the "residual structure" in the unfolded state. Another possible explanation would be presence of a partially structured intermediate state (such as a molten globular state) alongside the random coil, with the experiments providing the solvent-accessibility patterns "averaged" across the entire protein population.

An additional disadvantage of chemical labeling as a technique to study dynamic protein structures is that slow diffusion of a "bulky" labeling reagent to its target site may prevent efficient labeling. Even if the kinetics of the labeling reaction itself is very fast, the efficiency of a trapping reaction will be significantly limited by the slow diffusion of the reactant (covalent modifier) through the protein solution to the transiently "exposed" reactive site on the protein. However, there are three notable exceptions to this rule when the diffusion limit can be defeated and stable chemical modifications can be used to study conformational dynamics. The first two employ an omnipresent modifier (solvent), which either reacts selectively with non-native protein states (but is totally inert toward the native state) or else is totally unreactive in its ground state but can be activated on a very short time scale. The former of course, is the well-known and extremely popular technique of hydrogen–deuterium exchange, which will be discussed in detail in Section 5.3. The latter is a radiolytic footprinting, whose applications to the characterization of "static" protein structures have already been discussed in Chapter 4. Chance and co-workers have recently demonstrated that this experimental technique also has potential to provide structural information on partially unfolded protein states (47, 48). Another technique may not need external covalent modifiers at all, but rather relies on the internal chemically active groups within the protein that have an intrinsic propensity to form stable bonds in the native conformation [e.g., cysteinyl residues forming disulfide bridges (49)].

5.2.2. Exploiting Intrinsic Protein Reactivity: Formation and Scrambling of Disulfide Bonds

Monitoring the reactions of disulfide-bond formation can also be used to study protein conformation and dynamics. The reactions include oxidation, reduction, and reshuffling (Figure 5.4), which are all based on thiol/disulfide exchange, in which the thiolate anion R_1S^- displaces one sulfur of the disulfide bond R_2SSR_3 (49). Thiol/disulfide exchange reactions are attenuated by a variety of factors, including local electrostatic interactions and structural propensities. Stable tertiary structure is another important determinant, as it "locks in" the native disulfide bonds. Thus, regeneration of the disulfide bond pattern is often used to monitor the progress of protein folding (50). Traditional techniques of determining the disulfide bond pattern rely on proteases to cleave the polypeptide backbone between the cysteine residues. In some cases, a disulfide bond pattern within trapped folding intermediates can be determined using a combination of proteolysis and MS analysis (51). However, application of this procedure to proteins containing a large number of cysteine residues can be a very challenging task, particularly when the cysteine residues are located close to each other in the sequence.

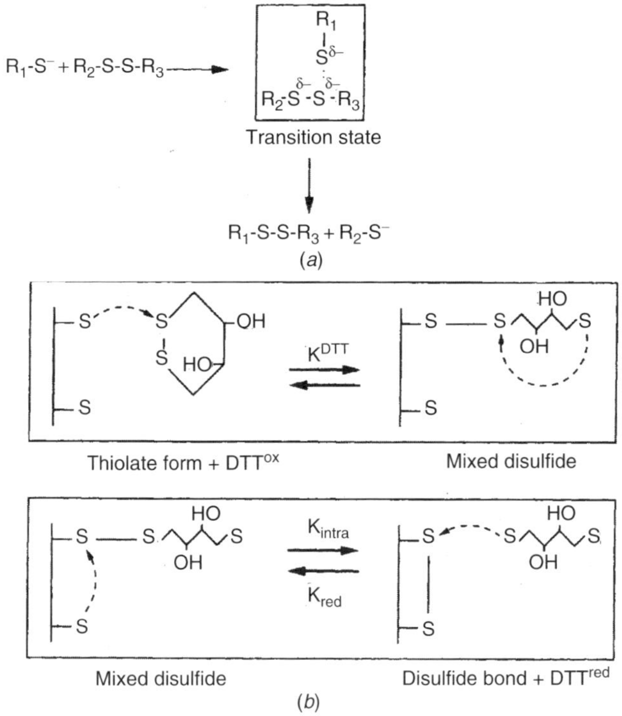

FIGURE 5.4. Chemistry of protein disulfide bond reactions. In thiol/disulfide exchange, a thiolate anion R_1S^- displaces one sulfur of a disulfide bond R_2SSR_3 (**a**). In the transition state, the negative charge of the thiolate appears to be delocalized among the three sulfur atoms (**a**). Protein disulfide bonds are formed and reduced by two such thiol/disulfide exchange reactions with a redox reagent, the first of which involves the formation of a *mixed disulfide bond* between the protein and the redox reagent. For completeness, the reactions are illustrated with cyclic (DTT^{red}/DTT^{ox}) and linear (GSSG/GSH) redox reagents (**b, c**). Thiol/disulfide exchange reactions can also occur intramolecularly; for example, a protein thiolate group may attack a disulfide bond of the same protein, leading to *disulfide reshuffling* (**d**). Reprinted with permission from (49). © 2000 American Chemical Society.

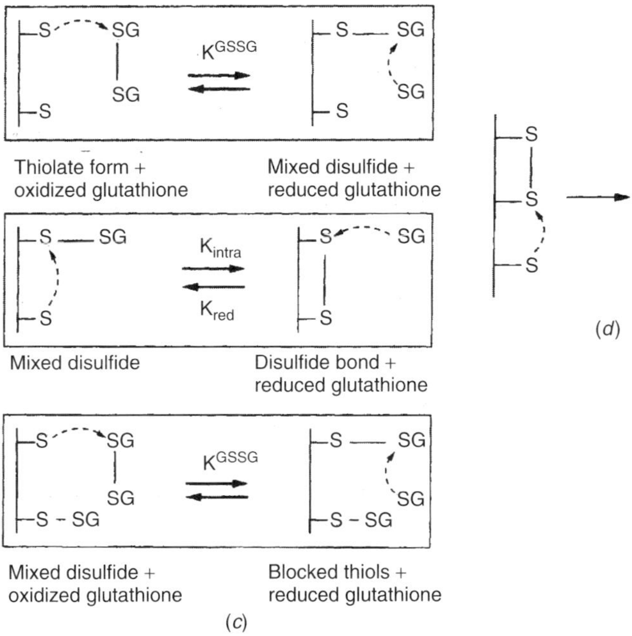

FIGURE 5.4. *(Continued)*

Straightforward attempts to map the disulfide bonds using tandem mass spectrometry are usually of limited use. Watson and co-workers have recently developed a method for determining the disulfide bond pattern in partially reduced proteins using cyanylation, which induces specific backbone cleavages at the N-terminal side of the modified (*S*-cyanocysteine) residues (52–54). The backbone cleavage step is followed by full reduction of the peptide mixture. The reduced peptides are separated by HPLC and analyzed by MALDI or ESI MS. This procedure can readily be applied to study late protein folding intermediates by quenching folding and trapping the disulfide intermediates under conditions that minimize disulfide scrambling (i.e., sulfhydryl–disulfide exchange) in solution (55, 56). One common problem that may arise in the course of data interpretation for cysteine-rich proteins is the correct assignment of the disulfide bonds. Watson and co-workers have recently introduced an efficient method for data processing and interpretation to support and extend disulfide mass-mapping methodology based on partial reduction and cyanylation-induced cleavage to proteins containing more than four cysteine residues (57). The method utilizes the concept of *negative signature mass** to identify the disulfide structure of a cystinyl protein given an input of mass spectral data and an amino acid sequence.

*This process is different from the conventional approaches to disulfide mapping in that it does not directly assign the linkages, but rather eliminates "candidate" linkages from a list of all possible Cys–Cys connections, aiming at ruling out enough linkages so that only a unique disulfide structure can be generated.

An alternative procedure for mapping proximal cysteine residues in partially unstructured proteins utilizes bi-thiol reagents (derivatives of arsenous acid, such as melarsen oxide and pyridinyl-3-arsonous acid) to link neighboring cysteine residues in reduced proteins (58, 59). Bis-thiol selective derivatization of intermediate states is principally different from mono-thiol trapping strategies, since bis-thiol modifications actually cross-link closely spaced cysteine residues pairwise. As a result, the chemical reduction does not result in protein unfolding. Two additional advantages of this experimental strategy are (i) relatively large protein mass increase upon single modification and (ii) absence of any side reactions even with a high molar excess of the reagent. As is the case with other chemical labeling techniques, the identity of the modified residues can be established using a combination of proteolysis, separation, and MS analysis.

5.3. STRUCTURE AND DYNAMICS OF INTERMEDIATE EQUILIBRIUM STATES: USE OF HYDROGEN EXCHANGE

5.3.1. Protein Dynamics and Hydrogen Exchange

Perhaps the major disadvantage of the experimental methods presented in the previous sections of this chapter is their inability to detect small-scale* and/or rare dynamic events in solution. As has been discussed in Chapter 1, even under native or near-native conditions proteins sample "non-native" states via either transient fluctuations or even large-scale unfolding events. Such large-scale motions (partial or complete loss of the native protein structure) are very rare events, unless, of course, the protein under consideration is an intrinsically unstructured one. The small-scale events do not alter the overall structure enough to induce changes in the charge state distributions, while their transient nature prevents efficient chemical trapping. Indeed, if the Boltzmann weight of the putative non-native states is negligible (as is often the case under native conditions), they would not be detectable using either protein ion charge state distributions or chemical labeling techniques.

One of the potent experimental tools that has been widely used to study protein dynamics is hydrogen–deuterium exchange (HDX) (60). HDX measurements detect little or no contribution from the protein in the "ground" states, providing an effective means to visualize transiently populated "activated" states (61, 62). While the majority of HDX studies still employ high-resolution NMR as a means to monitor the exchange kinetics (see Chapter 2 for a discussion of HDX NMR experiments), mass spectrometry is currently enjoying a dramatic surge in popularity in this field as well (63–65). The pioneering work of Katta and Chait first demonstrated the great potential of the HDX/ESI MS combination

*Here we use the term "small-scale dynamic events" as a general descriptor of local protein motions that alter the structure of a small portion of the protein without affecting the rest (an example of such an event would be a local conformational fluctuation).

as a tool to probe conformational dynamics of small proteins (66). In the following years, the number and scope of the applications of HDX/ESI MS (and, more recently, HDX/MALDI MS) methodology to probe dynamics of biomolecules has expanded dramatically, catalyzed by spectacular technological improvements in "soft" ionization methods. ESI MS offers several important advantages over NMR, namely, faster time scale, tolerance to paramagnetic ligands and cofactors, ability to monitor the exchange in a conformer-specific fashion, and much more forgiving molecular weight limitations. The ability of ESI MS to handle larger proteins and their complexes is particularly important when compared to high-field NMR, which still has limited application for proteins larger than approximately 30 kDa. The practical upper mass limit of ESI MS, on the other hand, has yet to be established, as the bar is being continuously raised (67). Another significant advantage offered by ESI MS is its superior sensitivity; hence many experiments can be carried out using only minute quantities of proteins, which in many cases enables the study of protein behavior at, or even below, endogenous levels. Importantly, because of the ESI MS ability to desorb biomolecular species directly from aqueous solutions, as well as high data acquisition rates afforded by most mass analyzers, the HDX/ESI MS measurements can often be carried out "on-line," enabling in many cases "real-time" monitoring of protein dynamics.

5.3.2. Hydrogen Exchange in Peptides and Proteins: General Considerations

In the preceding chapter we discussed briefly some basic aspects of HDX experiments relevant to probing "static" structures of proteins and protein assemblies. We noted that protein HDX involves two different types of reactions: (i) reversible protein unfolding that disrupts the H-bonding network and (ii) isotope exchange at individual unprotected amides. Since protein unfolding (either local or global) is a prerequisite for exchange at the sites that are protected* in the native conformation, HDX reactions serve as a reliable and sensitive indicator of the unfolding events. However, it is important to remember that the organization and dynamic features of the hydrogen-bonding network are not the only determinants of the HDX kinetics. Even in the absence of any protection, the exchange kinetics of any labile hydrogen atom will strongly depend on the nature of the functional group. (Thus, solvent-exposed η-hydrogen atoms of Arg guanidine group would have the fastest exchange rates at neutral pH, exceeding those of backbone amide and Trp indole hydrogen atoms by more than an order of magnitude and almost three orders of magnitude, respectively.) The pH dependencies of the "cumulative" intrinsic exchange rates for several types of labile hydrogen atoms, calculated based on the data compiled by Dempsey (68), are presented in Figure 5.5. The

*By "protection" we mean here one of the two things: (i) involvement in the hydrogen bonding network or (ii) sequestration from solvent in the core of the protein. In most cases, however, labile hydrogen atoms located in the protein core do participate in hydrogen bonding as well.

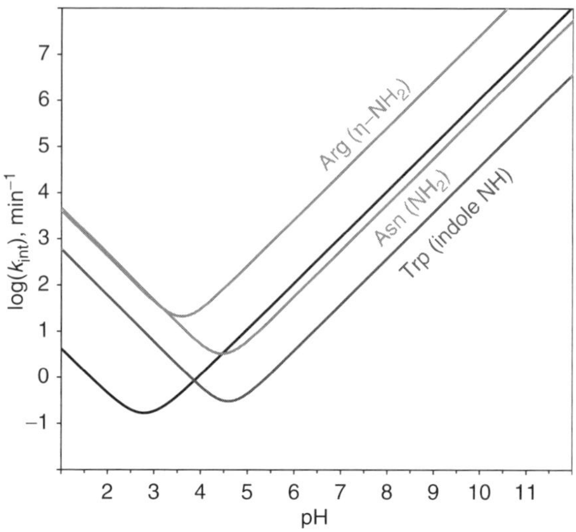

FIGURE 5.5. Intrinsic exchange rates of several types of labile hydrogen atoms as functions of solution pH calculated based on the data compiled in (68). The solid black line represents intrinsic exchange of backbone amides.

exchange rates are also influenced by the neighboring residues via both inductive and steric blocking effects (69).

Since hydrogen exchange at unprotected sites is both acid- and base-catalyzed, the overall exchange rate constant will depend on solution pH. Exchange through catalysis by water has also been reported, and the "intrinsic exchange" rate constant is usually presented as (68)

$$k_{\text{int}} = k_{\text{acid}}[\text{H}^+] + k_{\text{base}}[\text{OH}^-] + k_{\text{water}}. \tag{5-3-1}$$

For most labile hydrogen atoms, base catalysis is more effective, exceeding the acid-catalyzed rate constants by four to eight orders of magnitude (68). Isotopic composition of both solvent and the protein also exerts a certain influence on the intrinsic exchange rates. Rate constants for the acid-catalyzed exchange of amide groups NH, ND (N^2H), and NT (N^3H) in ^1H_2O are essentially identical, but a solvent isotope effect doubles the acid-catalyzed rate in ^2H_2O (70). Intrinsic exchange rate constants for base-catalyzed exchange in ^1H_2O decrease slowly in the order $N^1H > N^2H > N^3H$, while the alkaline rate constant in ^2H_2O appears to be very close to that in ^1H_2O after making corrections for glass electrode artifacts and differences in water autoprotolysis constant (70). The presence of small exchange catalysts, such as phosphate and carbonate, as well small organic molecules with carboxyl and/or amino groups, typically increases hydrogen exchange rates from both hydroxyl and amino groups of polypeptides (71). The presence of organic cosolvents also influences the intrinsic exchange rates, although this subject has not yet received much attention (68).

5.3.3. Global Exchange Kinetics: Mechanisms of Backbone Amide Hydrogen Exchange in a Two-State Model System

Backbone amide hydrogen atoms constitute a particularly interesting class of labile hydrogen atoms due to their uniform distribution throughout the protein sequence,* which makes them very convenient reporters of protein dynamics at the amino acid residue level. Therefore, it is not surprising that the majority of HDX experiments are concerned with the exchange of backbone amide hydrogen atoms, although in some cases it is rather difficult to separate contributions of the backbone amide hydrogen atoms from those of the side chains.

The mathematical formalism that is often used to describe HDX kinetics of backbone amides was introduced several decades ago and is based on a simple two-state kinetic model† (72):

$$\mathbf{ND}(\text{incompetent}) \underset{k_{cl}}{\overset{k_{op}}{\rightleftharpoons}} \mathbf{ND}(\text{competent}) \xrightarrow{k_{int}} \mathbf{NH}(\text{competent}) \rightleftharpoons \mathbf{NH}(\text{incompetent}), \quad (5\text{-}3\text{-}2)$$

where k_{op} and k_{cl} are the rate constants for the opening (unfolding) and closing (refolding) events that expose/protect a particular amide hydrogen to/from exchange with the solvent. The intrinsic exchange rate constants of amide hydrogen atoms from the exchange-competent state k_{int} can be estimated using short unstructured peptides (73). The **ND** $\rightarrow$ **NH** transition is essentially irreversible, as HDX experiments are carried out in significant excess (typically 10–100-fold) of exchange buffer.

In most HDX studies the exchange-incompetent state of the protein is considered to be its native state. The exchange-competent state is thought of as a non-native structure, which can be either fully unfolded (random coil) or partially unfolded (intermediate states). Alternatively, it can represent a structural fluctuation within the native conformation, which exposes an otherwise protected amide hydrogen to solvent transiently through *local unfolding* or *structural breathing* without unfolding (74, 75). Transitions between different non-native states under equilibrium conditions are usually ignored in mathematical treatments of HDX. One of the reasons is that the majority of HDX measurements are carried out under native or near-native conditions. The Boltzmann weight of non-native states for most proteins under these conditions is very low and the transitions among such states do not make any detectable contribution to the overall HDX kinetics.

Non-native HDX is now also experiencing a surge in popularity due to renewed interest in the structure and dynamics of intermediate protein states. An important

*Proline is the only naturally occurring amino acid that does not have an amide hydrogen atom when this residue is a part of a polypeptide chain.

†Most of the data presented in this chapter will have a fully deuterated protein as a starting point of the exchange reaction; the latter will be carried out by placing such a protein into a protiated buffer solution. Therefore, we have modified the original presentation of (5-3-2) by Hvidt and Nielsen, who used a fully protiated protein as a starting point.

advantage offered by ESI MS is its ability to directly visualize various protein states based on the difference in deuterium incorporation. As we will see below, a clear distinction between a native conformation and a (partially) unstructured state can be made under certain mildly denaturing conditions.

Unlike HDX NMR measurements, a typical HDX MS experiment provides information on *global protection patterns* by measuring the isotope content of the entire protein, rather than the exchange kinetics of individual amide hydrogen atoms. Nevertheless, interpretation of HDX MS data often utilizes the kinetic model (5-3-2) by making an implicit assumption that **ND**(incompetent) and **ND**(competent) represent *groups of amides*, rather than *individual amides* that become unprotected upon transition from one state to another. This simplistic view of protein behavior is illustrated in Figure 5.6, where the global energy minima (narrow deep potential wells) represent native protein conformations, while the diffuse shallow local minima represent the unstructured (exchange-competent) states. The potential wells representing global minima are narrow, allowing no conformational freedom for the protein (exchange-incompetent state). An exchange event can only occur if the protein molecule escapes the potential

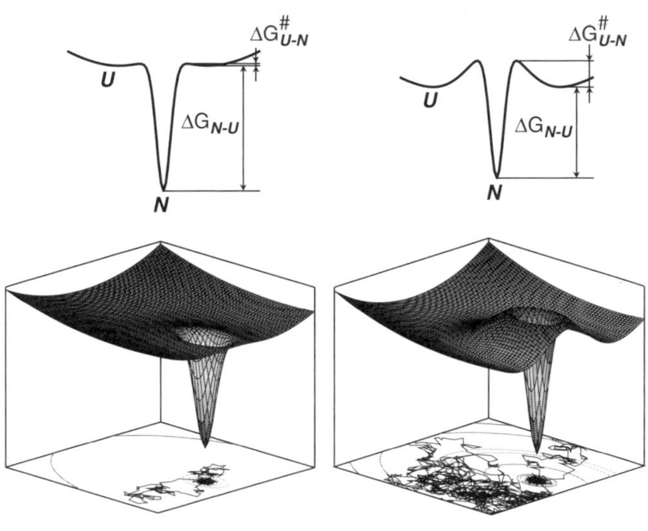

FIGURE 5.6. Minimalistic representations of a two-state protein system. The global energy minimum corresponds to an exchange-incompetent ("native") state of the protein. The diffuse local minima (concentric to the potential well of the "native" state) represent an exchange-competent state ("random coil"). The excursions from the global minimum are rare due to the significant difference in energy (Boltzmann statistics) and are very short-lived when the reverse activation energy barrier separating the local minima basin from the "native" potential is low (**A**). The protein molecules may become trapped in the "random coil" state for prolonged periods of time if the reverse activation energy barrier is significant (**B**).

well corresponding to the native (exchange-incompetent) state and spends some time in the local minimum basin, which is comprised of various microstates representing a random coil state of the protein. If the reverse activation energy barrier (separating the local free-energy minimum from the global one) is low, the protein will sample the unstructured state only transiently before returning back to the native state (Figure 5.6A). As a result, only a small fraction of all labile hydrogen atoms will be exchanged upon a single unfolding event.* Such a scenario (commonly referred to as the *EX2 exchange mechanism*) is realized if the residence time in the local minimum basin ($1/k_{cl}$) is much shorter than the characteristic time of exchange of an unprotected labile hydrogen atom ($1/k_{int}$). In this case, the probability of exchange for even a single amide during an unfolding event will be significantly less than one. The overall rate of exchange will be defined by both the frequency of unfolding events (k_{op}) and the probability of exchange during a single opening event:

$$k^{HDX} = k_{op} \cdot (k_{int}/k_{cl}) \quad (5\text{-}3\text{-}3)$$

or, after regrouping,

$$k^{HDX} = k_{int} \cdot K, \quad (5\text{-}3\text{-}4)$$

where K is an effective equilibrium constant for the unfolding reaction and is determined by the free-energy difference between the two states of the protein. While in NMR measurements k^{HDX} is simply a rate of depletion of a number of protein molecules labeled with ^1H at a specific amide (depending on the experiment, it may be a depletion of ^2H-labeled protein molecules), HDX MS measurements would regard this rate constant as a cumulative rate of exchange. In other words, k^{HDX} is an ensemble-averaged rate of loss of the entire ^1H content (for all amides combined). It can be shown that under these conditions both rate constants (measured by NMR and MS) would be numerically equal, although they actually have different meanings. Figure 5.7A shows a simulated pattern of HDX exchange under these conditions (EX2 limit for a two-state system with 55 identical amide hydrogen atoms).

Raising the reverse-activation energy barrier separating the local minimum basin from the potential well representing the global energy minimum, will decrease the refolding rate k_{cl}, leading to an increased residency time of the protein in the exchange-competent state (Figure 5.6B). This, of course, will lead to an increase of the exchange probability for each labile hydrogen atom during a single unfolding event (unless the intrinsic exchange rate also decreases substantially). If the barrier is sufficiently high (so that $k_{int} \gg k_{cl}$), the protein will become trapped in the unfolded state long enough to allow all labile hydrogen atoms to be exchanged during a single unfolding event. In this case (commonly

*An additional complication (which will be ignored here) arises from the fact that many microstates of a random coil state can have residual protection, making certain amides unavailable to exchange. In this case, the complete exchange will require adequate sampling of all microstates within the basin.

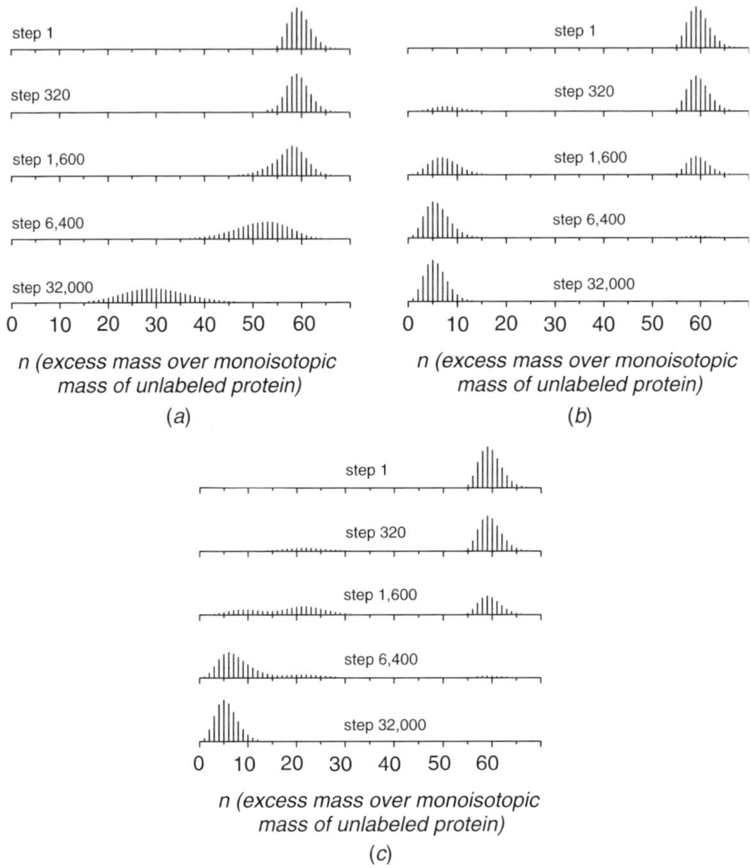

FIGURE 5.7. Simulated HDX MS patterns for minimalistic models of two-state protein systems whose energy surfaces are depicted on Figure 5.6. Proteins are assumed to be "fully deuterated" prior to exchange (the latter is initiated by infinite dilution in a protiated buffer solution). Simulation parameters correspond to EX2 (**A**), EX1 (**B**), and intermediate ($k_{\text{int}}/k_{\text{cl}} = 0.5$) EXX conditions (**C**).

referred to as the *EX1 exchange regime*), the exchange rate will be determined simply by the rate of protein unfolding:

$$k^{\text{HDX}} = k_{\text{op}}. \tag{5-3-5}$$

A pattern of HDX MS simulated under these conditions is presented in Figure 5.7B, which clearly shows distinct contributions from both states of the protein. In this situation the rate constant k^{HDX} defined by (5-3-5) is simply a depletion rate of the number of ^{1}H-labeled protein molecules. Therefore, the physical meaning of the MS-measured exchange rate in the EX1 regime is identical to that derived from the HDX NMR experiments.

Finally, in the intermediate exchange regime (when the values of k_{int} and k_{cl} are comparable) the residency time in the exchange-competent state is long enough to have one or more protons exchanged during each unfolding event, but too short to have the entire set of all labile hydrogen atoms exchanged at once. The k^{HDX} rate constant, as measured by NMR, will reflect a gradual decrease of the protein population with a certain amide retaining its ^{1}H label:

$$k^{\text{HDX}} = k_{\text{op}} \cdot (k_{\text{int}}/k_{\text{int}} + k_{\text{cl}}). \tag{5-3-6}$$

The situation with the HDX MS measurements under the same conditions will be much more complicated, as suggested by a convoluted appearance of the exchange pattern simulated under these conditions (Figure 5.7C). While the isotopic distribution is expected to exhibit a bimodal character, the distance (or mass difference) between the two clusters increases as the exchange progresses. As a result, two apparent rate constants will be measured in a single experiment, one describing the changes in the relative abundance of the two clusters and another one related to the shift of the lower m/z isotopic cluster. The former rate constant is determined solely by the protein unfolding rate, as is the case under EX1 exchange conditions (5-3-5). The second rate constant depends on both k_{int} and k_{cl}, as well as the frequency of unfolding events k_{op} in a fashion similar to (5-3-3).* Therefore, if the value of k_{int} is known, HDX MS measurements carried out under the intermediate conditions (for the lack of a better term, we will use *EXX* throughout this chapter) may allow the values of both k_{op} and k_{cl} to be determined in a single experiment. This contrasts with measurements conducted under the EX1 and EX2 conditions, which provide information only on k_{op} (EX1 regime) or the $k_{\text{op}}/k_{\text{cl}}$ ratio (EX2 regime).

The simplistic two-state model considered in this section is, of course, an overly idealistic representation of the protein dynamics, although it provides a good start in our discussion of how the dynamics of real protein systems is reflected in their HDX profiles. The following sections present a more detailed discussion of the HDX measurements carried out under the EX1 and EX2 exchange conditions as applied to more "sophisticated" systems.

5.3.4. Realistic Two-State Model System: Effect of Local Fluctuations on the Global Exchange Pattern Under EX2 Conditions

HDX of most proteins under near-native conditions (aqueous solutions maintained at or near neutral pH, moderate temperatures, and reasonable ionic strengths in the absence of denaturants) almost always follows EX2-type kinetics.† An example of such behavior is presented in Figure 5.8. However, EX2-type exchange can also be observed under denaturing conditions, as long as the intrinsic exchange rate is

*In this case the second "apparent" rate constant should be defined as a rate of shift of the monoisotopic peak in the low m/z cluster.

†The refolding rates are very high, since these conditions favor the natively folded protein conformation very strongly and the non-native states become populated only transiently.

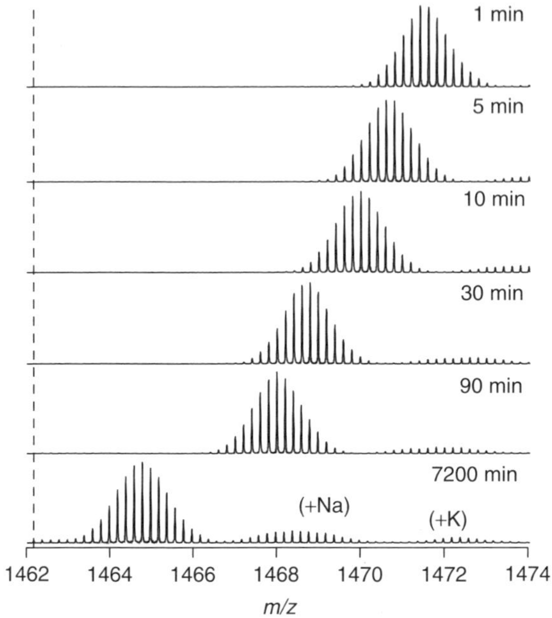

FIGURE 5.8. HDX of chymotrypsin inhibitor 2 under conditions favoring the EX2 exchange mechanism (the refolding rates are high, while the intrinsic exchange rates are moderate). The exchange was initiated by diluting a fully deuterated protein in a protiated solution (10 mM $CH_3CO_2NH_4$, pH adjusted to 7.0, temperature 23 °C; a 1:20 dilution, v:v). The position of the exchange "endpoint" is indicated with a dotted line.

significantly lower than the refolding rate, which can be achieved, for example, under mildly acidic conditions. Analysis of the HDX MS profiles recorded under near-native conditions reveals that real proteins almost never exhibit simple single-exponential kinetics, as could have been expected based on our consideration of a two-state model (Figure 5.7A). Even in the case of two-state protein systems, such as chymotrypsin inhibitor 2 (CI2), the exchange follows biphasic exponential kinetics (Figure 5.9). HDX MS measurements detect significantly higher initial protection compared to HDX NMR experiments carried out under similar conditions (76), as several amides (within the fast phase) exchange too fast to be measured on the time scale of a typical HDX NMR experiment. Increasing the solution temperature obviously accelerates the exchange kinetics (due to increase of both k_{int} and k_{op}), while keeping it biphasic (Figure 5.9). Further increase of the solution temperature (to 60 °C) results in monoexponential HDX kinetics with one apparent rate constant. However, in this case the fast exchange phase is simply too fast to be detected by MS without using more sophisticated time-resolved techniques. (A detailed discussion of HDX MS measurements on the subsecond time scale will be presented in Chapter 6.)

The most glaring deficiency of the simplistic two-state model discussed in the previous section is that it assigns zero conformational freedom to the protein in

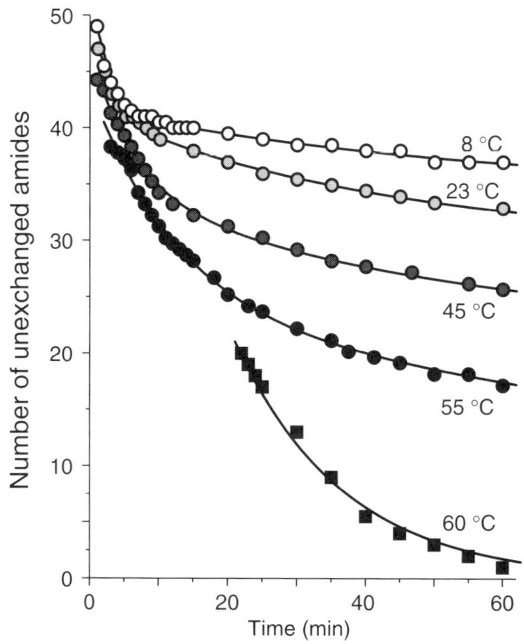

FIGURE 5.9. Hydrogen exchange kinetics of CI2 measured by HDX MS at different solution temperatures under "near-native" conditions (10 mM $CH_3CO_2NH_4$, pH adjusted to 7.0). Total number of amide hydrogen atoms is 59, only 30 of which are protected on the time scale of ^{1}H NMR experiments (76).

its native state (***N***). As we have seen in Chapter 1, this assumption ignores the dynamic character of native conformations. It is now commonly accepted that "native" protein conformations in solution are not static structures, but rather continuously sample various microstates. Although the $\boldsymbol{N} \rightleftharpoons \boldsymbol{U}$ transitions discussed above are important contributors to such dynamics, they are very rare (particularly under native conditions). Much more frequent dynamic events are the so-called local fluctuations, dynamic processes that result in transient (and localized) amide deprotection that appears to be uncooperative and denaturant-independent (77). Such amides usually exhibit fast, denaturant-independent exchange and typically reside at or near the protein surface; therefore, a solvent-penetration model can also be invoked to explain this behavior (78).

Our view of local fluctuations invokes the notion of an activated state $\boldsymbol{N}^*$, which is a collection of microstates, each having a structure (or, more precisely, amide protection pattern) identical to that of the native conformation with the exception of one (or several) amides at the protein surface. The reverse activation energy barrier separating this activated state from the ground state $\Delta G^{\ddagger}_{N*\rightarrow N}$ is close to zero, so that any transition from ***N*** to $\boldsymbol{N}^*$ would be very short-lived. The energy of this activated state G_{N*} is significantly lower than the activation energy for the $\boldsymbol{N} \rightarrow \boldsymbol{U}$ transition, so that the fluctuations occur much more frequently

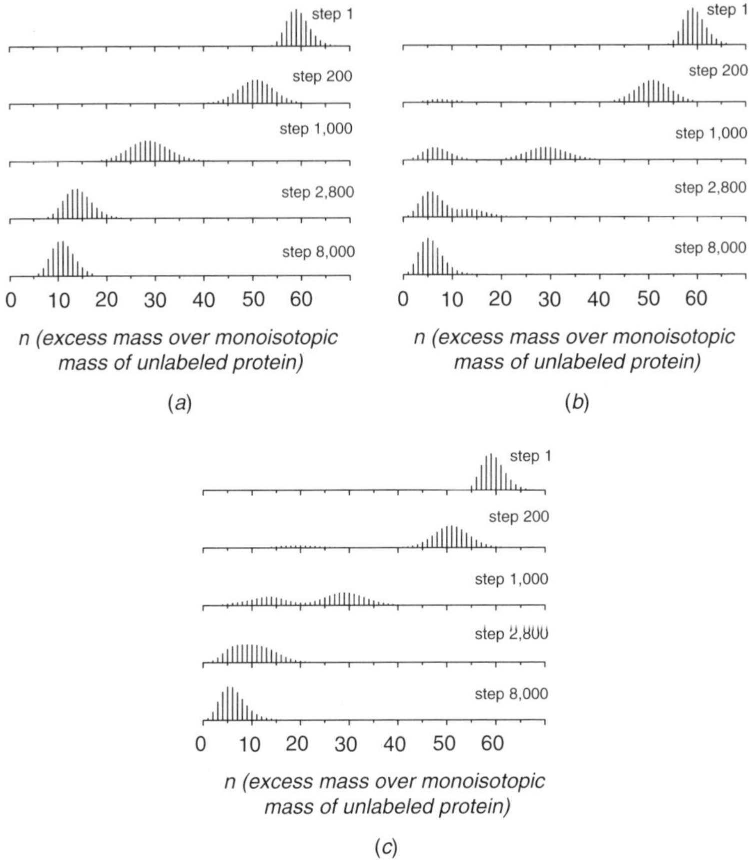

FIGURE 5.10. Simulated HDX MS patterns for a two-state protein model that accounts for transient local fluctuations (limited exchange competence is assigned to the "native" state). Proteins are assumed to be "fully deuterated" prior to exchange (infinite dilution in a protiated buffer solution). Simulation parameters correspond to EX2 (**A**), EX1 (**B**), and intermediate ($k_{int}/k_{cl} = 0.5$) EXX conditions (**C**).

than the global unfolding events. In most cases, however, there would be a set of amides that would never become accessible to solvent via a local fluctuation event, hence the limited amplitude of the exchange in the fast phase. Labile hydrogen atoms that remain protected during local fluctuations could only be exchanged via a global unfolding event (in the case of a two-state protein, such as CI2, it would be an $\boldsymbol{N} \rightarrow \boldsymbol{U}$ transition). It is these rare events that give rise to a slow exchange phase. Figure 5.10A shows a simulated HDX MS pattern of a two-state model protein under the EX2 conditions that takes into account local structural fluctuations in the native state.* Since the exchange of all labile amide

*To allow limited conformational freedom within the native state $\boldsymbol{N}$, we consider an activated state $\boldsymbol{N}^*$ whose structure and energetics are defined as follows. We assume $\boldsymbol{N}^*$ to be a heterogeneous state

hydrogen atoms is now accomplished in two steps, broadening of the isotopic cluster (which was so evident in Figure 5.7A) becomes much less significant.

Local structural fluctuations in real proteins are not all identical, as the protection factors exhibit significant variation among various sites. The primary determinants are the "rigidity" of local structure (β-sheets > α-helices > Ω-loops), the degree of burial (74), and the presence of stabilizing tertiary contacts. In many cases, however, structural fluctuations giving rise to exchange with similar kinetic characteristics can be grouped together, giving rise to a hierarchy of local fluctuations.

5.3.5. Effects of Local Fluctuations on the Global Exchange Pattern Under EX1 and Mixed (EXX) Conditions

Increasing the lifetime of the globally unfolded state ***U*** of CI2 (e.g., by selecting mildly denaturing conditions and/or increasing the intrinsic exchange rate) leads to switching the HDX kinetics to the EX1 regime, as suggested by the *bimodal* appearance of the isotopic cluster (Figure 5.11). We note, however, that there is an important difference between the experimentally measured HDX profile and the one simulated for a simplistic two-state system (Figure 5.7B). While both isotopic clusters are static (vis-à-vis their position on the m/z scale) in the simulated spectra, there is clearly a noticeable shift of the higher m/z cluster (corresponding to the native conformation) in the experimentally obtained HDX MS profiles of CI2. It seems natural to assume that the local conformational fluctuations affect protein dynamics under these mildly denaturing conditions as well, giving rise to the partial EX2-type character of the exchange reactions. The simulated HDX MS pattern that takes into account a possibility of local structural fluctuations (Figure 5.10B) is fully consistent with the experimentally measured HDX kinetics of CI2.

We have already seen in (5-3-5) that the rate of disappearance of the ionic signal representing the native state under the EX1 conditions is in fact the rate of protein

comprised of a large number of equienergetic structures, with a total number of amides exchangeable from ***N**** being L. The free energy of this state is significantly lower than that of the globally unfolded state (i.e., $G_U \gg G_{N^*} \gg G_N$), while the reverse activation energy barrier $\Delta G^{\ddagger}_{N^* \to N}$ is zero or very close to zero. The former feature ensures that ***N**** is sampled by a protein molecule much more often than the globally unfolded state. The latter feature ensures that such excursions are very short (i.e., these states are transient). Therefore, even though ***N**** is allowed to have a certain conformational freedom, the probability of exchanging even a single amide from ***N**** during its lifetime would be very small. As a result, the $\boldsymbol{N} \rightleftharpoons \boldsymbol{N}^*$ transitions will all have characteristics of local structural fluctuations, despite the fact that ***N**** is formally introduced as a third state of the protein with limited exchange competence. Such presentation of local fluctuations differs from a traditional view, which assumes that every local fluctuation leads to a unique conformation (i.e., such local minima are separated from one another by significant barriers, so that these microstates cannot be grouped together). The reverse activation energy barriers separating each of these states from the exchange-incompetent state may also be significant (in which case the exchange will actually proceed through an EX1 mechanism). It is easy to show, however, that the HDX patterns produced by the two systems would be identical, and the fast phase of the exchange would still have all characteristics of EX2-type kinetics (uncorrelated exchange).

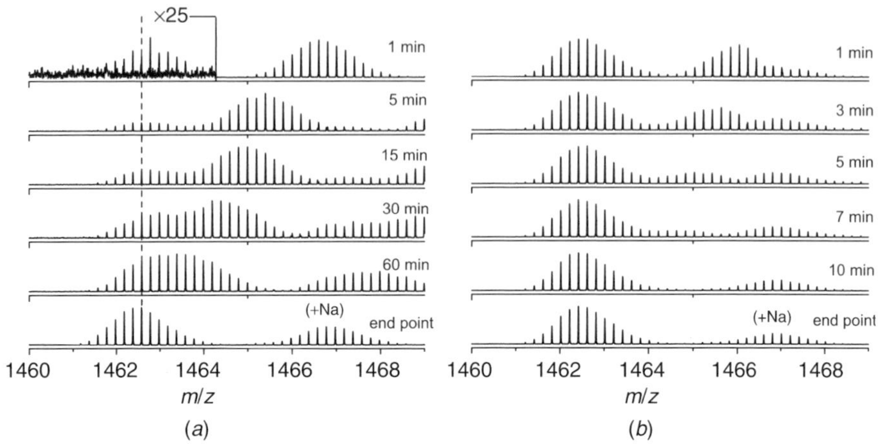

FIGURE 5.11. HDX of chymotrypsin inhibitor 2 under conditions favoring the EX1 exchange mechanism (the refolding rates are low compared to the intrinsic exchange rates). The exchange was initiated by diluting a fully deuterated protein in a protiated solution (10 mM $CH_3CO_2NH_4$, pH adjusted to 11.0 with NH_4OH, 60% CH_3OH). Solution temperature is 8 °C (**A**) and 35 °C (**B**).

unfolding k_{op} ($\boldsymbol{N} \rightarrow \boldsymbol{U}$ transition). Another quantitative characteristic of protein dynamics that can be extracted from these data relates to the local structural fluctuations within the native conformation.* The kinetics of both processes is greatly influenced by solution temperature (Figure 5.12), which may allow some useful thermodynamic data (e.g., unfolding entropy) to be extracted from such measurements.

The most complicated exchange pattern is observed in this two-state protein when the experiment is carried out under intermediate (EXX) exchange conditions. Three distinct features are evident in both experimentally measured (Figure 5.13) and simulated (Figure 5.10C) HDX MS profiles. These are the global unfolding and local fluctuations within the native state (which manifest themselves in the same ways, as under the EX1 regime). Additionally, a gradual mass shift of the lower m/z isotopic cluster is an indicator of the EXX regime, as discussed earlier (incomplete exchange during each global unfolding event). In principle, exchange under these conditions may provide a wealth of information on protein dynamics at all levels: k_{op} ($\boldsymbol{N} \rightarrow \boldsymbol{U}$) from changes in the relative abundance of the two isotopic clusters, $K = k^*_{op}/k^*_{cl}$ (local fluctuations within the native state) from the mass shift of the higher m/z isotopic cluster, and $K = k_{op}/k_{cl}$ ($\boldsymbol{N} \rightarrow \boldsymbol{U}$) from the mass shift of the lower m/z isotopic cluster (Figure 5.13B–D).

*The rate of a gradual "mass shift" of the higher m/z cluster will be determined by (5-3-3), although k_{op} and k_{cl} in this case relate to the local dynamic event rather than to global unfolding.

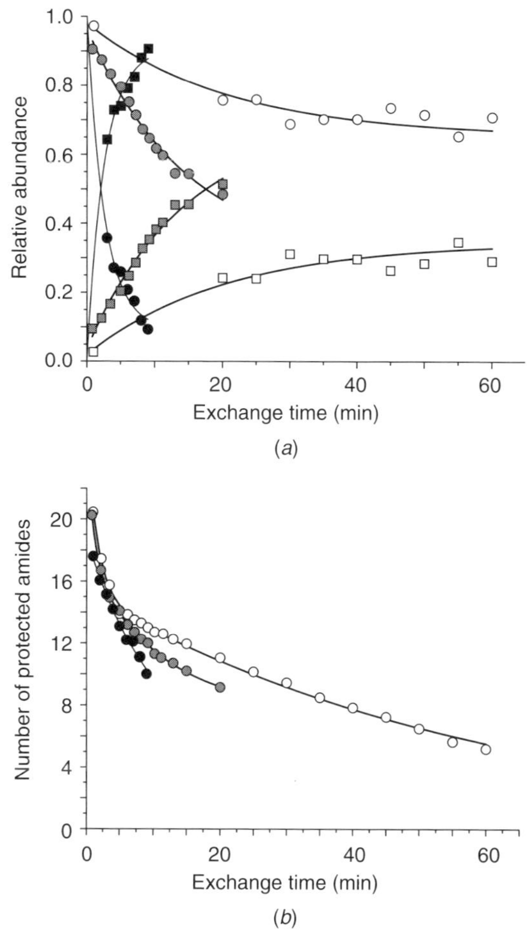

FIGURE 5.12. Hydrogen exchange kinetics of CI2 measured by HDX MS at different solution temperatures (8 °C, white; 23 °C, gray; and 35 °C, black) under conditions favoring EX1 exchange mechanism (see Figure 5.11 for details). (**A**) Changes in relative abundance of isotopic clusters corresponding to "protected" (circles) and "fully exchanged" (squares) protein molecules. (**B**) EX2-type kinetics reflecting local fluctuations within the "protected" conformation.

5.3.6. Exchange in Multistate Protein Systems: Superposition of EX1 and EX2 Processes and Mixed Exchange Kinetics

Most proteins are not simple two-state systems but possess one or more "intermediate" states that are usually only partially structured or else their structure may exhibit a significant degree of flexibility.* The presence of such intermediate states can be detected only indirectly by HDX MS under EX2 conditions

*A classical molten globular state is perhaps one of the most common examples of an intermediate state; see Chapter 1 for a more detailed discussion.

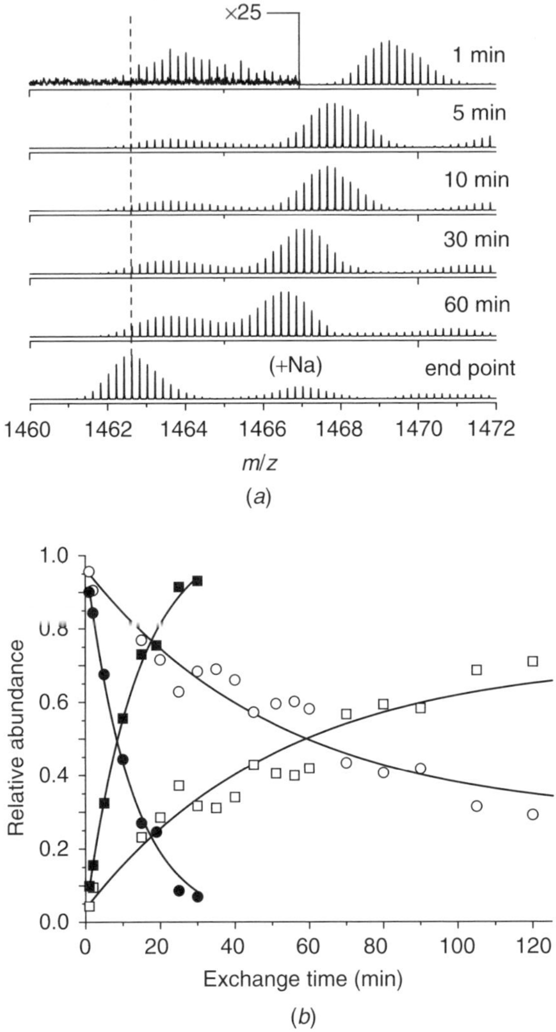

FIGURE 5.13. HDX of chymotrypsin inhibitor 2 under the intermediate exchange conditions (EXX regime, the refolding rates are comparable to the intrinsic exchange rates). The exchange was initiated by diluting a fully deuterated protein in a protiated solution (10 mM $CH_3CO_2NH_4$, pH adjusted to 10.0 with NH_4OH, 70% CH_3OH). (**A**) Evolution of the isotopic distribution as a function of exchange time at solution temperature 8 °C. (**B**) Changes in relative abundance of isotopic clusters corresponding to "protected" (circles) and "unprotected" (squares) protein molecules. (**C**) EX2-type kinetics reflecting local fluctuations within the "protected" conformation. (**D**) Mass shift of the isotopic cluster corresponding to the "unprotected" protein molecules. The two data sets in panels (**B–D**) correspond to solution temperature of 8 °C (white) and 23 °C (black).

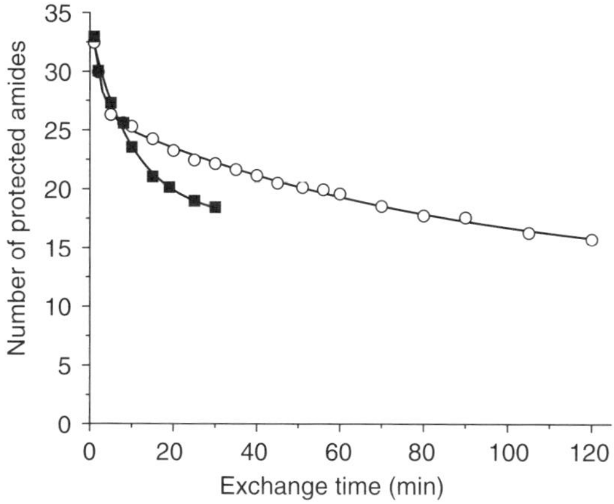

(*c*)

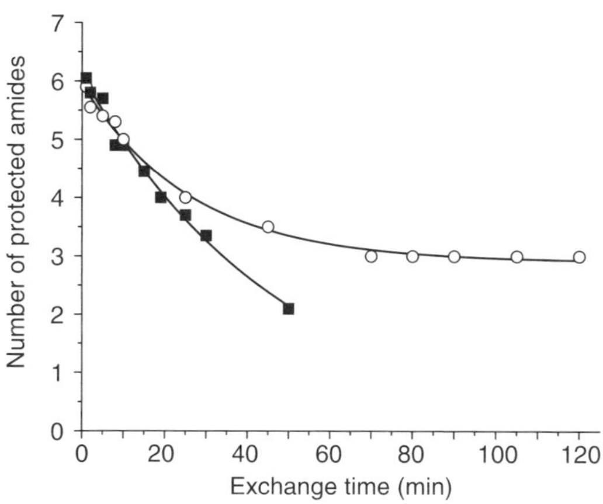

(*d*)

FIGURE 5.13. (*Continued*)

(as an intermediate phase in the overall exchange kinetics, which is faster than the global unfolding but slower than the local fluctuations). These intermediate states can be observed distinctly (at least, in principle) under EX1 conditions. However, this can only be achieved if the amide protection within the intermediate state is significant (to make it distinct from the fully unfolded state), but noticeably different from that of the native state. Another important requirement is that all state-to-state transitions in this system follow the EX1 mechanism. This latter requirement is not satisfied in many practically interesting cases (e.g., pH- or alcohol-induced unfolding). As a result, some states can escape direct detection (or else their residual protection can be misread) in a straightforward HDX MS experiment. One interesting example is presented in Figure 5.14, where HDX of a small protein ubiquitin (Ub) is carried out under conditions known

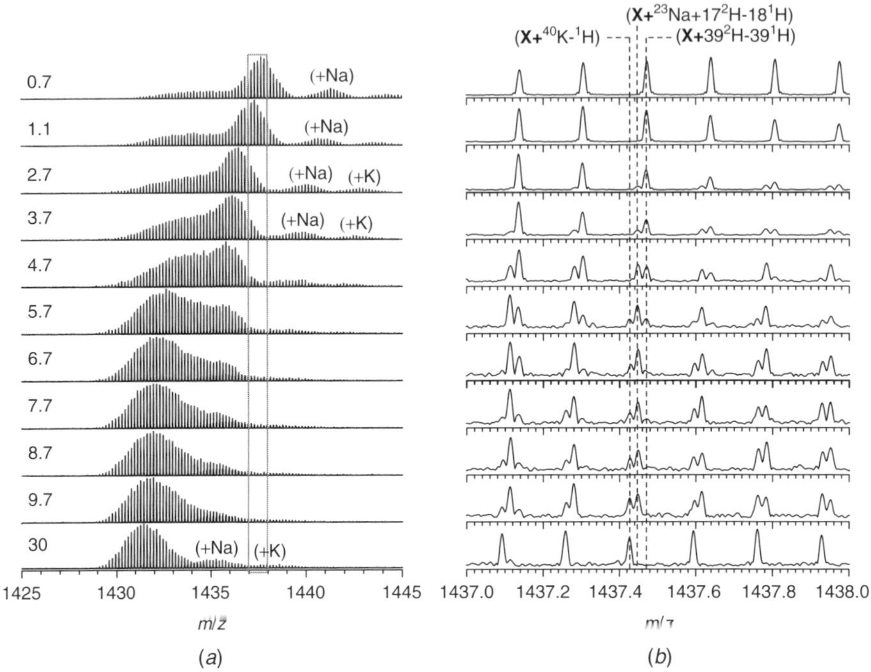

FIGURE 5.14. HDX of ubiquitin under conditions favoring EX1-type exchange (pH 7.0, 60% CH_3OH). Only $N \rightarrow A$ transition occurs under the EX1 regime, while the complete exchange occurs via fluctuations of the *A*-state and has all characteristics of uncorrelated (EX2-type) exchange (**A**). Zoomed regions of the spectra (**B**) show mass-resolved isobaric peaks (protonated and alkali metal ion-cationized ions having different ^{2}H content).

to cause a fraction of the protein molecules to assume a molten globular conformation (*A*-state), as well as a fully unstructured state (*U*) in addition to an abundant native conformation (*N*). However, the HDX profiles measured under such conditions are only bimodal (the fully unfolded state *U* escapes direct detection). The reason for such behavior is that while the $N \rightarrow A$ transition under these conditions appears to follow the EX1 exchange mechanism, sampling of the *U*-state occurs primarily through structural fluctuations of the *A*-state (direct $N \rightarrow U$ transitions are very rare), leading to an apparent EX2-type exchange in the *U*-state. Such behavior is illustrated with a three-state energy diagram presented in Figure 5.15. The reverse activation energy barrier separating the *A*-state from the native conformation is sufficiently high, so that each $N \rightarrow A$ transition lasts long enough to allow full exchange of all amides unprotected in the *A*-state. At the same time, the barrier separating the *U*-state from the *A*-state is very low, so that the *U*-state becomes populated only transiently, hence the EX2 character of the exchange kinetics corresponding to the $A \rightarrow U$ transition.

Despite the limitations discussed in this section, it is possible in some cases to observe truly multimodal isotopic distributions indicative of the presence of

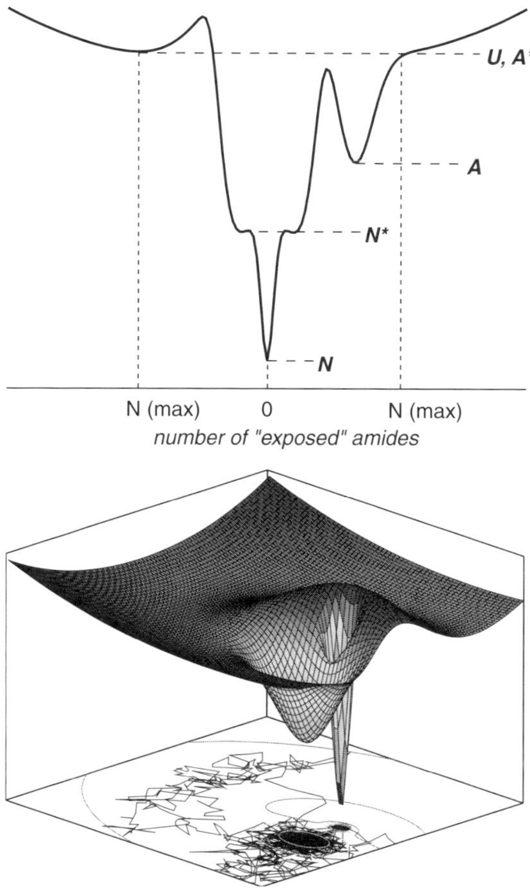

FIGURE 5.15. Minimalistic representation of a three-state protein system. Transition from the ***N***-state (global energy minimum) to a partially structured molten globule-like ***A***-state occurs under EX1 conditions. Transition from the ***A***-state to a fully unstructured ***U***-state occurs under EX2 conditions.

multiple (three or more) protein conformations in solution under equilibrium (Figure 5.16). More examples of detecting multiple (kinetic) intermediate states will be presented and discussed in Chapter 6. A final comment that should be made in connection with our discussion of global HDX patterns measured by MS relates to one particularly annoying artifact that often complicates data interpretation. As has been mentioned in Chapter 1, peptide and protein ions produced by ESI often contain one or more Na^+ and/or K^+ ions. These adduct ion peaks may overlap with and obscure the details of the convoluted bi- and multimodal isotopic distributions observed in the course of HDX MS measurements. Fortunately, high-resolution mass spectrometry offers an elegant way to solve this problem. Both Na and K have single stable isotopes with negative mass defect

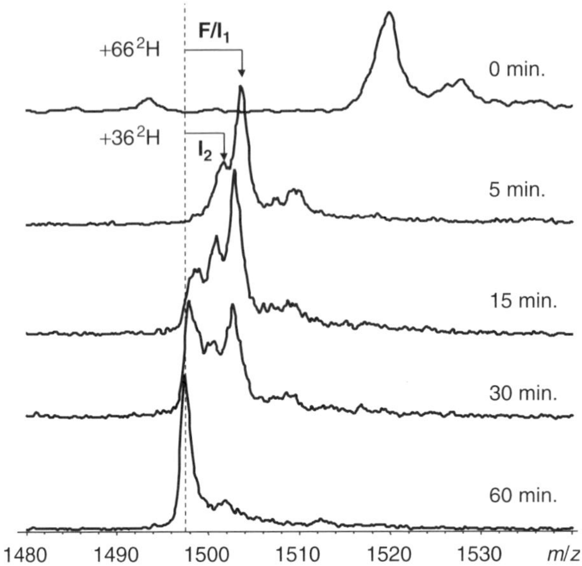

FIGURE 5.16. HDX of pseudo-wild-type cellular retinoic acid binding protein I under conditions favoring EX1 exchange regime (acid catalyzed exchange). Time evolution of a +12 ion peak profile throughout the course of the exchange reaction clearly shows the presence of at least three states under equilibrium differing by the degree of backbone protection. The dashed line on the left indicates the position of the fully exchanged protein ion peak. Reproduced with permission from (110).

(see Table 3.1). On the other hand, the mass difference between ^{2}H and ^{1}H is $+1.0063u$. Therefore, Na^+ and K^+ adducts can often be resolved from their deuterium-containing isobars (e.g., see Figure 5.14B).

The ability of HDX MS to reveal the presence and characterize the behavior of distinct intermediate states is quite unique, as ^{1}H NMR generates residue-specific exchange data averaged across the entire ensemble of states, thus complicating the detection of distinct conformations. Although in some instances certain information about the intermediates may be obtained by grouping amides with similar exchange kinetics into cooperative unfolding–refolding units, or *foldons* (62), this approach may result in artifacts (79, 80).

5.4. MEASUREMENTS OF LOCAL PATTERNS OF HYDROGEN EXCHANGE

As we have seen in the preceding section, HDX/ESI MS is a very powerful tool for characterizing the global behavior of proteins (by providing information on the overall bulk exchange pattern); but it is falling behind HDX NMR methodologies as far as local dynamic and structural detail. Such site-specific information is necessary for producing a detailed picture of the dynamic behavior of individual structural segments within a protein.

5.4.1. "Bottom-up" Approaches to Probing the Local Structure of Intermediate States

Mass spectrometry is unrivaled as far as being able to provide detailed structural information on proteins and peptides (i.e., amino acid sequence and post-translational modifications) using only minute quantities of the analyte and has become a primary experimental tool in proteomics (81, 82). However, using mass spectrometry to localize labile ^{1}H (or ^{2}H) atoms within a protein or a polypeptide is not a trivial task. Proteolytic fragmentation and separation of the fragments prior to MS analysis will inevitably alter both the overall isotopic content and its distribution across the polypeptide chain, even if carried out for short periods of time. This problem can be at least partially remedied for one particularly important class of labile hydrogen atoms, those belonging to the backbone amide groups. It was realized a quarter century ago that one ubiquitous proteolytic enzyme, pepsin, is active within the pH range 2.5–3, under which the intrinsic amide exchange rates are typically minimal, which makes it suitable for probing amide protection (83, 84). All side chain labile hydrogen atoms exchange fast in this pH range (see Figure 5.5), and the commonly used terms *quenched exchange* or *slow exchange* refer only to backbone amides. Even the amide hydrogen atoms continue to exchange under such slow exchange conditions, and the local exchange details can be maintained only if the sample handling is relatively fast and is performed at low solution temperature (typically 0–4 °C).

A decade ago Smith and co-workers combined HDX MS with peptic digest of the "labeled" proteins carried out under the "quenched exchange" conditions prior to mass analysis (63, 85). In the commonly used procedure, labeled proteins are digested in solution. Although this approach is very popular, it is not ideal, largely due to significant back-exchange during the proteolytic reaction, which typically takes several minutes. The large quantity of pepsin required for rapid digestion may also interfere with a chromatographic step (if separation of peptic fragments is carried out prior to mass analysis) and adversely affect the ESI process (86). These problems can be remedied by utilizing a recently introduced "on-line" approach (86, 87), which makes use of an immobilized pepsin column in tandem with an HPLC column to carry out fast and efficient digestion and desalting/preconcentration/separation of peptic fragments prior to their mass analysis (Figure 5.17). In many cases the separation step can be bypassed, as the different peptic fragments can be confidently resolved and identified even if their peaks in the mass spectra partially overlap (Figure 5.18). Any back-exchange occurring prior to mass analysis does not usually exceed 10% and may be accounted for by introducing a "back-exchange" correction factor (63, 85).

The spatial resolution offered by this technique is usually limited only by the extent of proteolysis. In general, a large number of fragments, particularly overlapping ones, would lead to greater spatial resolution, and hence more precise localization of the structural regions that have undergone exchange. Presently, pepsin remains the most popular proteolytic enzyme suitable for such applications (most other ubiquitous enzymes are inactive under the "slow exchange" conditions), although the search for suitable alternatives continues. Forest and co-workers recently reported that a combination of pepsin with

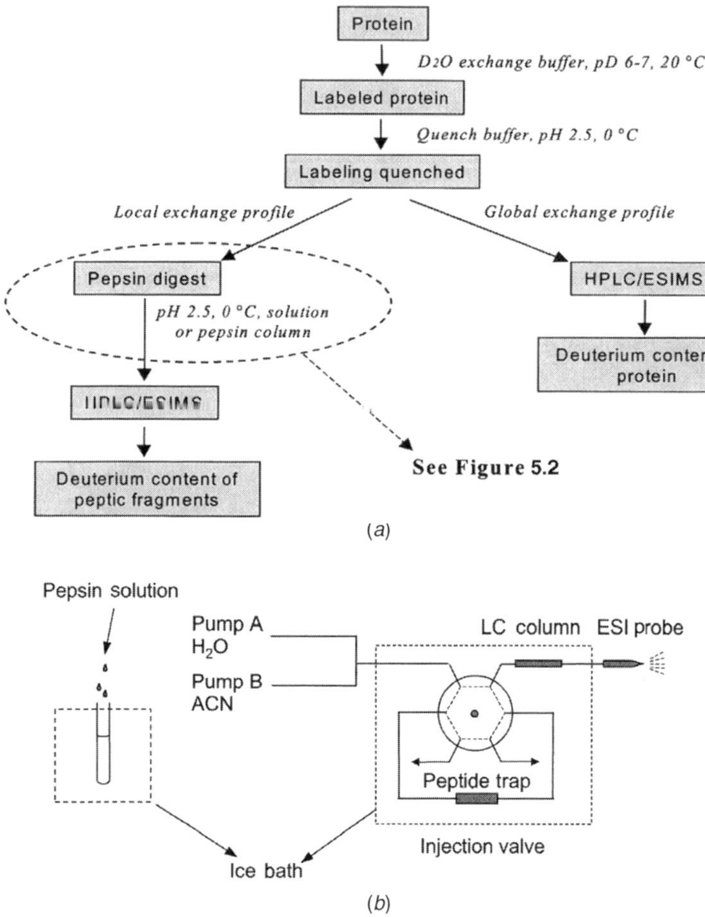

FIGURE 5.17. General procedures used to study backbone amide dynamics in a site-specific fashion by HDX MS (**A**). Labeled proteins may be analyzed either as intact proteins (global exchange) or peptic fragments (local exchange) by HPLC ESI MS under conditions that minimize deuterium loss at peptide amide groups. Two approaches can be used to digest the labeled proteins and mass-analyze the proteolytic fragments. (**B**) Proteolysis in solution is followed by desalting and separation of fragment peptides. (**C**) Proteolysis is carried out "on-line" using a column packed with immobilized pepsin and a peptide trap to concentrate the peptic fragments. The pepsin column does not have to be kept in the ice bath due to very short reaction time. The pH is always maintained at 2.5 to minimize back-exchange. Adapted with permission from (86).

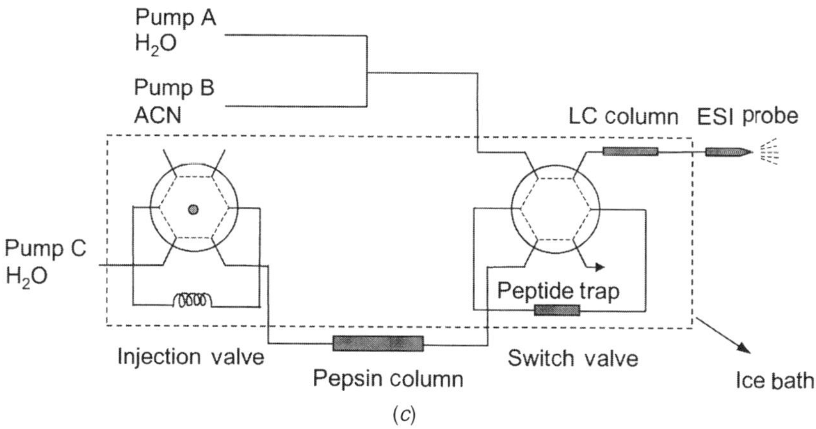

FIGURE 5.17. (*Continued*)

proteases from *Aspergillus satoi* (type XIII) and *Rhizopus* sp. (type XVIII) had increased sequence coverage of a 77 kDa protein under "slow exchange" conditions (0 °C, pH 2.5) (88). Like pepsin, neither of these proteases is specific, but there is good reproducibility in the protein digestion patterns. A significant number of other proteolytic agents (both endo- and exoproteases) retain their enzymatic activity in the pH range corresponding to the slow exchange conditions, although their commercial availability and/or cost remain an issue. An exception is a set of several commercially available carboxypeptidases, such as carboxypeptidase Y, which exhibits high enzymatic activity in acidic environments (89, 90).

A structural feature whose presence in proteins typically has a very significant negative impact on the spatial resolution achievable with "bottom-up" HDX MS measurements described in this section is disulfide bonding. Not only does it limit the action of pepsin under denaturing conditions (by helping to maintain compact structures, therefore reducing accessibility of the "candidate" cleavage sites to the enzyme), but it also prevents physical separation of peptic fragments connected by the cysteine–cysteine bridges even if the enzymatic reaction is successful. The common disulfide-reducing agents (such as dithiothreitol, DTT) are inactivated at acidic pH (91) and, therefore, cannot be used under the slow exchange conditions. The task of reducing disulfides under such conditions can be successfully carried out by another agent, tris(2-carboxyethyl)phosphine (92, 93), which has been shown to remain stable and retain its disulfide-reducing capacity at pH as low as 1.5 (91).

Potentially, spatial resolution can be improved further by introducing a gas phase fragmentation step during mass analysis (e.g., MS/MS). Viability of this approach was recently demonstrated by Smith and co-workers (94). In this work, cytochrome *c* amides were selectively labeled with deuterium at pD 7.0, followed by solution acidification to quench the exchange, and digestion with pepsin. Deuterium content and distribution within peptic fragments were determined by

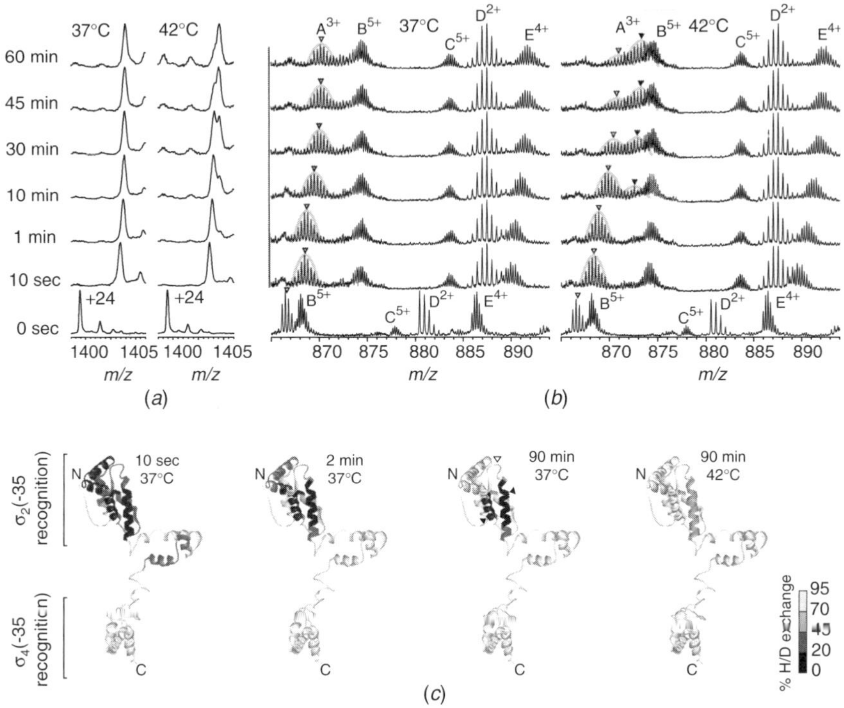

FIGURE 5.18. Exchange kinetics of peptic fragments of *E. coli* heat shock transcription factor $\sigma32$. (**A**) HDX of full-length $\sigma32$ at 37 and 42 °C (evolution of the charge state +24). After 30 min of exchange at 42 °C the bimodal distribution is clearly visible. (**B**) Exchange kinetics of peptic fragments of $\sigma32$ at 37 and 42 °C. Representative region of the mass spectra showing five peptic peptides (A–E) of $\sigma32$ after exposure to D_2O for different times. The isotopic natural abundance of the peptides is shown in the lowest trace (0 min, before amide hydrogen exchange). The traces above show the extent of deuterium incorporation of each peptide (0.2 to 60 min) at 37 °C (*left panel*) and 42 °C (*right panel*). For the triply charged peptide A, the isotope envelopes are indicated as well as the bimodal distribution (open and solid triangles). Peptic fragments are designated as follows: A, [79–101]; B, [121–155]; C, [257–294]; D, [2–18]; E, [19–49]. (**C**) Model of the $\sigma32$ structure colored according to the exchange behavior of various segments at 37 °C after 10 s, 2 min, and 90 min as well as at 42 °C after 90 min (left to right). In the third structure from the left (90 min, 37 °C), two helices with a connecting loop are marked by solid and open arrowheads, respectively, to demonstrate the precision with which the deuteron incorporation data fit this structural model. Reproduced with permission from (127).

carrying out MS and MS/MS (collision-induced dissociation, CID) experiments using a commercial ion trap instrument. Comparison of the experimental data with the results of published NMR experiments led to the conclusion that deuterium content of the *b*-series of fragment ions can serve as an accurate predictor of the exchange kinetics at individual amides. At the same time, deuterium content analysis of the y'' fragment ions yielded several discrepancies, suggesting the possibility of limited hydrogen migration during the dissociation process. Similar conclusions were reached more recently by Deinzer and co-workers, who studied acid unfolding of oxidized and reduced *E. coli* thioredoxins using a combination of HDX, peptic digest, and CID MS (95).

Although ESI MS is the most popular MS technique used in conjunction with HDX measurements, MALDI is now also rapidly gaining acceptance in the field. Komives and co-workers first demonstrated that the "standard" HDX/MS protocol (exchange–quench–proteolysis) can be implemented using MALDI MS to monitor deuterium incorporation in a site-specific fashion (96). Only minor modifications of established MALDI sample preparation protocols are required to carry out such experiments. Quenching the amide HDX and peptic digest are followed by chilled, rapid drying of the samples mixed with a matrix solution whose pH is adjusted to 2.5 on the MALDI target to ensure minimal loss of deuterons from peptide amide groups. Although the sample remains completely dry and is kept under high vacuum during the analysis, the deuterium content of the peptides decreases slowly (single-curve exponential decay). Therefore, a correction for the deuterium loss prior to mass analysis can easily be made (96). Although the HDX/MALDI MS experiments usually provide kinetic information comparable to that obtained using ESI, some discrepancies do exist. In their studies of unfolding and aggregation of a surfactant protein, Gustafsson and co-workers noted that the measured HDX rates were consistently higher when ESI MS was used as a detector, as compared to MALDI (97). This was interpreted as a result of the polypeptide unfolding in the ESI source leading to greater exposure of amide protons to the solvent.

Use of MALDI offers superior sensitivity (at subpicomolar level) and eliminates the need for preconcentration and separation steps prior to MS analysis of the deuterium content of peptic fragments. As is the case with HDX/ESI MS experiments, the spatial resolution offered by HDX/MALDI MS experiments may be enhanced even further by using peptide ion dissociation in the gas phase, for example, postsource decay in time-of-flight (TOF) analyzers (98).

5.4.2. "Top-down" Approaches to Probing the Local Structure of Intermediate States

An alternative method to probe amide HDX kinetics locally in smaller polypeptides with ESI MS was introduced by Anderegg and co-workers (99). The method relied on the ability of mass spectrometers to produce a wealth of structural information in tandem (MS/MS or MS^n) experiments. In this scheme, collision-induced dissociation (CID) was employed to fragment relatively small polypeptides (<4 kDa) in the gas phase in order to obtain residue-specific information on the location of labile hydrogen atoms. More recent applications include the 23-residue N-terminal domain of HIV-1 glycoprotein gp41 (100), model fibril-forming peptides (98), and model trans-membrane peptides (101, 102). The experimental scheme is very attractive, as it offers the simplicity of an "on-line" HDX/ESI MS experiment and yet produces detailed information on amide protection at the residue level. It has been also shown that dissociation of peptic fragments in the gas phase can improve spatial resolution of the HDX MS measurements that utilize proteolysis under slow exchange conditions prior to mass analysis (94).

Extensive fragmentation of the polypeptide ions in the gas phase often allows protection of every backbone amide to be measured; this latter advantage, however, diminishes dramatically as the polypeptide size increases. The practical upper mass limit of a polypeptide for which abundant structurally diagnostic CID fragment ions can be produced using most types of mass analyzers is about 5 kDa. Such a severe mass limitation (much less forgiving than that imposed by ^{1}H NMR spectroscopy) inevitably narrowed the scope of systems to which this new methodology could be applied.

The mass range of biopolymers amenable to sequencing in the gas phase has expanded very dramatically in the past several years following spectacular progress in Fourier transform ion cyclotron resonance (FT ICR) technology. It is now possible in many cases to obtain a wealth of structural information (both sequence and post-translational modifications) on large proteins without carrying out extensive proteolytic fragmentation and separation of proteins prior to MS and MS/MS analysis (103). This work was stimulated in large part by the ever-increasing demands of proteomics, where the necessity to obtain maximal sequence coverage within minimal time using only minute protein quantities is now shifting the attention from the traditional "bottom-up" approach (104) to "top-down" methods (105). FT ICR MS has now truly become a practical tool to study protein structure, displaying an impressive array of unique analytical capabilities. Perhaps the most important ones within the context of the subject of this chapter are the unsurpassed mass resolution provided by FT ICR analyzers and an expanded arsenal of ion fragmentation methods. Application of infrared multiphoton dissociation (IRMPD) (106) and electron capture dissociation (ECD) (107) for polypeptide and protein sequencing has greatly expanded the range of biopolymers for which structurally diagnostic fragment ions can be produced. In addition, even "classical" ion fragmentation techniques (e.g., CID) result in significantly higher yields of fragment ions (as compared to other types of mass analyzers) due to longer ion trapping times afforded by FT ICR MS. Furthermore, FT ICR MS versatility easily allows several ion activation techniques to be combined in one experiment, further increasing fragmentation yields and sequence coverage (108, 109). Interpretation of such fragmentation spectra can pose a significant challenge, however, due to a plethora of possible fragment ion peaks confined to a relatively narrow m/z range (typically <2500 u). This task is greatly simplified when FT ICR is used for mass analysis, as the high resolving power provides a means to resolve the isotopic clusters of overlapping peaks of ions differing in their charge states.

These advances have had a major impact on the development of the HDX/ESI/CID MS methodology. The protein ion fragmentation efficiency offered by commercial FT ICR mass spectrometers has become sufficiently high to have enabled extensive characterization of local dynamic events for a protein exceeding 15 kDa (pseudo-wild-type cellular retinoic acid binding protein I, *wt**-CRABP I) under mildly denaturing conditions (110, 111). Protein ion fragmentation was carried out directly in the ESI source using nozzle-skimmer collisional activation, that is, without prior mass selection. This approach (fragmentation of the entire protein ion population) also has been adopted in other work (112), since it maximizes the yield of fragment ions without having any detrimental effect on the measurements. Mass evolution of each protein fragment ion (Figure 5.19) provides a measure of local deuterium content of the protein as a function of the exchange time in solution and, therefore, its backbone dynamics.

The HDX/ESI/CID MS measurements of the *wt**-CRABP I dynamics were carried out on-line (by continuously infusing the protein solution from the exchange reactor to the ESI source). Off-line experiments (i.e., taking aliquots from the exchange reactor after certain periods of time and quenching the exchange, followed by CID MS analysis) present another choice (113, 114). This might be a preferred method when longer data acquisition time is needed to improve signal-to-noise ratio, for instance, in a situation where the protein ion fragmentation efficiency is low.

It should probably be mentioned that the validity of data obtained in HDX/ESI/CID MS experiments can be questioned based on the possibility of hydrogen scrambling in the gas phase (prior to protein ion dissociation) (115). Johnson and co-workers were perhaps the first to identify this problem (116). Subsequently, McLafferty and colleagues reported extensive hydrogen scrambling when fragmenting gaseous cytochrome *c* ions in the cell of an FT ICR MS (117). On the other hand, the groups of Anderegg (99) and Waring (100) reported that scrambling was minimal for short helical peptides under typical CID conditions in a collision cell of a triple quadrupole mass spectrometer. More recently, Kraus and co-workers reported on the use of ESI MS/MS (ion trap) to monitor deuterium content of short peptides in their studies of amyloid fibril formation in solution, and the results appeared to be unaffected by scrambling (98). As mentioned earlier in this chapter, Smith and co-workers (94) and later Deinzer and co-workers (95) observed some hydrogen scrambling within the *y*-type (but not *b*-type) fragment ions produced by CID of small peptic fragments in a quadrupole ion trap MS.

We have examined the scrambling issue within larger protein ions ($>$18 kDa) by monitoring deuterium content of an intrinsically unstructured His-tag (111). Protein ion dissociation was achieved by using CID in the ESI interface region of an FT ICR mass spectrometer. We argued that the absence of scrambling in the gas phase would render the fragment ions derived from the His-tag region essentially label-free (due to lack of any protection of this segment in solution). On the other hand, even limited scrambling was expected to introduce some labels into that segment prior to protein ion fragmentation. No evidence of the latter was found, leading us to conclude that scrambling did not occur or else was very limited (111). Finally, Heck and co-workers have examined proton mobility within shorter model transmembrane peptides and reported detectable scrambling whose extent could be linked to the nature of the charge carrier (i.e., protonated and alkali metal-cationized peptide ions behaved differently) (118).

A lack of consensus in these and other (119) studies has led us to suggest that there may be several factors that govern proton mobility in peptide and protein ions (115). Outcome of any particular experiment would then depend on a combination of such factors. Two factors have been suggested as prime candidates as major determinants of scrambling, namely, collision energy and flexibility of the polypeptide chain in the gas phase (115). Higher collision energy (or, more generally, rapid ion activation) favors direct cleavages, while slow heating methods (120) often lead to extensive internal rearrangements prior to dissociation. It follows then that the slow heating processes, such as SORI (120), should be expected to result in extensive hydrogen scrambling prior to fragmentation, in line with recent observations (117). Most other mass analyzers, however, employ significantly higher activation energy (120). Limited scrambling associated with formation of *y*-ions (but not *b*-ions) in quadrupole ion traps most likely reflects the differences in the mechanisms of their formation (94, 95).

Chain flexibility in the gas phase should also be important, since the "intrinsic" proton mobility is rather limited. Flexibility may enhance proton mobility by allowing two distant segments of the protein to come into close contact, thus enabling intersegment transfer of labile hydrogen atoms. In order for the internal exchange to occur, the two exchanging sites must come into direct contact with one another (akin to a bimolecular H/D exchange reaction in the gas phase). Lack of such flexibility due to either (partial) retention of the solution structure in the gas phase (121) or the "cementing" role of metal cations (122) would limit the extent of scrambling. These simple considerations suggest that the best chance of avoiding hydrogen scrambling in the gas phase would result from employing fast (high-energy) ion fragmentation methods and using solution conditions that allow proteins to remain (partially) structured. We now have evidence that confirms the correctness of this notion (123).

5.4.3. Further Modifications and Improvements of HDX MS Measurements

One other difficulty facing HDX MS measurements in general (both "global" and "local" measurements) is that even under EX1 exchange conditions the isotopic clusters corresponding to different intermediate states can overlap. This overlap would be particularly significant if the extent of protection does not differ greatly among the intermediate states involved. Structural fluctuations within these intermediate states would result in a partial EX2-like exchange component of the observed HDX pattern (vide supra), thus broadening the isotopic clusters and further exacerbating the problem. Even without any exchange the isotopic clusters of protein ions have substantial width due to the contributions of naturally occurring "minor" isotopes of carbon, nitrogen, oxygen, and sulfur (^{13}C, ^{15}N, ^{18}O, ^{33}S, and ^{34}S).* The latter problem can be addressed by removing the background isotopic distributions. Depletion of ^{13}C alone is expected to give a narrower isotopic cluster for CRABP I ions (Figure 5.19A). This can be achieved by

*Contribution of the naturally occurring deuterium is negligible and can safely be ignored despite the large number of hydrogen atoms in a typical protein.

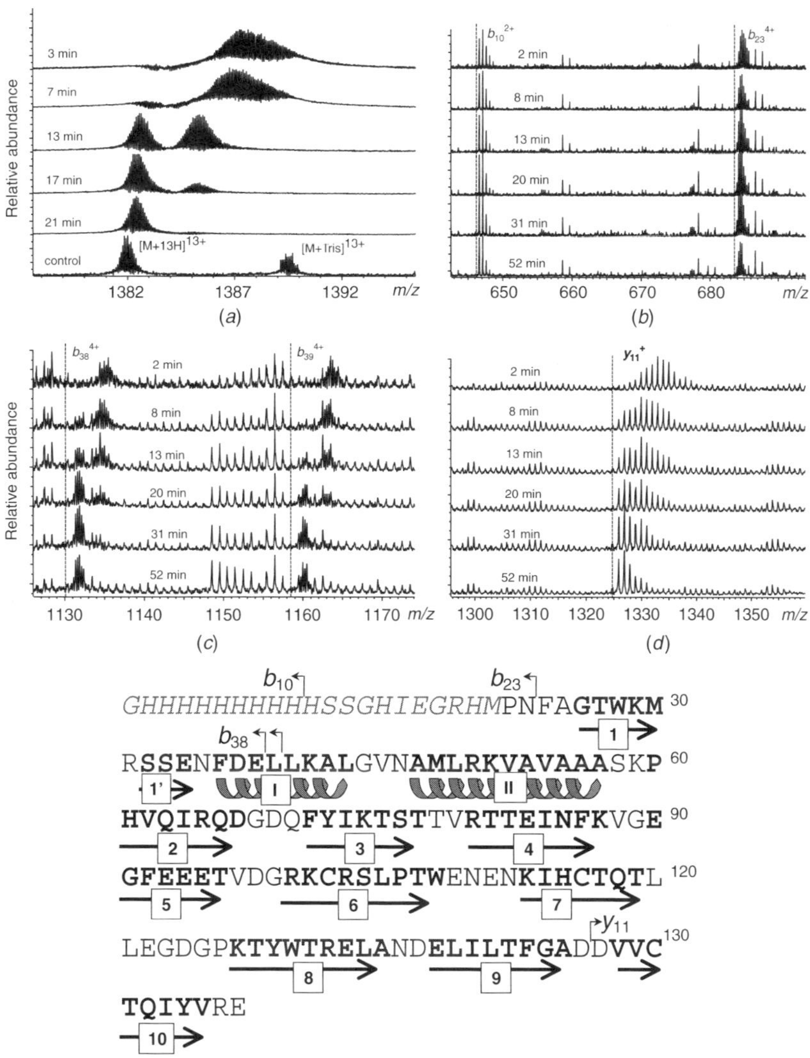

FIGURE 5.19. Use of HDX ESI MS and protein ion fragmentation in the gas phase (CAD) to probe local conformational dynamics. Concentrated solution of a fully deuterated protein (pseudo-wild-type cellular retinoic acid binding protein I) in D_2O was diluted 1:50 (v:v) in a protiated buffer (^1H_2O). HDX was monitored globally and locally by following protein mass shift due to replacement of labile ^2H atoms with ^1H atoms. Panel (**A**) shows time evolution of the isotopic distribution of intact protein ions (+13 charge state). Panels (**B**–**D**) show time evolution of fragment ion peaks representing three different protein segments: slow uncooperative exchange (EX2-type) within a "frayed" region of the protein (**D**), very fast exchange within the unstructured His-tag (**B**), and slow cooperative exchange (EX1-type) within a helical region (**C**). The diagram at the bottom shows primary and secondary structure of the protein and positions of the backbone cleavages giving rise to fragment ions (**B**–**D**). Adapted with permission from (111). © 2000 American Chemical Society.

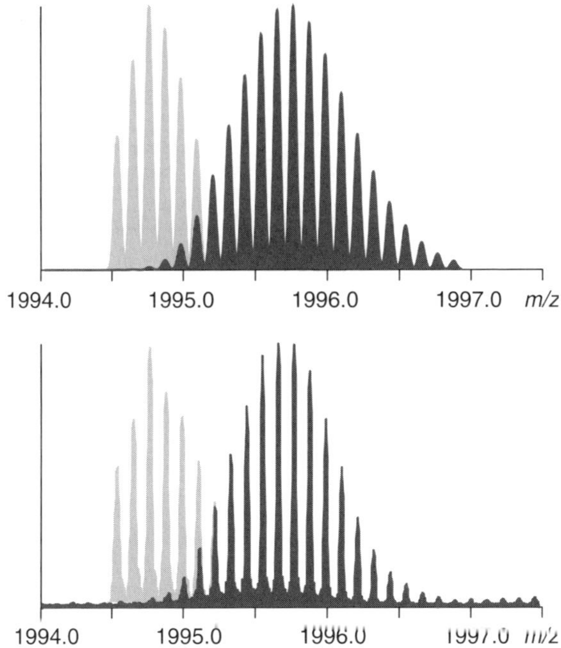

FIGURE 5.20. The effect of ^{13}C depletion on the isotopic distribution of a pseudo-wild-type cellular retinoic acid binding protein I ions. Calculated (*top*) and experimentally measured (*bottom*) isotopic distributions corresponding to intact ions (charge state +9) of proteins expressed in *E. coli* fed with growth media containing regular (dark) and ^{12}C-enriched (99.9%) glucose (light gray). Adapted with permission from (112).

expressing proteins in cells grown on "isotopically depleted" media. An example of the isotopic distribution of CRABP I "grown" on 99.9% ^{12}C glucose is shown on Figure 5.20. In addition to "collapsing" the broad natural distribution to a few isotopic peaks, one can also clearly see the monoisotopic peak of the intact protein ion. This allows the interpretation of the mass spectra to be dramatically simplified (124) and should enable more precise local measurements of deuterium occupancies in the HDX MS experiments that utilize either proteolytic or gas phase protein fragmentation prior to mass analysis.

Alternatively, the convoluted isotopic distributions in some cases can be decomposed to yield contributions from each protein conformation (125). Perhaps it will be possible in the future to streamline such deconvolution by using various chemometric techniques, similar to processing convoluted charge state distributions of protein ions using factor analysis (24).

Another important modification of HDX MS measurements that greatly improves the temporal resolution afforded by the technique utilizes a rapid mixing apparatus (126) interfaced to the ESI source and will be discussed in the next chapter.

REFERENCES

1. Chowdhury S. K., Katta V., Chait B. T. (1990). Probing conformational changes in proteins by mass spectrometry. *J. Am. Chem. Soc.* **112**: 9012–9013.
2. Loo J. A., Loo R. R., Udseth H. R., Edmonds C. G., Smith R. D. (1991). Solvent-induced conformational changes of polypeptides probed by electrospray-ionization mass spectrometry. *Rapid Commun. Mass Spectrom.* **5**: 101–105.
3. Fenn J. B. (1993). Ion formation from charged droplets: roles of geometry, energy and time. *J. Am. Soc. Mass Spectrom.* **4**: 524–535.
4. Yu X., Wojciechowski M., Fenselau C. (1993). Assessment of metals in reconstituted metallothioneins by electrospray mass spectrometry. *Anal. Chem.* **65**: 1355–1359.
5. Wang F., Tang X. (1996). Conformational heterogeneity of stability of apomyoglobin studied by hydrogen/deuterium exchange and electrospray ionization mass spectrometry. *Biochemistry* **35**: 4069–4078.
6. Konermann L., Douglas D. J. (1997). Acid-induced unfolding of cytochrome *c* at different methanol concentrations: electrospray ionization mass spectrometry specifically monitors changes in the tertiary structure. *Biochemistry* **36**: 12296–12302.
7. Konermann L., Douglas D. J. (1998). Equilibrium unfolding of proteins monitored by electrospray ionization mass spectrometry: distinguishing two-state from multistate transitions. *Rapid Commun. Mass Spectrom.* **12**: 435–442.
8. Konermann L., Douglas D. J. (1998). Unfolding of proteins monitored by electrospray ionization mass spectrometry: a comparison of positive and negative ion modes. *J. Am. Soc. Mass Spectrom.* **9**: 1248–1254.
9. Konermann L., Collings B. A., Douglas D. J. (1997). Cytochrome *c* folding kinetics studied by time-resolved electrospray ionization mass spectrometry. *Biochemistry* **36**: 5554–5559.
10. Lee V. W., Chen Y. L., Konermann L. (1999). Reconstitution of acid-denatured holomyoglobin studied by time-resolved electrospray ionization mass spectrometry. *Anal. Chem.* **71**: 4154–4159.
11. Kolakowski B. M., Simmons D. A., Konermann L. (2000). Stopped-flow electrospray ionization mass spectrometry: a new method for studying chemical reaction kinetics in solution. *Rapid Commun. Mass Spectrom.* **14**: 772–776.
12. Kolakowski B. M., Konermann L. (2001). From small-molecule reactions to protein folding: studying biochemical kinetics by stopped-flow electrospray mass spectrometry. *Anal. Biochem.* **292**: 107–114.
13. Gumerov D. R., Kaltashov I. A. (2001). Dynamics of iron release from transferrin N-lobe studied by electrospray ionization mass spectrometry. *Anal. Chem.* **73**: 2565–2570.
14. van den Bremer E. T., Jiskoot W., James R., Moore G. R., Kleanthous C., Heck A. J., Maier C. S. (2002). Probing metal ion binding and conformational properties of the colicin E9 endonuclease by electrospray ionization time-of-flight mass spectrometry. *Protein Sci.* **11**: 1738–1752.
15. Merrifield M. E., Huang Z., Kille P., Stillman M. J. (2002). Copper speciation in the α and β domains of recombinant human metallothionein by electrospray ionization mass spectrometry. *J. Inorg. Biochem.* **88**: 153–172.

16. Gumerov D. R., Mason A. B., Kaltashov I. A. (2003). Interlobe communication in human serum transferrin: metal binding and conformational dynamics investigated by electrospray ionization mass spectrometry. *Biochemistry* **42**: 5421–5428.
17. Kashiwagi T., Yamada N., Hirayama K., Suzuki C., Kashiwagi Y., Tsuchiya F., Arata Y., Kunishima N., Morikawa K. (2000). An electrospray-ionization mass spectrometry analysis of the pH-dependent dissociation and denaturation processes of a heterodimeric protein. *J. Am. Soc. Mass Spectrom.* **11**: 54–61.
18. Hanefeld U., Stranzl G., Straathof A. J., Heijnen J. J., Bergmann A., Mittelbach R., Glatter O., Kratky C. (2001). Electrospray ionization mass spectrometry, circular dichroism and SAXS studies of the (*S*)-hydroxynitrile lyase from *Hevea brasiliensis*. *Biochim. Biophys. Acta* **1544**: 133–142.
19. Guy P., Remigy H., Jaquinod M., Bersch B., Blanchard L., Dolla A., Forest E. (1996). Study of the new stability properties induced by amino acid replacement of tyrosine 64 in cytochrome C553 from *Desulfovibrio vulgaris* Hildenborough using electrospray ionization mass spectrometry. *Biochem. Biophys. Res. Commun.* **218**: 97–103.
20. Griffith W. P., Kaltashov I. A. (2003). Highly asymmetric interactions between globin chains during hemoglobin assembly revealed by electrospray ionization mass spectrometry. *Biochemistry* **42**: 10024–10033.
21. Kamadurai H. B., Subramaniam S., Jones R. B., Green-Church K. B., Foster M. P. (2003). Protein folding coupled to DNA binding in the catalytic domain of bacteriophage lambda integrase detected by mass spectrometry. *Protein Sci.* **12**: 620–626.
22. Zhang Y. H., Yan X., Maier C. S., Schimerlik M. I., Deinzer M. L. (2001). Structural comparison of recombinant human macrophage colony stimulating factor beta and a partially reduced derivative using hydrogen deuterium exchange and electrospray ionization mass spectrometry. *Protein Sci.* **10**: 2336–2345.
23. Dobo A., Kaltashov I. A. (2001). Detection of multiple protein conformational ensembles in solution via deconvolution of charge state distributions in ESI MS. *Anal. Chem.* **73**: 4763–4773.
24. Mohimen A., Dobo A., Hoerner J. K., Kaltashov I. A. (2003). A chemometric approach to detection and characterization of multiple protein conformers in solution using electrospray ionization mass spectrometry. *Anal. Chem.* **75**: 4139–4147.
25. Iavarone A. T., Jurchen J. C., Williams E. R. (2000). Effects of solvent on the maximum charge state and charge state distribution of protein ions produced by electrospray ionization. *J. Am. Soc. Mass Spectrom.* **11**: 976–985.
26. Iavarone A. T., Williams E. R. (2002). Supercharging in electrospray ionization: effects on signal and charge. *Int. J. Mass Spectrom.* **219**: 63–72.
27. Schnier P. D., Gross D. S., Williams E. R. (1995). On the maximum charge state and proton transfer reactivity of peptide and protein ions formed by electrospray ionization. *J. Am. Soc. Mass Spectrom.* **6**: 1086–1097.
28. Gumerov D. R., Dobo A., Kaltashov I. A. (2002). Protein-ion charge-state distributions in electrospray ionization mass spectrometry: distinguishing conformational contributions from masking effects. *Eur. J. Mass Spectrom.* **8**: 123–129.

29. Kim M. Y., Maier C. S., Reed D. J., Deinzer M. L. (2002). Conformational changes in chemically modified *Escherichia coli* thioredoxin monitored by H/D exchange and electrospray ionization mass spectrometry. *Protein Sci.* **11**: 1320–1329.
30. Yan X., Watson J., Ho P. S., Deinzer M. L. (2003). Mass spectrometric approaches using electrospray ionization charge states and hydrogen–deuterium exchange for determining protein structures and their conformational changes. *Mol. Cell. Proteomics* **3**: 10–23.
31. Kebarle P., Ho Y. (1997). On the mechanism of electrospray mass spectrometry. In: *Electrospray Ionization Mass Spectrometry: Fundamentals, Instrumentation and Applications.*, ed. R. B. Cole, pp. 3–63. Hoboken, NJ. John Wiley & Sons.
32. Constantopoulos T. L., Lackson G. S., Enke C. G. (2000). Challenges in achieving a fundamental model for ESI MS. *Anal. Chim. Acta* **406**: 37–52.
33. Pan P., McLuckey S. A. (2003). Electrospray ionization of protein mixtures at low pH. *Anal. Chem.* **75**: 1491–1499.
34. Kaltashov I. A., Fenselau C. C. (1995). A direct comparison of first and second gas-phase basicities of the octapeptide RPPGFSPF. *J. Am. Chem. Soc.* **117**: 9906–9910.
35. Kaltashov I. A., Fenselau C. (1996). Thermochemistry of multiply charged melittin in the gas phase determined by the modified kinetic method. *Rapid Commun. Mass Spectrom.* **10**: 857–861.
36. Grandori R. (2003). Origin of the conformation dependence of protein charge-state distributions in electrospray ionization mass spectrometry. *J. Mass Spectrom.* **38**: 11–15.
37. Read R. J. (1996). As MAD as can be. *Structure* **4**: 11–14.
38. Kataoka M., Goto Y. (1996). X-ray solution scattering studies of protein folding. *Fold. Design* **1**: R107–R114.
39. Murphy R. M. (1997). Static and dynamic light scattering of biological macromolecules: what can we learn? *Curr. Opin. Biotechnol.* **8**: 25–30.
40. Perkins S. J., Ashton A. W., Boehm M. K., Chamberlain D. (1998). Molecular structures from low angle X-ray and neutron scattering studies. *Int. J. Biol. Macromol.* **22**: 1–16.
41. Svergun D. I. (1999). Restoring low resolution structure of biological macromolecules from solution scattering using simulated annealing. *Biophys. J.* **76**: 2879–2886.
42. de la Mora J. F. (2000). Electrospray ionization of large multiply charged species proceeds via Dole's charged residue mechanism. *Analyt. Chim. Acta* **406**: 93–104.
43. Kebarle P., Peschke M. (2000). On the mechanisms by which the charged droplets produced by electrospray lead to gas phase ions. *Anal. Chim. Acta* **406**: 11–35.
44. Cole R. B. (2000). Some tenets pertaining to electrospray ionization mass spectrometry. *J. Mass. Spectrom.* **35**: 763–772.
45. Kebarle P. (2000). A brief overview of the present status of the mechanisms involved in electrospray mass spectrometry. *J. Mass Spectrom.* **35**: 804–817.
46. Craig P. O., Ureta D. B., Delfino J. M. (2002). Probing protein conformation with a minimal photochemical reagent. *Protein Sci.* **11**: 1353–1366.
47. Maleknia S. D., Downard K. M., Chance M. R. (2000). Synchrotron X-ray footprinting reveals unfolding of apomyoglobin helices. *Biophys. J.* **78**: 44a.

48. Chance M. R. (2001). Unfolding of apomyoglobin examined by synchrotron footprinting. *Biochem. Biophys. Res. Comm.* **287**: 614–621.
49. Narayan M., Welker E., Wedemeyer W. J., Scheraga H. A. (2000). Oxidative folding of proteins. *Acc. Chem. Res.* **33**: 805–812.
50. Creighton T. E. (1997). Protein folding coupled to disulphide bond formation. *Biol. Chem.* **378**: 731–744.
51. van den Berg B., Chung E. W., Robinson C. V., Dobson C. M. (1999). Characterisation of the dominant oxidative folding intermediate of hen lysozyme. *J. Mol. Biol.* **290**: 781–796.
52. Wu J., Watson J. T. (1997). A novel methodology for assignment of disulfide bond pairings in proteins. *Protein Sci* **6**: 391–398.
53. Wu J., Watson J. T. (1998). Optimization of the cleavage reaction for cyanylated cysteinyl proteins for efficient and simplified mass mapping. *Anal. Biochem.* **258**: 268–276.
54. Yang Y., Wu J., Watson J. T. (1998). Disulfide mass mapping in proteins containing adjacent cysteines is possible with cyanylation/cleavage methodology. *J. Am. Chem. Soc.* **120**: 5834–5835.
55. Wu J., Yang Y., Watson J. T. (1998). Trapping of intermediates during the refolding of recombinant human epidermal growth factor (hEGF) by cyanylation, and subsequent structural elucidation by mass spectrometry. *Protein Sci.* **7**: 1017–1028.
56. Yang Y., Wu J., Watson J. T. (1999). Probing the folding pathways of long R(3) insulin-like growth factor-I (LR(3)IGF-I) and IGF-I via capture and identification of disulfide intermediates by cyanylation methodology and mass spectrometry. *J. Biol. Chem.* **274**: 37598–37604.
57. Qi J., Wu W., Borges C. R., Hang D., Rupp M., Torng E., Watson J. T. (2003). Automated data interpretation based on the concept of "negative signature mass" for mass-mapping disulfide structures of cystinyl proteins. *J. Am. Soc. Mass Spectrom.* **14**: 1032–1038.
58. Happersberger H. P., Przybylski M., Glocker M. O. (1998). Selective bridging of bis-cysteinyl residues by arsonous acid derivatives as an approach to the characterization of protein tertiary structures and folding pathways by mass spectrometry. *Anal. Biochem.* **264**: 237–250.
59. Happersberger H. P., Cowgill C., Glocker M. O. (2002). Structural characterization of monomeric folding intermediates of recombinant human macrophage-colony stimulating factor β (rhM-CSF) by chemical trapping, chromatographic separation and mass spectrometric peptide mapping. *J. Chromatogr. B* **782**: 393–404.
60. Englander S. W., Mayne L., Bai Y., Sosnick T. R. (1997). Hydrogen exchange: the modern legacy of Linderstrøm-Lang. *Protein Sci.* **6**: 1101–1109.
61. Englander S. W. (2000). Protein folding intermediates and pathways studied by hydrogen exchange. *Annu. Rev. Biophys. Biomol. Struct.* **29**: 213–238.
62. Englander S. W., Mayne L., Rumbley J. N. (2002). Submolecular cooperativity produces multi-state protein unfolding and refolding. *Biophys. Chem.* **101–102**: 57–65.
63. Engen J. R., Smith D. L. (2001). Investigating protein structure and dynamics by hydrogen exchange MS. *Anal. Chem.* **73**: 256A–265A.
64. Kaltashov I. A., Eyles S. J. (2002). Studies of biomolecular conformations and conformational dynamics by mass spectrometry. *Mass Spectrom. Rev.* **21**: 37–71.

65. Hoofnagle A. N., Resing K. A., Ahn N. G. (2003). Protein analysis by hydrogen exchange mass spectrometry. *Annu. Rev. Biophys. Biomol. Struct.* **32**: 1–25.

66. Katta V., Chait B. T. (1991). Conformational changes in proteins probed by hydrogen-exchange electrospray-ionization mass spectrometry. *Rapid Commun. Mass Spectrom.* **5**: 214–217.

67. Sobott F., Robinson C. V. (2002). Protein complexes gain momentum. *Curr. Opin. Struct. Biol.* **12**: 729–734.

68. Dempsey C. E. (2001). Hydrogen exchange in peptides and proteins using NMR-spectroscopy. *Progr. Nucl. Magn. Res. Spectrosc.* **39**: 135–170.

69. Bai Y., Milne J. S., Mayne L., Englander S. W. (1993). Primary structure effects on peptide group hydrogen exchange. *Proteins* **17**: 75–86.

70. Connelly G. P., Bai Y. W., Jeng M. F., Englander S. W. (1993). Isotope effects in peptide group hydrogen exchange. *Proteins* **17**: 87–92.

71. Liepinsh E., Otting G. (1996). Proton exchange rates from amino acid side chains—implications for image contrast. *Magn. Reson. Med.* **35**: 30–42.

72. Hvidt A., Nielsen S. O. (1966). Hydrogen exchange in proteins. *Adv. Protein Chem.* **21**: 287–386.

73. Bai Y., Milne J. S., Mayne L., Englander S. W. (1994). Protein stability parameters measured by hydrogen exchange. *Proteins* **20**: 4–14.

74. Maity H., Lim W. K., Rumbley J. N., Englander S. W. (2003). Protein hydrogen exchange mechanism: local fluctuations. *Protein Sci.* **12**: 153–160.

75. Qian H., Chan S. I. (1999). Hydrogen exchange kinetics of proteins in denaturants: a generalized two-process model. *J. Mol. Biol.* **286**: 607–616.

76. Itzhaki L. S., Neira J. L., Fersht A. R. (1997). Hydrogen exchange in chymotrypsin inhibitor 2 probed by denaturants and temperature. *J. Mol. Biol.* **270**: 89–98.

77. Milne J. S., Mayne L., Roder H., Wand A. J., Englander S. W. (1998). Determinants of protein hydrogen exchange studied in equine cytochrome *c*. *Protein Sci.* **7**: 739–745.

78. Miller D. W., Dill K. A. (1995). A statistical mechanical model for hydrogen exchange in globular proteins. *Protein Sci.* **4**: 1860–1873.

79. Arrington C. B., Teesch L. M., Robertson A. D. (1999). Defining protein ensembles with native-state NH exchange: kinetics of interconversion and cooperative units from combined NMR and MS analysis. *J. Mol. Biol.* **285**: 1265–1275.

80. Arrington C. B., Robertson A. D. (2000). Correlated motions in native proteins from MS analysis of NH exchange: evidence for a manifold of unfolding reactions in ovomucoid third domain. *J. Mol. Biol.* **300**: 221–232.

81. Aebersold R. (2003). A mass spectrometric journey into protein and proteome research. *J. Am. Soc. Mass Spectrom.* **14**: 685–695.

82. Patterson S. D., Aebersold R. H. (2003). Proteomics: the first decade and beyond. *Nat. Genet.* **33** Suppl: 311–323.

83. Rosa J. J., Richards F. M. (1979). An experimental procedure for increasing the structural resolution of chemical hydrogen-exchange measurements on proteins: application to ribonuclease S peptide. *J. Mol. Biol.* **133**: 399–416.

84. Englander J. J., Rogero J. R., Englander S. W. (1985). Protein hydrogen exchange studied by the fragment separation method. *Anal. Biochem.* **147**: 234–244.

85. Smith D. L., Deng Y., Zhang Z. (1997). Probing noncovalent structure of proteins by amide hydrogen exchange and mass spectrometry. *J. Mass Spectrom.* **32**: 135–146.

86. Wang L., Pan H., Smith D. L. (2002). Hydrogen exchange-mass spectrometry: optimization of digestion conditions. *Mol. Cell. Proteomics* **1**: 132–138.

87. Ehring H. (1999). Hydrogen exchange/electrospray ionization mass spectrometry studies of structural features of proteins and protein/protein interactions. *Anal. Biochem.* **267**: 252–259.

88. Cravello L., Lascoux D., Forest E. (2003). Use of different proteases working in acidic conditions to improve sequence coverage and resolution in hydrogen/deuterium exchange of large proteins. *Rapid Commun. Mass Spectrom.* **17**: 2387–2393.

89. Stennicke H. R., Mortensen U. H., Breddam K. (1996). Studies on the hydrolytic properties of (serine) carboxypeptidase Y. *Biochemistry* **35**: 7131–7141.

90. Jung G. M., Ueno H., Hayashi R. (1999). Carboxypeptidase Y: structural basis for protein sorting and catalytic triad. *J. Biochem. (Tokyo)* **126**: 1–6.

91. Han J. C., Han G. Y. (1994). A procedure for quantitative determination of tris(2-carboxyethyl)phosphine, an odorless reducing agent more stable and effective than dithiothreitol. *Anal. Biochem.* **220**: 5–10.

92. Levison M. E., Josephson A. S., Kirschenbaum D. M. (1969). Reduction of biological substances by water-soluble phosphines–gamma-globulin (IgG). *Experientia* **25**: 126–127.

93. Burns J. A., Butler J. C., Moran J., Whitesides G. M. (1991). Selective reduction of disulfides by tris(2-carboxyethyl)phosphine. *J. Org. Chem.* **56**: 2648–2650.

94. Deng Y. Z., Pan H., Smith D. L. (1999). Selective isotope labeling demonstrates that hydrogen exchange at individual peptide amide linkages can be determined by collision-induced dissociation mass spectrometry. *J. Am. Chem. Soc.* **121**: 1966–1967.

95. Kim M. Y., Maier C. S., Reed D. J., Deinzer M. L. (2001). Site-specific amide hydrogen/deuterium exchange in *E. coli* thioredoxins measured by electrospray ionization mass spectrometry. *J. Am. Chem. Soc.* **123**: 9860–9866.

96. Mandell J. G., Falick A. M., Komives E. A. (1998). Measurement of amide hydrogen exchange by MALDI-TOF mass spectrometry. *Anal. Chem.* **70**: 3987–3995.

97. Gustafsson M., Griffiths W. J., Furusjo E., Johansson J. (2001). The palmitoyl groups of lung surfactant protein C reduce unfolding into a fibrillogenic intermediate. *J. Mol. Biol.* **310**: 937–950.

98. Kraus M., Janek K., Bienert M., Krause E. (2000). Characterization of intermolecular beta-sheet peptides by mass spectrometry and hydrogen isotope exchange. *Rapid Commun. Mass Spectrom.* **14**: 1094–1104.

99. Anderegg R. J., Wagner D. S., Stevenson C. L., Borchardt R. (1994). The mass spectrometry of helical unfolding in peptides. *J. Am. Soc. Mass Spectrom.* **5**: 425–433.

100. Waring A. J., Mobley P. W., Gordon L. M. (1998). Conformational mapping of a viral fusion peptide in structure-promoting solvents using circular dichroism and electrospray mass spectrometry. *Proteins* **S2**: 38–49.

101. Demmers J. A., Haverkamp J., Heck A. J., Koeppe R. E. II, Killian J. A. (2000). Electrospray ionization mass spectrometry as a tool to analyze hydrogen/deuterium exchange kinetics of transmembrane peptides in lipid bilayers. *Proc. Natl. Acad. Sci. U.S.A.* **97**: 3189–3194.

102. Demmers J. A., van Duijn E., Haverkamp J., Greathouse D. V., Koeppe R. E. II, Heck A. J., Killian J. A. (2001). Interfacial positioning and stability of transmembrane peptides in lipid bilayers studied by combining hydrogen/deuterium exchange and mass spectrometry. *J. Biol. Chem.* **276**: 34501–34508.

103. Reid G. E., McLuckey S. A. (2002). "Top down" protein characterization via tandem mass spectrometry. *J. Mass Spectrom.* **37**: 663–675.

104. Aebersold R., Goodlett D. R. (2001). Mass spectrometry in proteomics. *Chem. Rev.* **101**: 269–295.

105. McLafferty F. W. (2001). Tandem mass spectrometric analysis of complex biological mixtures. *Int. J. Mass Spectrom.* **212**: 81–87.

106. Little D. P., Speir J. P., Senko M. W., O'Connor P. B., McLafferty F. W. (1994). Infrared multiphoton dissociation of large multiply charged ions for biomolecule sequencing. *Anal. Chem.* **66**: 2809–2815.

107. Zubarev R. A., Kelleher N. L., McLafferty F. W. (1998). Electron capture dissociation of multiply charged protein cations. *J. Am. Chem. Soc.* **120**: 3265–3266.

108. Hofstadler S. A., Sannes-Lowery K. A., Griffey R. H. (1999). Infrared multiphoton dissociation in an external ion reservoir. *Anal. Chem.* **71**: 2067–2070.

109. Horn D. M., Ge Y., McLafferty F. W. (2000). Activated ion electron capture dissociation for mass spectral sequencing of larger (42 kDa) proteins. *Anal. Chem.* **72**: 4778–4784.

110. Eyles S. J., Dresch T., Gierasch L. M., Kaltashov I. A. (1999). Unfolding dynamics of a β-sheet protein studied by mass spectrometry. *J. Mass Spectrom.* **34**: 1289–1295.

111. Eyles S. J., Speir P., Kruppa G., Gierasch L. M., Kaltashov I. A. (2000). Protein conformational stability probed by Fourier transform ion cyclotron resonance mass spectrometry. *J. Am. Chem. Soc.* **122**: 495–500.

112. Kaltashov I. A., Eyles S. J., Xiao H. (2004). Combination of protein hydrogen exchange and tandem mass spectrometry as an emerging tool to probe protein structure, dynamics and function. In: *Focus on Protein Research*, ed. J. W. Robinson, pp. 191–218. Hauppauge, NY: Nova Science Publishers, Inc.

113. Akashi S., Naito Y., Takio K. (1999). Observation of hydrogen–deuterium exchange of ubiquitin by direct analysis of electrospray capillary-skimmer dissociation with Fourier transform ion cyclotron resonance mass spectrometry. *Anal. Chem.* **71**: 4974–4980.

114. Akashi S., Takio K. (2000). Characterization of the interface structure of enzyme–inhibitor complex by using hydrogen–deuterium exchange and electrospray ionization Fourier transform ion cyclotron resonance mass spectrometry. *Protein Sci.* **9**: 2497–2505.

115. Kaltashov I. A., Eyles S. J. (2002). Crossing the phase boundary to study protein dynamics and function: combination of amide hydrogen exchange in solution and ion fragmentation in the gas phase. *J. Mass Spectrom.* **37**: 557–565.

116. Joshnson R. S., Krylov D., Walsh K. A. (1995). Proton mobility within electrosprayed peptide ions. *J. Mass. Spectrom.* **30**: 386–387.

117. McLafferty F. W., Guan Z. Q., Haupts U., Wood T. D., Kelleher N. L. (1998). Gaseous conformational structures of cytochrome *c*. *J. Am. Chem. Soc.* **120**: 4732–4740.

118. Demmers J. A., Rijkers D. T., Haverkamp J., Killian J. A., Heck A. J. (2002). Factors affecting gas-phase deuterium scrambling in peptide ions and their implications for protein structure determination. *J. Am. Chem. Soc.* **124**: 11191–11198.

119. Buijs J., Hagman C., Hakansson K., Richter J. H., Hakansson P., Oscarsson S. (2001). Inter- and intra-molecular migration of peptide amide hydrogens during electrospray ionization. *J. Am. Soc. Mass Spectrom.* **12**: 410–419.

120. McLuckey S. A., Goeringer D. E. (1997). Slow heating methods in tandem mass spectrometry. *J. Mass Spectrom.* **32**: 461–474.

121. Kaltashov I. A., Fenselau C. (1997). Stability of secondary structural elements in a solvent-free environment: the alpha helix. *Proteins* **27**: 165–170.

122. Vonhelden G., Wyttenbach T., Bowers M. T. (1995). Conformation of macromolecules in the gas phase. *Science* **267**: 1483–1485.

123. Hoerner J. K., Xiao H., Dobo A., Kaltashov I. A. (2004). Is there hydrogen scrambling in the gas phase? Energetic and structural determinants of proton mobility within protein ions. *J. Am. Chem. Soc.* **126**: 7709–7717.

124. Marshall A. G., Senko M. W., Li W., Li M., Dillon S., Guan S., Logan T. M. (1997). Protein molecular mass to 1 Da by ^{13}C, ^{15}N double-depletion and FT ICR mass spectrometry. *J. Am. Chem. Soc.* **119**: 433–434.

125. Tsui V., Garcia C., Cavagnero S., Siuzdak G., Dyson H. J., Wright P. E. (1999). Quench-flow experiments combined with mass spectrometry show apomyoglobin folds through an obligatory intermediate. *Protein Sci.* **8**: 45–49.

126. Konermann L., Simmons D. A. (2003). Protein-folding kinetics and mechanisms studied by pulse-labeling and mass spectrometry. *Mass Spectrom. Rev.* **22**: 1–26.

127. Rist W., Jorgensen T. J. D., Roepstorff P., Bukau B., Mayer M. P. (2003). Mapping temperature-induced conformational changes in the *Escherichia coli* heat shock transcription factor σ 32 by amide hydrogen exchange. *J. Biol. Chem.* **278**: 51415–51421.

6

KINETIC STUDIES BY MASS SPECTROMETRY

In the previous chapter we demonstrated the power of MS for the study of biomolecules at equilibrium. For instance, we can allow a sample to equilibrate in the solution conditions of choice and then acquire a spectrum at leisure. By varying the solvent composition, it becomes possible to infer some of the properties of alternate conformations transiently sampled in the equilibrium population, yielding information about the structural features of the states that are visited. Analysis of charge state distributions exhibited in ESI mass spectra provides valuable information about the degree of "foldedness" of these conformational ensembles, and from hydrogen exchange experiments it is possible to measure the exposure to solvent of these species.

Kinetic studies are crucial to a complete understanding of biochemical reaction mechanisms, including protein folding. Kinetic investigations enable rate constants and activation energies to be derived, and by monitoring the progress of the reaction it may be possible to observe transient intermediates that give key mechanistic information. Transient intermediates detected during protein folding can provide insight into the dominant pathway of folding; likewise short-lived intermediate species during the catalytic cycle of enzyme activity give valuable information about the reaction mechanism. However, these intermediates are, by their very nature, short-lived so, unlike equilibrium measurements, there is the added challenge of time scale. Once a reaction is initiated it must be rapidly transferred to an observation point in order to be followed kinetically. In this chapter we discuss some of the methodologies developed to study these kinetic events.

Mass Spectrometry in Biophysics: Conformation and Dynamics of Biomolecules
By Igor A. Kaltashov and Stephen J. Eyles
ISBN 0-471-45602-0

6.1. KINETICS OF PROTEIN FOLDING

Perhaps the simplest way to measure the kinetics of a reaction is to mix the reactants and then monitor change with time of a spectroscopic signal. In the case of protein folding this involves removing chemical denaturant by dilution into conditions permissive for refolding, or adjustment of pH, for instance. Proteins fold to their native state rapidly under renaturing conditions, however, generally on the order of only a few seconds. Instrumentation for *continuous flow* and *stopped-flow* measurement of enzyme kinetics was originally developed in the 1930s (1, 2).

6.1.1. Stopped-Flow Measurement of Kinetics

The basic stopped-flow experiment requires rapid mixing of reactants in a controlled ratio with subsequent placement of the reaction mixture in a flow cell, allowing the reaction to be monitored by optical methods. In the simplest folding experiment a syringe is loaded with protein denatured chemically by high concentrations of chaotrope. A second syringe contains buffer and the two are connected to a mixing device (Figure 6.1A). Upon activation of the two syringes, denatured protein is mixed with refolding buffer, effecting dilution of the denaturant so that the protein can begin to fold. The reaction mixture is then rapidly transferred to an observation cell and the flow is stopped. The reaction can then be followed as a function of time by a variety of spectroscopic methods (see Chapter 2), each of

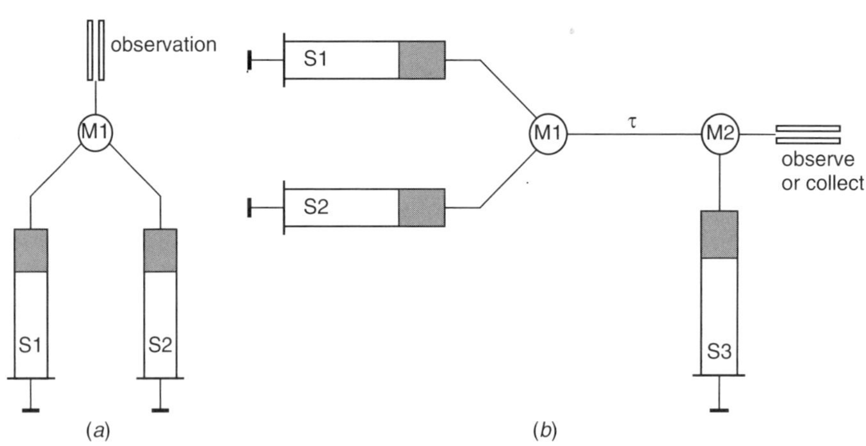

FIGURE 6.1. (**A**) Schematic diagram of a stopped-flow device. Reactants contained in the two drive syringes (S1 and S2) are mixed in mixer M1 when the syringes are activated, and the reaction mixture is then transferred to an observation cell where the reaction kinetics can be monitored in real time spectroscopically. (**B**) In a quenched flow device the reaction mixture emerging from M1 is allowed to react for a variable time period, τ, before quenching by mixing at M2 with either a quenching agent or by injecting into liquid nitrogen.

which monitors different properties of the folding protein. Commonly, circular dichroism is employed to track secondary (in the far-UV wavelength region) or tertiary (near-UV) structure formation; fluorescence is employed to measure the environment of tryptophan residues or extrinsic fluorophores; optical absorption measurements can be used at a wavelength where there is a change in extinction coefficient upon folding.

Commercial stopped-flow instrumentation that ensures accurate mixing ratios by employing syringes controlled by stepping motors or pneumatic drive systems has been available for some time. Dead times are minimized by use of high-efficiency mixers that provide turbulent flow and minimal flow paths between mixer and observation cell. Additionally, stop syringes or valves provide a precise stop of the flow within the cell so that turbulence effects during observation are minimized. Typically, dead times in the millisecond range are readily achievable, enabling folding reactions that typically have time scales from several milliseconds to several seconds to be monitored. Even more rapid folding kinetics can be measured using high-efficiency mixers coupled to continuous flow measurement devices (3) as described later in the chapter.

Stopped-flow measurements by optical spectroscopy require the presence of a chromophore whose properties change as the reaction occurs (see Chapter 2). An alternative and in some cases much more tractable technique involves quenching the reaction at some point during its progress, followed by off-line analysis of the species present in the quenched mixture. Analysis can then be by a variety of different methods that are not dependent on a chromophore. This effectively allows a snapshot to be taken of the species present at that precise time. While this does not allow kinetics to be measured directly, a kinetic curve can be produced by determining the species present in the quenched reaction at different time points. A *quenched flow apparatus* involves not one but two mixing chambers linked by a delay loop (Figure 6.1B). Reactants are introduced into the first mixing chamber (M1) and the reaction is allowed to occur in the delay line for varying time periods, followed by quenching of the reaction by mixing in the second chamber (M2). Alternatively, the reaction may be quenched by injecting the reaction mixture into liquid nitrogen. The species present at that time point can then be measured at leisure under conditions where no further reaction occurs. By varying the flow rate and/or volume of the delay loop, the reaction time allowed before quenching can be adjusted, enabling a kinetic time course to be reconstructed from individual reaction time points.

This type of methodology has been employed for many decades to study enzymatic reactions by using, for instance, radiolabeled reagents in the initial mix followed by quenching with an excess of the unlabeled analogue. The classic *pulse chase* experiment, for instance, can be used for a variety of biochemical studies from localization of proteins within cells to investigation of protein translation in the ribosome, or to determine how metabolites are processed in response to different external factors (4–6). For example, protein synthesis under a certain cellular condition can be monitored by allowing cells to incubate for varying periods of time in the presence of ^{35}S-methionine, incorporating this radioisotope into newly translated peptide chains. After the "pulse" period the labeled amino acid is rinsed out and replaced by a "chase" buffer containing an excess of unlabeled methionine, effectively preventing further incorporation of radiolabel into proteins translated beyond that point. Subsequent separation of the cellular components allows the identification and localization of those proteins produced during the labeling pulse by detection of radioactivity.

6.1.2. Kinetic Measurements with Hydrogen Exchange

The methods of isotope exchange at labile sites on proteins have been used for many years to investigate protein structures. From the previous chapters we have seen that labile protons (particularly amides) are useful probes of structure and dynamics, since those that are buried or involved in structure exchange more slowly with solvent than those that are exposed. This was first demonstrated using tritium exchange experiments (7–9). More recently, hydrogen–deuterium exchange (HDX) labeling has proved extremely powerful for studying the kinetics of protein folding (10, 11). As an unfolded polypeptide forms structure during the folding reaction, labile amide protons become protected against exchange with bulk solvent due to burial or formation of intramolecular hydrogen bonds (i.e., as part of secondary structure). In a *competition experiment*, unfolded protein with all amides protiated* is allowed to refold by rapid dilution into deuterated buffer. During this time, amides that become rapidly protected against exchange will remain protiated, whereas those that remain exposed to solvent will exchange with the buffer deuterons. After a certain time period, the reaction may be quenched by dropping the pH to a level where the intrinsic exchange rate is minimal (pH 2–3) and/or by reducing the temperature to 0 °C. Refolding then continues under quenched conditions, where no further labeling can occur, thus enabling the regions of structure that become protected to be probed. In practice, it is a simpler matter to adjust the pH for the labeling reaction and allow for competition between folding and exchange rather than controlling the time of exposure to the labeling buffer. Since the intrinsic exchange rate changes, as a rule of thumb, by an order of magnitude for every pH unit, varying the pH of the labeling buffer during the folding reaction allows one to probe a variety of folding rates.† In this manner, no special equipment is required since the intrinsic rate of exchange can be matched to the folding rate, provided that the native structure is protected

*Of course, the reverse is also possible. All labile protons can be replaced with deuterium by incubating the protein for extended periods at elevated temperature in ^2H_2O. Generally, a temperature and pH that do not completely denature the protein are preferred to avoid potential aggregation. Lyophilization and a further round of deuteration are generally sufficient to completely exchange all labile sites. Deuteration of the protein enables subsequent steps to be carried out in protiated buffers, which can often be more cost effective when large volumes are involved.

†The exchange process of labile protons with solvent is catalyzed by general acid, general base, and water. The overall intrinsic rate is given by $k_{int} = k_a[H^+] + k_b[OH^-] + k_w[H_2O]$. Rate constants for these reactions and activation energies have been measured in detail using polyalanine model peptides (12), enabling the intrinsic rate at a given pH to be calculated. For instance, at pH 9.5, 25 °C, $t_{1/2} \sim 1$ ms.

against exchange and folded regions in intermediate states are sufficiently stable at the chosen pH such that no further exchange occurs from these states. The folding reaction can be allowed to go more or less to completion before quenching and subsequent sample workup for analysis.

To add a further level of complexity and even more power to probe folding pathways and kinetics, the *pulse labeling experiment* employs a short labeling pulse after a variable refolding reaction time, followed by a third mixing step to quench exchange. This prevents further labeling, so that even marginally stable intermediates remain protected, and allows folding to continue to completion (Figure 6.2). This method enables labeling to occur at given points during the reaction rather than continuously from initiation. The setup has become extremely popular as a tool for measuring protein folding kinetics by hydrogen/deuterium labeling. The method takes advantage of the fact that exchange of labile hydrogens within proteins is extremely sensitive to pH. Thus, a high pH pulse will rapidly exchange all exposed amides, while if the labeling pulse is kept short it generally does not have a significant effect on folding kinetics. At low pH the intrinsic exchange rate is minimal so that over the course of a folding reaction no significant further exchange with bulk solvent occurs, at least at amide hydrogens. On the other hand, at pH 9.5 the half-time for exchange is of the order of 1 ms.

Denatured protiated protein is first diluted by mixing with renaturation buffer at a pH where folding can occur. Refolding is allowed to proceed for a time τ_1, governed by the first delay loop (generally this delay can be varied from one or two milliseconds up to several seconds), followed by a short isotope labeling pulse

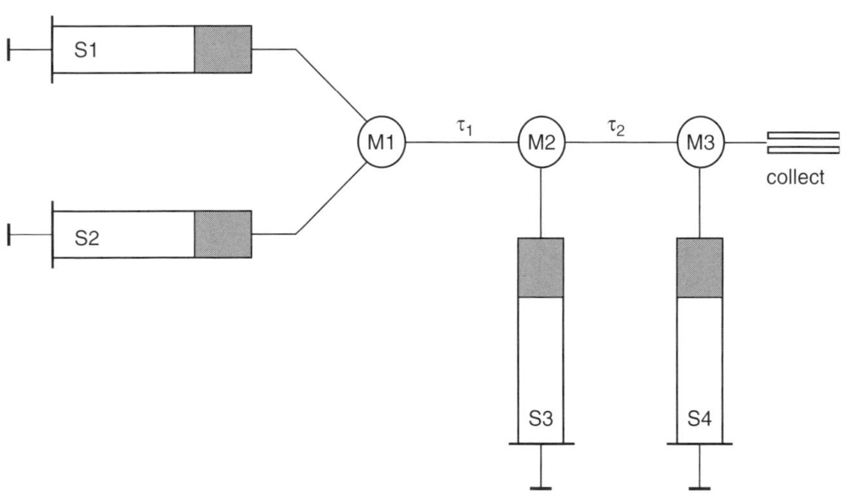

FIGURE 6.2. In the pulse labeling refolding experiment, reactants from S1 and S2 are mixed to initiate the folding reaction for time τ_1. The labeling reagent from S3 is injected and reacted for time τ_2 before quenching in M3. The quenched reaction mixture is then collected for off-line measurement.

(τ_2) with deuterated buffer at high pH. This pulse is used to label with deuterium any amides that are not sequestered from solvent by burial or hydrogen bonding in secondary structure.* The length of time of the labeling pulse is controlled by the second delay loop, followed by quenching in the third mixing chamber of any further HDX by reducing pH to a level close to the minimum for intrinsic exchange (pH 3). By varying τ_1 a kinetic picture can be built up of the folding process based on the regions of structure that become protected against exchange.

The most common modern method to monitor deuterium incorporation from the pulse labeling experiment described above is NMR, from which the average proton occupancy at each amino acid can be measured as a function of time τ_1 during the folding process. By measuring peak intensities, a kinetic curve can be constructed for each residue and the structural details of folding intermediates can be inferred from correlation of amides with similar protection rates. Over

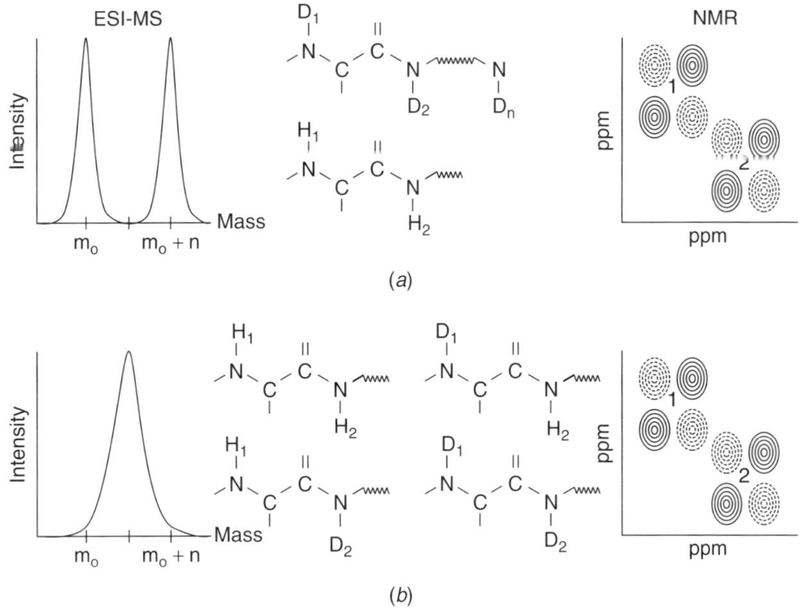

FIGURE 6.3. Distinguishing exchanging populations by MS. Simulated mass spectra and NMR spectra for a population of n amides with average 50% H/D exchange. (**A**) In 50% of the sample population exchange for deuterium is complete, whereas for the remaining 50% no exchange has occurred (EX1 mechanism). The mass spectrum shows two peaks representing exchanged and unexchanged, and NMR shows 50% exchange at all amide sites. (**B**) The 50% random exchange at each amide site (EX2 mechanism) produces an identical NMR spectrum, whereas mass spectrometry now exhibits a single broad peak at average mass representing 50% HDX. Reproduced with permission from (16). © 1993 AAAS.

*A 10 ms labeling pulse at pH 9.5 is equivalent to ten half-lives of exchange and is sufficient to completely label all exposed labile sites.

the past two decades this method has been applied extensively to mechanistic studies of the folding of a number of small proteins (11, 13, 14). The technique is limited, however, by the requirement for assignment of the NMR resonances in order to obtain meaningful structural information (see Chapter 2).

Another significant limitation arises from the fact that each NMR signal from a native protein sample corresponds to the sum of resonances from each molecule in the sample. For instance, the peak corresponding to the amide at position x in a protein sequence is the summation of resonance intensities of every amide at position x in every protein molecule in the NMR tube. In the case of a 0.5 mL sample containing 1 mM protein, this corresponds to $\sim 3 \times 10^{17}$ molecules! Thus, for an HDX experiment the measured NMR resonance intensity corresponds only to the average proton occupancy within the sample. As pointed out by Miranker et al. (16), NMR cannot distinguish between the different populations of amide protection present in a solution of native protein (Figure 6.3). For instance, a sample containing molecules in which 50% are fully protonated and the remaining 50% fully deuterated could not be distinguished from a sample in which all the molecules in the sample are 50% randomly deuterated. In both cases, the NMR peak for a given amide would simply appear to have half the intensity as that of the control protonated sample.

6.2. KINETICS BY MASS SPECTROMETRY

6.2.1. Pulse Labeling Mass Spectrometry

Since the pioneering work of Katta and Chait (15) demonstrated that differences in deuterium content could be measured in intact proteins by mass spectrometry, there has been an upsurge in the use of ESI MS in the study of protein folding kinetics. The beauty of the technique and what makes it complementary to the information obtained by NMR methods is that individual populations with differing deuterium contents can be distinguished by mass spectrometry by virtue of the difference in molecular weight. This was first demonstrated by Miranker and co-workers for the protein hen egg white lysozyme (16). Prior studies of this small protein using kinetic methods showed that a transient intermediate state populates during refolding under the conditions employed (14). Pulse labeling experiments revealed that residues in the α-helical domain become protected more rapidly during folding than in the β-sheet domain, but also, more interestingly, the kinetic traces for all of the amides monitored by NMR exhibited biphasic kinetics with a fast phase rate constant similar for all probed amides. This suggested that there may be an alternative pathway of folding that rapidly protects the entire structure against exchange, effectively bypassing the intermediate. By adjusting the intensity (labeling time or pH) of the labeling pulse, the authors inferred the presence of parallel folding pathways, but no direct evidence could be obtained by NMR alone due to the aforementioned averaging of cross-peak intensities over the entire sample.

Mass spectrometry, on the other hand, can clearly distinguish these parallel folding events. If a population of folding molecules attains native-like structure and hence complete protection against exchange, then it will have a different molecular weight from the unfolded state that becomes fully exchanged during the labeling pulse, or indeed any partially folded intermediate states. In the case of lysozyme folding in a pulse labeling experiment, deuterated unfolded protein was allowed to refold for varying times before labeling with a $^{1}H_2O$ pulse. As shown in Figure 6.4, even after only 20 ms of folding three peaks were observable in

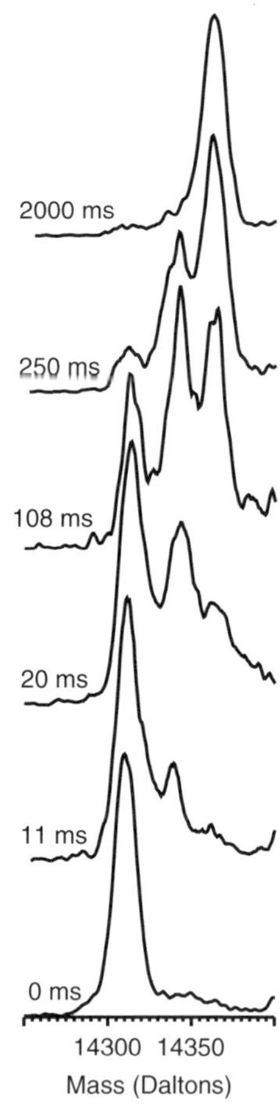

FIGURE 6.4. Time course for refolding of hen lysozyme measured by pulse labeling HDX MS. At the earliest labeling time point (0 ms), no protection exists so all protein molecules are fully labeled with ^{1}H. By contrast, after long refolding time (2000 ms) the protein becomes fully protected before the labeling pulse and so remains fully deuterated. At intermediate time points, several conformational populations coexist, including fully protected and fully unprotected states, and a conformation with intermediate protection (degree of exchange). Adapted with permission from (16). © 1993 AAAS.

the mass spectrum corresponding to different degrees of exchange in the folding population. The low mass peak represents protein that remains unfolded and thus all labile sites were exchanged with protons. The two other peaks correspond to populations that have 22 and 50 amides protected, the former representing the folding intermediate while the latter arises from a parallel folding population in which complete protection is afforded to the entire molecule. At longer refolding times as folding continues, the populations of the unfolded and intermediate states diminish as they are converted to native protected protein.

These elegant experiments demonstrated the power of MS to distinguish populations based on isotope incorporation during refolding. While unable to give high-resolution information as to the structural regions of the protein that become protected first, MS is unique in its ability to identify parallel folding pathways and other details about populations in an HDX sample that would be unattainable by NMR. Another huge advantage of ESI MS over NMR is that very little protein is required (routinely $<1\%$ of that needed for NMR). In fact, the low concentration requirements ($<10\ \mu M$) for ESI MS would, in principle, allow pulse labeled samples to be directly infused into the mass spectrometer without further concentration provided that the buffers used to effect the pH changes are compatible. Alternatively, rapid column desalting could be used, in any case drastically reducing the time for workup and measurement, and thus minimizing the possible problems of back-exchange. More discussion of the synergy of different techniques in biophysics will be included in Chapter 8.

An interesting aspect of pulse labeling in combination with MS is the ability to distinguish not only different folding populations, but also altered kinetics in different proteins, if they are sufficiently different in mass that there is no overlap of peaks. Site-directed mutagenesis is often used to probe the influence of single amino acid residues on the folding and stability of proteins, but it can often be quite difficult to ensure absolutely identical refolding conditions are employed for each individual protein. Hooke and co-workers noted that hen and human lysozyme, which have highly homologous amino acid sequences but apparently quite different folding kinetics, could be refolded together as a mixture and then their HDX behavior monitored in the same experiment (17). This technique avoids the potential ambiguities that could arise from slightly different denaturation concentrations or mixing ratios during the refolding experiments and allows very small changes in folding behavior to be assigned unequivocally. Of course, this option is not available to standard stopped-flow optical techniques unless extrinsic chromophores are added to the protein in order to distinguish them spectroscopically.

More recent studies have employed the HDX labeling strategy in conjunction with ESI MS to investigate the controversies of parallel pathways during protein folding. Experimental folding studies have clearly demonstrated in many cases the existence of intermediate species during the renaturation reactions for many proteins (18–22), and for many years it was assumed that intermediates may be essential precursors to the native structure. Characterization of these states has been considered for some time a key aspect for the better understanding of the processes by which a protein folds. Theoretical treatments and the development of the protein folding process in terms of the "new view" or landscape model have since brought into question the relevance of these transient states (see Chapter 1 for a more detailed discussion of folding pathways and landscapes). It has been suggested that these partially folded states may in fact merely be nonproductive kinetic traps that are off pathway and not important to the folding process to the native state (23, 24). A number of small proteins have been identified that appear to fold via a pure two-state mechanism involving no observable intermediates, suggesting that folding is a much more concerted process (25–29). In fact, experimental data for proteins that previously had been shown to have well-characterized intermediates have since suggested that proteins such as lysozyme or cytochrome *c* may in fact fold without kinetically detected intermediates under certain conditions (30, 31).

The protein interleukin-1β is a small all β-sheet protein that folds very slowly (13), making it readily amenable to hydrogen exchange studies. The protein refolds via a well-characterized intermediate based on HDX NMR and stopped-flow optical methods. However, the question of whether this intermediate is an essential step during the folding of this protein was answered recently using mass spectrometry (32). Pulse labeling experiments were performed using a rapid mixing apparatus to monitor protection at the earliest time points, or even manual mixing for later points (the time constant for folding of this protein was measured to be ~40 s). The degree of protection as a function of folding time was then measured by ESI MS (Figure 6.5). This experiment demonstrated that there was no evidence for formation of any folded structure with native state protection at early time points. Instead, only after a significant accumulation of the partially folded intermediate was later conversion to native state observed. These experiments clearly showed that, for this protein at least, the native state does not become rapidly populated but instead there is a lag phase during which an intermediate species forms before formation of the native folded structure can occur. These mass spectrometric data suggest that even in the landscape view of protein folding there may be a preferred pathway of folding. Although the unfolded state samples a large amount of conformational space, there are only a limited number of ways to get to the native structure, and the intermediate structure detected is obligatory for correct folding, at least in this case.

A similar study was performed to investigate the folding of apomyoglobin. This helical protein has been the subject of many folding studies by a variety of biophysical methods and the general consensus is that an intermediate state forms during folding in which three of the native α-helices (the N-terminal A helix and the two C-terminal G and H helices) fold and dock together prior to formation of the native structure. Once again quenched flow HDX experiments coupled to ESI MS analysis were able to show that no alternative route to the native state was observed (33). In fact, for this protein within the first measured time point (6 ms) no unfolded protein was present, but had all been converted to an intermediate state. Only at later times, however, was a mass spectral signal corresponding

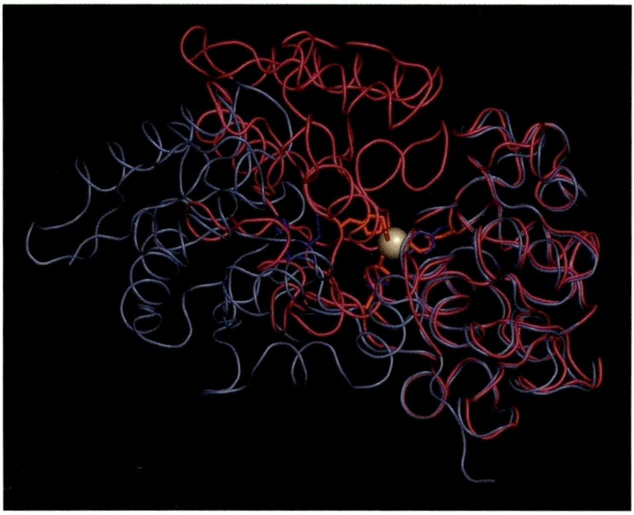

FIGURE 1.8. Superimposed crystal structures of the apo- and holo-forms of the N-lobe of human serum transferrin. Created with PyMol. © 2004 Delano Scientific LLC.

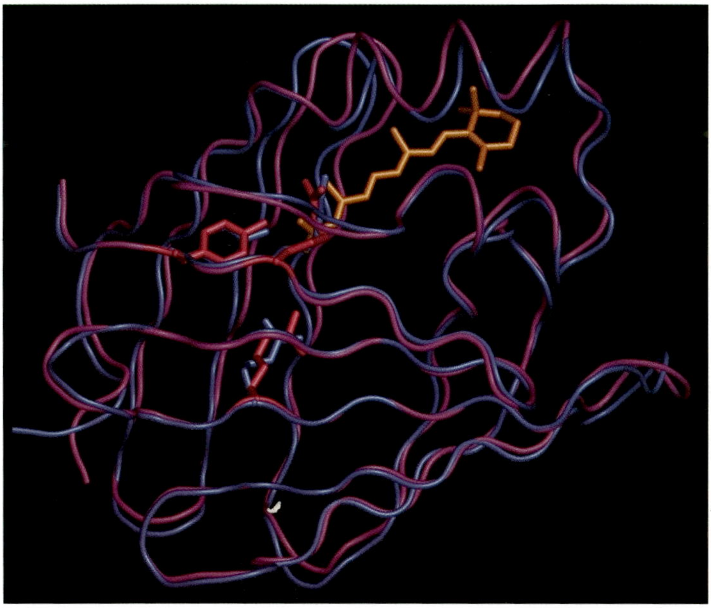

FIGURE 1.9. Crystal structures of the apo- and holo-forms of cellular retinoic acid binding protein I. Created with PyMol. © 2004 Delano Scientific LLC.

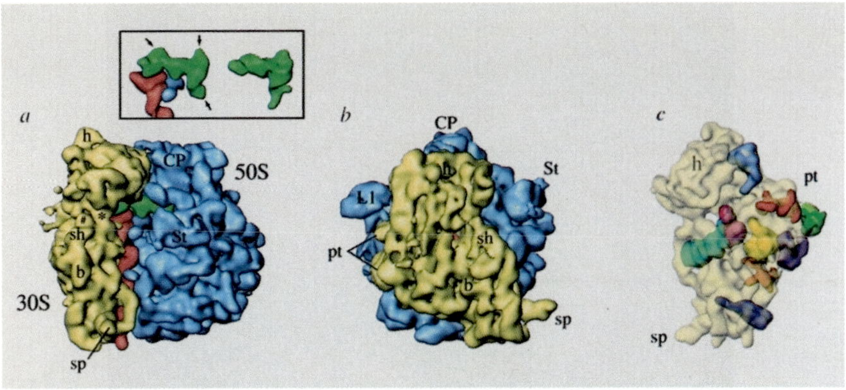

FIGURE 11.6. *Escherichia coli* initiation complex, with fMet-tRNA bound to the ribosome in the presence of mRNA with an AUG codon in its middle. (The 30S subunit is shown in yellow, the 50S subunit in blue, the helix 44 of 16S rRNA in red, and the tRNA in green.) (*a*, *b*) Two views of the 70S ribosome. Inset in (*a*) shows on the left the experimental mass attributed to tRNA in contact with the decoding center, and on the right, atomic structure of initiator tRNA, filtered to 10 Å resolution. Arrows point to contacts between tRNA and ribosome. (*c*) The 30S portion of the map, with fitted X-ray structures of proteins. Adapted with permission from (200) as presented in (201).

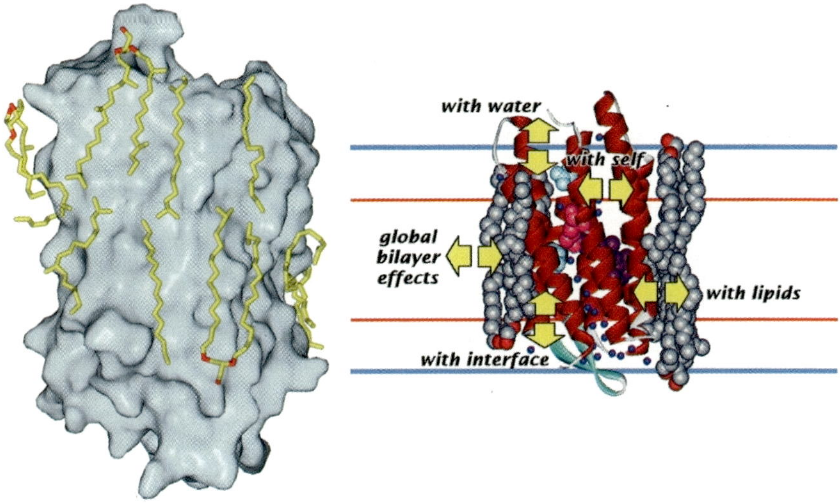

FIGURE 11.10. (*Left*) A view of membrane protein interactions with lipids. Native lipids are seen bound to the surface of bacteriorhodopsin and suggest intimate and specific interactions between the protein and lipids. Reproduced with permission from (79). (*Right*) A schematic representation of the polypeptide interactions that determine the structure and stability of membrane proteins. (The blue lines represent schematically the total thickness of the lipid bilayer, and the red lines the central hydrocarbon core bounded by the interfacial regions.) Besides interactions of the polypeptide chain with itself, water, neighboring lipids, and the membrane interface, the thermodynamic and electrostatic properties of the lipid bilayer itself are important. The lipid bilayer, like proteins, resides in a free energy minimum resulting from numerous interactions. This equilibrium can be disturbed by the introduction of proteins or other solutes, resulting in so-called bilayer effects, which also include solvent properties peculiar to bilayers that arise from motional anisotropy and chemical heterogeneity. Reproduced with permission from (202).

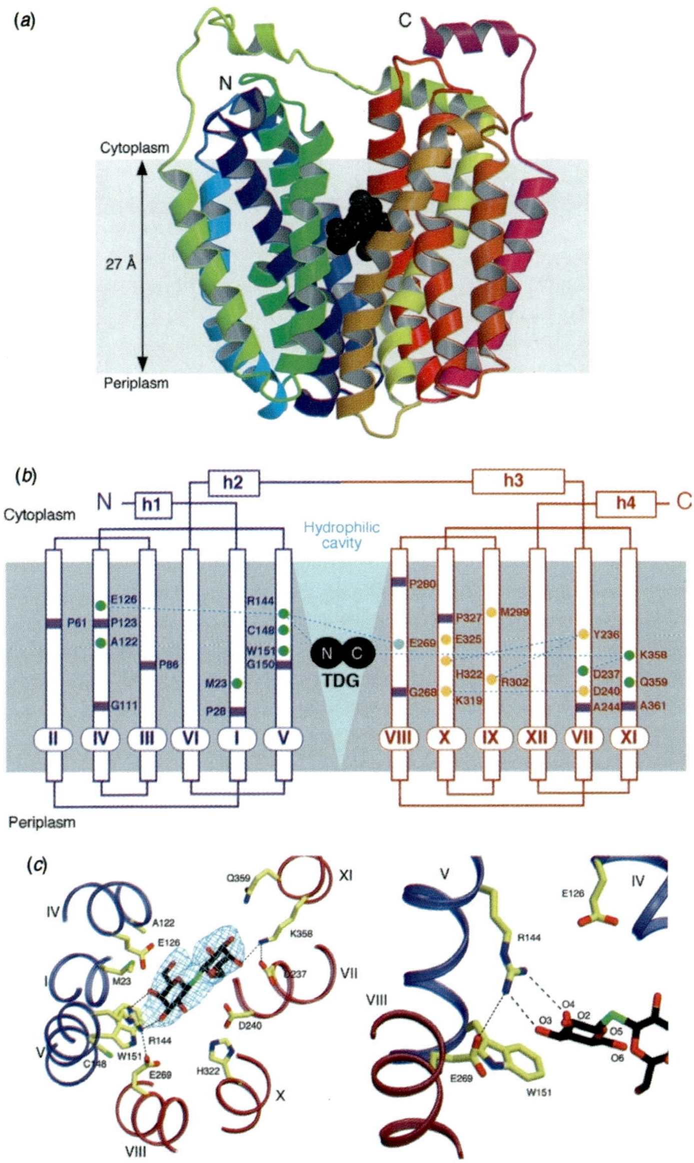

FIGURE 11.11. Overall structure of lactose permease from *E. coli* (LacY) with a bound substrate homologue D-galactopyranosyl-β-D-thio-galactopyranoside (TDG). (**A**) Ribbon representation of LacY viewed parallel to the membrane. (The twelve transmembrane helices are colored from the N-terminus in purple to the C-terminus in pink and TDG is represented by black spheres.) (**B**) Secondary structure schematic of LacY. (The N- and C-terminal domains of the transporter are colored blue and red, respectively; residues marked with green and yellow circles are involved in substrate binding and proton translocation, respectively; residue Glu^{269} marked by a light blue circle, is involved in both substrate binding and proton transfer; and the hydrophilic cavity is designated by a light blue triangle.) h1 to h4 denote surface helices. (**C**) Substrate-binding site of LacY. (Transmembrane helices in the N- and C-terminal domains are colored blue and red, respectively.) Reproduced with permission from (94). © 2003 AAAS.

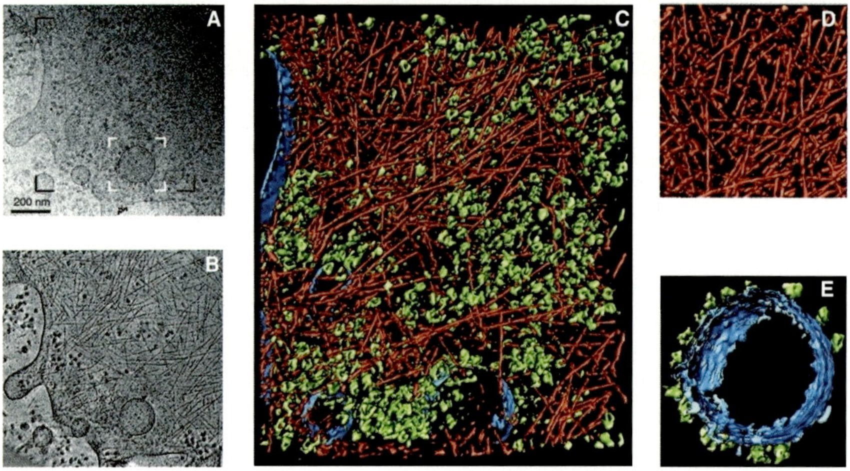

FIGURE 11.24. Macromolecular crowding of eukaryotic cytoplasm visualized by cryo-electron tomography. Shown are a peripheral region of a vitrified *Dictyostelium* imaged with conventional transmission electron microscopy, cell thickness 200–350 nm (**A**), and a representative tomographic slice of 60 nm thickness, revealing a cortical actin network (**B**). Visualization (surface rendering) of actin network, membranes, and cytoplasmic macromolecular complexes (**C**) in a 815 × 870 × 97 nm volume corresponds to the area in panel (**A**) framed in black. [Actin filaments are colored in red, other macromolecular complexes (mostly ribosomes) in green, and membranes in blue.] An image of an actin layer (**D**) taken from the upper left portion of (**C**) reveals a crowded network of branched and cross-linked filaments. Surface rendering of a ribosome-decorated portion of endoplasmic reticulum segmented into membrane and ribosomes is shown in (**E**). The densely packed structures in the lumen are omitted for clarity. Adapted with permission from (171). © 2002 AAAS.

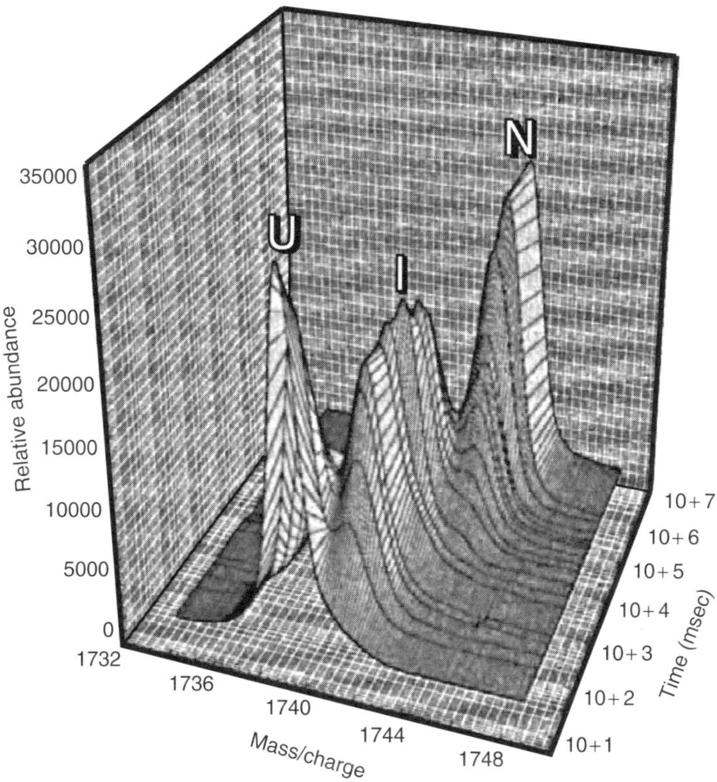

FIGURE 6.5. Kinetic refolding of the protein interleukin-1β monitored by hydrogen exchange MS. Evolution of the +10 charge state is shown as a function of refolding time. At early refolding times only the unfolded state peak is observed, and as time progresses an intermediate species becomes significantly populated. Since no native state forms until the intermediate becomes significantly populated, this indicates that the partially folded species is an obligatory state on the folding pathway of IL-1β under these refolding conditions. Reproduced with permission from (32). © 1997 Nature Publishing Group.

to fully protected native protein detectable. It has been suggested that partially folded states may not be on the preferred folding pathway and are in fact kinetic traps (31). If this were the case, then one would expect to see formation of the native state in parallel with any nonproductive intermediates. In both of the above studies this is not the case, providing some evidence that for these proteins intermediates are necessary and on the pathway of productive folding.

In the case of lysozyme and interleukin-1β described above, the labeled masses of the intermediate states are sufficiently different from those of either the native or unfolded states so that they can readily be resolved in the mass spectra. In the experiments performed on myoglobin, however, the rapidly formed intermediate could not be completely resolved (33, 34). In this case the data were fit to a sum of Gaussian curves representing contributions from each of the three states

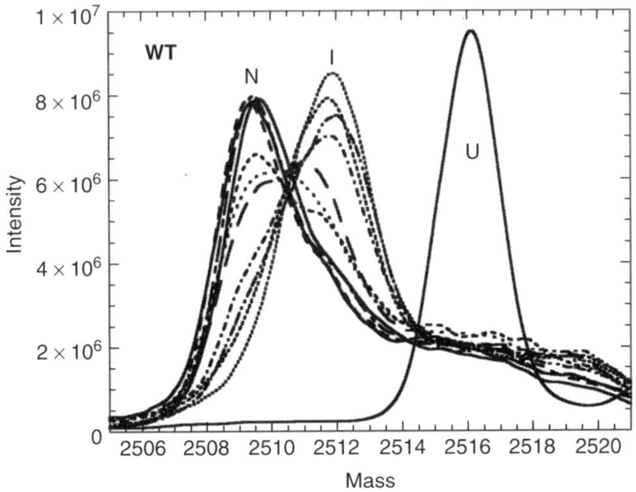

FIGURE 6.6. HDX MS traces for refolding of apomyoglobin. No peak immediately arises corresponding to native-like protection, suggesting an obligate intermediate species. In the case of apomyoglobin, the intermediate state is fully populated within the dead time of measurement (6 ms). The m/z values for native and intermediate state protections are very close and not fully resolved, but they can nevertheless be deconvoluted. Reproduced with permission from (34). © 2003 Elsevier.

(native, N, intermediate, I, and unfolded, U), based on complementary data from NMR experiments (35–37). It was found that satisfactory fits could be achieved for data at all time points, from which the time course of evolution of the intermediate and native states could be obtained (Figure 6.6). Interestingly, the centroid masses and widths of the Gaussians were kept constant to fit the data at each time point. This suggests a remarkable homogeneity and stability in protection of the intermediate state throughout the time course of the folding reaction, since no further changes in the exchange profile of this species were detectable, at least by the fitting procedure employed. More recent studies by NMR have sought to further investigate structure in the kinetic intermediate and suggest a more heterogeneous ensemble that may be masked by the out-exchange during the labeling pulse (38).

As mentioned above, one of the assumptions of the pulse labeling technique is that the folding pathway is not altered by the high pH labeling pulse. Marginally stable regions may well be destabilized to a sufficient extent that they too become labeled and hence are not detected. Alternatively, if the folding rate is altered by a pH change then this may give rise to incorrect apparent protection rates due to incomplete labeling (39). Thus, a great deal of care must be taken in the interpretation of pulse labeling experiments. However, if carefully controlled and if all factors are taken into consideration, the effects of varying pulse length and pulse pH can be valuably applied to investigate the conformational stability of partially folded intermediate species (11, 40, 41). A potential solution to these inherent difficulties has been provided by using continuous flow mass spectrometry (42, and see below).

A further enhancement of the pulse labeling method, namely, subsequent fragmentation of the labeled protein prior to mass spectrometry, enables more specific information to be obtained about the regions of the polypeptide chain protected during folding. The techniques of Smith and co-workers described in the previous chapter may readily be applied to pulse labeled samples. After refolding, labeling, and quenching, the reaction mixture is subjected to peptic digest and the resulting proteolytic fragments are analyzed by HPLC coupled to ESI MS (Figure 6.7). This method was applied to samples of cytochrome *c* labeled after

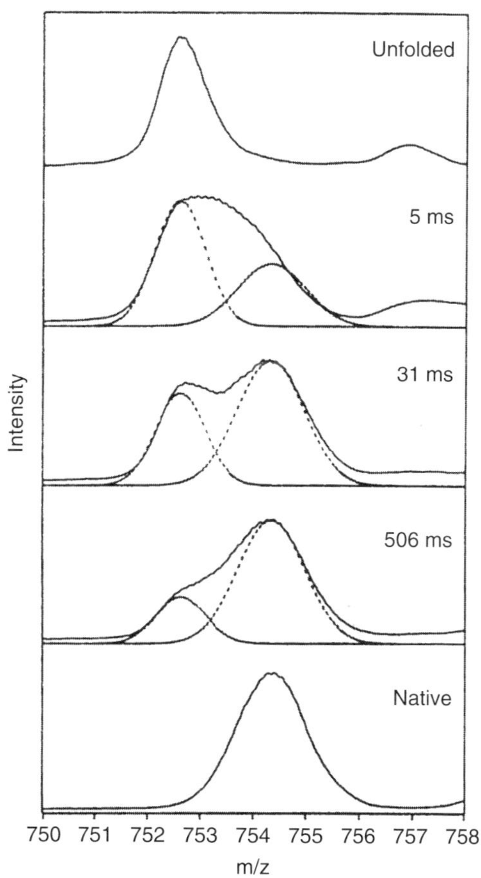

FIGURE 6.7. Proteolytic digestion after pulse labeling enables hydrogen exchange protection information to be gleaned from the separate fragments. For cytochrome *c* the N-terminal fragment, residues 1–21, becomes cooperatively protected during folding, with 50% of the population completely protected in this region by 30 ms. Adapted with permission from (43). © 1997 American Chemical Society.

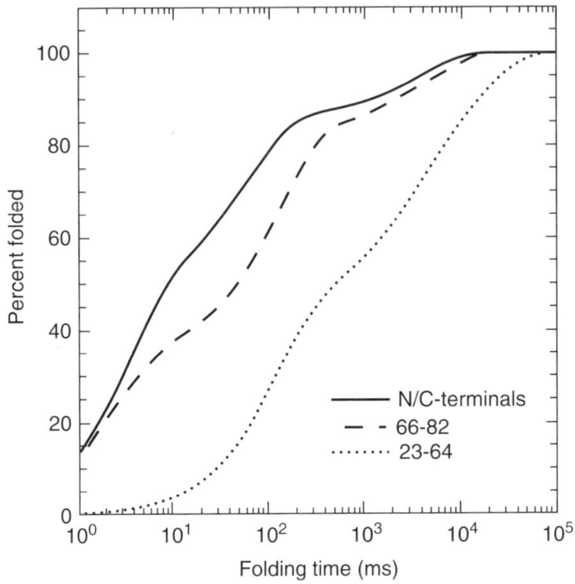

FIGURE 6.8. Time course of protection for various segments of cytochrome *c* during folding. The N- and C-termini become rapidly protected in the intermediate state encompassing the A, G, and H helices. In contrast, the 23–64 segment folds significantly more slowly and the 66–82 region exhibits kinetics between the two. Reproduced with permission from (43). © 1997 American Chemical Society.

varying refolding times (43). The authors found that each peptic fragment was either completely labeled or unlabeled, indicating that folding in each segment was a highly cooperative process. The N- and C-terminal regions, however, were formed into stable structure protective against exchange at earlier times than the remainder of the molecule (Figure 6.8). These results were in accord with previous studies by NMR, but again MS can provide valuable information about structural heterogeneity in the folding intermediates from the measured isotope pattern. In addition, this method could be extended to proteins of higher molecular weight, beyond the current capabilities of NMR methods.

One example of pulse labeling applied to the study of larger proteins is the case of rabbit muscle aldolase. This α-/β-barrel protein forms a homotetramer of molecular mass 157 kDa, far beyond the current capabilities of NMR investigation. Pan and Smith (44) employed a dialysis method to initiate refolding from the guanidine hydrochloride denatured state of this protein followed by pulse labeling of aliquots to study the refolding and reconstitution of this protein as a function of time. Using peptic digest to identify structural regions protected against exchange, the authors were able to delineate three structural domains (each consisting of approximately 110 residues) folded sequentially within each monomer. This is consistent with the current notion that larger proteins may fold initially to form subdomains of approximately this size (45, 46), although the folding domains in this protein consist of noncontiguous segments. Interestingly, the authors observed that the domains differed in the folding and unfolding directions, suggesting that different intermediates might be more favored, although they attributed this primarily to stabilizing quaternary interactions in the folded state that may locally stabilize structural regions only in the unfolding pathway.

In principle, the "top-down" approaches discussed in Chapter 5 should also be applicable to the structural characterization of pulse labeled samples. Measurement of deuterium content in the intact protein by mass spectrometry could be followed by subsequent localization to specific regions using the gas phase fragmentation methods previously described. An important aspect of HDX studied by MS is that the protein does not have to be in its native state for measurement of deuteration by MS. For the NMR experiments, generally only the native state resonance assignments are available so the final folded state must be formed prior to subsequent analysis, and under conditions for which NMR assignments have been made (NMR chemical shifts are highly dependent on pH and temperature). On the other hand, provided that the intrinsic exchange rate is minimal, deuterium labeling can be measured by MS under any conditions. Thus, even though many proteins are only marginally stable at pH 2–3, they can readily be subjected to mass spectrometric analysis, and these conditions may in fact be more favorable for promoting efficient gas phase fragmentation.*

6.2.2. Continuous Flow Mass Spectrometry

The beauty and relative simplicity of the pulse labeling method is that all of the folding and labeling are performed off-line before introduction of the sample into the mass spectrometer. Thus, in the absence of back-exchange processes that can be relatively easily minimized by maintaining samples frozen until used, samples can be analyzed at leisure. However, even under carefully controlled conditions, some information about residues protected during folding can be lost in the concentration and desalting processes due to exchange with solvent. For this reason alternative methods have been sought that enable direct interface to the electrospray source to enable monitoring of exchange on-line. In principle, this could not only improve the kinetic resolution but also allow more amide probes to be monitored, even those that are only marginally stable under quench conditions.

The technique of *time-resolved* ESI MS was initially developed by Henion et al. (48), who used manual mixing followed by infusion of the reaction mixture into the spectrometer, and Sam et al. (49, 50) to study reactions of small organic molecules. The latter setup is in many ways similar to the stopped-flow device employed for spectroscopic kinetic studies except that the observation cell is replaced by a connection to the ESI interface. Since standard electrospray requires a continuous flow of analyte, it is not possible to stop the flow of liquid for observation (although see Section 6.2.3 for techniques in which this effectively can be achieved). Instead, the syringes are pushed continuously and the

*For a given molecular weight, a molecule with a higher charge state will have a greater collision energy at the same electrical potential, thus giving rise to more efficient fragmentation (47).

two reactants mix and then pass into a delay line. The total flow rate from the two syringes and the volume of the delay line between the mixer and the tip of the electrospray needle determine the age of the reaction as the mixture enters the mass spectrometer (the reaction time t_{rxn} = volume/flow rate). Thus, in the continuous flow experiment, the reaction mixture reaching the electrospray emitter is of a constant age, enabling the reaction progress at a specific time point to be investigated. Adjustment of flow rate or delay loop volume can then be used to monitor different time points during the reaction.

A dramatic improvement to the time resolution of these continuous flow devices was attained by Konermann et al. (51). Previously, only times in the range of several seconds to minutes were accessible, but by miniaturizing the mixing apparatus and ESI source regions these researchers were able to access reaction times of $\leq$100 ms. Use of narrow bore capillary tubing, a low dead volume mixing device, and a very short ESI needle (7 mm) enabled this time resolution to be achieved. Other folding times were accessed by varying the length of tubing between mixer and ESI needle.

As a model system the authors used cytochrome *c* and investigated the evolution of charge state distributions resulting from folding (51). Native proteins tend to exhibit narrow distributions with a relatively low net charge, as a result of burial of basic sites and/or low surface area (see Chapter 5). In contrast, denatured proteins adopt significantly more extended conformations and can thus accommodate many more positive charges. Unfolded state mass spectra are characterized by broad charge state distributions centered around high charges. By acquiring data at each time point and then plotting intensity of the signal from each charge state as a function of time, one can construct a kinetic profile (Figure 6.9). In the case of cytochrome *c*, the disappearance of unfolded state signal and concomitant appearance of signal at lower charge state corresponding to the native state could be adequately fit to two time constants. This is in agreement with stopped flow spectroscopic experiments that showed approximately 90% of the population folds fast with a 10% slow folding population rate-limited by proline *cis–trans* isomerization (52).

The continuous flow method has also been applied to study the kinetics of folding and reconstitution of myoglobin (53, 42), and also its unfolding under acidic conditions (54). In these experiments not only could the charge state distributions be followed to show loss of structure, but also the change in mass associated with loss of the noncovalently bound heme group also served as a probe of unfolding upon acidification. Further studies indicated that an intermediate could be detected kinetically during unfolding but only at low pH (55), whereas at basic pH the unfolding reaction was highly cooperative.

HDX reactions can also be measured by continuous flow methods (42, 56). In this experimental setup, the variable delay loop is separated from the ESI needle by a second mixing chamber. Protein folding is initiated by mixing with buffer and then passed through the delay loop to a second mixer where the refolding species are mixed with deuterated buffer that effects exchange at all unprotected sites. After mixing, the protein is immediately injected on-line into the mass

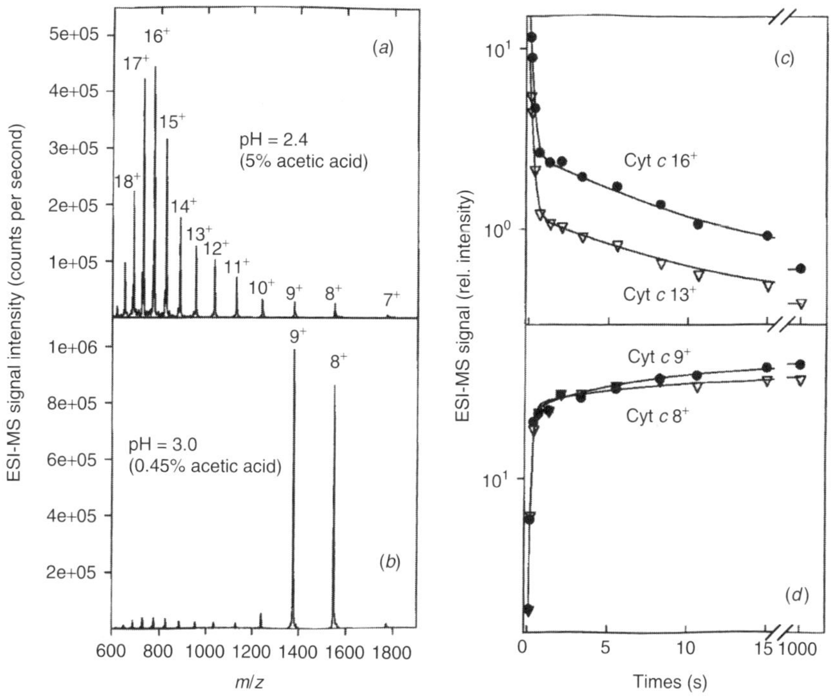

FIGURE 6.9. Time resolved refolding of cytochrome *c* by measuring changes in charge state intensity. The unfolded state at pH 2.4 is characterized by maximum intensity of the +16 charge state (**A**), whereas for the folded state the charge state maximum is +9 (**B**). During refolding the disappearance of the unfolded state (**C**) and appearance of native state (**D**) peaks in the mass spectrum are measured. Reproduced with permission from (51). © 1997 American Chemical Society.

spectrometer, where the extent of deuterium label incorporation is measured. The labeling time is controlled by the volume of the transfer line from this second mixer to the ESI emitter and must be kept short to avoid exchanging amide sites that are only marginally stable. In the work described, reconstitution and refolding of myoglobin was investigated on a time scale from 90 ms to 1.9 s using a labeling pulse time of 7 ms at pD 10.2. At different time points intermediate species were detected that corresponded to a largely unstructured state capable of binding heme, and also a state that bound a dimeric form of heme. In the latter case the charge state distribution suggested a compact state similar to the native state but with a significantly perturbed structure as judged by protection against HDX.

One of the major advantages of this technique is that the material reaching the electrospray emitter is of constant reaction age for a given flow rate and length of tubing. In stopped-flow analysis the kinetics of a reaction can be measured directly, but any loss of signal during acquisition would render the experiment useless. Additionally, to improve signal-to-noise ratios, often multiple data sets must be acquired and averaged, requiring precise timing to enable reproducible overlay of the time axis. On the other hand, the continuous flow method allows time for optimization of instrument parameters, and acquisition times are limited only by sample availability. The disadvantage, on the other hand, is that flow rates or loop lengths must be adjusted to acquire data for each time point.

6.2.3. Stopped-Flow Mass Spectrometry

The above mass spectrometric experimental techniques are all indirect methods to measure kinetics. Either a reaction is quenched and then the extent of reaction measured, or else in the continuous flow setup a single reaction age is monitored directly. By changing the reaction time in each case, a picture of the complete kinetic profile can be built up. But a much more attractive scheme would be the possibility of directly measuring the folding reaction on-line, as can be achieved using optical spectroscopy by stopped-flow methods.

The obvious problem here is that a stopped flow is, by definition, not directly compatible with ESI MS since the electrospray interface requires a continuous flow of analyte solution to function. Additionally, the very high flow rates generally used to achieve efficient mixing and shortest dead times in a stopped-flow apparatus are generally incompatible with the ESI source conditions. An interesting possibility of switching flow rates has been developed for the measurement of subfemtomolar quantities of analyte using nanoflow LC/MS (57, 58). With this technique, material is separated using microcapillary columns equipped with nanospray emitters. A switching valve allows the rate of infusion into the spectrometer to be switched from *high* (120 nL/min) to *low* (<10 nL/min) flow by adjusting the split ratio of the mobile phase between column and waste. This switch effectively allows the chromatography to be paused in a data-dependent manner when a peak of interest elutes from the column and enables the eluting compound to be analyzed for an extended period of time. This method of *peak parking* is extremely useful for species at very low concentration that may otherwise be missed or insufficiently sampled under normal flow conditions. Clearly, it has great potential for usage in proteomics and for other techniques requiring chromatographic separation, but it has yet to be applied to protein folding studies.

Devices that couple stopped flow with MS have been designed and employed to look at chemical reactions including protein folding. The first of these employed a rotating ball inlet coupled to an electron ionization source to measure the kinetics of organic reactions between small molecules (59). Such an inlet system was required to maintain a sufficient vacuum in the source region. Fortunately, ESI is an atmospheric pressure ionization technique so one could in principle mix the reactants and then infuse directly into the ESI source. This would have the same effect as the continuous mixing method described above, leading not to kinetic measurement but monitoring the reaction at a constant time point. Alternatively, the reactants could be premixed and then infused, but this would yield very long dead times, potentially missing all of the reaction kinetics.

Kolakowski and Konermann (60, 61) circumvented these limitations by elegantly redesigning a stopped-flow setup in a manner that first allows the reaction to be rapidly initiated, and then infused into the source as a second step using a series of valves and a third syringe. In the first stage, two syringes are activated at high flow rate to mix the two reactants, flush old solution from the reaction chamber, and then deliver fresh reaction mixture into the reaction tube. A valve switching system then shuts off flow from these syringes and the reaction is delivered into the ESI source by a third syringe running at a low flow rate compatible with electrospray. Thus, relatively short dead times are achievable (100 ms) and the amount of reactant delivered corresponds to an observation time window of ~40 s before dilution artifacts lead to loss of signal. Careful optimization of the wash phases and solvent flow rates ensured reproducibility between individual stopped-flow "shots."

In this manner, the authors were able to monitor in real time using ESI MS the acid induced unfolding of myoglobin, a process that is complete within 10 s (Figure 6.10). Monitoring the ion intensity of the +14 charge state of

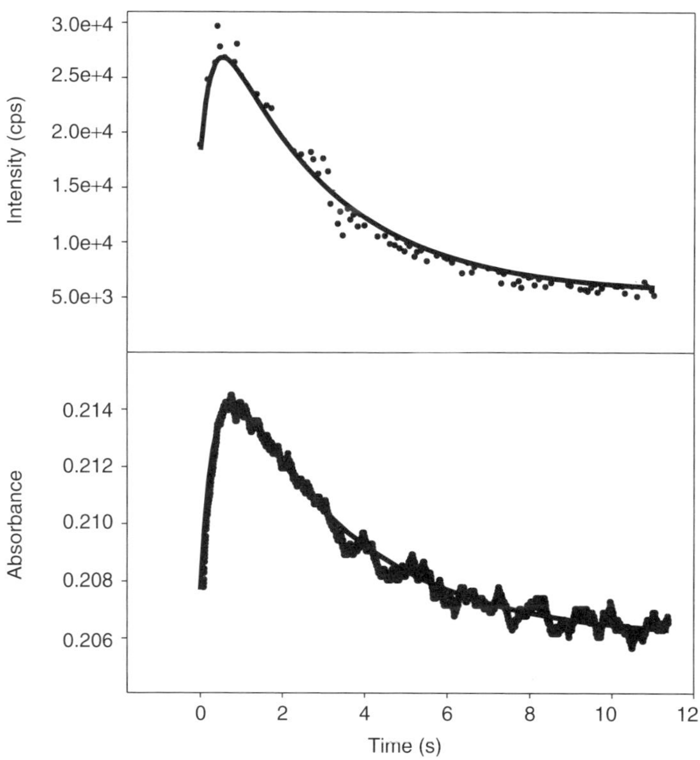

FIGURE 6.10. Unfolding kinetics of myoglobin at pH 3.0 monitored by stopped-flow MS. The evolution of the +14 charge state during unfolding (*top*) corresponds closely with the change in absorption of the Soret band at 441 nm (*bottom*). Reproduced with permission from (61). © 2001 Elsevier.

holomyoglobin resolved two kinetic phases with time constants of 0.29 and 2.8 s. These are in very good agreement with stopped-flow optical measurements and are thought to correspond to unfolding of the holoprotein, with subsequent loss of the noncovalent heme group in the slower phase.

While still not yet on a par with the kinetic resolution of stopped-flow optical devices, this technology is clearly rapidly advancing and further design improvements will surely lead to shorter dead times and enable faster folding reactions to be measured. Furthermore, MS offers several major advantages over other spectroscopic techniques as previously discussed. Lower detection limits not only allow for studies to be performed on samples only available in limited supply, but the lower concentrations (2.5 μM) also avoid the potential problems of sample aggregation during folding. Natural proteins can be used without the need to introduce extrinsic chromophores into the structure that may interfere with the folding reaction. Use of ESI TOF instrumentation would also enable a full spectral scan very rapidly, thus allowing evolution of the entire charge state envelope to be measured, and subsequently direct detection of folding intermediates.

With the modern need for high-throughput instrumentation, major technological advances have been made in the field of microfluidic devices. Already, devices have been designed capable of miniaturizing both quenched flow (62) and electrospray sources (63). Further development will surely lead to chip-based systems that can be used to rapidly yield kinetic data with extremely short dead times and using minute sample quantities.

6.2.4. Kinetics of Disulfide Formation During Folding

An alternative to hydrogen–deuterium exchange as a probe of protein folding is to monitor the change of other characteristics by MS. In the case of proteins that contain disulfide bridges, the order of their formation provides an indication of the folding pathway (64, 65). As secondary and tertiary structural elements form, they orient cysteine residues correctly to enable disulfide formation. An elegant study by Deinzer and co-workers employed the combination of HDX MS with measurement of disulfide bond formation and evolution of charge state distribution during the oxidative refolding of the protein human macrophage colony stimulating factor β (rhm-CSFβ) (66). This protein forms a homodimer containing three intramolecular and three intersubunit disulfide bridges. Dilution of the reduced denatured protein into buffer containing a mixture of oxidized and reduced glutathione enabled formation of the disulfide bridges as the protein refolded. Aliquots of the refolding mixture were removed at various time points during the reaction and any free cysteinyl residues not involved in disulfide bonds were carboxyamidomethylated by reaction with iodoacetamide. The mixture of partially refolded proteins was then separated by chromatography, digested with pepsin, and then analyzed by HPLC ESI MS. Those fragments that contained blocked cysteine residues were identified by the mass difference resulting from alkylation. In the case of more than one cysteine occurring in the same peptic fragment, tandem MS was employed to locate the modified residue.

In this manner, the authors were able to determine not only the order of formation of structure within the monomeric subunits but also the interactions that enable formation of the homodimer.

As we have seen previously, the charge state distribution exhibited by a protein is exquisitely sensitive to structure, at least in terms of solvent accessible surface area. In addition to mapping the location of disulfide bonds described above, the partially refolded mixture was also analyzed by ESI MS (see Chapter 5). At the earliest time points only high charge states were detected, indicating the presence of predominantly unstructured chains. As oxidative refolding continued, the spectra obtained became more complex due to the presence of more compact intermediate species, and eventually also the formation of dimeric species (Figure 6.11).

A novel technique involving chemical labeling has been developed recently to study kinetics of protein folding. The groups of Russell and Baldwin used

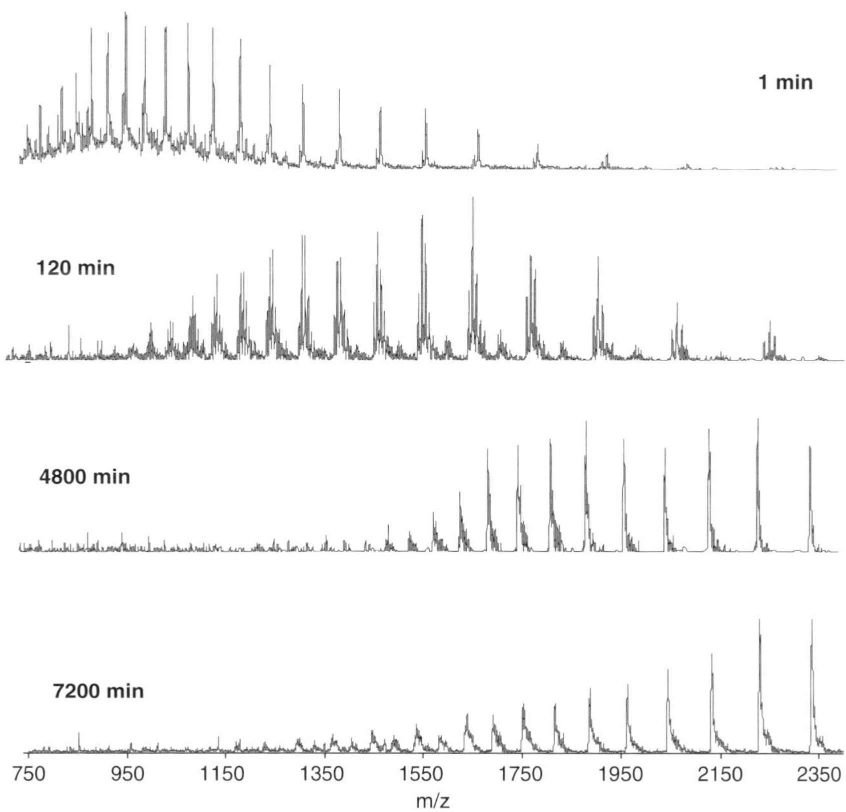

FIGURE 6.11. Evolution of charge state distribution during the oxidative refolding of rhm-CSFβ. As folding proceeds and native disulfide bonds form, the protein becomes more compact and the charge state shifts from unfolded to native-like with time. Adapted with permission from (66). © 2002 American Chemical Society.

a method termed pulsed alkylation mass spectrometry (PA/MS) as an attractive alternative to HDX measurements (67). Alkylation has the huge advantage over isotope exchange reactions that it is effectively irreversible; hence there is no complication from the possibility of back-exchange. The method relies on the specific reactivity of N-alkylmaleimides with free cysteine residues (i.e., those that are not involved in disulfide linkages) that are exposed to the reagent in the bulk solution. Bacterial luciferase is a 76 kDa heterodimer that contains 14 cysteinyl residues distributed throughout the α- and β-subunits, providing a set of reporters of accessibility across the structure. The authors used a pulse for 5 s with alkylating reagent followed by quenching of the reaction by addition of β-mercaptoethanol to scavenge any remaining reagent. The protein was then digested with the proteolytic enzyme AspN and the resulting fragments were identified using MALDI TOF mass spectrometry. By varying the solvent conditions, cysteine accessibility was probed either in the native state or in a partially folded species populated in 2 M urea (Figure 6.12).

While this method was used to investigate the structure and dynamics of the various conformational states at equilibrium, the technique could, in principle, be applied to kinetic studies. One of the most intriguing aspects is the wide range of reactivity of the maleimide derivative depending on chain length of the alkyl group. This would allow tailoring of the pulsed alkylation reaction depending on the protein refolding rates, enabling shorter labeling times for faster folding reactions. In addition, the broad diversity of chemical functionality of amino acid residues allows the possibility of using alternative reagents, or even mixed reagents, to label different residues and enable even more detailed probing of accessibility.

Thus, one could envisage an experimental scheme analogous to the deuterium exchange pulse labeling experiment where protein folding is first initiated by diluting into solvent conditions favoring renaturation. After a variable refolding time, a labeling reagent is pulsed into the system for a short time followed by quenching of the reaction and completion of refolding. Protection of amino acid side chains against labeling during refolding can thus be probed at various time points in the reaction. Mass spectrometric techniques can then be applied at leisure to identify the labeled sites either by subsequent proteolysis or gas phase fragmentation strategies as described in Chapter 5. A similar methodology has been applied to locate the metal centers within metalloenzymes by taking

---

FIGURE 6.12. Pulsed alkylation mass spectrometry to study kinetic refolding of bacterial luciferase. As the protein refolds, disulfides are formed that prevent alkylation in folded regions. After modification the protein is proteolytically digested and the fragments are analyzed by mass spectrometry to determine the sites of alkylation. (**A**) Digest of unmodified luciferase. (**B**) After modification the identified peptide masses shift by 125 Da corresponding to alkylation at a single cysteinyl residue (annotated by *). Panels (**C**) and (**D**) show superimpositions of digests before and after alkylation with the number of asterisks corresponding to the number of cysteine residues modified. Reproduced with permission from (67). © 2001 American Chemical Society.

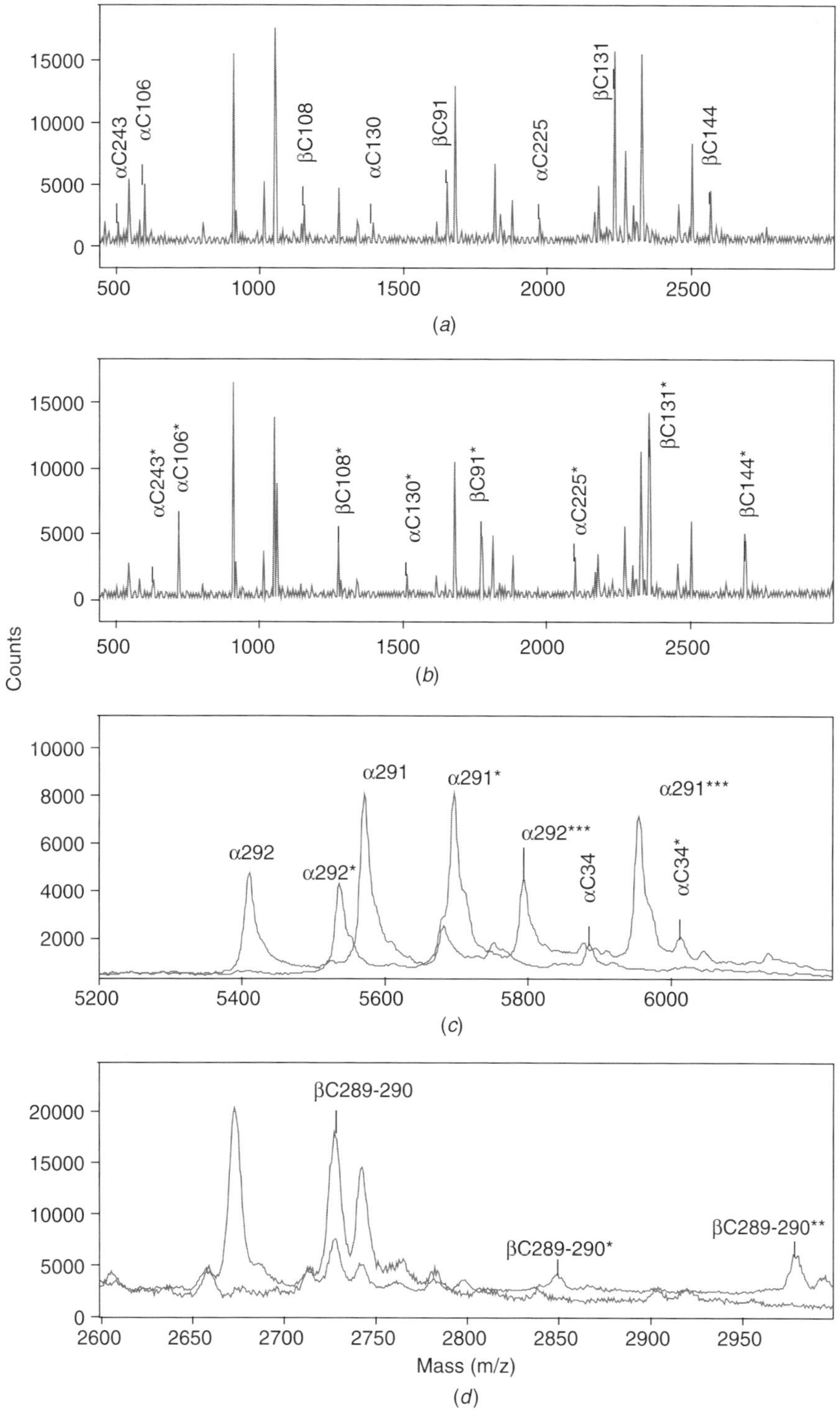
Counts
αC243
αC106
βC108
αC130
βC91
αC225
βC131
βC144
(a)
αC243*
αC106*
βC108*
αC130*
βC91*
αC225*
βC131*
βC144*
(b)
α292
α292*
α291
α291*
α292***
αC34
α291***
αC34*
(c)
βC289-290
βC289-290*
βC289-290**
Mass (m/z)
(d)

advantage of the metal catalyzed oxidation reactions that can occur at residues in close proximity to the bound metal in the presence of an oxidizing reagent (68).

One caveat, of course, is that the labeling reagents must be carefully chosen such that modification of the residues of interest does not alter the efficacy of folding. Nevertheless, a range of compounds are already available for such purposes. Another important factor is that the reagent must be sufficiently reactive to completely label all exposed residues within the time of the labeling pulse. However, it would seem that this scheme may be more tractable than traditional hydrogen exchange pulse labeling experiments since it could be achieved without drastic changes in pH. The latter must be carefully controlled in many cases since many proteins may actually denature at the extreme pH applied for the labeling.

6.2.5. Kinetics of Protein Assembly

Another fascinating aspect of protein dynamics and folding is how proteins assemble to form multimeric complexes. For instance, some proteins are not functional as monomers but require assembly. The chaperone GroEL in its active form consists of a homotetradecamer, arranged as two 7-membered rings stacked on top of each other. Kinetics of assembly is also an interesting aspect of protein folding.

In order to detect macromolecular complexes in their native state, instrumentation is required that can achieve measurement in the high m/z range. The extended mass range afforded by time-of-flight mass spectrometry is an essential tool for the measurement of protein assembly and disassembly. These types of experiments have been pioneered recently by the Robinson group and have provided valuable insight into the assembly processes of large multiprotein complexes (see Chapter 4). One of the first examples of kinetic measurement of the assembly process was a study of the chaperone complex MtGimC (69). The active form of this protein is a heterohexamer ($\alpha_2\beta_4$) formed from two subunits MtGimα and MtGimβ. The intact hexamer is 87 kDa and the charge states adopted by the native state complex give rise to peaks in the range 4000–5000 m/z. Using very gentle source conditions, including elevated pressures within both the source and analyzer regions of an ESI-TOF mass spectrometer, the authors were able to observe the intact complex and various partially dissociated species. In addition, manual mixing of the α- and β-subunits in a 1:2 ratio at 37 °C followed by infusion into the mass spectrometer allowed the assembly process to be monitored in real time. Interestingly, the assembly process appeared to be a highly cooperative event involving no observable intermediates (Figure 6.13). Over the time course measured, only peaks corresponding to the intact hexamer were detected, with consequent depletion of the monomer populations. By contrast, controlled dissociation revealed a variety of dimers, trimers, and tetramers, all consistent with a $(\beta\alpha\beta)_2$ architecture in which the trimers appear to be linked primarily through the α-subunit.

Another intriguing study of protein assembly by mass spectrometry involves the kinetics of subunit exchange in small heat shock proteins (sHSPs) (70). These

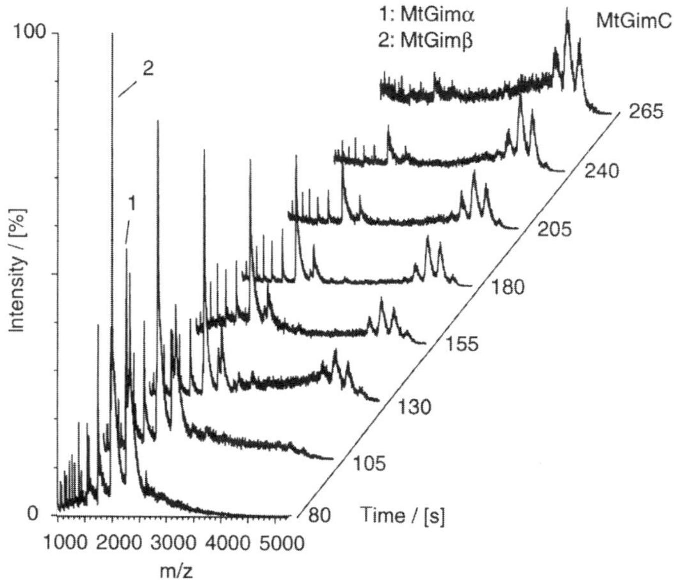

FIGURE 6.13. Assembly kinetics in MtGimC from subunits MtGimα and MtGimβ. As time progresses peaks arise from the intact complex and the monomer becomes depleted. Reproduced with permission from (69). © 2000 National Academy of Sciences, USA.

proteins combine to form complexes of defined stoichiometries in plants, but in mammalian cells consist of polydisperse oligomers ranging from 9 to 24 monomer subunits, making this system very difficult to characterize by other methods. The heterogeneous nature leads to difficulties in crystallographic structure determination and even characterization by electrophoresis. Once again the power of mass spectrometry is evident: using nano-ESI and measurement by TOF mass spectrometry the polydispersity and dynamic nature of these species can be probed. An artificial mixture of homologous sHSPs obtained from different species (pea and wheat) were shown to interact and form dynamic dodecamers (Figure 6.14). Time resolved studies enabled the kinetics of subunit exchange to be measured, revealing that although exchange is relatively facile, it is predominantly dimers that are the exchanging species. On a slower time scale monomeric subunits also exchange. The envelope comprising peaks from the various species becomes very complex but could be well-fitted by simulations to reveal the nature of these oligomeric species.

6.3. KINETICS OF ENZYME CATALYSIS

In the previous section we examined some of the mass spectrometric methods used to investigate the folding and unfolding reactions of proteins. Armed with this information, we can gain insight into how a protein achieves its native folded

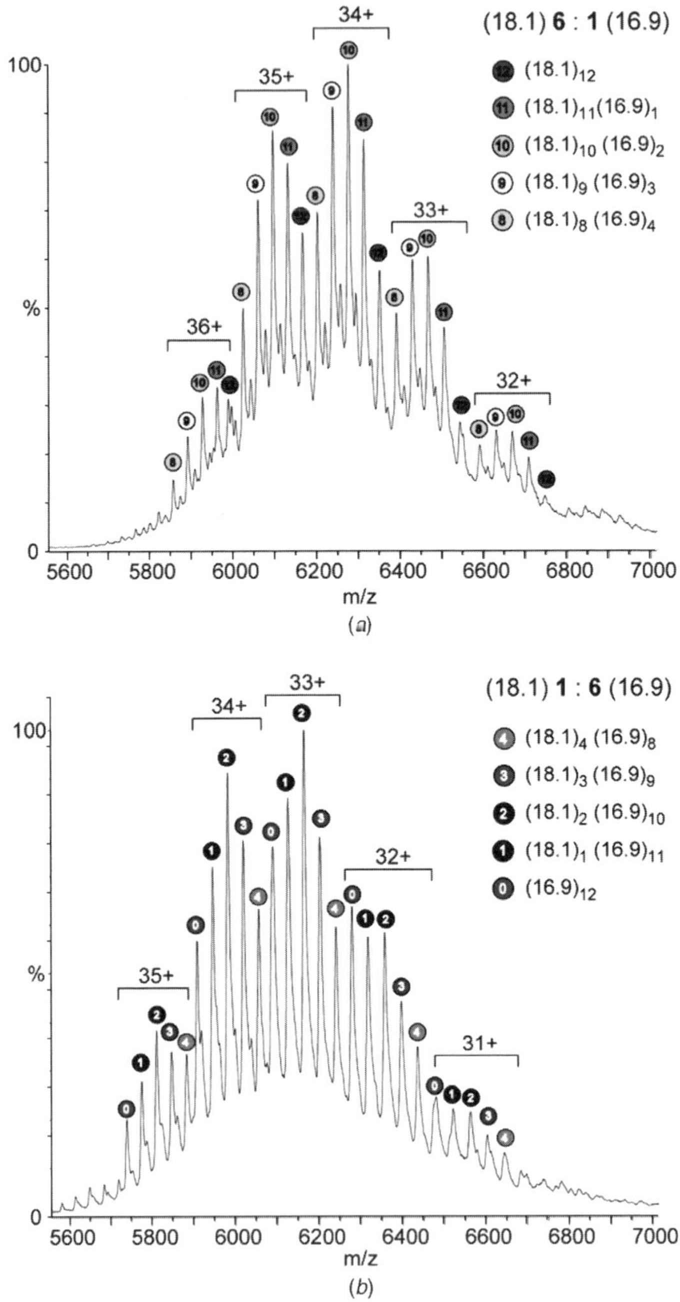

FIGURE 6.14. Subunit exchange in the sHSPs. Mixing PsHSP18.1 from pea with TaHSP16.9 from wheat allows subunit exchange between the two to occur, producing a variety of different stoichiometries over time that can be characterized by MS. Reproduced with permission from (70).

and functional structure. Equally important though is an understanding of how the folded protein functions. In Chapter 5 the importance of dynamics was discussed and the many ways we can look at the dynamic events occurring in the native structure. One particular class of proteins is also involved in the catalysis of complex chemical reactions. Enzymes are capable of transforming reactants into products in a highly efficient and specific manner. Biophysicists are extremely interested in the kinetics of these reactions and how nature has optimized the structure and function of proteins to best achieve their genetically destined goals. In this section we briefly discuss some of the mass spectrometric approaches to study the kinetics of enzyme catalysis.

Enzymes are extremely efficient biological machines, converting substrates to products on time scales similar, or often faster, than the process of protein folding. Their action is catalytic so they are capable of performing multiple turnovers provided that there are sufficient reactants present. For this reason it is often easier to obtain at least limited kinetic information about the reaction process since it is continuous. In the case of protein folding, once the denatured protein is exposed to conditions that are permissive for folding, the entire population proceeds to fold toward the native state, albeit potentially at different rates depending on the pathway(s) available. For enzyme catalysis one can employ a limiting concentration of catalyst in the present of excess substrate and observe the rate of reaction.

The simplest method to measure enzyme kinetics is at the "steady state." In excess of substrate, a rapid equilibrium is set up, in which the concentration of intermediate species builds up until they are effectively at a steady state, being formed and depleted at the same rate. Substrate is gradually depleted over time, but in such an excess that for a sufficiently short time window of measurement this too can be considered constant. Under such conditions the reaction rate is more or less constant so that the rate of formation of products and the dissociation constant for enzyme–substrate complex can be measured.

The theory of enzyme catalysis was developed over several years and proposed by Michaelis and Menten (71), based on the formation of a noncovalent complex of enzyme (E) and substrate (S), at equilibrium, from which the reaction is catalyzed to yield product (P). The general reaction scheme may be represented as

$$\mathrm{E} + \mathrm{S} \xrightleftharpoons{K_\mathrm{S}} \mathrm{ES} \xrightarrow{k_\mathrm{cat}} \mathrm{E} + \mathrm{P}, \tag{6-3-1}$$

assuming substrate is present in vast excess over the enzyme concentration. $K_\mathrm{S} = [\mathrm{ES}]/[\mathrm{E}][\mathrm{S}]$ represents the dissociation constant for the enzyme–substrate complex (also known as K_M, the Michaelis constant). The chemical reaction that converts reactants to products then has a reaction velocity $v = k_\mathrm{cat}[\mathrm{ES}]$, which leads to the classic Michaelis–Menten equation:

$$v = \frac{[\mathrm{E}]_0[\mathrm{S}]k_\mathrm{cat}}{K_\mathrm{M} + [\mathrm{S}]} \tag{6-3-2}$$

The equation is valid with the assumption that substrate is present in vast excess over the enzyme. Generally, kinetic experiments are performed under conditions where the substrate concentration is saturating, leading to a limiting maximum reaction velocity, $V_{\max} = k_{\text{cat}}[\text{E}]_0$.

Traditionally, measurement of steady-state kinetics can be achieved spectroscopically using a chromogenic substrate that undergoes a color change, thus enabling different parts of the reaction to be probed using different artificial substrates (72). However, this method is generally not applicable to natural substrates and is thus necessarily limited in scope. Another problem with modified substrates is that the very presence of the chromophore may alter the kinetics of catalysis (73). For this reason there has been a great deal of interest in using mass spectrometric methods for the analysis of enzyme kinetics, since it effectively serves as a "universal" detector (74). One of the great powers of mass spectrometry is that with a sufficient time resolution, not only can kinetics be monitored as the change in intensity of spectral peaks, but also the intermediate species present can be determined, provided the reaction gives rise to a resolvable change in mass. If the catalytic reaction occurs within a time scale amenable to measurement on-line, then one can monitor the kinetics of formation of products and any intermediate species simultaneously.

One of the earliest direct measurements of enzyme catalysis was described by Smith and Caprioli (75, 76), who used FAB ionization to measure the hydrolysis of peptides by trypsin. This method, however, requires high concentrations of glycerol as liquid matrix, which can rapidly cause inactivation of enzymes. Electrospray ionization provides a much more friendly solvent environment as shown by Lee et al. (48) in which continuous flow MS was employed to measure the kinetics of a variety of enzymatic reactions including the hydrolysis of *O*-nitrophenyl β-D-galactopyranoside by lactase and the proteolytic degradation of a short peptide, dynorphin 1–8, by α-chymotrypsin. These reactions occurred on a sufficiently slow time scale that manual mixing followed by infusion into the mass spectrometer was sufficient to enable reaction kinetics to be measured directly as depletion of reactant and formation of product. Quantitation of kinetic parameters can be achieved by use of internal standards, and by separating the various species by HPLC. This avoids the potential problems of suppression of ionization that would give rise to nonlinear concentration responses. Henion and co-workers demonstrated the applicability of HPLC MS to the measurement of kinetics of hydrolysis by β-galactosidase of its cognate substrate lactose, using a disaccharide as an internal standard for quantitation (77).

Steady-state kinetic measurement is a useful but limited technique. While it can be used to directly measure the overall rate of reaction (k_{cat}) and the Michaelis constant (K_{M}) for the reaction, it reveals very little detail about the reaction mechanism. Only by determining the rates of formation and depletion of intermediate states can a full description of the enzymatic reaction be made. Using MS, it has proved possible to detect both covalently and noncovalently bound intermediate enzyme–substrate complexes that, when identified, can provide quite detailed mechanistic information. In addition, with rapid mixing, detection of the rates of the processes that interconvert these intermediate species allows a more complete kinetic description of the processes.

As with studies of protein folding, pushing the time limits to shorter and shorter times can provide more useful details about enzyme catalysis. For instance, the observation of transient intermediate species can provide insight into the mechanism of the enzymatic reaction and the species that become populated during the process. Additionally, measurement of the rates of formation of these species enables a more complete kinetic picture to be drawn. Various approaches have been taken in an attempt to address these questions using mass spectrometry, by employing many of the methodologies similar to those described above for protein folding studies. Chemical quench techniques followed by off-line detection using radiolabeling techniques or NMR, for instance, can be used to detect intermediate species but only if they are stable to the quench conditions employed (78–80). However, even in the case where the transient intermediates are labile, they still may be detectable using MS. *Pulsed flow* mixing with millisecond resolution enabled detection of a tetrahedral intermediate in the conversion of shikimate-3-phosphate to phosphoenol pyruvate by the enzyme 5-enoylpyruvyl-shikimate-3-phosphate synthase (EPSP-synthase) by ESI MS (81). This setup was effectively the same as a continuous flow mixing device that was able to achieve a time resolution of the order of 50 ms. Although no kinetic measurements were made, the observation of a number of peaks corresponding to intermediates in the reaction pathway of this enzyme were observed without the need to quench the reaction before detection.

An elegant example of the application of MS to the study of much more rapid enzyme kinetics was given by Li et al. (82), who used a continuous flow mixing device with variable flow rates to monitor the reaction of 3-deoxy-D-manno-2-octulosonate-8-phosphate synthase (KDO8PS). This enzyme produces KDO8P by condensation of arabinose-5-phosphate (A5P) with phospho*enol*pyruvate (PEP). Using ESI TOF and flow rates that allowed a time resolution from 7 to 160 ms, the authors were able to detect a variety of enzyme complexes not only with substrate and product, but also with an intermediate species that had previously only been postulated. Since each of these species differ in mass, they could all be readily observed and resolved from each other. Presumably with improved time resolution, it may also be possible to extract kinetic constants to define the catalytic reaction of this enzyme in more detail. The beauty of this technique is the ability not only to monitor concentrations of reactant and product moieties, but also to detect any intermediate species present as long as the reaction involves a change in molecular weight.

By scanning the mass spectrometer over short m/z ranges, or indeed monitoring only single or multiple ions, significantly higher sensitivity can be achieved (83), which can be useful particularly for quantitation. Tandem MS techniques can be extremely useful in situations where impurities or other contaminants are present that may otherwise complicate the analysis. For instance, all the interference from contaminant ions can be effectively eliminated if a specific product ion derived from the parent of interest is measured. Norris and co-workers employed multiple reaction monitoring to quantitate substrate and product concentrations and that of an internal standard as a function of reaction time for the enzyme fucosyltransferase V (84). Potentially this technique could be applicable to the study of enzyme kinetics in a crude reaction mixture, perhaps even approximating physiological conditions.

The continuous flow and stopped-flow instrumentation described above were initially developed many years ago to monitor enzyme kinetics (1, 2). Subsequent improvements in mixing efficiency and reproducibility of flow rates and stop conditions have made these instruments invaluable for the study of reaction kinetics and have allowed the time range to be shortened such that pre-steady-state kinetics can be measured. In the earliest steps of a reaction there is initial buildup of intermediate species at rates that can be measured by rapid mixing experiments. In this pre-steady-state regime it becomes possible, if one can measure the concentration of each of the intermediates as they form, to extract each of the microscopic rate constants. Here the power of MS really comes into its own, since the vast majority of intermediates will differ in mass from the original reactants, so that with sufficient mass resolution the evolution of each species can be measured kinetically.

Continuous flow MS may also be used to obtain kinetic constants of an enzymatic reaction. If the time resolution of the mixer and the interface to the mass spectrometer are sufficient, then it becomes possible to measure directly the rates of formation of enzyme intermediate species (85). Pre-steady-state kinetic measurements were performed on a single point mutant of the protein xylanase, which hydrolyses β-1,4-glycosidic bonds in xylan to form disaccharides, at a rate sufficiently slow to resolve transient glycosyl enzyme intermediate species by continuous flow MS. Using an artificial substrate 2,5-dinitrophenyl β-xylobioside (2,5-DNPX$_2$), the authors were able to follow the hydrolysis reaction photometrically from the release of 2,5-dinitrophenol (2,5-DNP), and also mass spectrometrically by observation of the intermediate xylanase-X$_2$. The observed rates of formation of the two species measured by the two different methods were in close accord, indicating that the mass spectrometric technique can be equally well applied.

Other studies have demonstrated the power of MS for the identification of intermediate species and for mechanistic clarification. One such study of TEM-2 β-lactamase, a well-studied enzyme, and its inhibition by clavulanic acid, called into question the mechanism of inhibition proposed by previous research (86). Previous studies had suggested that a single serine (Ser70) was responsible for binding the carboxylate group of the inhibitor with subsequent opening of the β-lactam ring. By observation of the time course of inhibition, proteolytic digestion of the enzyme–inhibitor complex, and subsequent mass analysis of the digest products, the authors were able to show that not only was Ser70 an important binding residue but also a second site, Ser130, was also involved in cross-linking to the inhibitor during the reaction. We will discuss applications of MS to investigate more of the mechanistic aspects of protein function in the next chapter.

The observation of enzyme kinetics by MS has been dominated by electrospray ionization. In the case of quenched flow measurements, however, it is also possible to investigate enzyme kinetics by other ionization methods. Recently,

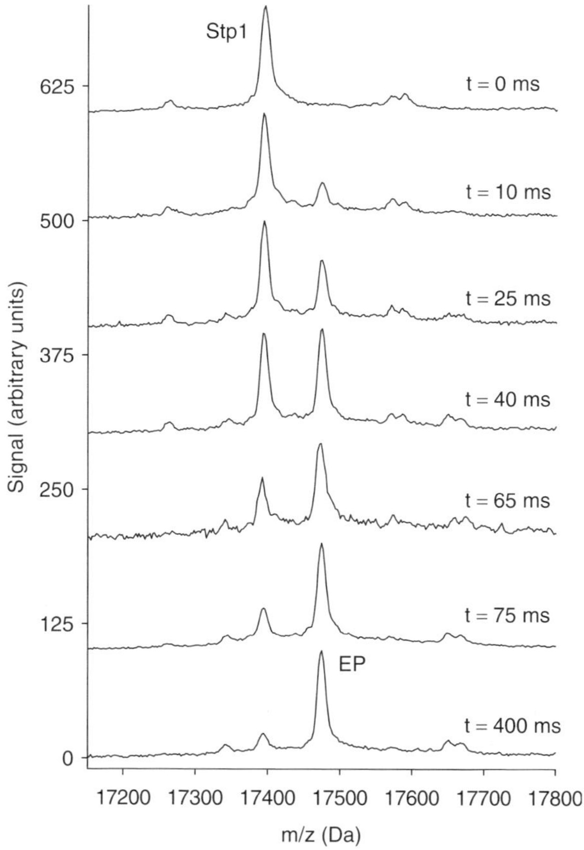

FIGURE 6.15. Time course for production of the covalent enzyme phosphate intermediate (EP) from the protein tyrosine phosphatase enzyme Stp1 during the turnover of *p*-nitrophenylphosphate (*p*-NPP). Beyond 400 ms of reaction, steady-state kinetics are achieved such that [EP] remains effectively constant as *p*-NPP is turned over. Reproduced with permission from (87). © 2000 American Chemical Society.

MALDI-TOF mass spectrometry has also been used for the investigation of pre-steady-state kinetics (87). The protein tyrosine phosphatase (PTPase) catalyzes the dephosphorylation reaction of phosphotyrosine via a mechanism involving a covalent phosphoenzyme intermediate species with mass 80 Da higher than the native enzyme. This same intermediate species is formed regardless of substrate, making quantitation of this species as a function of time an attractive avenue for the screening of substrate reactivity and/or inhibition studies. In this study, the reaction progress was followed at various times from 4 m to 1 s and the relative abundances of enzyme and phosphoenzyme intermediate at each time point were quantitatively measured using MALDI-TOF mass spectrometry. In this manner, pre-steady-state kinetics of formation of the intermediate were measured, and

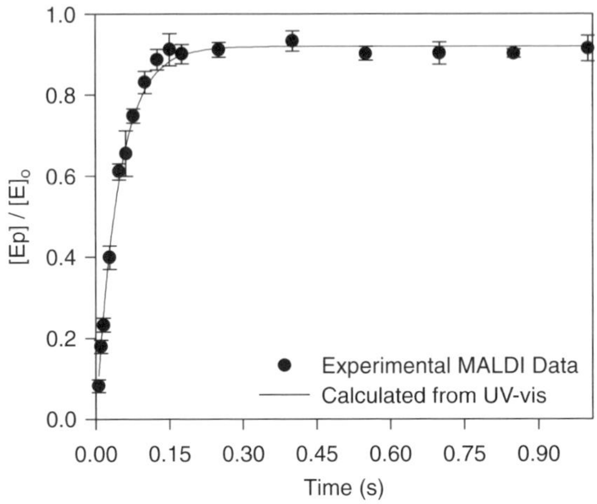

FIGURE 6.16. Reaction rates obtained from MALDI-TOF analysis can be confirmed by comparison with the rate of release of *p*-nitrophenol as measured by UV-Vis spectrophotometry upon dephosphorylation of *p*-NPP. Reproduced with permission from (87). © 2000 American Chemical Society.

moreover the approach to steady state could be directly observed (Figure 6.15). Excellent correlation with optical spectroscopy confirms the applicability of this technique (Figure 6.16). Although MALDI has not become routine thus far for measurement of enzyme catalysis, coupled to quench flow methods it too will no doubt become a very valuable biophysical tool.

In this chapter we have shown how kinetic measurements can be employed to elucidate mechanistic details of protein folding and function. Coupled with studies of dynamics at equilibrium described in the previous chapter, this shows how MS is becoming an extremely powerful tool in the biophysical arena.

REFERENCES

1. Roughton F. J. W. (1934). The kinetics of haemoglobin IV—general methods and theoretical basis for the reactions with carbon monoxide. *Proc. R. Soc. London B* **115**: 451–464.
2. Chance B. (1940). The accelerated-flow method for rapid reactions. *J. Franklin Inst.* **229**: 455–766.
3. Shastry M. C. R., Luck S. D., Roder H. (1998). A continuous-flow capillary mixing method to monitor reactions on the microsecond time scale. *Biophys. J.* **74**: 2714–2721.
4. Alberts B., Bray D., Lewis J., Raff M., Roberts K., Watson J. (1983). *Molecular Biology of the Cell*. New York: Grandland Publishing.
5. Dyson R. (1974). *Cell Biology*. Boston: Allyn and Bacon.
6. Darnell J., Lodish H., Baltimore D. (1990). *Molecular Cell Biology*. New York: W. H. Freeman.

7. Englander S. W., Englander J. J. (1978). Hydrogen–tritium exchange. *Methods Enzymol.* **49**: 24.
8. Englander S. W., Kallenbach N. R. (1984). Hydrogen exchange and the dynamic structures of proteins. *Q. Rev. Biophys.* **16**: 521–655.
9. Englander J. J., Rogero J. R., Englander S. W. (1985). Protein hydrogen exchange studied by the fragment separation method. *Anal. Biochem.* **147**: 234–244.
10. Jacobs M. D., Fox R. O. (1994). Protein folding intermediates characterized by pulsed hydrogen exchange. *Tech. Protein Chem.* **5**: 447–454.
11. Roder H., Elöve G. A., Englander S. W. (1988). Structural characterisation of folding intermediates in cytochrome *c* by hydrogen exchange labelling and proton NMR. *Nature* **335**: 700–704.
12. Connelly G. P., Bai Y. W., Jeng M. F., Englander S. W. (1993). Isotope effects in peptide group hydrogen exchange. *Proteins: Structure, Function and Genetics* **17**: 87–92.
13. Varley P., Gronenborn A. M., Christensen H., Wingfield P. T., Pain R. H., Clore G. M. (1993). Kinetics of folding of the all β-sheet protein interleukin-1-β. *Science* **260**: 1110–1113.
14. Radford S. E., Dobson C. M., Evans P. (1992). The folding of hen lysozyme involves partially structured intermediates and multiple pathways. *Nature* **358**: 302–307.
15. Katta V., Chait B. T. (1991). Conformational changes in proteins probed by hydrogen-exchange electrospray-ionization mass spectrometry. *Rapid Commun. Mass Spectrom.* **5**: 214–217.
16. Miranker A., Robinson C. V., Radford S. E., Aplin R. T., Dobson C. M. (1993). Detection of transient protein folding populations by mass spectrometry. *Science* **262**: 896–900.
17. Hooke S. D., Eyles S. J., Miranker A., Radford S. E., Robinson C. V., Dobson C. M. (1995). Cooperative elements in protein folding monitored by electrospray ionisation mass spectrometry. *J. Am. Chem. Soc.* **117**: 7548–7549.
18. Matthews C. R. (1993). Pathways of protein folding. *Annu. Rev. Biochem.* **62**: 653–683.
19. Kuwajima K. (1989). The molten globule state as a clue for understanding the folding and cooperativity of globular protein structure. *Proteins: Structure, Function and Genetics* **6**: 87–103.
20. Baldwin R. L. (1994). Finding intermediates in protein folding. *BioEssays* **16**: 207–210.
21. Bai Y., Sosnick T. R., Mayne L., Englander S. W. (1995). Protein folding intermediates: native state hydrogen exchange. *Science* **269**: 192–196.
22. Zitzewitz J. A., Matthews C. R. (1993). Protein engineering strategies in examining protein folding intermediates. *Curr. Opin. Struct. Biol.* **3**: 594–600.
23. Dill K. A., Chan H. S. (1997). From Levinthal to pathways to funnels. *Nat. Struct. Biol.* **4**: 10–19.
24. Wolynes P. G., Luthey-Schulten Z., Onuchic J. N. (1996). Fast-folding experiments and the topography of protein folding energy landscapes. *Chem. Biol.* **3**: 425–432.
25. Jackson S. E., Fersht A. R. (1991). Folding of chymotrypsin inhibitor 2. 1. Evidence for a two-state transition. *Biochemistry* **30**: 10428–10433.

26. Jackson S. E. (1998). How do small single-domain proteins fold? *Folding and Design* **3**: R81–R91.
27. Kragelund B. B., Hojrup P., Jensen M. S., Schjerling C. K., Juul E., Knudsen J., Poulsen F. M. (1996). Fast and one-step folding of closely and distantly related homologous proteins of a four-helix bundle family. *J. Mol. Biol.* **256**: 187–200.
28. Ferguson N., Capaldi A. P., James R., Kleanthous C., Radford S. E. (1999). Rapid folding with and without populated intermediates in the homologous four-helix proteins Im7 and Im9. *J. Mol. Biol.* **286**: 1597–1608.
29. Gunasekaran K., Eyles S. J., Hagler A. T., Gierasch L. M. (2001). Keeping it in the family: folding studies of related proteins. *Curr. Opin. Struct. Biol.* **11**: 83–93.
30. Kiefhaber T. (1995). Kinetic traps in lysozyme folding. *Proc. Natl. Acad. Sci. U.S.A.* **92**: 9029–9033.
31. Sosnick T. R., Mayne L., Hiller R., Englander S. W. (1994). The barriers in protein folding. *Nat. Struct. Biol.* **1**: 149–156.
32. Heidary D. K., Gross L. A., Roy M., Jennings P. A. (1997). Evidence for an obligatory intermediate in the folding of interleukin 1β. *Nat. Struct. Biol.* **4**: 726–731.
33. Tsui V., Garcia C., Cavagnero S., Siuzdak G., Dyson H. J., Wright P. E. (1999). Quench-flow experiments combined with mass spectrometry show apomyoglobin folds through an obligatory intermediate. *Protein Sci.* **8**: 45–49.
34. Nishimura C., Wright P. E., Dyson H. J. (2003). Role of the B helix in early folding events in apomyoglobin: evidence from site-directed mutagenesis for native-like long range interactions. *J. Mol. Biol.* **334**: 293–307.
35. Jennings P. A., Wright P. E. (1993). Formation of a molten globule intermediate early in the kinetic folding pathway of apomyoglobin. *Science* **262**: 892–896.
36. Hughson F. M., Wright P. E., Baldwin R. L. (1990). Structural characterization of a partly folded apomyoglobin intermediate. *Science* **249**: 1544–1548.
37. Cavagnero S., Theriault Y., Narula S. S., Dyson H. J., Wright P. E. (2000). Amide proton hydrogen exchange rates for sperm whale myoglobin obtained from ^{15}N-^{1}H NMR spectra. *Protein Sci.* **9**: 186–193.
38. Nishimura C., Dyson H. J., Wright P. E. (2002). The apomyoglobin folding pathway revisited: structural heterogeneity in the kinetic burst phase intermediate. *J. Mol. Biol.* **322**: 483–489.
39. Bieri O., Kiefhaber T. (2001). Origin of apparent fast and non-exponential kinetics of lysozyme folding measured in pulsed hydrogen exchange experiments. *J. Mol. Biol.* **310**: 919–935.
40. Deng Y., Zhang Z., Smith D. L. (1999). Comparison of continuous and pulsed labeling amide hydrogen exchange/mass spectrometry for studies of protein dynamics. *J. Am. Soc. Mass Spectrom.* **10**: 675–684.
41. Udgaonkar J. B., Baldwin R. L. (1990). Early folding intermediate of ribonuclease A. *Proc. Natl. Acad. Sci. U. S. A.* **87**: 8197–8201.
42. Simmons D. A., Konermann L. (2002). Characterization of transient protein folding intermediates during myoglobin reconstitution by time-resolved electrospray mass spectrometry with on-line isotopic pulse labeling. *Biochemistry* **41**: 1906–1914.
43. Yang H., Smith D. L. (1997). Kinetics of cytochrome *c* folding examined by hydrogen exchange and mass spectrometry. *Biochemistry* **36**: 14992–14999.

44. Pan H., Smith D. L. (2003). Quaternary structure of aldolase leads to differences in its unfolding and unfolding intermediates. *Biochemistry* **42**: 5713–5721.

45. Wolynes P. G. (1997). Folding funnels and energy landscapes of larger proteins within the capillary approximation. *Proc. Natl. Acad. Sci. U. S. A.* **94**: 6170–6175.

46. Bryngelson J. D., Onuchic J. N., Socci N. D., Wolynes P. G. (1995). Funnels, pathways, and the energy landscape of protein folding: a synthesis. *Proteins: Structure, Function and Genetics* **21**: 167–195.

47. Rockwood A. L., Busman M., Smith R. D. (1991). Coulombic effects in the dissociation of large highly charged ions. *Int. J. Mass Spectrom. Ion Proc.* **111**: 103–129.

48. Lee E. D., Muck W., Henion J. D., Covey T. R. (1989). Real time reaction monitoring by continuous-introduction ion spray tandem mass spectrometry. *J. Am. Chem. Soc.* **111**: 4600–4604.

49. Sam J. W., Tang X.-J., Peisach J. (1994). Electrospray mass spectrometry of iron bleomycin: demonstration that activated bleomycin is a ferric peroxide complex. *J. Am. Chem. Soc.* **116**: 5250–5256.

50. Sam J. W., Tang X.-J., Magliozzo R. S., Peisach J. (1995). Electrospray mass spectrometry of iron bleomycin II: investigation of the reaction of Fe(III)-bleomycin with iodosylbenzene. *J. Am. Chem. Soc.* **117**: 1012–1018.

51. Konermann L., Collings B. A., Douglas D. J. (1997). Cytochrome c folding kinetics studied by time-resolved electrospray ionization mass spectrometry. *Biochemistry* **36**: 5554–5559.

52. Brandts J. F., Halvorson H. R., Brennan M. (1975). Consideration of the possibility that the slow step in protein denaturation reactions is due to cis-trans isomerism of proline residues. *Biochemistry* **14**: 4953–4963.

53. Lee V. W. S., Chen Y.-L., Konermann L. (1999). Reconstitution of acid-denatured holomyoglobin studied by time resolved electrospray ionization mass spectrometry. *Anal. Chem.* **71**: 4154–4159.

54. Konermann L., Rosell F. I., Mauk A. G., Douglas D. J. (1997). Acid-induced denaturation of myoglobin studied by time-resolved electrospray ionization mass spectrometry. *Biochemistry* **36**: 6448–6454.

55. Sogbein O. O., Simmons D. A., Konermann L. (2000). Effects of pH on the kinetic reaction mechanism of myoglobin unfolding studied by time-resolved electrospray ionization mass spectrometry. *J. Am. Soc. Mass Spectrom.* **11**: 312–319.

56. Simmons D. A., Dunn S. D., Konermann L. (2003). Conformational dynamics of partially denatured myoglobin studied by time-resolved electrospray mass spectrometry with online hydrogen-deuterium exchange. *Biochemistry* **42**: 5896–5905.

57. Martin S. E., Shabanowitz J., Hunt D. F., Marto J. A. (2000). Subfemtomole MS and MS/MS peptide sequence analysis using nano-HPLC micro-ESI Fourier transform ion cyclotron resonance mass spectrometry. *Anal. Chem.* **72**: 4266–4274.

58. Settlage R. E., Christian R. E., Hunt D. F. 2000. *U.S. Patent No. 6139734*.

59. Ørsnes H., Graf T., Degn H. (1998). Stopped flow mass spectrometry with rotating ball inlet: application to the ketone–sulfite reaction. *Anal. Chem.* **70**: 4751–4754.

60. Kolakowski B. M., Simmons D. A., Konermann L. (2000). Stopped-flow electrospray ionization mass spectrometry: a new method for studying chemical reaction kinetics in solution. *Rapid Commun. Mass Spectrom.* **14**: 772–776.

61. Kolakowski B. M., Konermann L. (2001). From small-molecule reactions to protein folding: studying biochemical kinetics by stopped-flow electrospray mass spectrometry. *Anal. Biochem.* **292**: 107–114.
62. Bökenkamp D., Desai A., Yang X., Tai Y.-C., Marzluff E. M., Mayo S. L. (1998). Microfabricated silicon mixers for submillisecond quench-flow analysis. *Anal. Chem.* **70**: 232–236.
63. Oleschuk R. D., Harrison D. J. (2000). Analytical microdevices for mass spectrometry. *Trends Anal. Chem.* **19**: 379–388.
64. Creighton T. E. (1992). *Protein Folding*. New York: W. H. Freeman.
65. Creighton T. E. (1984). Disulfide bond formation in proteins. *Methods Enzymol.* **107**: 305–329.
66. Zhang Y. H., Yan X., Maier C. S., Schimerlik M. I., Deinzer M. L. (2002). Conformational analysis of intermediates involved in the in vitro folding pathways of recombinant human macrophage colony stimulating factor β by sulfhydryl group trapping and hydrogen/deuterium pulse labeling. *Biochemistry* **41**: 15495–15504.
67. Apuy J. L., Park Z.-Y., Swartz P. D., Dangott L. J., Russell D. H., Baldwin T. O. (2001). Pulsed-alkylation mass spectrometry for the study of protein folding and dynamics: development and application to study of a folding/unfolding intermediate of bacterial luciferase. *Biochemistry* **40**: 15153–15163.
68. Lim J., Vachet R. W. (2003). Development of a methodology based on metal-catalyzed oxidation reactions and mass spectrometry to determine the metal binding sites in copper metalloproteins. *Anal. Chem.* **75**: 1164–1172.
69. Fändrich M., Tito M. A., Leroux M. R., Rostom A. A., Hartl F. U., Dobson C. M., Robinson C. V. (2000). Observation of the noncovalent assembly and disassembly pathways of the chaperone complex MtGimC by mass spectrometry. *Proc. Natl. Acad. Sci. U.S.A.* **97**: 14151–14155.
70. Sobott F., Benesch J. L. P., Vierling E., Robinson C. V. (2002). Subunit exchange of multimeric protein complexes. Real time monitoring of subunit exchange between small heat shock proteins by using electrospray mass spectrometry. *J. Biol. Chem.* **277**: 38921–38929.
71. Michaelis L., Menten M. L. (1913). Die kinetik der invertinwirkung. *Biochem. Z.* **49**: 333–339.
72. Huggett A. S. G., Nixon D. A. (1957). Enzymic determination of blood glucose. *Biochem. J.* **66**: 12P.
73. Wallenfels K. (1962). Beta-galactosidase (crystalline). *Methods Enzymol.* **5**: 212–219.
74. Northrop D. B., Simpson F. B. (1997). Beyond enzyme kinetics: direct determination of mechanisms by stopped-flow mass spectrometry. *Bioorg. Med. Chem. Lett.* **5**: 641–644.
75. Smith L. A., Caprioli R. M. (1983). Following enzyme catalysis in real time inside a fast atom bombardment mass spectrometer. *Biomed. Mass Spectrom.* **10**: 98–102.
76. Smith L. A., Caprioli R. M. (1984). Enzyme reaction rates determined by fast atom bombardment mass-spectrometry. *Biomed. Mass Spectrom.* **11**: 392–395.
77. Hsieh F. Y. L., Tong X., Wachs T., Ganem B., Henion J. (1995). Kinetic monitoring of enzymatic reactions in real time by quantitative high performance liquid chromatography mass spectrometry. *Anal. Biochem.* **229**: 20–25.

78. Anderson K. S., Sikorski J. A., Johnson K. A. (1988). A tetrahedral intermediate in the EPSP synthase reaction observed by rapid quench kinetics. *Biochemistry* **27**: 7395–7406.

79. Anderson K. S., Sikorski J. A., Johnson K. A. (1988). Evaluation of 5-*enol*pyruvoyl-shikimate-3-phosphate synthase substrate and inhibitor binding by stopped-flow and equilibrium fluorescence measurements. *Biochemistry* **27**: 1604–1610.

80. Brown E. D., Marquardt J. L., Lee J. P., Walsh C. T., Anderson K. S. (1994). Detection and characterization of a phospholactoyl-enzyme adduct in the reaction catalyzed by UDP-*N*-acetylglucosamine *enol*pyruvoyl transferase, MurZ. *Biochemistry* **33**: 10638–10645.

81. Paiva A. A., Tilton R. F., Crooks G. P., Huang L. Q., Anderson K. S. (1997). Detection and identification of transient enzyme intermediates using rapid mixing, pulsed-flow electrospray mass spectrometry. *Biochemistry* **36**: 15472–15476.

82. Li Z., Sau A. K., Shen S., Whitehouse C., Baasov T., Anderson K. S. (2003). A snapshot of enzyme catalysis using electrospray ionization mass spectrometry. *J. Am. Chem. Soc.* **125**: 9938–9939.

83. Bothner B., Chavez R., Wei J., Strupp C., Phung Q., Schneemann A., Siuzdak G. (2000). Monitoring enzyme catalysis with mass spectrometry. *J. Biol. Chem.* **275**: 13455–13459.

84. Norris A. J., Whitelegge J. P., Faull K. F., Toyokumi T. (2001). Analysis of enzyme kinetics using electrospray ionization mass spectrometry and multiple reaction monitoring: fucosyltransferase V. *Biochemistry* **40**: 3774–3779.

85. Zechel D. L., Konermann L., Withers S. G., Douglas D. J. (1998). Pre-steady state kinetic analysis of an enzymatic reaction monitored by time-resolved electrospray ionization mass spectrometry. *Biochemistry* **37**: 7664–7669.

86. Brown R. P. A., Aplin R. T., Schofield C. T. (1996). Inhibition of TEM-2β-lactamase from *Escherichia coli* by clavulanic acid: observation of intermediates by electrospray ionization mass spectrometry. *Biochemistry* **35**: 12421–12432.

87. Houston C. T., Taylor W. P., Widlanski T. S., Reilly J. P. (2000). Investigation of enzyme kinetics using quench-flow techniques with MALDI-TOF mass spectrometry. *Anal. Chem.* **72**: 3311–3319.

7

PROTEIN INTERACTION: A CLOSER LOOK AT THE "STRUCTURE–DYNAMICS–FUNCTION" TRIAD

In the preceding three chapters we have discussed various MS-based techniques to study biomolecular higher order structure and dynamics under a variety of conditions. Characterization of protein structure and dynamics is often key to understanding how these macromolecules carry out their "duties" via a sophisticated web of interactions with each other, other biopolymers, and small organic and inorganic ligands. We have already discussed several qualitative examples of how both structure and dynamics govern protein–protein interaction. In this chapter we consider several MS-based methods that can be used to extract quantitative information on protein–ligand interaction. We first consider several approaches to measuring protein–small ligand binding energy. Particular attention is paid to complexes that are unstable in the gas phase. We then proceed to discuss various methods that are used to obtain quantitative information on protein–protein interactions. The remainder of the chapter discusses advanced uses of MS to characterize the dynamics of multiprotein assemblies and its role in modulating protein function. Two specific examples that are considered include both large-scale dynamics (such as transient unfolding, as well as intrinsic disorder) and relatively small-scale "long-distance" motions (allosteric effects).

7.1. PROTEIN–LIGAND INTERACTIONS: CHARACTERIZATION OF NONCOVALENT COMPLEXES USING DIRECT ESI MS MEASUREMENTS

Although the earliest applications of ESI MS focused on measuring mass and obtaining sequence information of single-chain biopolymers, the apparent

Mass Spectrometry in Biophysics: Conformation and Dynamics of Biomolecules
By Igor A. Kaltashov and Stephen J. Eyles
ISBN 0-471-45602-0

"gentleness" of this ionization technique prompted several researchers to explore if the structural integrity of protein–small ligand complexes can be maintained throughout the desorption process. As early as 1991, Ganem, Li, and Henion observed intact complexes formed by a small (~12 kDa) immunosuppressive binding protein and small immunosuppressive agents rapamycin and its analogue (1). This report was soon followed by two other important studies in which intact noncovalent complexes of lysozyme and its natural hexasaccharide substrate (2), as well as of the myoglobin–heme group complex (3) were observed. The list of ligands that bind to proteins strongly enough to survive the transition from solution to the gas phase has been expanded dramatically in the following couple of years to include metal ions (4, 5), nucleotides (6), oligonucleotides (7, 8), and peptides and proteins (9). Concurrently, utilization of ESI MS for direct observation of specific associations of other biopolymers, such as DNA duplexes (10), and quadruplexes (11), were reported.

The spectacular success of these early studies provided motivation for further exploration of the utility of ESI MS for characterization of such interactions in solution. Several groups evaluated the possibility to correlate binding properties in solution and complex ion stability in the gas phase for a variety of systems (12, 13). This issue was systematically explored by Douglas and co-workers, who used a set of myoglobin and cytochrome b_5 variants with mutations affecting the protein–heme group binding strength in solution (14). A good correlation was found to exist between the solution and gas phase binding properties within the set of mutants in which the number of hydrogen bonds between the heme propionate groups and the protein surface residues was systematically reduced. However, removal of the hydrogen bonds linking the heme to the proximal histidine residue did not result in detectable destabilization of the heme–protein complex ion in the gas phase (14).

This observation and other reports detailing frequent deviations of binding characteristics in solution from those in the gas phase (15, 16) indicate that it may be dangerous to simply equate the solution and gas phase properties of noncovalent complexes. As noted by Ganem and Henion, "noncovalent binding of macromolecules usually involves a broad range of weak interactions, ranging from hydrogen bonding and electrostatic effects to ... π-stacking and hydrophobic effects [with] each type of interaction responding differently to changes in solvation, as would occur during ionspray or electrospray ionization" (17). Nevertheless, if the protein–ligand interaction involves an extensive array of hydrogen bonds [or any other intermolecular interactions that are not weakened by the removal of solvent (e.g., salt bridges)], solution binding characteristics can be obtained from straightforward ESI MS measurements. Ganem and Henion demonstrated that monitoring protein–small ligand interaction in solution with ESI MS in some instances can provide quantitative information on the energetics of the binding process (17). Direct monitoring of the relative abundance of complex ions formed by a model glycopeptide and antibiotics from the vancomycin family was shown to provide a correct order of magnitude for complex association constants (K_A). K_A values were obtained by constructing a Scatchard plot whose parameters were derived from ESI MS measurements. Similar conclusions have been reached in a number of other studies, where MS was shown to be capable of providing correct quantitative assessment of binding affinity in solution (18–21).

Although this approach continues to be used to evaluate binding in solution in both qualitative (i.e., ranking ligands according to binding affinity) and quantitative fashion for select systems (22–26), its limitations have also become apparent (27–29). One particularly striking example has recently been reported by Heck, who studied gas phase dissociation of a ternary complex ion composed of two proteins and a Zn^{2+} ion (30). The protein–protein complex of colisin E9 DNase and its cognate immunity protein Im9 has one of the lowest equilibrium dissociation constants reported so far for a protein complex ($K_D \sim 10^{-16}$ M), while the binding affinity of Zn^{2+} for the E9 DNase is seven orders of magnitude lower. Nevertheless, in the gas phase dissociation experiments, dissociation of the protein constituents occurs before the release of the metal ion. A detailed analysis of the protein–protein interface suggests that hydrophobic interactions and hydrogen bonds mediated by interfacial water molecules are key contributors in defining the stability and specificity of the complex. The apparent weakening of the protein–protein interaction in the gas phase was attributed to diminishing complex stabilization by van der Waals forces and disruption of the hydrogen-bonding network as a result of ion desolvation. By contrast, the electrostatic interaction was not diminished as a result of solvent removal, hence the remarkable stability of the protein–metal ion bond in the gas phase (30).

An interested reader is referred to an excellent recent review by Zenobi and co-workers, who provide a detailed discussion and multiple examples of how quantitative data on noncovalent binding can be extracted from direct ESI MS measurements (23).

7.2. INDIRECT CHARACTERIZATION OF NONCOVALENT INTERACTIONS: MEASUREMENTS UNDER NATIVE CONDITIONS

The gas phase behavior of the Zn^{2+}–E9 Dnase–Im9 ternary complex ion considered in the previous section of this chapter presents an example of hydrophobic interactions, however strong in solution, "losing out" in a competition with electrostatic interactions in the gas phase. Following removal of the solvent, hydrophobic interactions are reduced to weak van der Waals and London dispersion forces, leading to significant destabilization of the complex ion in the gas phase. Electrostatic interactions, on the other hand, are usually enhanced in a solvent-free environment (due to dramatic decrease of the dielectric constant).*

*Although the dielectric constant ε of water exceeds that of vacuum by almost two orders of magnitude, the actual decrease of ε upon protein transition from an aqueous solution to the gas phase will not be as dramatic, as the dielectric constant within the protein interior is usually much smaller compared to its surface, which is smaller still than that of the surrounding solvent (due to limited mobility of the polar groups within the protein as opposed to solvent molecules).

Since hydrophobic interaction is in many cases an important, but not the only, factor in stabilizing protein–ligand complexes in solution, stability of the complex ion in the gas phase may have some correlation with binding energetics in solution, but the deviations may be very significant. Whenever hydrophobic interactions play a significant role in stabilizing noncovalent associations in solution, relative abundance of noncovalently bound complex ions in ESI mass spectra cannot be used as a reliable indicator of the binding strength in solution. This section presents several methods that are used to probe noncovalent interactions indirectly; that is, they do not rely on "survival" of the complex ions upon their transition from solution to the gas phase.

7.2.1. Assessment of Ligand Binding by Monitoring Dynamics of "Native" Proteins with HDX MS

One particularly interesting example of such limited correlation between protein–ligand binding properties in solution and the complex stability in the gas phase comes from our own work with cellular retinoic acid binding protein I (CRABP I), a transporter of *all-trans* retinoic acid (RA).* The X-ray crystal structure suggests that RA–CRABP I complex is stabilized primarily through a network of hydrophobic interactions, although there is also an intermolecular salt bridge that makes a significant contribution toward stability of this complex (K_D in the subnanomolar range). In addition to RA, CRABP I also binds (albeit with significantly lower affinity) stereoisomers of RA, 9-cis and 13-cis (both are noncognate ligands to CRABP I). Although the ESI mass spectra of the CRABP I/RA mixture acquired under mild conditions in the ESI interface do show prominent signals corresponding to protein–ligand complex ions (Figure 7.1A), their relative intensities do not agree with ranking of binding affinities in solution (31).

As we have already discussed in Chapter 1, it is now realized that in many cases protein folding and function should not be considered separately from each other (32). This notion is often used when studying protein–protein interactions (33, 34); however, it can also be very useful when protein interaction with small organic ligands is considered. A common assumption here is that binding reduces flexibility, which has been proved experimentally in the case of several retinoid carriers (35–37). Therefore, changes in protein backbone dynamics upon ligand introduction can be used as indicators of the protein–ligand interaction. This hypothesis was tested by measuring HDX kinetics of CRABP I under near-native conditions (Figure 7.2). Under these conditions HDX clearly follows the EX2 regime (see Chapter 5 for a detailed discussion of hydrogen exchange under the EX2 regime), and the overall kinetics exhibit three phases (fast, intermediate, and slow). As we have discussed in Chapter 5, the slow exchange phase usually corresponds to global unfolding events that occur very rarely under the native conditions. According to (5-3-4), the apparent rate constant corresponding to this phase would be related to both intrinsic exchange rate k_{int} and the equilibrium

*RA is a major physiologically active metabolite of vitamin A.

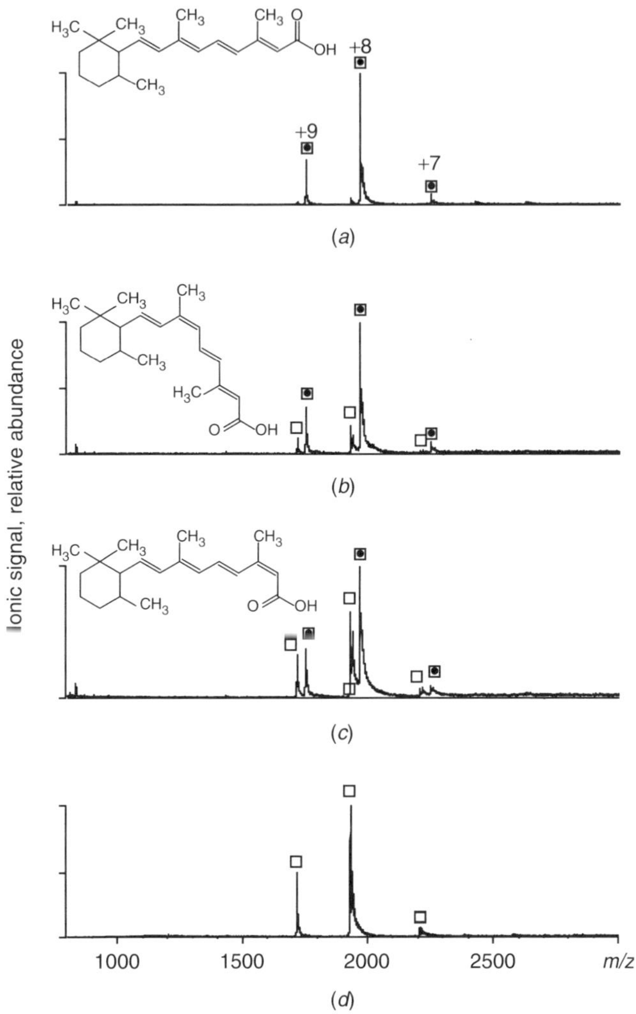

FIGURE 7.1. ESI mass spectra of CRABP I mixed with RA isomers and analogues (all-trans, **A**; 13-cis, **B**; 9-cis, **C**; no ligand added, **D**). The protein-to-ligand ratio (where appropriate) is 1:5. All spectra are acquired under "near-native" solution conditions. ESI interface settings were adjusted to minimize collisional activation/dissociation. Adapted with permission from (31). © 2003 American Society for Mass Spectrometry.

constant for the unfolding process K_{unfold}:

$$k^{\text{HDX}} = k_{\text{int}} \cdot K_{\text{unfold}} \tag{7-2-1}$$

Since the equilibrium constant K_{unfold} relates to a free energy of the native conformation in the following fashion:

$$\Delta G_N = -RT \cdot \ln(K_{\text{unfold}}), \tag{7-2-2}$$

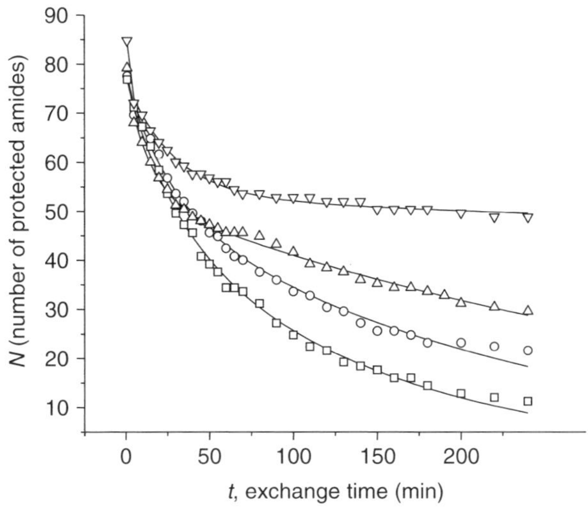

FIGURE 7.2. HDX kinetics of CRABP I at 36 °C under "near-native" conditions in the absence of ligands (squares) and in the presence of RA isomers (13-cis, circles; 9-cis, triangles; and all-trans, inverted triangles). Adapted with permission from (31). © 2003 American Society for Mass Spectrometry.

and the protein–ligand binding energy can be estimated as a free energy change of the native conformation induced by the ligand (the asterisk indicates the presence of ligand in the system):

$$\Delta G_{\text{binding}} = \Delta G_{N*} - \Delta G_N = \Delta\Delta G_{N^*-N}, \tag{7-2-3}$$

one can estimate the binding energy simply by comparing the apparent HDX rate constant measured in the presence and in the absence of the ligand:

$$\Delta G_{\text{binding}} = RT \cdot \ln(k_{\text{HDX}}/k^*_{\text{HDX}}). \tag{7-2-4}$$

An implicit assumption used here is that there is no significant conformational difference between the apoprotein and its holo-form, so that ligand-induced structure stabilization would be the only factor contributing to the free energy change $\Delta\Delta G_{N^*-N}$ upon ligand binding by the protein (7-2-3). In other words, this approach can safely be used only if the protein–ligand binding process conforms to the "lock-and-key" mechanism.* Using this approach to evaluate CRABP I affinity to various retinoids not only provides correct ranking of RA stereoisomers, but also gives an estimation of the protein–RA binding energy, which is reasonably close to that obtained by calorimetric measurements (31).

*There may be significant conformational changes during the various stages of the binding process. However, as long as the endpoints of the binding reaction (apo- and holo-forms of the protein) are structurally similar, the thermodynamic treatment presented here can safely be used.

A discussion of this method would probably be incomplete without mentioning the fact that the assumption that "binding reduces flexibility" is not universally true. It is often anticipated that specific, high-affinity binding proceeds through formation of well-defined interactions between a protein and its ligand, leading to a loss of conformational entropy, which would need to be more than offset by favorable enthalpic interactions for the binding to be thermodynamically favored. However, an increase of the conformational entropy of the residues interacting with the ligand may also promote the binding process, as "positive" ΔS would lead to "more negative" (and, therefore, more favorable) ΔG even if enthalpic contribution ΔH is rather insignificant ($\Delta G = \Delta H - T\Delta S \approx -T\Delta S$) (38). Indeed, there are several documented examples when binding actually leads to a detectable increase in protein flexibility (39–41). These findings, however, are likely to have little bearing on the validity of the method discussed in this section. Indeed, the observed increases of protein flexibility upon ligand binding refer to local events, typically at or near the binding site(s) (38). Such behavior would manifest itself as an increase of the HDX rates in the fast (local fluctuations) and medium (partially unstructured intermediate states) phases of the overall HDX pattern. None of these measurements is used to determine the binding energy, whose calculation (7-2-4) is based solely on the parameters of the slow phase (rare global unfolding events).

7.2.2. PLIMSTEX: Binding Assessment Via Monitoring Conformational Changes with HDX MS in Titration Experiments

Another method that uses HDX to quantify protein–ligand interaction (PLIMSTEX, short for **p**rotein–**li**gand **i**nteractions in solution by **MS**, **t**itration, and H/D **ex**change) was introduced recently by Gross and co-workers (42). PLIMSTEX requires that ligand binding result in a change in HDX (e.g., due to different solvent exposure of the apo- and holo-forms of the protein). The method is analogous to titration monitoring by spectroscopic methods. The approach, although indirect, overcomes a major difficulty of ESI MS to measure, without discrimination, solution concentrations at any point, including at equilibrium.

The protein is first equilibrated with different concentrations of ligand in aqueous buffer, and HDX reactions are then initiated by diluting these protein solutions in D_2O without changing any other solution parameters (salt content, buffer compositions, etc.). After reaching a near-steady state (after at least an hour), the exchange is quenched, following by quick desalting under slow exchange conditions and mass analysis (see Chapter 5 for a more detailed discussion of HDX quenching). Since quenching and desalting typically cause the ligand(s) to dissociate, mass measurements simply reveal the number of deuterium atoms on solvent-accessible amides (corrected for back-exchange during quench and desalting). A plot of these numbers (deuterium uptake) as a function of the total ligand concentration gives the PLIMSTEX curve. The curve is then fit by using a 1:n (protein/ligand) sequential binding model (where n is the number of binding sites per protein molecule) to produce cumulative equilibrium binding constants

(β_i, $i = 1$ to n)* and "deuterium uptake shifts" ΔD_i (differences between the average deuterium level of each binding species and that of the apoprotein). A positive ΔD_i indicates that binding of i ligand(s) to the protein leads to more protection (less solvent exposure and/or conformational flexibility) compared to the apo-form. A negative ΔD_i points to the formation of a more open structure (or more flexible structure) relative to the ligand-free form of the protein. As such, PLIMSTEX can be applied to determine both binding stoichiometry and affinity associated with a wide range of protein–ligand interactions, as well as to characterize conformational changes accompanying such binding events. Figure 7.3 shows an application of PLIMSTEX to probe Ca^{2+} binding to porcine calmodulin (43).

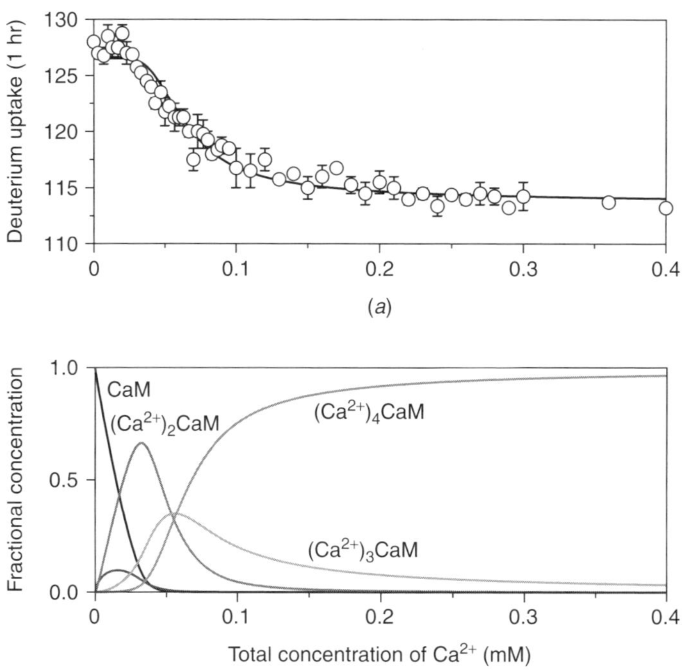

FIGURE 7.3. Use of PLIMSTEX to probe Ca^{2+} binding to porcine calmodulin. (**A**) Ca^{2+} titration of 15 μM porcine calmodulin (CaM) in 50 mM HEPES (pH 7.4, 21.5 °C, 90% D_2O). (**B**) Fractional concentrations of $(Ca^{2+})_i$CaM species as a function of total Ca^{2+} concentration in solution. The PLIMSTEX titration data (**A**) are nearly a mirror image of those of the $(Ca^{2+})_4$CaM fractional concentration (**B**), suggesting that $(Ca^{2+})_4$CaM formation plays a major role in determining the titration curve shape. None of the nonspecific binding of more than four Ca^{2+} ions is registered by PLIMSTEX, indicating that if further Ca^{2+}–CaM binding occurs, it does not cause any significant conformational changes in proteins. Adapted with permission from (43). © 2003 American Chemical Society.

*Binding constants K_i can easily be calculated from the cumulative constants, since $\beta_i = \Pi K_i$.

7.2.3. Other Titration Methods Utilizing HDX MS Under Native Conditions

Engen recently used HDX MS to estimate a dissociation constant K_D of a heterodimer formed by a ubiquitin homologue SUMO-1 and a ubiquitin conjugating enzyme UBC9 (44). Changes in deuterium levels upon formation of the complex were probed by comparing the amount of deuterium incorporated in each protein alone and in the presence of its binding partner. There was no change in UBC9 deuterium incorporation at times ranging from 10 s to 8 h in the presence of excess SUMO-1, indicating that there was no solvent shielding or conformational change in UBC9 upon complex formation. On the other hand, SUMO-1 displayed a significant reduction in the amount of deuterium incorporation when UBC9 was present. Since the reduction in SUMO-1 deuterium levels occurred mostly after 1 min of incubation in D_2O, it would seem unlikely that solvent shielding played a significant role (amide hydrogen atoms at the surface of proteins ordinarily exchange quite rapidly). In order to determine how much UBC9 relative to SUMO-1 is required to elicit a detectable change in SUMO-1 deuterium levels and to estimate the K_D for the complex, a constant molar amount of SUMO-1 was titrated with UBC9. The titration curve suggests that the ratio of UBC9/SUMO-1 required for efficient binding is at least 5, which is equivalent to a UBC9 concentration of 60–100 μM, thus providing an estimate of K_D.

7.2.4. Binding Revealed by Changes in Ligand Mobility

Another creative approach to characterizing protein–ligand interaction in solution with mass spectrometry was introduced recently by Clark and Konermann (45). The technique is applicable in situations when the ligand's physical dimensions are much smaller compared to those of the ligand–protein complex, so that diffusion coefficients of the two species in solution will differ very significantly from one another. The diffusion coefficients are determined by measuring the spread of an initially sharp boundary between two solutions of different concentration in a laminar flow tube due to Taylor dispersion (46). In the absence of the protein–ligand interaction, the measured ESI MS dispersion profiles show a gradual transition for the protein and a steep transition for the small ligand (Figure 7.4). However, if a noncovalent complex is formed in solution, the dispersion profiles of the two species will be very similar, since diffusion of the small ligand is now limited by the slow Brownian motions of the protein. Unlike the direct ESI MS-based methods of probing noncovalent complexes, this technique does not rely on the complex "survival" upon its transition from solution to the gas phase. On the contrary, "harsh" conditions in the ESI interface are required to dissociate the complex ion in the gas phase, such that the dispersion profiles of its constituents can be monitored separately. Figure 7.5 illustrates application of this technique to study noncovalent heme–protein interactions in myoglobin under both native and denaturing conditions.

This method has recently been used to screen a small library of low molecular weight saccharides to identify those that bind to a protein target (47). Mixtures of

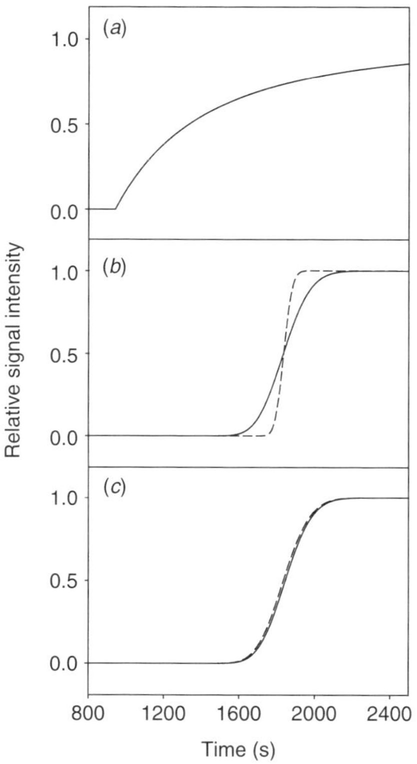

FIGURE 7.4. Simulated dispersion profiles representing signal intensity of selected analytes at the end of a laminar flow tube under different conditions. (**A**) A dispersion profile expected for a single analyte in the absence of diffusion. (**B**) Dispersion profiles expected for a protein ($D = 10^{-10}$ m^2/s, solid curve), and a small ligand ($D = 10^{-9}$ m^2/s, dashed curve) that do not interact in solution. (**C**) Dispersion profiles as in (**B**), but under the assumption of tight noncovalent binding between the two analytes (the profiles measured for the two species will be identical under these conditions). Flow parameters used in simulations: tube length, 3 m; tube radius, 129.1 μm; flow rate 5 μL/min (corresponding to $v_{max} = 3.18 \times 10^{-3}$ m/s). Reproduced with permission from (45). © 2003 American Society for Mass Spectrometry.

six different saccharides were screened for noncovalent binding to lysozyme. Of the six "candidate" ligands, only N,N',N''-triacetylchitotriose (NAG_3) is known to bind to the protein. In "direct" binding tests, NAG_3 showed a significantly reduced diffusion coefficient in the presence of the protein, while no changes were observed for any of the other five candidates. In a second set of experiments, the candidate ligands were tested for their ability to displace a reference ligand (N, N',N'', N'''-tetraacetylchitotetraose, NAG_4) from the target. Only addition of NAG_3 significantly increased the diffusion coefficient of the reference ligand, while mixtures that did not contain NAG_3 had no effect on NAG_4 mobility in the presence of the target protein. Competition screening may also be

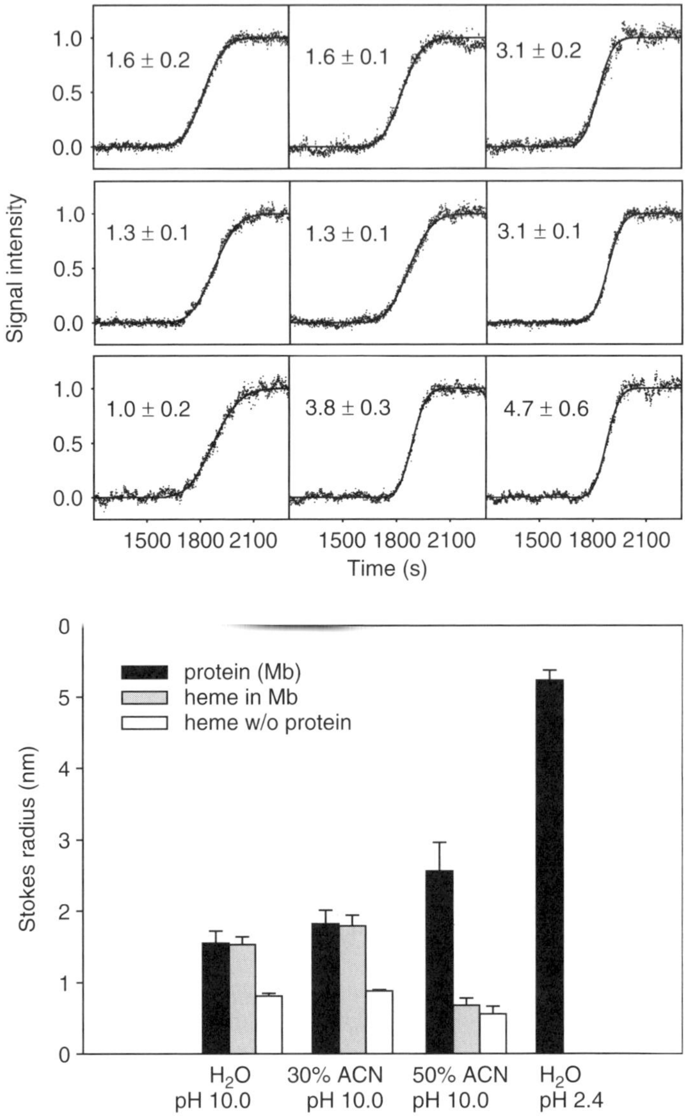

FIGURE 7.5. (**A**) Experimental dispersion profiles recorded for protein (first column) and heme (second column) in solutions of myoglobin at pH 10. The third column shows dispersion profiles recorded for heme in protein-free solutions. The experiments were carried out in the presence of 0% acetonitrile (*top row*), 30% acetonitrile (*middle row*), and 50% acetonitrile (*bottom row*). Solid lines are fits to the experimental data from which diffusion coefficients (shown in each panel in units of 10^{-9} m^2/s). (**B**) Stokes radii of myoglobin and heme, calculated based on the experimentally determined diffusion coefficients (**A**). Problems with heme solubility precluded the determination of heme Stokes radius at pH 2.4. Adapted with permission from (45). © 2003 American Society for Mass Spectrometry.

useful in cases when mixtures of potential ligands include analytes that are not amenable to "direct" ESI MS detection or when studying noncovalent interactions involving two or more macromolecules. Dispersion measurements are not necessarily restricted to water-soluble proteins; studies on nucleic acids, solubilized membrane receptors, or entire organelles may also be feasible. Another interesting aspect of this technique that still awaits careful consideration is the possibility of determination of dissociation constants directly from the measured dispersion profiles.

7.3. INDIRECT CHARACTERIZATION OF NONCOVALENT INTERACTIONS: EXPLOITING PROTEIN DYNAMICS UNDER DENATURING CONDITIONS

One of the methods of indirect assessment of protein–ligand binding considered in the previous section of this chapter is based on the assumption that ligand binding leads to greater stabilization of the native conformation of the protein. Such stabilization could only be detected with HDX, since both apo- and holo-forms of the protein were predominantly in the native conformation, with rare excursions to either partially or fully unfolded state. Ligand binding decreased the (already extremely low) frequency of unfolding events, thus making it possible to estimate binding energy as a difference between ΔG_{N*} and ΔG_N (31). Under some circumstances the stabilizing effect of ligand binding is much more pronounced, particularly if the apo-form of the protein fails to maintain the native conformation in the absence of ligand (i.e., the protein is intrinsically disordered in the absence of the ligand). In such cases the stabilizing effect of the ligand is so obvious that it does not require laborious HDX measurements, since the dramatic changes in large-scale protein dynamics induced by the ligand can be seen directly in the charge state distributions of the protein ions in ESI mass spectra. Examples of such behavior (Figure 7.6) abound particularly among metalloproteins, such as metallothioneins (4), endonucleases (30), or nuclear hormone receptor DNA-binding domains (48).

7.3.1. Ligand-Induced Protein Stabilization Under Mildly Denaturing Conditions: Charge State Distributions Reveal the Presence of "Invisible" Ligands

In most cases, however, both apo- and holo-forms of the protein are natively folded. Even in such circumstances the stabilizing effect of the ligand can often be "felt" indirectly as it delays the onset of protein unfolding (induced by mild chaotropes) when compared to the unfolding of the apo-form of the protein. We have seen such effects in iron-binding proteins from the transferrin family, where presence of the ferric ion delays the acid-induced unfolding (as indicated by the protein ion charge state distributions) by at least one pH unit (49, 50). Importantly, metal ion dissociation from the holo-form of the protein in the absence of

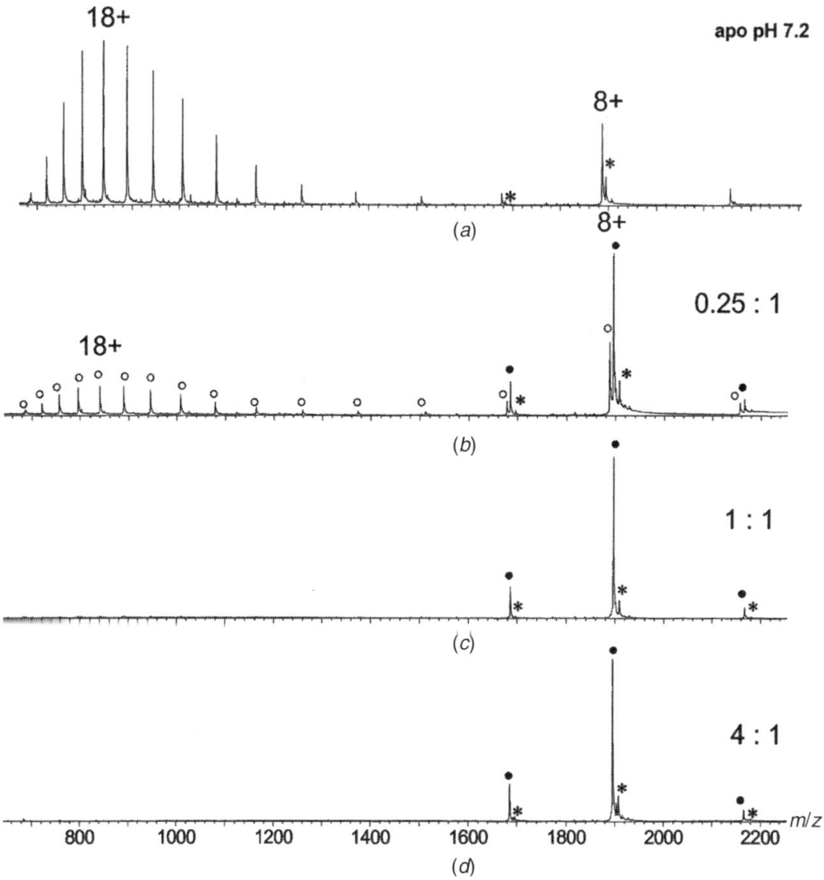

FIGURE 7.6. Stabilizing effect of Zn^{2+} on colicin E9 endonuclease (E9 DNase) conformation. Nano-ESI mass spectra of E9 DNase at pH 7.2 recorded in the absence (**A**) and in the presence (**B**–**D**) of Zn^{2+} in solution. Metal-to-protein molar ratios are 0.25:1 (**B**), 1:1 (**C**), and 4:1 (**D**). Ion peaks of Zn^{2+}-bound and metal-free E9 DNase are labeled with solid and open circles, respectively. Adapted with permission from (30). © 2002 Cold Spring Harbor Laboratory Press.

chelating agents is accompanied by a significant loss of tertiary structure (as indicated by a dramatic change in the protein ion charge state distributions), strongly suggesting that there is an intimate link between metal binding and large-scale conformational dynamics.

The primary function of transferrins is sequestration and transport of iron in bodily fluids, but these proteins are also known to bind a variety of other metals. While for many of these metals binding to transferrin clearly manifests itself in ESI mass spectra as a mass increase, there is at least one metal ion (Bi^{3+}) whose binding to transferrin is known to occur, but cannot be verified in a straightforward ESI MS experiment (Figure 7.7). This is a rather intriguing

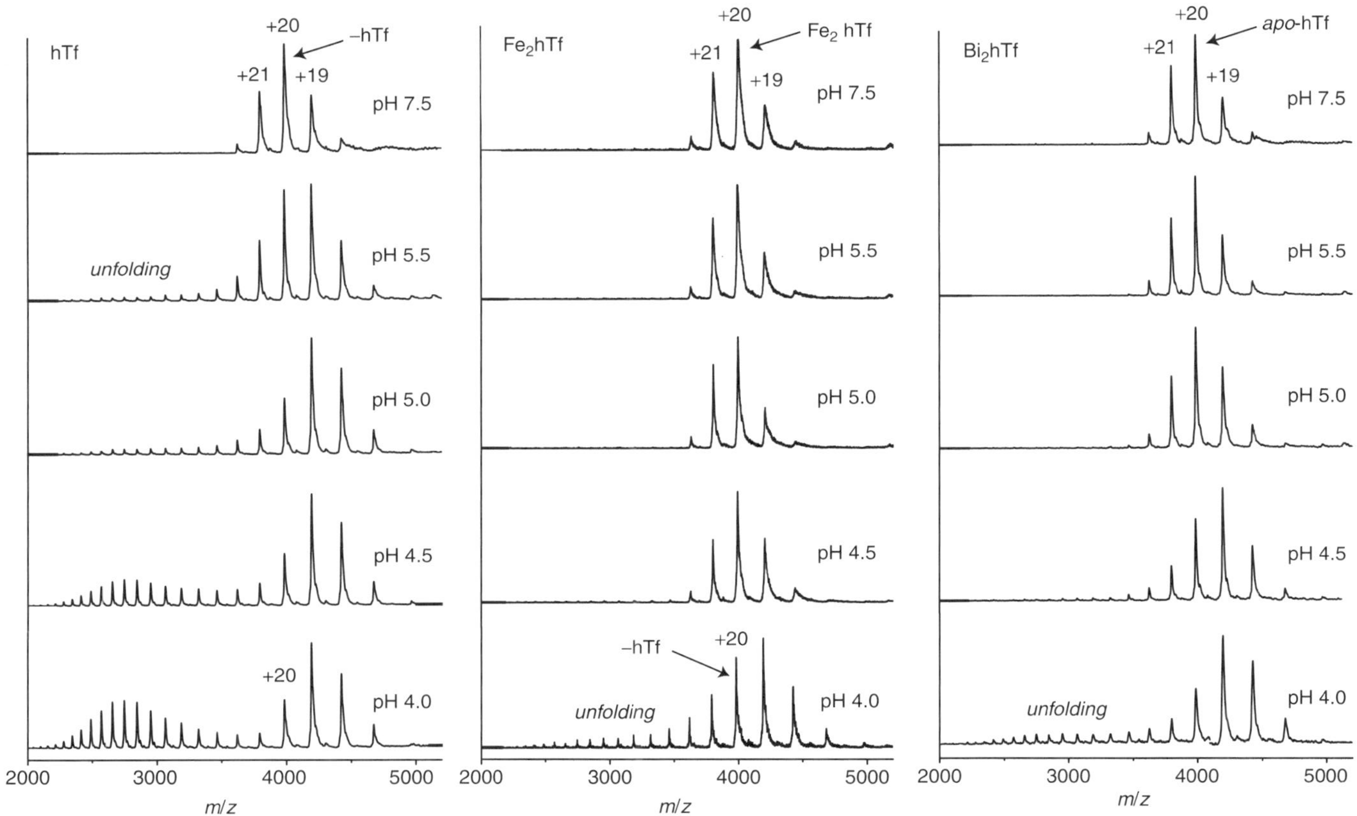

FIGURE 7.7. Acid-induced unfolding of human serum transferrin (*h*Tf) monitored by ESI MS: apo-hTf (*left column*), Fe^{3+}-saturated *h*Tf (*center column*), and Bi^{3+}-saturated *h*Tf (*right column*).

observation, since the Bi^{3+}—transferrin binding energy in solution is second only to that of Fe^{3+} (the cognate ligand). However, Bi^{3+}—transferrin binding can easily be detected indirectly by monitoring the evolution of charge state distributions of transferrin ions desorbed from a solution saturated with Bi^{3+} (Figure 7.7). The onset of transferrin unfolding in metal-saturated solution occurs at the same pH 4.5, regardless of whether the metal is Fe^{3+}, In^{3+}, or Bi^{3+}. On the other hand, unfolding of the apoprotein occurs at significantly higher pH 5.5, providing a clear indication that Bi^{3+} does bind to transferrin in solution just like Fe^{3+} and In^{3+}. Apparently it dissociates from the protein very easily in the gas phase (unlike Fe^{3+} and In^{3+}), which makes direct detection impossible.

Although this indirect approach to detection of protein–ligand interaction has not been used widely as yet, it clearly has a great potential. It is not as technically demanding as other indirect methods and should be applicable to a wide spectrum of protein–ligand interactions. The major shortcoming is that such measurements only establish the fact of protein–ligand interactions taking place in solution without providing any information on binding stoichiometry or energetics.*

7.3.2. SUPREX: Utilizing HDX Under Denaturing Conditions to Discern Protein–Ligand Binding Parameters

This latter disadvantage (inability to provide quantitative affinity data) can be addressed in some cases using another indirect method of probing protein–ligand interactions, which relies on measurements carried out under denaturing conditions. This method, termed SUPREX (short for **s**tability of **u**npurified **p**roteins from **r**ates of H/D **ex**change), estimates protein stability in the presence of increasing amount of denaturant by measuring the extent of HDX at a certain time (51). Protein samples are subjected to HDX by dilution into a series of deuterated exchange buffers containing different concentrations of a denaturant (typically guanidinium chloride, GdmCl). After a specified exchange time, the deuterium content of each protein sample is determined with MALDI MS. The change in mass (relative to the fully protonated sample) is plotted as a function of denaturant concentration to generate a SUPREX curve.

SUPREX curves can be used to extract thermodynamic parameters for protein unfolding, provided the equilibrium unfolding behavior conforms to a two-state process and that the exchange occurs under the EX2 regime. As we have seen in Chapter 5, the exchange rate in a two-state system (where all labile amides are identical and no local fluctuations occur) can be expressed simply as

$$k^{\mathrm{HDX}} = k_{\mathrm{int}} \cdot K, \qquad (7\text{-}3\text{-}1)$$

where k_{int} is the "intrinsic" exchange rate (from the unprotected state) and K is the equilibrium constant for protein unfolding. Switching from K to the equilibrium constant of the reverse process $K_{\mathrm{fold}} = 1/K$, expression (7-3-1) can be

*It is conceivable, though, that at least some qualitative data on the binding energetics (e.g., ligand ranking according to their affinity) can be obtained in such measurements, as stronger binding is expected to shift the onset of unfolding to lower pH or higher chaotrope concentration.

rewritten as

$$k^{\mathrm{HDX}} = k_{\mathrm{int}}/(1+K). \tag{7-3-2}$$

As we have discussed in Chapter 5, k^{HDX} as expressed in (7-3-2) is defined in NMR terms (i.e., it is simply a "rate of depletion" of a number of protein molecules labeled with ^{1}H at a specific amide). On the other hand, HDX MS measurements regard this rate constant as a "cumulative" rate of exchange (an ensemble-averaged rate of loss of the entire ^{1}H content for all amides). If the basic SUPREX assumptions (as stated in the beginning of this paragraph) are correct, then both rate constants (measured by NMR and MS) would be numerically equal, although they actually have different meanings. HDX progress in a two-state system is measured by MS simply as a protein mass gain ΔM:

$$\Delta M = \Delta M_\infty + (\Delta M_0 - \Delta M_\infty)\cdot e^{-k^{\mathrm{HDX}}\cdot t}, \tag{7-3-3}$$

where ΔM_0 is mass gain due to the exchange that does not require global unfolding, ΔM_∞ is a total mass gain after the exchange is complete, and t is the exchange time. Combining (7-3-2) and (7-3-3), one can link the extent of exchange at any time t and the folding equilibrium constant K_{fold} as

$$\Delta M = \Delta M_\infty + (\Delta M_0 - \Delta M_\infty)\cdot e^{-[k_{\mathrm{int}}/(1+K_{\mathrm{fold}})]\cdot t}. \tag{7-3-4}$$

The folding free energy of the folded state ($\Delta G = -RT\cdot \ln K_{\mathrm{fold}}$) in the presence of denaturant D is often estimated as

$$\Delta G = \Delta G_0 + m\cdot[D], \tag{7-3-5}$$

where $[D]$ is concentration of the denaturant, $m = \partial(\Delta G)/\partial[D]$, and ΔG_0 is the free energy of the folded state in the absence of the denaturant. Therefore, if the parameters m and k_{int} are somehow known (or can be estimated), then the SUPREX curve (ΔM as a function of $[D]$) can be used to extract the value of ΔG_0.

Typically, a best fit (7-3-4) and (7-3-5) is found for a set of experimental data points (Figure 7.8), from which a value of ΔG_0 can be deduced by extrapolation to "zero" denaturant concentration. Although the SUPREX curves have a familiar look of conventional denaturation curves, they are actually rather different. The position of the midpoint in conventional denaturation curves $C_{1/2}$ is a function of ΔG_0 and m, while in the SUPREX curves the position of the midpoint $C_{1/2}^{\mathrm{SUPREX}}$ also depends on the exchange time t and the intrinsic exchange rate k_{int} (51):

$$C_{1/2}^{\mathrm{SUPREX}} = C_{1/2} - \frac{RT}{m}\cdot \ln\left(\frac{k_{\mathrm{int}}\cdot t}{0.693} - 1\right), \tag{7-3-6}$$

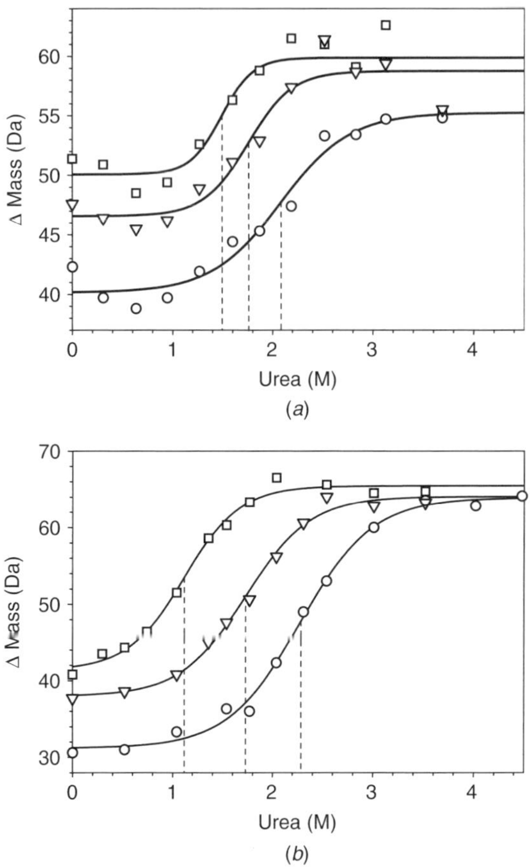

FIGURE 7.8. Representative SUPREX curves for S-protein (10 μM) in the absence (**A**) and in the presence of 67 μM peptide 7 (**B**). (**A**) Exchange times of 10 (circles), 20 (triangles), and 30 min (squares) resulted in $C_{1/2}^{\text{SUPREX}}$ values of 2.08, 1.76, and 1.49 M urea concentration, respectively. (**B**) Exchange times of 25 (circles), 85 (triangles), and 312 min (squares) resulted in $C_{1/2}^{\text{SUPREX}}$ values of 2.28, 1.73, and 1.12 M urea concentration, respectively. The solid lines represent the best fits for each set of data. The vertical dotted lines mark the values of $C_{1/2}^{\text{SUPREX}}$. Adapted with permission from (55). © 2003 American Chemical Society.

where $C_{1/2}$ can be used to calculate free energy of folding simply as $\Delta G_0 = -m \cdot C_{1/2}$. An extension of this method takes into account protein oligomerization (which "locks" the protein in the exchange-incompetent state) in solution (52):

$$\Delta G_0 = -m \cdot C_{1/2} + RT \cdot \ln\left(\frac{n^n}{2^{n-1}} \cdot [P]^{n-1}\right), \tag{7-3-7}$$

where $[P]$ is the protein concentration and n is the number of the protein subunits in the oligomer.

Since ligand binding usually stabilizes proteins, the ΔG_0 values obtained in SUPREX measurements with and without ligands can be used to estimate protein–ligand binding energies (as $\Delta\Delta G_0$) and dissociation constants (53–55), as illustrated in Figure 7.8. One of the greatest advantages of the SUPREX methodology is its sensitivity and tolerance to a wide spectrum of salts, buffers, and other environmental conditions under which the HDX reactions and MALDI MS measurements can be carried out. These advantages are exemplified in one particularly interesting application of this methodology to study protein stability in vivo (the in vivo environment used in this work was *E. coli* cytoplasm) (56), which showed a great promise of this technology to probe protein stability in the cell.

The major caveat of the quantitative analysis of SUPREX data (as is the case with most denaturation experiments) is that the protein must unfold in a cooperative, two-state process without any intermediate states present.* Another serious limitation of SUPREX is the requirement that HDX conform to the EX2 exchange regime, which is often impossible at medium or high denaturant concentrations or reasonably high pH, as has been discussed in Chapter 5.

7.4. UNDERSTANDING PROTEIN ACTION: MECHANISTIC INSIGHTS FROM THE ANALYSIS OF STRUCTURE AND DYNAMICS UNDER NATIVE CONDITIONS

The examples of using MS to characterize protein–ligand interaction considered so far have focused on identifying the players and extracting quantitative information regarding the strength of such interaction in solution. Although such information is very important, in most cases it is not nearly sufficient to understand how proteins work, for example, to describe a mechanism of protein–ligand binding and identify structural and dynamic features of the protein–ligand system serving as a "driving force" of the interaction. MS-based methods can often provide such mechanistic insights, allowing a diverse set of questions to be answered, ranging from catalytic activity of enzymes to "long-distance" signal transfer through allosteric interactions to "forced" protein folding. Specifically, exploitation of HDX under near-native conditions to locate and quantify functionally sensitive dynamic events often provides valuable mechanistic insights.

7.4.1. Dynamics at the Ligand Binding Site and Beyond: Understanding Enzymatic Mechanisms

Soon after the introduction of HDX MS by Smith (57–59), its utility as a tool to probe functionally important structural and dynamic features of proteins was recognized. For example, mechanisms of enzymatic action at the atomic level

*A multiphasic titration curve is observed (or else the transition is broadened) in such cases and any extrapolated stability measurement becomes impossible. However, the midpoint of the curve may still serve as a qualitative indicator of protein stability.

are often deduced from the structures of the end-points of the reaction, that is, enzyme–substrate and enzyme–product complexes. Characterization of these complexes by HDX MS provides an important new dimension, as it often allows the distribution of protein flexibility throughout the entire protein to be determined. As we have discussed in Chapter 1, it is the conformational dynamics (not the structure alone) and the way it is distributed throughout the protein that determines its biological activity. Therefore, probing enzyme dynamics in the presence and in the absence of its substrates and/or products often provides invaluable information on the molecular mechanism of its catalytic activity.

Tang and co-workers used HDX MS to identify a catalytically essential domain movement in *E. coli* dihydrodipicolinate reductase (one of seven enzymes in the succinylase pathway of bacterial L-lysine biosynthesis). The binding of NADH (substrate) and 2,6-pyridinedicarboxylate (inhibitor) to the enzyme led to substantial changes (reduction) in the rates of deuterium incorporation. The most affected regions of the enzyme were those located in the substrate binding pocket, as well as in the segment whose movements are crucial for the enzyme activation (60). In a similar study, Angeletti and co-workers used HDX MS to explore the effects of NADPH binding on the dynamics of *C. glutamicum* mesodiaminopimelate dehydrogenase (another enzyme of the L-lysine biosynthetic pathway in bacteria) (61). Binding of both NADPH and diaminopimelate to the enzyme was found to reduce the extent of exchange. Eight peptic fragments were identified whose deuterium exchange slowed considerably upon substrate binding, representing the regions known or thought to bind NADPH and diaminopimelate. One of the peptic fragments is located at the interdomain hinge region of the protein and was proposed to be exchangeable in the catalytically inactive "open" conformation, while remaining inaccessible to solvent in the catalytically active "closed" form of the enzyme formed after substrate binding and domain closure (61).

Structural changes remote from the active site were also found in another enzyme by Forest and co-workers, who used HDX MS to understand the mechanism of Mg^{2+} and NADPH-induced activation of acetohydroxy acid isomeroreductase, the second enzyme of the parallel branched chain amino acid pathway (62). The "long-distance" conformational changes were interpreted as structural effects of the ligand binding process. Conformational changes not directly involved in the binding of substrates observed by the same group in another system (creatine kinase isoenzyme) using HDX MS were interpreted as a movement that allows the enzyme to properly align the substrates for optimal catalysis (63). Important mechanistic insights on the mode of enzymatic action can also be obtained by monitoring ligand-induced changes in HDX kinetics of enzyme analogues where mutations affect ligand binding sites (64).

Angeletti and co-workers evaluated dynamics of purine nucleoside phosphorylase (PNP) at various stages of the enzymatic reaction by monitoring the differences of HDX of this protein induced by a substrate analogue, products, and a transition state analogue (enzyme inhibitor). Site-specific HDX measurements (peptic digestion under "slow exchange" conditions) identified regions whose flexibility was decreased in all protein ligand complexes. More importantly, several protein segments were identified whose dynamics were decreased only in the complex with the inhibitor. The segments most highly influenced by the inhibitor surround the catalytic site, providing evidence for reduced protein dynamic motion caused by the transition state analogue (65).

Adams and co-workers have used HDX MS to probe conformational changes in the COOH-terminal Src kinase* (Csk) with the aim of identifying dynamic events that influence regulatory motions in the catalytic pathway (66). The exchange was carried out in the absence and presence of nucleotides ADP and ATP analogues. A comparison of the ATP- and ADP-induced dynamic changes in the protein reveals unique structural changes induced by the γ-phosphate group of the nucleotide and provide a structural framework for understanding phosphorylation-driven conformational changes. Dynamics of various protein segments was monitored by measuring deuterium content of corresponding peptic fragments, allowing identification of segments that exchange differentially with solvent (Figure 7.9). Both ATP and ADP were found to protect similar regions of the protein but the extent of such protection varied markedly in several crucial areas, such as activation loop and helix G in the kinase domain, as well as several interdomain regions (Figure 7.9). The results of this study clearly demonstrate that delivery of the γ-phosphate group of ATP induces unique local and long-range conformational changes in Csk, which may have significant implications for regulatory motions in the catalytic pathway (66). A similar approach was later used to identify "long-distance" conformational changes transmitted through an adenosine 3′,5′-cyclic monophosphate (cAMP)-dependent protein kinase upon its activation with cAMP (67, 68).

Although most HDX studies of protein function utilize enzymatic degradation of the labeled protein under slow exchange conditions to identify "ligand-sensitive" sites, utilization of alternative methods of protein fragmentation is also possible. Akashi and Takio used HDX MS in combination with protein ion fragmentation in the gas phase (see Chapter 5 for a detailed discussion of this technique) to investigate the interaction between a thiol protease inhibitor, cystatin, and its target enzyme, papain (69). Deuterium incorporation into cystatin was analyzed at different time points both in the presence and in the absence of papain. Although fewer fragments were generateded for the cystatin–papain complex ions, significant reductions in initial deuterium content were recognized throughout the entire sequence of cystatin, with the N-terminal region (flexible in the apo-form of cystatin) affected the most. Therefore, although complex formation restricts the flexibility of the entire cystatin molecule, its flexible N-terminal region is particularly tightly fixed in the binding pocket, suggesting the mode of inhibitor action (69).

Very often HDX MS allows detection of rather small (but functionally very important) conformational changes within large proteins. An example of such

*The Src family of nonreceptor protein tyrosine kinases phosphorylate protein targets associated with discrete signaling pathways. All Src enzymes are regulated through a single kinase, Csk (which downregulates kinase activity by phosphorylating a single tyrosine residue in the C terminus of the Src enzymes).

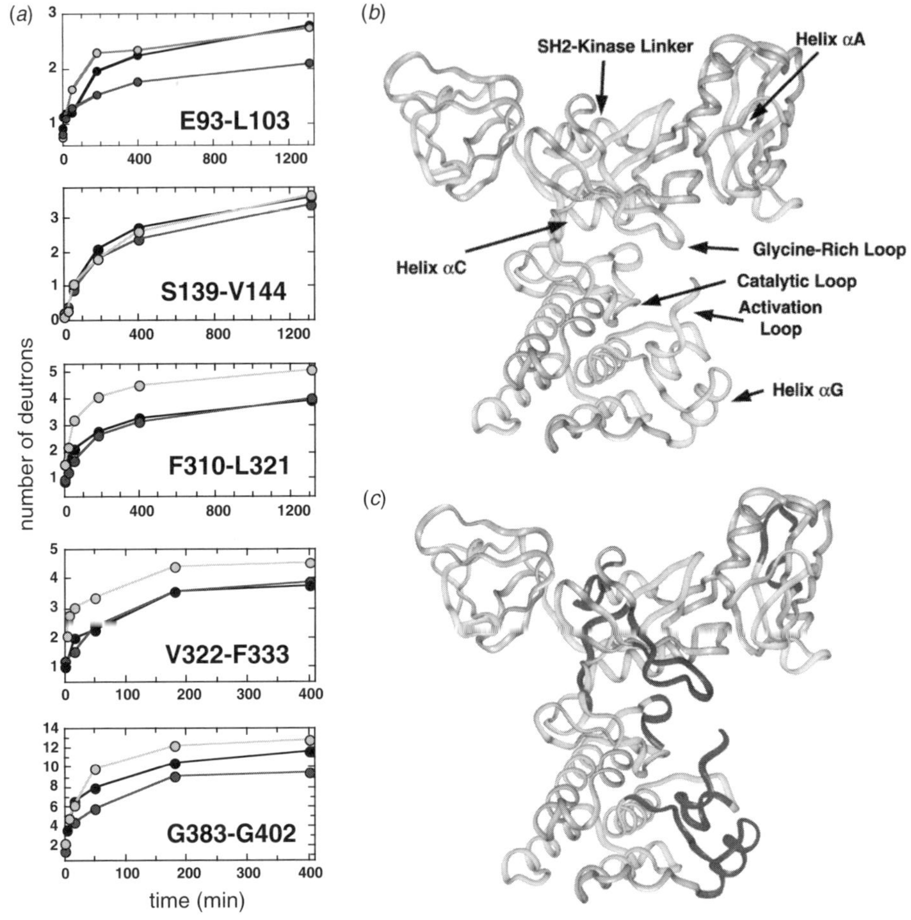

FIGURE 7.9. Probing phosphorylation-driven motions in the COOH-terminal Src kinase (Csk) with HDX MS. (**A**) Evolution of deuterium content of several peptic fragments of CsK in the absence (light gray) and presence of nucleotides: darker shade of gray, ATP analog AMPPNP and black, ADP. Coverage of Csk sequence with peptic fragments detected in HDX MS experiments (mapped to the X-ray structure for Csk) is shown in gray (**B**). The fragment peptides shown on panel **A** correspond to the following structural elements of Csk: E93-L103, αA in SH2; S139-V144, SH2-kinase linker; F310-L321, catalytic loop; V322-F333, activation loop; and G383-G402, αG. (**C**) HDX map of Csk showing effects of nucleotide binding: segments shown in darker shades of gray identify the regions whose protection is variably affected by ADP (as compared to ATP analog AMPPNP). Consult panels **A** and **B** for specific changes induced by each of the two ligands. Adapted with permission from (66). Copyright 2002 Elsevier.

sensitive measurements was presented by Marshall and co-workers, who were able to detect and assign minute changes in the solvent exposure of troponin C (TnC, a calcium-binding protein of the thin filament of muscle, which plays a regulatory role in skeletal and cardiac muscle contraction) induced by Ca^{2+} binding, total exposed surface area increases from 3090 ± 86 Å^2 to 3108 ± 71 Å^2 upon Ca^{2+} binding (70). In many cases, however, the significant background of all exchangeable hydrogen atoms may interfere with the identification of those that are functionally relevant. This problem can be circumvented by using a combination of HDX MS and functional labeling, as discussed in the following section of this chapter.

7.4.2. Allosteric Effects Probed by HDX MS

Englander and co-workers used tetrameric hemoglobin as a model system to study how protein molecules manage intramolecular signal transduction processes (71). Mammalian hemoglobins function by transducing a part of the binding energy of its initially bound O_2 ligands into structure-change energy. The energy is carried through the protein to distant heme sites in the form of structure changes, and there transduced back into binding energy. The initial relatively low binding energy and the later enhanced binding provide the foundation of cooperative oxygen binding by hemoglobin. Identification of individual allosterically important structural changes, their individual energetic contributions, as well as orchestration of such individual changes resulting in the allosteric function were studied with HDX MS using a *functional labeling* technique (71, 72). The idea of functional labeling, as an extension of HDX measurements, was initially discussed by Englander (73), namely, to focus only on those sites that change their exchange rates as a result of interactions. The experimental strategy is similar to the pulse chase experiments commonly used in biology and results in the isotope labeling only changing at sites affected by the folding or binding interaction. This eliminates a background of exchangeable amide hydrogen atoms that are located on the allosterically insensitive sites. In this manner the HDX data can be much simplified since complicating exchange processes due to local fluctuations are effectively removed from the measurements.

In the case of hemoglobin, amide exchange at an unresolved whole-molecule level does show significant differences between the oxy- (***R***-state) and deoxy- (***T***-state) forms. However, the number of allosterically sensitive amides and their exchange rates in either protein form cannot be determined because of the large background of unresolved sensitive and insensitive amides that exchange over a wide range of time scales (71). This difficulty is circumvented by using a functional labeling method, which focuses HDX labeling on only the sites that are affected by the $\boldsymbol{T} \rightarrow \boldsymbol{R}$ transition. The fast-exchanging ***R***-state is partially labeled by exchange-in for a limited time period. Both allosterically sensitive and insensitive sites that exchange in this time period become labeled with ^2H (or ^3H). The hemoglobin sample is then switched to the slow-exchanging ***T***-state and moved into protiated solvent, allowing the bound label (^2H or ^3H) to exchange-out. After some exchange-out (chase) time, label on allosterically insensitive sites (same exchange rate in both forms) is largely lost, but it is retained on allosterically sensitive sites, which are more protected in the ***T***-state. The exchange-in/exchange-out sequence produces a sample with the ^{2}H label selectively placed on functionally sensitive sites. Site-specific HDX measurements then reveal the number of allosterically sensitive residues and their exchange rates in both forms of hemoglobin. This procedure allowed three sets of allosterically sensitive residues to be identified, which were then associated with known important structural changes occurring in hemoglobin upon the $\boldsymbol{T} \rightarrow \boldsymbol{R}$ transition (71).

7.4.3. Protein Activation by Physical "Stimulants"

The examples of "functionally important" protein dynamics considered so far have centered on protein activation induced by ligand binding. In many protein systems physiologically important conformational changes can be induced by other means, both chemical (e.g., pH change during endocytosis) and physical (e.g., pressure-, heat-, and light-induced activation).

Smith and co-workers explored conformation and dynamics of a small heat shock protein from wheat using HDX MS (74). Although this protein exists as a dodecameric assembly in solution, the regions that form extensive and essential intersubunit contacts in the assembly show little or no protection after incubation in D_2O for 5 s (25 °C). Such behavior is clearly indicative of large conformational fluctuations in solution and breaking intersubunit contacts. The intersubunit contact breaking occurs on a time scale between 10 ms and 5 s, as suggested by the results of pulse labeling. At 42 °C, the protein exists in a suboligomeric form. When the temperature dependence of "intrinsic" HDX is taken into account, exchange patterns at 25 and 42 °C are identical within experimental error, suggesting that the conformation of individual protein subunits is the same in both the dodecameric and subdodecameric forms. Significant protection is seen in regions that form the dimeric interface, suggesting that the stable suboligomeric form is a dimer. A model of heat activation is proposed that invokes a shift in the dodecamer $\rightleftharpoons$ dimer equilibrium in favor of dimers without any conformational changes of the dimers (74).

A different approach to characterizing thermal dissociation of a small heat shock protein was recently described by Robinson and co-workers (75), who used a nano-ESI source equipped with a novel probe that allows the thermocontrol of the solution in the electrospray capillary. Thermal dissociation of a 200 kDa complex of the small heat shock protein monitored by ESI MS revealed both dissociation into suboligomeric species and an increase in its size and polydispersity at elevated temperatures (75).

We have already mentioned in Chapter 5 a recent study by Rist and colleagues, who probed heat-induced conformational changes in *E. coli* heat shock transcription factor* σ^{32} using HDX MS (76). The study identified a protein region exhibiting temperature-dependent decrease in stabilization energy (see Figure 5.17 for more detail on site-specific HDX MS measurements in σ^{32}),

*Upon temperature up-shift, σ^{32} out-competes the housekeeping sigma-factor, σ^{70}, for binding to core RNA polymerase and initiates heat shock gene transcription.

acting as a thermosensor, which may be important for the heat shock regulation (76).

Hellingwerf and co-workers used HDX (monitored by both ESI MS and FTIR spectroscopy) to probe dynamic events triggered by light activation of photoactive yellow protein (PYP), a eubacterial photosensor (77). In the presence of an adequate hydration shell, large structural changes in the protein backbone, involving both solvent accessible and core regions, were detected using FTIR difference spectroscopy. A significant part (23%) of the amide groups, which are protected in the "resting state" of PYP, become exposed to the solvent in the light-activated (putative signaling) state (as measured by HDX MS and HDX FTIR), while the HDX kinetics of the inactive (apo) form of PYP was not affected by the continuous illumination of the protein sample. A model for photoactivation of PYP that involves large-scale conformational changes was proposed based on the results of this study (77).

7.5. UNDERSTANDING PROTEIN ACTION: MECHANISTIC INSIGHTS FROM THE ANALYSIS OF STRUCTURE AND DYNAMICS UNDER DENATURING CONDITIONS

The majority of the "mechanistic" studies focusing on protein function discussed so far utilize HDX under native conditions. For most proteins, amide exchange under such conditions occurs under the EX2 regime, meaning that HDX MS experiments produce a picture of protein dynamics averaged across the entire protein population, just like HDX NMR. The existence of the functionally important intermediate (partially structured) states (e.g., participating directly or indirectly in the ligand binding process) can be inferred from such experiments indirectly, that is, by grouping the amides exhibiting similar exchange rates. Switching the amide exchange to the EX1 regime allows in many cases the distinct protein states to be identified, based on their respective residual levels of backbone protection. (See Chapter 5 for a more detailed discussion of the two exchange regimes and the information they can provide.) Unfortunately, switching exchange to the EX1 regime would also mean switching to denaturing conditions, raising doubts vis-à-vis relevance of any partially unstructured state(s) populated under such conditions for the protein–ligand binding under native conditions. In this section we use CRABP I, a protein that has already been mentioned in the beginning of this chapter, as a test case for addressing these concerns.

CRABP I is a member of a family of small soluble intracellular proteins that bind hydrophobic ligands such as fatty acids, lipids, and retinoids. This protein contains 136 residues, which form two 5-stranded β-sheets. The first two strands are connected by a helix–turn–helix motif, and all of the others by reverse turns. The two β-sheets are packed orthogonally to form a solvent-filled β-barrel. The ligand binding pocket, which physiologically accommodates all-trans retinoic acid (RA), is located inside the barrel (Figure 7.10). The ligand binding/release mechanism by CRABP I has not been clearly established. A "portal" model has been postulated, which invokes the notion of a highly dynamic segment within the protein that serves as an opening, thus allowing the entry of ligand into the

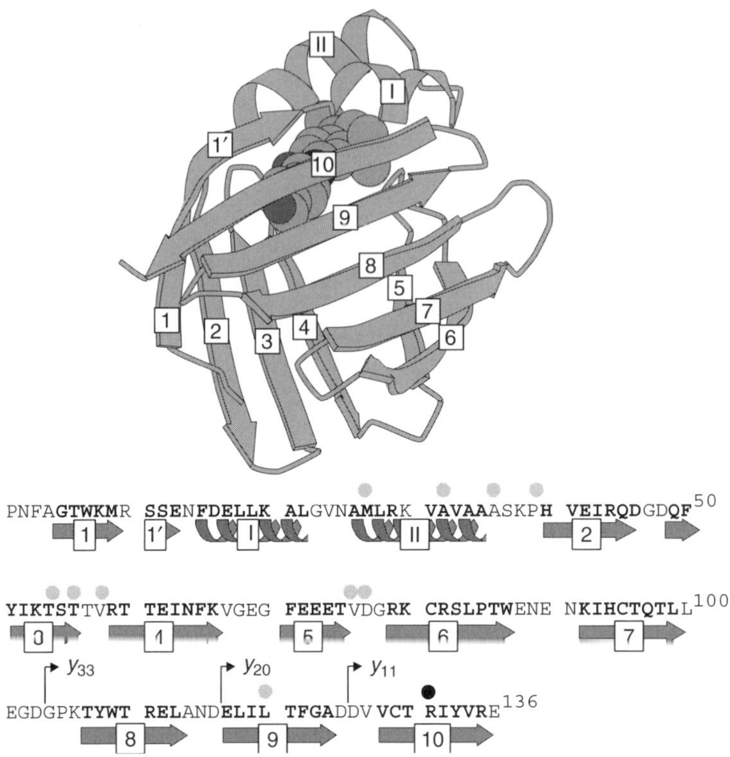

FIGURE 7.10. Ribbon diagram of CRABP I showing RA in the ligand binding pocket (*top*) and the amino acid sequence of the protein with the X-ray derived ligand contact sites and secondary structural elements shown above and below the sequence, respectively (*bottom*). Grey circles indicate hydrophobic contacts; salt bridge-forming Arg residue is marked with a black circle.

cavity(35). In CRABP I this portal region constitutes the helix–turn–helix motif and two flanking β-hairpins. Reduced dynamics within this region in the holo-form of the protein effectively inhibits dissociation of the complex in solution.

We have seen earlier in this chapter that measurements of HDX kinetics of CRABP I carried out in aqueous solution at neutral pH indicate that the protein backbone dynamics are reduced dramatically in the presence of the cognate ligand (31), consistent with earlier NMR measurements. Although the HDX kinetics under native conditions can be used to obtain fairly accurate estimates of the RA–CRABP I binding energy, such measurements alone are not particularly revealing as far as the mechanism of protein–ligand binding is concerned. However, native HDX MS experiments do tell us that there are several intermediate states of the protein whose Boltzmann weight is affected by the presence of the ligand (31). We have previously postulated that at least some of these intermediate states [which are often viewed as "partial replicas of the native structure at lesser or greater degree of advancement" (78)] are essential for ligand binding, as they provide easy access to the protein cavity. The transient nature of such intermediate states would prevent their direct observation under native conditions. Their presence, of course, can be detected in HDX MS and HDX NMR experiments. However, exchange follows the EX2 mechanism, making it impossible to differentiate among several possible activated states of the protein. To "visualize" the putative intermediate states of CRABP I that may play a role in ligand binding, we have to find a way to either increase their Boltzmann weight (which would affect the protein ion charge state distributions) and/or increase the potential energy barrier separating these activated states of the protein from the native conformation (which would affect the protein HDX).* Both objectives can be achieved by inducing mildly denaturing conditions (e.g., mild acidification of the protein solution). Even though partial denaturation disrupts the protein–ligand interaction (ligand retention efficiency diminishes due to a decrease of the Boltzmann weight of the native state of the protein), some residual binding is observed until the pH is lowered to 3.0 (31).

The HDX kinetics clearly exhibit partial EX1 character within the pH range 3.0–3.5. The exchange pattern gives a clear indication that the presence of RA in the exchange buffer appears to stabilize the more structured states of the protein at pH as low as 3.2 (an example of such behavior is shown in Figure 7.11). In order to gain an insight into the details of the protein–ligand interaction, the isotope content of various fragment ions (produced by fragmenting CRABP I ions in the gas phase) was measured as a function of the exchange time in solution. Protein ion fragmentation (CID) was carried out "in-source" (i.e., without mass selection of the precursor ion prior to dissociation) and was induced by increasing the capillary exit potential and ion residence time in the hexapole ion guide prior to injection to the ICR cell. Although full sequence coverage of CRABP I is not achieved under these conditions, the CID spectrum does contain an extended series of abundant structurally diagnostic fragment ions ranging in size from several amino acid residues to well over half of the protein. Figure 7.12 shows an example of the time evolution of the isotopic clusters corresponding to three C-terminal fragment ions of CRABP I with and without RA present in the exchange buffer. All three fragment ions (y_{11}^{+}, y_{20}^{2+}, and y_{33}^{3+}) contain Arg^{131}, a residue whose side chain forms a salt bridge with the carboxy group of the ligand in the "native" complex according to the available crystallographic data (79). Consideration of the protein–ligand distances in the native complex suggests that Leu^{119} (contained within the y_{20}^{2+} and y_{33}^{3+} fragments) may form a hydrophobic contact with the ligand, while other amino acid residues within the segments under consideration do not appear to interact directly with RA. Isotopic clusters corresponding to all three fragments clearly exhibit bimodal character following short exposure to protiated solvent (top two traces on Figure 7.12), indicating presence of at least two protein states. This bimodal distribution is particularly clear in the case of the smallest fragment, y_{11}^{+}, with the two maxima separated by nine mass units. This separation gives a numerical value of the amide

*This would result in longer "residence times" of the protein molecules within the potential wells corresponding to the intermediate states of the protein and switch the HDX kinetics to the EX_1 type.

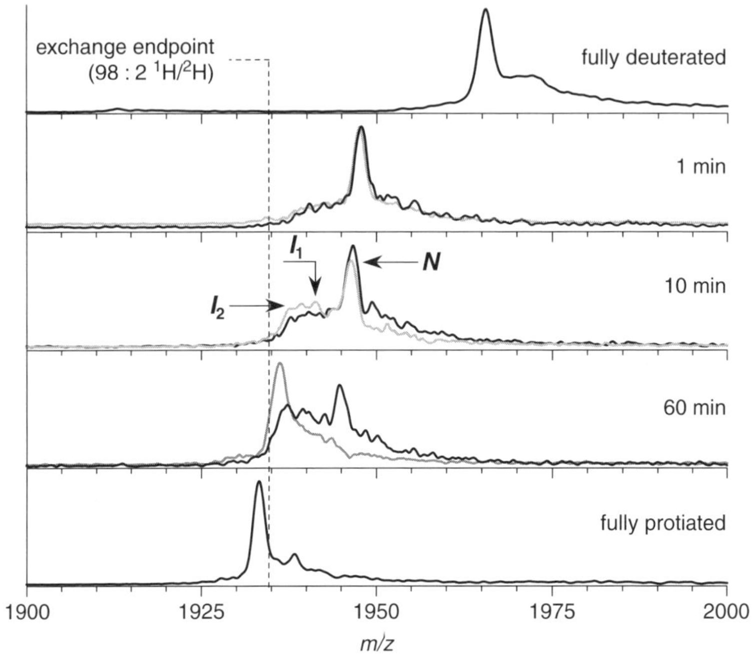

FIGURE 7.11. Global backbone dynamics of CRABP I probed with HDX/ESI MS under conditions favoring EX1 exchange mechanism (a 100 μM solution of a fully deuterated protein in D_2O was diluted 1:50, v:v in the exchange buffer solution, $H_2O/NH_4CH_3CO_2$, pH adjusted to 3.5 with CH_3CO_2H). Traces correspond to the fully deuterated (*top*) and unlabeled (*bottom*) protein; black and gray traces correspond to the exchange in the presence and in the absence of all-trans retinoic acid (5-molar excess) in the exchange buffer. All profiles correspond to a +8 charge state of CRABP I ions. Adapted with permission from (80).

protection difference between the two conformations. Remarkably, this number is very close to a full number of the backbone amides within the segment corresponding to the $y_{11}{}^+$ fragment ion (i.e., nine out of ten amides remain deuterated following 1 min exposure to the protiated solvent). The exchange pattern of this segment has familiar features of the EX1 (cooperative) exchange overlaid with the exchange reactions due to uncooperative dynamic events, such as local structural fluctuations. (See Chapter 5 for a detailed discussion of "mixed" exchange regimes.) Importantly, presence of the ligand clearly stabilizes the more structured conformation of the protein. Even after 30 min of the protein exposure to protiated solvent under these denaturing conditions, the $y_{11}{}^+$ fragment retains 2.5 ± 0.5 labile deuterium atoms when RA is present in the exchange buffer. All the while, the isotopic pattern of this fragment ion in the absence of the ligand is indistinguishable from that of a "fully exchanged" $y_{11}{}^+$ fragment.

Similar analysis of the isotopic distribution of the $y_{20}{}^{2+}$ and $y_{33}{}^{3+}$ fragment ions as a function of exchange time in solution indicates that the fraction of

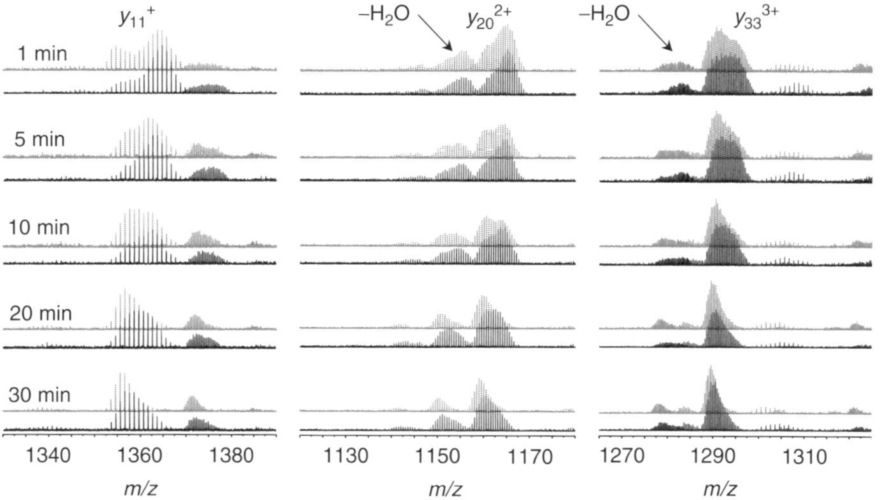

FIGURE 7.12. Effect of all-trans retinoic acid (RA) on the local backbone dynamics of CRABP I in aqueous acidic (pH 3.2) solution probed with HDX/ESI/CAD MS. The three columns represent evolution of the isotopic distributions of three C-terminal fragments (y_{11}^+, y_{20}^{2+}, and y_{33}^{3+}) throughout 30 min of exposure of the deuterated protein to a protiated buffer (10 mM $NH_4CH_3CO_2$, pH adjusted to 3.2 with CH_3CO_2H). Less abundant clusters on the left-hand sides of the main peaks correspond to secondary fragments generated by a neutral loss of H_2O from the primary fragment ions.

protected backbone amides actually diminishes as the size of the protein segment under consideration increases. Thus, the "stable" protein form retains only 13 (out of possible 19) labile deuterium atoms within the [$Glu^{117} \rightarrow Glu^{136}$] segment following 1 min of exposure to the protiated solvent (as evidenced by the spacing between the two maxima in the isotopic clusters of the y_{20}^{2+} fragment ion). The increase in flexibility is even more dramatic within the larger segment, [$Gly^{104} \rightarrow Glu^{136}$] (only 17 out of 31 labile deuterium atoms are retained by the y_{33}^{3+} fragment ion).* In both cases, presence of the ligand stabilizes the "more protected" form of the protein. Interestingly, the extent of measured protection at longer exchange times appears to be invariant of the segment size. Thus, both y_{20}^{2+} and y_{33}^{3+} fragment ions retain only 3.0 ± 0.5 labile deuterium atoms in the presence of RA following 30 min exposure to the protiated buffer. This ligand-induced protection remains essentially the same (within experimental error) as for the much shorter segment corresponding to the y_{11}^+ fragment ion.

*The isotopic clusters of the y_{33}^{3+} fragment ions appear to contain contributions from more than two species that cannot be clearly resolved at the present time (see comments on isotopic depletion strategies in the final section of this chapter). In our analysis we take into account what appears to be the two extremes, the most and the least protected species.

These results clearly indicate that the acid-induced intermediate form of CRABP I does interact with the ligand at the "native" binding site located within the β-strand **10**. This conclusion is very important, as it proves that the equilibrium intermediate states of the protein populated under mildly denaturing conditions bind the ligand at the "native" site and, therefore, are likely to bear significant resemblance to the transient intermediates that facilitate ligand binding under native conditions. Similar analysis of the local exchange pattern deduced from the N-terminal (b-type) fragments representing strands **1**, **1′**, and **2**, and helices **I** and **II** (data not shown), also reveals uneven distribution of protein flexibility within the [$\text{Pro}^1 \rightarrow \text{Asp}^{48}$] segment of the protein. Once again, the presence of RA in the exchange buffer dramatically affects local exchange kinetics, revealing intimate details of the ligand interaction with the activated (partially structured) states of the protein (80).

In this chapter we have considered only a limited number of examples of various mass spectrometry-based approaches to study quantitative and mechanistic aspects of protein function. Because of the space limitations, we could not possibly provide an exhaustive account of all work carried out in this field, so we focused on the methodological aspects of such work. More examples of functional studies of various systems will be presented in Chapter 11.

REFERENCES

1. Ganem B., Li Y. T., Henion J. D. (1991). Detection of noncovalent receptor ligand complexes by mass spectrometry. *J. Am. Chem. Soc.* **113**: 6294–6296.
2. Ganem B., Li Y.-T., Henion J. D. (1991). Observation of noncovalent enzyme–substrate and enzyme–product complexes by ion-spray MS. *J. Am. Chem. Soc.* **113**: 7818–7819.
3. Katta V., Chait B. T. (1991). Observation of the heme globin complex in native myoglobin by electrospray-ionization mass-spectrometry. *J. Am. Chem. Soc.* **113**: 8534–8535.
4. Yu X., Wojciechowski M., Fenselau C. (1993). Assessment of metals in reconstituted metallothioneins by electrospray mass spectrometry. *Anal. Chem.* **65**: 1355–1359.
5. Hu P., Ye Q. Z., Loo J. A. (1994). Calcium stoichiometry determination for calcium binding proteins by electrospray ionization mass spectrometry. *Anal. Chem.* **66**: 4190–4194.
6. Ganguly A. K., Pramanik B. N., Tsarbopoulos A., Covey T. R., Huang E., Fuhrman S. A. (1992). Mass-spectrometric detection of the noncovalent Gdp-bound conformational state of the human h-Ras protein. *J. Am. Chem. Soc.* **114**: 6559–6560.
7. Cheng X. H., Harms A. C., Goudreau P. N., Terwilliger T. C., Smith R. D. (1996). Direct measurement of oligonucleotide binding stoichiometry of gene V protein by mass spectrometry. *Proc. Natl. Acad. Sci. U.S.A.* **93**: 7022–7027.
8. Cheng X., Morin P. E., Harms A. C., Bruce J. E., Ben-David Y., Smith R. D. (1996). Mass spectrometric characterization of sequence-specific complexes of DNA and transcription factor PU.1 DNA binding domain. *Anal. Biochem.* **239**: 35–40.
9. Smith R. D., Lightwahl K. J., Winger B. E., Loo J. A. (1992). Preservation of noncovalent associations in electrospray ionization mass-spectrometry—multiply charged polypeptide and protein dimers. *Org. Mass Spectrom.* **27**: 811–821.

10. Ganem, Bruce, Li Y.-T., Henion, Jack D. (1993). Detection of oligonucleotide duplex forms by ion-spray mass spectrometry. *Tetrahedron Lett.* **34**: 1445–1448.

11. Goodlett D. R., Camp D. G., Hardin C. C., Corregan M., Smith R. D. (1993). Direct observation of a DNA quadruplex by electrospray ionization mass-spectrometry. *Biol. Mass Spectrom.* **22**: 181–183.

12. Huang E. C., Pramanik B. N., Tsarbopoulos A., Reichert P., Ganguly A. K., Trotta P. P., Nagabhushan T. L., Covey T. R. (1993). Application of electrospray mass-spectrometry in probing protein–protein and protein–ligand noncovalent interactions. *J. Am. Soc. Mass Spectrom.* **4**: 624–630.

13. Light-Wahl K. J., Schwartz B. L., Smith R. D. (1994). Observation of the noncovalent quaternary associations of proteins by electrospray-ionization mass-spectrometry. *J. Am. Chem. Soc.* **116**: 5271–5278.

14. Hunter C. L., Mauk A. G., Douglas D. J. (1997). Dissociation of heme from myoglobin and cytochrome b_5: comparison of behavior in solution and the gas phase. *Biochemistry* **36**: 1018–1025.

15. Aplin R. T., Robinson C. V., Schofield C. J., Westwood N. J. (1994). Does the observation of noncovalent complexes between biomolecules by electrospray-ionization mass-spectrometry necessarily reflect specific solution interactions. *J. Chem. Soc. Chem. Comm.* **1994**: 2415–2417.

16. Ding J., Anderegg R. J. (1995). Specific and nonspecific dimer formation in the electrospray ionization mass spectrometry of oligonucleotides. *J. Am. Soc. Mass Spectrom.* **6**: 159–164.

17. Lim H. K., Hsieh Y. L., Ganem B., Henion J. (1995). Recognition of cell-wall peptide ligands by vancomycin group antibiotics—studies using ion-spray mass-spectrometry. *J. Mass Spectrom.* **30**: 708–714.

18. Greig M. J., Gaus H., Cummins L. L., Sasmor H., Griffey R. H. (1995). Measurement of macromolecular binding using electrospray mass-spectrometry—determination of dissociation-constants for oligonucleotide–serum albumin complexes. *J. Am. Chem. Soc.* **117**: 10765–10766.

19. Loo J. A., Hu P. F., McConnell P., Mueller W. T., Sawyer T. K., Thanabal V. (1997). A study of Src SH2 domain protein–phosphopeptide binding interactions by electrospray ionization mass spectrometry. *J. Am. Soc. Mass Spectrom.* **8**: 234–243.

20. Jorgensen T. J. D., Roepstorff P., Heck A. J. R. (1998). Direct determination of solution binding constants for noncovalent complexes between bacterial cell wall peptide analogues and vancomycin group antibiotics by electrospray ionization mass spectrometry. *Anal. Chem.* **70**: 4427–4432.

21. Jorgensen T. J. D., Staroske T., Roepstorff P., Williams D. H., Heck A. J. R. (1999). Subtle differences in molecular recognition between modified glycopeptide antibiotics and bacterial receptor peptides identified by electrospray ionization mass spectrometry. *J. Chem. Soc. Perkin Trans. 2* **1999**: 1859–1863.

22. Rostom A. A., Tame J. R. H., Ladbury J. E., Robinson C. V. (2000). Specificity and interactions of the protein OppA: partitioning solvent binding effects using mass spectrometry. *J. Mol. Biol.* **296**: 269–279.

23. Daniel J. M., Friess S. D., Rajagopalan S., Wendt S., Zenobi R. (2002). Quantitative determination of noncovalent binding interactions using soft ionization mass spectrometry. *Int. J. Mass Spectrom.* **216**: 1–27.

24. Daniel J. U. M., McCombie G., Wendt S., Zenobi R. (2003). Mass spectrometric determination of association constants of adenylate kinase with two noncovalent inhibitors. *J. Am. Soc. Mass Spectrom.* **14**: 442–448.

25. Nousiainen M., Derrick P. J., Lafitte D., Vainiotalo P. (2003). Relative affinity constants by electrospray ionization and fourier transform ion cyclotron resonance mass spectrometry: calmodulin binding to peptide analogs of myosin light chain kinase. *Biophys. J.* **85**: 491–500.

26. Zhang S., Van Pelt C. K., Wilson D. B. (2003). Quantitative determination of noncovalent binding interactions using automated nanoelectrospray mass spectrometry. *Anal. Chem.* **75**: 3010–3018.

27. Prinz H., Lavie A., Scheidig A. J., Spangenberg O., Konrad M. (1999). Binding of nucleotides to guanylate kinase, p21ras, and nucleoside-diphosphate kinase studied by nano-electrospray mass spectrometry. *J. Biol. Chem.* **274**: 35337–35342.

28. Nesatyy V. J. (2002). Mass spectrometry evaluation of the solution and gas-phase binding properties of noncovalent protein complexes. *Int. J. Mass Spectrom.* **221**: 147–161.

29. Douglas D. J., Collings B. A., Numao S., Nesatyy V. J. (2001). Detection of noncovalent complex between alpha-amylase and its microbial inhibitor tendamistat by electrospray ionization mass spectrometry. *Rapid Commun. Mass Spectrom.* **15**: 89–96.

30. van den Bremer E. T., Jiskoot W., James R., Moore G. R., Kleanthous C., Heck A. J., Maier C. S. (2002). Probing metal ion binding and conformational properties of the colicin E9 endonuclease by electrospray ionization time-of-flight mass spectrometry. *Protein Sci.* **11**: 1738–1752.

31. Xiao H., Kaltashov I. A., Eyles S. J. (2003). Indirect assessment of small hydrophobic ligand binding to a model protein using a combination of ESI MS and HDX/ESI MS. *J. Am. Soc. Mass Spectrom.* **14**: 506–515.

32. Baker D., Lim W. A. (2002). From folding towards function. *Curr. Opin. Struct. Biol.* **12**: 11–13.

33. Tsai C.-J., Kumar S., Ma B., Nussinov R. (1999). Folding funnels, binding funnels, and protein function. *Protein Sci.* **8**: 1181–1190.

34. Verkhivker G. M., Bouzida D., Gehlhaar D. K., Rejto P. A., Freer S. T., Rose P. W. (2002). Complexity and simplicity of ligand–macromolecule interactions: the energy landscape perspective. *Curr. Opin. Struct. Biol.* **12**: 197–203.

35. Krishnan V. V., Sukumar M., Gierasch L. M., Cosman M. (2000). Dynamics of cellular retinoic acid binding protein I on multiple time scales with implications for ligand binding. *Biochemistry* **39**: 9119–9129.

36. Lu J., Lin C. L., Tang C., Ponder J. W., Kao J. L., Cistola D. P., Li E. (2000). Binding of retinol induces changes in rat cellular retinol-binding protein II conformation and backbone dynamics. *J. Mol. Biol.* **300**: 619–632.

37. Franzoni L., Lucke C., Perez C., Cavazzini D., Rademacher M., Ludwig C., Spisni A., Rossi G. L., Ruterjans H. (2002). Structure and backbone dynamics of apo- and holo-cellular retinol-binding protein in solution. *J. Biol. Chem.* **4**: 4.

38. Stone M. J. (2001). NMR relaxation studies of the role of conformational entropy in protein stability and ligand binding. *Acc. Chem. Res.* **34**: 379–388.

39. Yu L., Zhu C. X., Tse-Dinh Y. C., Fesik S. W. (1996). Backbone dynamics of the C-terminal domain of *Escherichia coli* topoisomerase I in the absence and presence of single-stranded DNA. *Biochemistry* **35**: 9661–9666.
40. Zidek L., Novotny M. V., Stone M. J. (1999). Increased protein backbone conformational entropy upon hydrophobic ligand binding. *Nat. Struct. Biol.* **6**: 1118–1121.
41. Fayos R., Melacini G., Newlon M. G., Burns L., Scott J. D., Jennings P. A. (2003). Induction of flexibility through protein–protein interactions. *J. Biol. Chem.* **278**: 18581–18587.
42. Zhu M. M., Rempel D. L., Du Z., Gross M. L. (2003). Quantification of protein–ligand interactions by mass spectrometry, titration, and H/D exchange: PLIMSTEX. *J. Am. Chem. Soc.* **125**: 5252–5253.
43. Zhu M. M., Rempel D. L., Zhao J., Giblin D. E., Gross M. L. (2003). Probing Ca^{2+}-induced conformational changes in porcine calmodulin by H/D exchange and ESI-MS: effect of cations and ionic strength. *Biochemistry* **42**: 15388–15397.
44. Engen J. R. (2003). Analysis of protein complexes with hydrogen exchange and mass spectrometry. *Analyst* **128**: 623–628.
45. Clark S. M., Konermann L. (2003). Diffusion measurements by electrospray mass spectrometry for studying solution-phase noncovalent interactions. *J. Am. Soc. Mass Spectrom.* **14**: 430–441.
46. Clark S. M., Leaist D. G., Konermann L. (2002). Taylor dispersion monitored by electrospray mass spectrometry: a novel approach for studying diffusion in solution. *Rapid Commun. Mass Spectrom.* **16**: 1454–1462.
47. Clark S. M., Konermann L. (2004). Screening for non-covalent ligand–receptor interactions by electrospray ionization mass spectrometry-based diffusion measurements. *Anal. Chem.* **76**: A–G.
48. Low L. Y., Hernandez H., Robinson C. V., O'Brien R., Grossmann J. G., Ladbury J. E., Luisi B. (2002). Metal-dependent folding and stability of nuclear hormone receptor DNA-binding domains. *J. Mol. Biol.* **319**: 87–106.
49. Gumerov D. R., Kaltashov I. A. (2001). Dynamics of iron release from transferrin N-lobe studied by electrospray ionization mass spectrometry. *Anal. Chem.* **73**: 2565–2570.
50. Gumerov D. R., Mason A. B., Kaltashov I. A. (2003). Interlobe communication in human serum transferrin: metal binding and conformational dynamics investigated by electrospray ionization mass spectrometry. *Biochemistry* **42**: 5421–5428.
51. Ghaemmaghami S., Fitzgerald M. C., Oas T. G. (2000). A quantitative, high-throughput screen for protein stability. *Proc. Natl. Acad. Sci. U.S.A.* **97**: 8296–8301.
52. Powell K. D., Wales T. E., Fitzgerald M. C. (2002). Thermodynamic stability measurements on multimeric proteins using a new H/D exchange- and matrix-assisted laser desorption/ionization (MALDI) mass spectrometry-based method. *Protein Sci* **11**: 841–851.
53. Powell K. D., Ghaemmaghami S., Wang M. Z., Ma L. Y., Oas T. G., Fitzgerald M. C. (2002). A general mass spectrometry-based assay for the quantitation of protein–ligand binding interactions in solution. *J. Am. Chem. Soc.* **124**: 10256–10257.
54. Ma L., Fitzgerald M. C. (2003). A new H/D exchange- and mass spectrometry-based method for thermodynamic analysis of protein–DNA interactions. *Chem. Biol.* **10**: 1205–1213.

55. Powell K. D., Fitzgerald M. C. (2003). Accuracy and precision of a new H/D exchange- and mass spectrometry-based technique for measuring the thermodynamic properties of protein–peptide complexes. *Biochemistry* **42**: 4962–4970.

56. Ghaemmaghami S., Oas T. G. (2001). Quantitative protein stability measurement in vivo. *Nat. Struct. Biol.* **8**: 879–882.

57. Zhang Z., Smith D. L. (1993). Determination of amide hydrogen exchange by mass spectrometry: a new tool for protein structure elucidation. *Protein Sci.* **2**: 522–531.

58. Smith D. L., Zhang Z. (1994). Probing noncovalent structural features of proteins by mass spectrometry. *Mass Spectrom. Rev.* **13**: 411–429.

59. Smith D. L., Deng Y., Zhang Z. (1997). Probing noncovalent structure of proteins by amide hydrogen exchange and mass spectrometry. *J. Mass Spectrom.* **32**: 135–146.

60. Wang F., Blanchard J. S., Tang X. J. (1997). Hydrogen exchange/electrospray ionization mass spectrometry studies of substrate and inhibitor binding and conformational changes of *Escherichia coli* dihydrodipicolinate reductase. *Biochemistry* **36**: 3755–3759.

61. Wang F., Scapin G., Blanchard J. S., Angeletti R. H. (1998). Substrate binding and conformational changes of *Clostridium glutamicum* diaminopimelate dehydrogenase revealed by hydrogen/deuterium exchange and electrospray mass spectrometry. *Protein Sci.* **7**: 293–299.

62. Halgand F., Dumas R., Biou V., Andrieu J. P., Thomazeau K., Gagnon J., Douce R., Forest E. (1999). Characterization of the conformational changes of acetohydroxy acid isomeroreductase induced by the binding of Mg^{2+} ions, NADPH, and a competitive inhibitor. *Biochemistry* **38**: 6025–6034.

63. Mazon H., Marcillat O., Forest E., Vial C. (2003). Changes in MM-CK conformational mobility upon formation of the ADP-Mg(2+)-NO(3)(−)-creatine transition state analogue complex as detected by hydrogen/deuterium exchange. *Biochemistry* **42**: 13596–13604.

64. Wang F., Li W., Emmett M. R., Hendrickson C. L., Marshall A. G., Zhang Y. L., Wu L., Zhang Z. Y. (1998). Conformational and dynamic changes of *Yersinia* protein tyrosine phosphatase induced by ligand binding and active site mutation and revealed by H/D exchange and electrospray ionization Fourier transform ion cyclotron resonance mass spectrometry. *Biochemistry* **37**: 15289–15299.

65. Wang F., Miles R. W., Kicska G., Nieves E., Schramm V. L., Angeletti R. H. (2000). Immucillin-H binding to purine nucleoside phosphorylase reduces dynamic solvent exchange. *Protein Sci.* **9**: 1660–1668.

66. Hamuro Y., Wong L., Shaffer J., Kim J. S., Stranz D. D., Jennings P. A., Woods V. L., Adams J. A. (2002). Phosphorylation driven motions in the COOH-terminal Src kinase, Csk, revealed through enhanced hydrogen–deuterium exchange and mass spectrometry (DXMS). *J. Mol. Biol.* **323**: 871–881.

67. Hamuro Y., Zawadzki K. M., Kim J. S., Stranz D. D., Taylor S. S., Woods V. L., (2003). Dynamics of cAPK type IIβ activation revealed by enhanced amide H/^{2}H exchange mass spectrometry (DXMS). *J. Mol. Biol.* **327**: 1065–1076.

68. Zawadzki K. M., Hamuro Y., Kim J. S., Garrod S., Stranz D. D., Taylor S. S., Woods V. L. Jr. (2003). Dissecting interdomain communication within cAPK regulatory subunit type IIβ using enhanced amide hydrogen/deuterium exchange mass spectrometry (DXMS). *Protein Sci.* **12**: 1980–1990.

69. Akashi S., Takio K. (2000). Characterization of the interface structure of enzyme–inhibitor complex by using hydrogen–deuterium exchange and electrospray ionization Fourier transform ion cyclotron resonance mass spectrometry. *Protein Sci.* **9**: 2497–2505.

70. Wang F., Li W., Emmett M. R., Marshall A. G., Corson D., Sykes B. D. (1999). Fourier transform ion cyclotron resonance mass spectrometric detection of small Ca(2+)-induced conformational changes in the regulatory domain of human cardiac troponin C. *J. Am. Soc. Mass Spectrom.* **10**: 703–710.

71. Englander J. J., Del Mar C., Li W., Englander S. W., Kim J. S., Stranz D. D., Hamuro Y., Woods V. L. Jr. (2003). Protein structure change studied by hydrogen–deuterium exchange, functional labeling, and mass spectrometry. *Proc. Natl. Acad. Sci. U.S.A.* **100**: 7057–7062.

72. Rogero J. R., Englander J. J., Englander S. W. (1986). Individual breathing reactions measured by functional labeling and hydrogen exchange methods. *Methods Enzymol.* **131**: 508–517.

73. Englander S. W., Englander J. J. (1994). Structure and energy change in hemoglobin by hydrogen exchange labeling. *Methods Enzymol.* **232**: 26–42.

74. Wintrode P. L., Friedrich K. L., Vierling E., Smith J. B., Smith D. L. (2003). Solution structure and dynamics of a heat shock protein assembly probed by hydrogen exchange and mass spectrometry. *Biochemistry* **42**: 10667–10673.

75. Benesch J. L., Sobott F., Robinson C. V. (2003). Thermal dissociation of multimeric protein complexes by using nanoelectrospray mass spectrometry. *Anal. Chem.* **75**: 2208–2214.

76. Rist W., Jorgensen T. J. D., Roepstorff P., Bukau B., Mayer M. P. (2003). Mapping temperature-induced conformational changes in the *Escherichia coli* heat shock transcription factor {sigma}32 by amide hydrogen exchange. *J. Biol. Chem.* **278**: 51415–51421.

77. Hoff W. D., Xie A., Van Stokkum I. H., Tang X. J., Gural J., Kroon A. R., Hellingwerf K. J. (1999). Global conformational changes upon receptor stimulation in photoactive yellow protein. *Biochemistry* **38**: 1009–1017.

78. Englander S. W., Mayne L., Rumbley J. N. (2002). Submolecular cooperativity produces multi-state protein unfolding and refolding. *Biophys. Chem.* **101–102**: 57–65.

79. Kleywegt G. J., Bergfors T., Senn H., Le Motte P., Gsell B., Shudo K., Jones T. A. (1994). Crystal structures of cellular retinoic acid binding proteins I and II in complex with all-trans-retinoic acid and a synthetic retinoid. *Structure* **2**: 1241–1258.

80. Kaltashov I. A., Eyles S. J., Xiao H. (2004). Combination of protein hydrogen exchange and tandem mass spectrometry as an emerging tool to probe protein structure, dynamics and function. In: *Focus on Protein Research*, ed. J. W. Robinson, pp. 191–218. Hauppauge, NY: Nova Science Publishers.

8

SYNERGISM BETWEEN BIOPHYSICAL TECHNIQUES

No single biophysical method can sufficiently and completely describe a biological process. Each technique provides different types of information that, when assembled together, enable a more complete picture to be drawn of the system under scrutiny and the processes that make the biological machine function. In this chapter we select two systems that have been studied in great detail and demonstrate the synergy of biophysical techniques. Some methods provide overlapping information to confirm the evidence, while others provide completely unique details.

8.1. HEN EGG WHITE LYSOZYME

8.1.1. Folding of Hen Lysozyme

The protein hen lysozyme has become the paradigm for protein folding studies since it has been studied by so many different techniques (1, 2). Its structure has been known for forty years—it was the first enzyme for which a high-resolution X-ray crystal structure was obtained (3). It constitutes a large proportion of the proteinaceous material in egg white, and its function, although still not completely understood, is to catalyze the hydrolysis of β-1,4-glycosidic linkages in certain gram-positive bacterial cell walls and chitin (4). Homologous *c*-type lysozymes are found in the milk and lachrymal fluids of mammals and a variety of other secretions, and it is thought to act as a general bactericidal agent. Subsequent

Mass Spectrometry in Biophysics: Conformation and Dynamics of Biomolecules
By Igor A. Kaltashov and Stephen J. Eyles
ISBN 0-471-45602-0

structural studies of this protein bound to a trisaccharide inhibitor [NAG_3, tri-(N-acetyl-β-1,4-glucosamine)] led to one of the first detailed mechanistic proposals for enzyme function (5).

The structure of lysozyme makes it an attractive model for protein folding studies. It is a readily available, relatively small monomeric enzyme consisting of 129 residues (14 kDa) and contains many of the common structural elements of proteins. The single chain structure can be divided into two structural subdomains cross-linked by four disulfide bridges (Figure 8.1). The α-domain is constructed from four α-helices and a short 3_{10} helix, while the β-domain, in the center of the sequence, is made up of a three-stranded antiparallel β-sheet and a second 3_{10} helix. A second short β-sheet region links the N-terminus to the β-subdomain. The interface between the two domains forms the substrate binding cleft into which up to six N-acetylglucosamine units can be accommodated.

Folding of lysozyme has been the subject of intense scrutiny for several decades. Some of the earlier studies focused on the order of disulfide formation during oxidative refolding from the reduced state (6–8). These studies involved allowing denatured reduced lysozyme to refold and measuring the kinetics of recovery of enzymatic activity. This process occurs over a period of hours so it can be monitored in real time without sophisticated equipment. The order of formation of disulfides has also been investigated (7, 9). In the past this

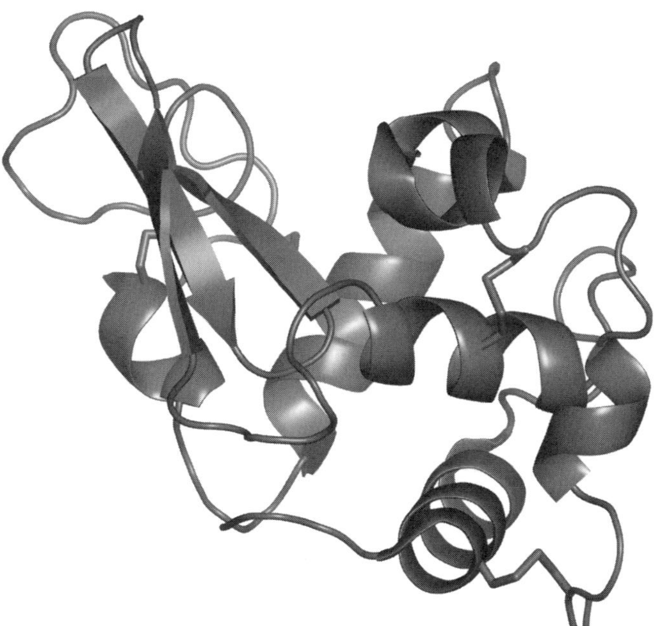

FIGURE 8.1. Ribbon diagram of hen lysozyme showing the α-helical (right lobe) and β-sheet (left lobe) folding domains. Figure produced from PDB code 1HEL using PyMol™, DeLano Scientific LLC.

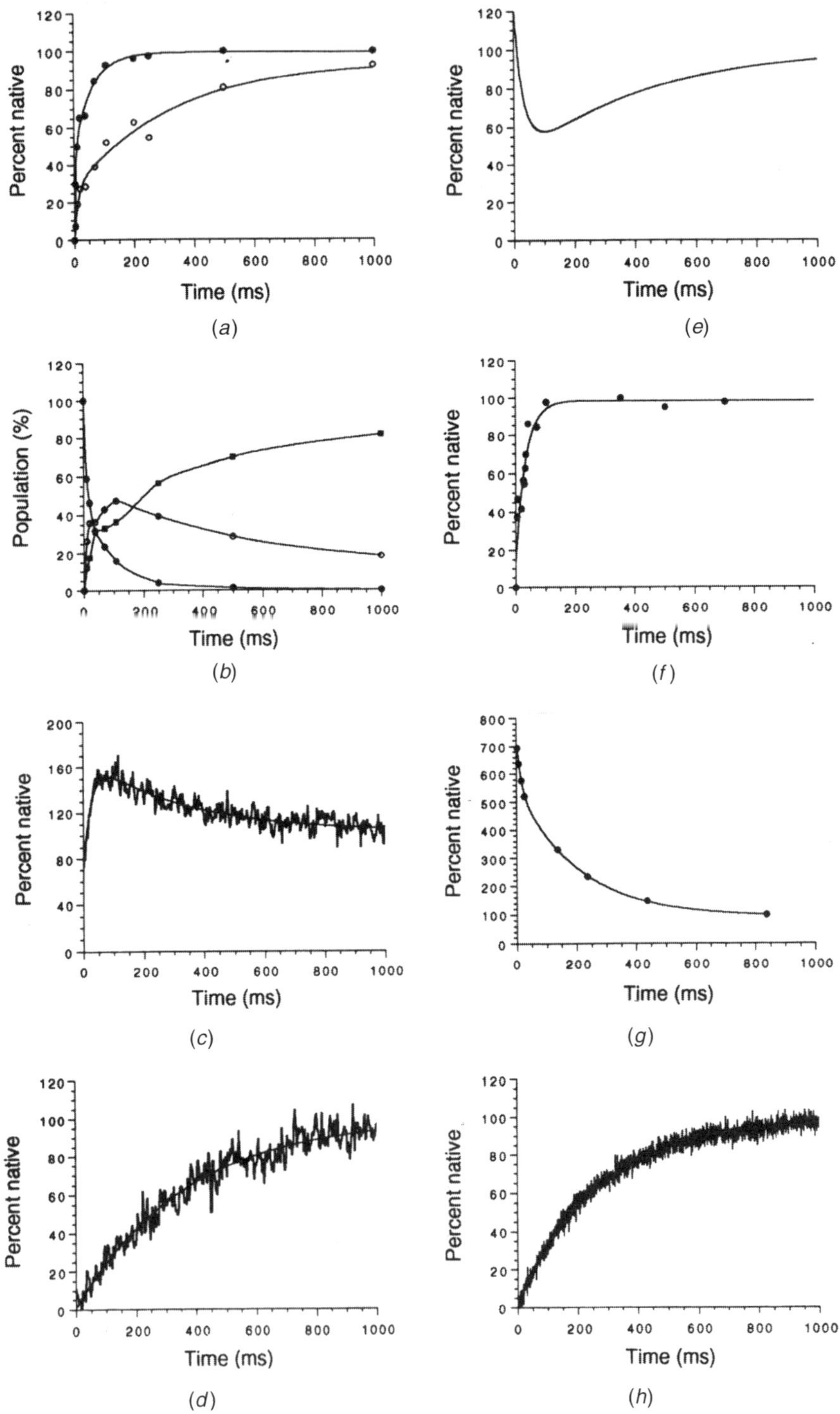
Percent native
Time (ms)
(a)
Population (%)
Time (ms)
(b)
Percent native
Time (ms)
(c)
Percent native
Time (ms)
(d)
Percent native
Time (ms)
(e)
Percent native
Time (ms)
(f)
Percent native
Time (ms)
(g)
Percent native
Time (ms)
(h)

was a painstaking task involving blocking of free sulfhydryls with a thiol specific reagent, such as iodoacetamide, and subsequent proteolysis followed by identification of the sequences containing the alkylated residue by amino acid sequencing or N-terminal sequencing. Today MS has made this task much easier since the chemical modification gives rise to a mass shift for each cysteine residue alkylated. (See Chapter 5 for a more detailed discussion of this technique.) A recent study used MS to quantify the number of disulfides formed during oxidative refolding as a function of time by measuring intensities of peaks in the mass spectrum corresponding to species containing one or more disulfides (10). Due to the number of possible isomers from formation of disulfide bridges between eight possible cysteine residues in lysozyme, disulfide mapping was not performed in this study but could, in principle, be used to determine if there is a specific pathway of disulfide formation during folding. Earlier studies concluded that the disulfide linkages in the β-domain of the protein (Cys64–Cys80 and Cys76–Cys94) were the first to form. These cysteine residues are, incidentally, closest to each other in the primary sequence, compared to those in the α-domain that bring the N- and C-termini together (Cys6–Cys127 and Cys30–Cys115). A recent study has indicated that a dominant folding intermediate that accumulates during the oxidative refolding of lysozyme contains three native disulfide bridges but lacks the Cys76–Cys94 linkage. The location of the free cysteine residues was determined by peptic digest and analysis of the fragments using ESI Q-TOF and FT-ICR MS measurement (11). This highly native-like species presumably accumulates due to burial of Cys94 during folding that prevents forming of the fourth disulfide bridge.

In recent years more emphasis has been put on other aspects of the folding process, as instrumentation has become available to couple rapid mixing devices to spectrophotometers. As discussed in Chapter 2, different spectroscopic methods probe different aspects of the molecular structure, and together the assembled body of evidence allows detailed mechanistic conclusions to be drawn. One of the most detailed studies of lysozyme folding was performed by Radford et al. (12), in which a number of biophysical probes were employed to monitor folding. Protein was denatured chemically in 6 M GuHCl (guanidine hydrochloride) and then refolded by dilution to a final concentration of 0.55 M, pH 5.2. In these experiments all the native disulfide bridges were left intact throughout unfolding and refolding. Here we will discuss in more detail the contributions from each technique to the more complete understanding of the folding of lysozyme.

FIGURE 8.2. Folding of lysozyme monitored by a variety of complementary biophysical techniques. (**A**) Hydrogen exchange pulse labeling monitored by NMR in the α-(●) and β-(○) domains. (**B**) Evolution of unprotected/unfolded (●), intermediate (○), and native-like state (■) as monitored by ESI MS. Refolding monitored by spectroscopic methods: (**C**) secondary structure from far-UV CD at 225 nm, (**D**) tertiary structure from near-UV CD at 289 nm, (**E**) intrinsic tryptophan fluorescence, (**F**) quenching of fluorescence by iodide ions, (**G**) binding of the hydrophobic dye 1,8-anilinonaphthalenesulfonate, and (**H**) binding of fluorescent inhibitor MeU-diNAG. Reproduced with permission from (1). © 1994 Elsevier Science Ltd.

Circular dichroism measures the extent of secondary and tertiary structure formation in proteins. In the far-UV region of the spectrum, the formation of helical structure can be monitored by change in CD signal intensity at 225 nm.* Refolding of lysozyme under these conditions exhibits biphasic kinetics in the far-UV region CD (Figure 8.2). Initially, a significant complement of secondary structure is developed within the 2 ms dead time of the stopped-flow instrumentation, followed by an apparent excess of helicity that forms within 25 ms. Subsequently, this signal relaxes back to that of the native state with a time constant of ~300 ms. This overshoot in the CD signal could be the result of formation of non-native helical structure but was attributed to spectral contributions in this region of the spectrum from the presence of disulfide bridges (12).

Tertiary structure formation can be monitored spectroscopically by CD in the near-UV region since this protein contains six intrinsic tryptophan residues located throughout the structure. Stopped-flow measurements showed formation of a signal characteristic of native structure with a time constant of 300 ms, indicating that the aromatic residues adopt their native chiral environments on a time scale consistent with the slow phase of secondary structure formation (12). Fluorescence measurements indicated an initial very rapid phase within the dead time of measurement suggesting that the tryptophan residues at first collapse into an environment where local non-native interactions cause intramolecular quenching of the fluorescence intensity (13). This effect is somewhat unusual in protein folding, since tryptophan fluorescence generally increases upon collapse due to the sequestering of hydrophobic residues away from the quenching effects of proximal solvent molecules. Further quenching occurs with a time constant of 25 ms, followed by recovery of the fluorescence signal characteristic of the native state with $\tau \sim 300$ ms, again consistent with CD measurements. Additional stopped-flow fluorescence measurements using iodide as a quenching solute indicate, however, that the fast phases do indeed result from rapid hydrophobic collapse, since iodide ions cannot penetrate to effect further fluorescence quenching. Rearrangement to form the native structure occurs without significant changes in the solvent environment of the tryptophan residues.

The fluorescent hydrophobic dye 1-anilinonaphthalenesulfonic acid (ANS) has been used extensively to detect exposed hydrophobic patches in kinetic and equilibrium folding intermediates since its fluorescence is substantially enhanced upon binding (14). ANS binding to lysozyme during folding exhibited two kinetic phases remarkably similar to those measured by far-UV CD, suggesting that secondary structural formation is accompanied by changes in exposure of hydrophobic residues. The dead time phase was accompanied by a substantial increase in ANS fluorescence followed by a biphasic decrease as continued folding to the native state resulted in burial of hydrophobic patches.

*These refolding experiments are generally performed by dilution from a high concentration of chemical denaturant. Even under renaturing conditions, the residual concentration of GuHCl is sufficiently high that its absorbance in the far-UV region below 225 nm renders stopped-flow CD experiments at these wavelengths impractical. With other denaturants, such as pH, it is often possible to monitor CD at much shorter wavelengths and hence gain even more information about secondary structural formation during refolding.

Finally, binding of a fluorescent disaccharide substrate analogue was employed to monitor formation of the active site cleft (13). 4-Methylumbelliferyl-N,N'-diacetyl-β-chitobiose (MeU-diNAG) fluorescence developed on a time scale of 350 ms, in good agreement with the slow phases measured by CD and tryptophan fluorescence techniques, indicating that this phase does indeed represent formation of a native-like structure competent to bind the inhibitor molecule. Together, these spectroscopic data provide highly complementary information as to the nature of the folding mechanism. They indicate an initial very rapid hydrophobic collapse followed by rearrangement of the molecule in two subsequent kinetic phases to the native state, the first involving primarily secondary structure formation, and the second slower phase creating native tertiary contacts.

As with all spectroscopic methods, these data provide primarily global information. We know that secondary structure formation gives rise to CD signal but from these results alone we cannot determine which regions of structure are forming. Similarly, with six tryptophan residues in the lysozyme molecule, the fluorescence data can only give a measure of the average environment experienced by these side chains. One approach to obtaining more site-specific information would be to remove all but one of the intrinsic tryptophan residues by site-directed mutagenesis as has been performed with, for example, CRABP I (15, 16). In this manner the contribution to the total fluorescence from individual tryptophan residues may be dissected to yield a more detailed picture of local interactions formed during folding.

Time resolved X-ray scattering measurements have been used to indicate how the radius of gyration (R_g) of lysozyme molecules changed kinetically during refolding (17). At pH 2 folding is significantly slower than at pH 5.2 used for the experiments described above. Chen and co-workers determined that the early collapsed state formed by the first time point of measurement is significantly more compact than expected for a random coil but yet much more expanded than the native structure. This is consistent with nonspecific hydrophobic collapse. Additionally, these authors have studied the equilibrium unfolding of lysozyme as a function of urea concentration. Other spectroscopic methods have concluded that the denaturation process is highly cooperative, involving no intermediate species. However, by X-ray scattering measurements, a stable intermediate species was detected with $R_g \sim 19.8$ Å, compared to 15.3 Å for the native state and 22 Å for the fully denatured state (18). Although the interpretation of X-ray scattering data can be extremely complex, it clearly has complementary value as a biophysical technique that can detect intermediate species not observed by other techniques.

In order to gain more detailed structural information about the folding intermediate of lysozyme, pulse labeling HDX experiments monitored by NMR were performed (12). Intriguingly, stable hydrogen bonding that protected against hydrogen exchange was found to form significantly more rapidly in the α-helical domain of the protein relative to the remainder of the molecule. The time scale for protection of these residues is ~60 ms, whereas for those residues in the β-domain, protection occurs with an average time constant of 350 ms. Armed with this knowledge, we can now see that the folding of hen lysozyme can be divided into two folding domains. Initially, helical structure forms as evidenced from far-UV CD stopped flow measurements and then the α-domain becomes stabilized by formation of hydrogen bonds protecting this region against exchange (19). In a subsequent slower kinetic phase, the remainder of the molecule, the β-domain, forms stable structure, at the same time organizing tryptophan residues into their native orientations and forming the saccharide binding cleft. However, even this is not the complete story. Kinetics of hydrogen exchange protection were observed to be biphasic for all residues, with each residue becoming 20–50% protected within 10 ms. This could result from partial protection or else alternative folding pathways (12). The insensitivity of the level of protection to the pH of the labeling pulse, however, strongly suggested the latter.

The power of MS really comes into its own to address the question of parallel folding pathways. If the alternate folding pathways lead to species with different levels of protection against exchange, then this will manifest itself as peaks of different mass that can be measured by MS.* By contrast, partial protection against exchange would result in a broadening of the mass spectral peak since only partial exchange occurs at each residue. As demonstrated by Miranker et al. (20), for lysozyme folding the former case occurs. Three distinct peaks are observed in the ESI mass spectrum of pulse labeled samples (see Figure 6.4), the proportions of which change with refolding time. The initial spectrum, with no time allowed before labeling with isotope, corresponds to the completely unfolded protein that becomes totally labeled by the pulse. As refolding time progresses, two other species are observed, one that has a mass close to that expected from protection upon folding of the α-domain, and another peak corresponding to fully protected molecules that is present even within 100 ms of folding. This result clearly demonstrates that a portion of the folding population can attain native-like protection very rapidly, whereas other molecules must fold via the α-domain intermediate. MS is the only technique that can unequivocally show this to be the case, hence its deserved place in the biophysical toolkit (21).

MS has also proved useful to study the folding of lysozyme variants, which may have only small differences in folding characteristics. Comparison of kinetics between lysozymes obtained from different species shows small but significant differences in kinetics that can be detected by spectroscopic methods. However, to ensure that refolding conditions are absolutely identical, pulse labeling experiments can be performed on mixtures of proteins providing that they differ sufficiently in mass to be resolved by MS. In this manner the rates can be compared directly in the same experiment. For instance, avian lysozymes from Japanese quail and Lady Amherst pheasant differ from the hen protein by only six amino acid residues, but these could be refolded in mixtures with uniformly ^{15}N-labeled hen lysozyme to directly compare differences in cooperativity of

*Strictly speaking, this is only useful under EX1 exchange conditions since for EX2 individual peaks cannot be resolved. Fortunately, for pulse labeling experiments, the high pH labeling pulse is assumed to only label unfolded proteins and unstructured regions of intermediate states. For these species the EX1 regime may safely be assumed. See Chapter 5 for a more detailed discussion of HDX theory.

folding by MS (22). Slight differences in rates of formation of the α-domain intermediate could readily be distinguished by this method using small quantities of protein and with complete certainty that the differences observed did not result from slight changes in folding conditions (Figure 8.3).

In a similar manner, a chemically modified variant containing only three of the native disulfide bridges, but lacking the Cys6–Cys127 cross-link in the α-domain ($CM^{6,127}$-lysozyme), was shown to fold via a somewhat different mechanism, where in this case the lack of stabilizing disulfide in the α-domain prevented the accumulation of the helical intermediate (23). Nevertheless, MS was used to confirm that even in this variant the parallel "fast folding" track was still available to this protein. Studies in which hen lysozyme was refolded at elevated temperatures similarly showed that under conditions where the α-domain intermediate is not stable the intermediate is not detected, although the effect of temperature can be reversed by the addition of stabilizing salts (24). Another investigation of the effect of the chaperone protein GroEL on lysozyme folding indicated that the mechanism remains unchanged; that is, the α-domain intermediate is still populated, but the overall folding rate was accelerated by the presence of the chaperone protein, presumably by providing an environment more permissive for efficient folding (25).

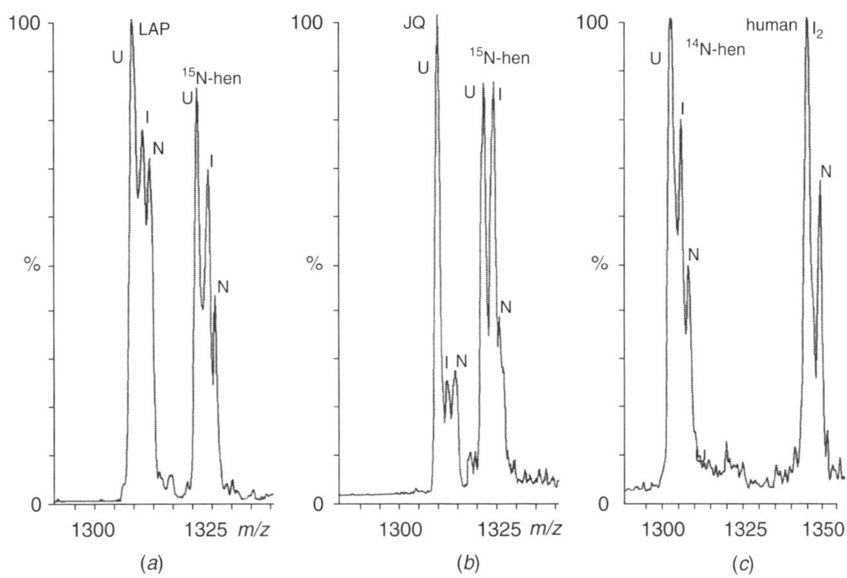

FIGURE 8.3. Folding of variant lysozymes. By mixing equimolar amounts of two variant lysozymes and performing pulse labeling experiments on the mixture, a direct and simultaneous comparison can be made of refolding kinetics for the two proteins. In the case where the two proteins have very similar mass, they can be resolved by instead using an isotopically labeled version. Comparison of the folding of lysozymes from hen, human, Lady Amherst pheasant (LAP), and Japanese quail (JQ) sources. Reproduced with permission from (22). © 1994 American Chemical Society.

Protein misfolding has become a recent topic of great interest, since when this complex process goes awry it can lead to a variety of disease states. (In Chapter 11 we will discuss emerging methodologies to investigate protein misfolding by MS.) Aggregation or amyloid formation has been associated with such pathogenic conditions as Alzheimer's and Creutzfeld–Jakob diseases. Single point mutations in human lysozyme have been identified as the origin of hereditary familial amyloidosis (26), in which biologically active lysozyme molecules can form amyloid fibrils (27). Studies using techniques including MS have shown that these misfolding events may be due to the dramatically increased unfolding rates relative to the wild type protein, and hence structural plasticity is likely a major determinant in the ability of these variants to form misfolded amyloidogenic states (28).

8.1.2. Substrate Binding to Lysozyme

The catalytic mechanism of lysozyme has been the subject of study for many years. The hydrolysis of β-1,4-linked oligosaccharides is known to involve two catalytic carboxylates from residues Glu35 and Asp52, although the rate of hydrolysis in the wild type protein is too fast for the direct observation of intermediate states. Two mechanisms have been proposed, one involving a covalent glycosyl–enzyme intermediate (29) in which Asp52 stabilizes the oligosaccharide during hydrolysis (Figure 8.4). Subsequent to solving the crystal structure of the enzyme alone and cocrystallized with NAG_3, an alternative mechanism was proposed that suggested noncovalent attachment, but instead stabilization of a carbonium ion intermediate by the proximity of the carboxylate (5, 30). Various other studies have been performed in which chemical modification or mutagenesis of the key catalytic residues resulted in virtually inactive enzymes (31–34), but none has been able to distinguish mechanistic detail.

A number of studies aimed at elucidating the mechanism of lysozyme have been hampered by the high activity of this enzyme. In order to study intermediate species, conditions must be found under which any intermediate states are resident for sufficiently long times to be studied. However, even at low temperature it has proved difficult to find conditions for the wild-type protein under which intermediate states accumulate (35, 36). The crystal structure of an inactive mutant form of lysozyme (D52S) cocrystallized in the presence of tri- and hexasaccharides demonstrated the six binding sites for each of the glycosyl moieties but, in spite, of extremely low activity no evidence for any intermediate structures (37). NMR studies were used to measure changes in the chemical environment of aromatic residues upon saccharide binding (38) but again only limited information about the nature of binding was obtained.

MS has been an extremely valuable tool for the investigation of the nature of the lysozyme–saccharide interaction. Ganem and co-workers demonstrated that noncovalent adducts of lysozyme with NAG_6 can be observed by MS, and also various enzyme complexes with hydrolysis products, using ion spray under aqueous conditions (39). A similar experiment was performed using electrospray ionization in 50:50 water/methanol with the D52S inactive mutant (38).

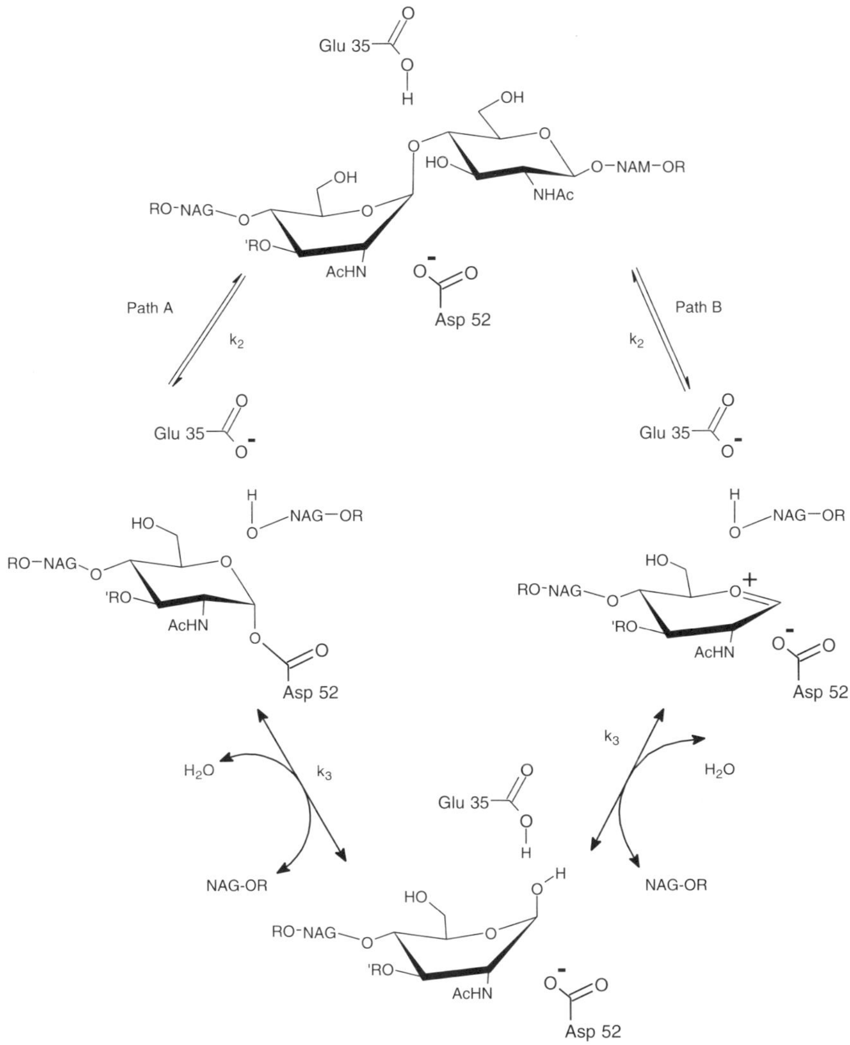

FIGURE 8.4. Alternative postulated catalytic mechanisms for the hydrolysis of β1,4-glycosidic linkages by hen lysozyme. Path A (*left*) in which a covalent acyl enzyme intermediate is formed involving the side chain of Asp52 (29). In Path B (*right*) the developing acylium ion intermediate is stabilized electrostatically by the proximal aspartate anion, although no covalent species is formed (30). Reproduced with permission from (40). © 2001 Nature Publishing Group.

This study also revealed the existence of enzyme–substrate associations, even under conditions where one might expect dissociation of noncovalent complexes, suggesting the possibility that the complex may in fact be covalent. A more recent study of another inactive mutant (E35Q) by ESI MS under denaturing solvent conditions in combination with X-ray structures from crystals diffracting to 1.46 Å clearly demonstrated the presence of a covalent enzyme–substrate intermediate (40). This most recent result argues strongly that enzymatic hydrolysis of glycosidic linkages by lysozyme involves a covalent glycosyl–enzyme intermediate, the mechanism originally proposed by Koshland (29). This shows how no single biophysical method is sufficient; but with evidence amassed from a variety of biophysical techniques, a complete picture can be drawn that fully represents the process.

8.2. MOLECULAR CHAPERONES

Studying protein folding in the test tube is a very convenient method, but there has always been a significant amount of controversy as to its physiological relevance. In the intracellular environment that proteins are exposed to as they are translated and released from the ribosome, they are by no means isolated species allowed to refold under their own devices at low concentration. Instead, they are released into an extremely crowded intracellular environment containing a variety of other proteins, nucleotides and ions, and at a significant osmotic pressure. Thus, there is significant interest in determining not only the mechanistic details of protein folding in vitro, but also the similarities and differences in vivo. One of the major differences is that a class of helper proteins, known as molecular chaperones, are thought to mediate the process of folding in the cell, in some cases by maintaining the chains in an extended conformation to prevent premature folding, and in other cases to provide an environment conducive to efficient renaturation without aggregation. Which of the proteins that are expressed interact with these chaperones is still somewhat unknown, but many have been identified by proteomics-type approaches (41), and in other cases likely candidates are those proteins that do not fold well in isolation in vitro. The latter class of proteins have been found almost always to fold more efficiently in the presence of chaperones.

A number of different classes of chaperones are known, first identified as *heat shock proteins* (HSPs) due to their increased expression under stress conditions such as elevated temperature, where newly expressed proteins may need more assistance in correct and efficient folding. They are classified according to their approximate molecular weight: thus, HSP70s are approximately 70 kDa and so forth. The HSP70s bind to extended hydrophobic regions in partially unfolded proteins under stress conditions and upon ATP hydrolysis release them for translocation to other cellular compartments or else to continue folding. This class of proteins also appears essential for the correct folding and export of certain proteins even under normal physiological conditions. HSP20s bind to the surface of non-native proteins under heat shock, HSP90s bind to signal transduction proteins, and there are a variety of other classes that have more specific functions including targeting proteins that are beyond repair for degradation.

One of the most interesting classes of chaperone proteins, at least in terms of studies by MS is HSP60, specifically GroEL in bacterial cells. This protein, as

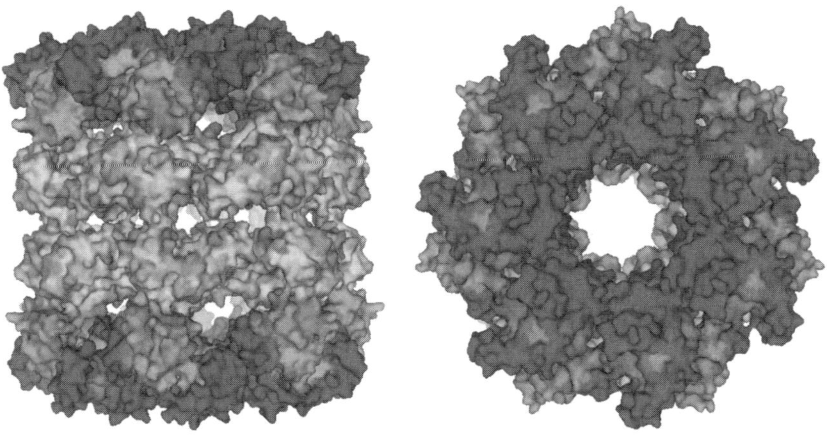

FIGURE 8.5. Structure of the GroEL tetradecamer based on cryo-EM data. The stacked heptameric rings can be seen in the upper panel, and the internal central cavity formed by the tetradecameric assembly can be clearly seen in the end-on view below. Picture created from PDB code 1GR5 (76) using PyMol™, DeLano Scientific LLC.

its name suggests, is comprised of 60 kDa monomers, but these are assembled into a toroidal structure, the active form of the protein being a macromolecular assembly of two stacked heptameric rings (Figure 8.5), with an overall molecular weight of ~800 kDa. The function of this structure is to provide a hydrophobic cavity within which partially unfolded proteins can be sequestered, unfolded, and then allowed to refold in an environment where aggregation is disfavored. This complex process requires large conformational changes in the GroEL machine and is known to be assisted by the cochaperone protein GroES, a single ring heptamer of HSP10 proteins, with energy provided by hydrolysis of Mg-ATP.

The earliest physical pictures of the GroEL molecular machine were obtained by electron microscopy and showed the toroidal structure of the protein (42, 43). Subsequent cryo-EM studies demonstrated binding of GroES to one or both ends of the structure to form an enclosed cavity (44–47), and that polypeptides sequestered within the cavity appeared to adopt a non-native structure within this cavity (48, 49). These observations led to speculation that the GroEL assembly was more of a passive element in protein folding, allowing misfolded proteins to experience a hydrophobic environment unlike the external solvent, in which they could unfold and disentangle prior to release for another attempt at correct folding. Further biophysical characterization was necessary to refute that possibility.

Shortly after this, a high-resolution crystal structure of GroEL was solved to 2.8 Å (50), which shed light on the atomic details of the toroidal structure previously observed. Each 60 kDa subunit could be separated into three structural domains, termed the apical, intermediate, and equatorial. The helical equatorial domain contains the ATPase domain and the majority of the intersubunit contact residues, whereas the apical domain is less well ordered and constitutes most of the residues at the opening of the central cavity. Studies by site-directed mutagenesis have identified several hydrophobic residues in the apical domain that are important for binding of substrate polypeptides (51), suggesting that recognition of exposed hydrophobic patches in partially folded proteins may be an important factor in the GroEL-mediated rescue process.

Further extensive studies, mainly by crystallography and cryo-EM, have been able to show various conformational changes that occur during the reaction cycle of GroEL and clearly demonstrate that the role of this chaperone in promoting refolding is in fact a very active and complicated one (52, 53, and references therein). Upon binding substrate polypeptide and ATP, GroEL undergoes a dramatic conformational change involving movement of the apical domains on one side of the toroid structure. This has the effect of increasing the affinity for GroES and also changing the shape and hydrophobicity of the interior cavity. Once GroES is bound, the unfolded polypeptide is now completely enclosed and in a larger and more hydrophilic environment than previously, thus promoting refolding. In the next step of the cycle, ATP binds to the opposite (*trans*) ring of the chaperone complex, and this promotes release of the GroES capping structure, nucleotide, and also the substrate peptide from the *cis* ring (Figure 8.6). This latter step also leaves the trans ring in a high affinity state for unfolded polypeptides, thus allowing the cycle to continue but on the opposite side of the molecule.

Crystallography sheds light on each of the phases of the reaction cycle by trapping individual states using, for instance, nonhydrolyzable ATP analogues. Cryo-EM, on the other hand, takes advantage of freezing out molecules at

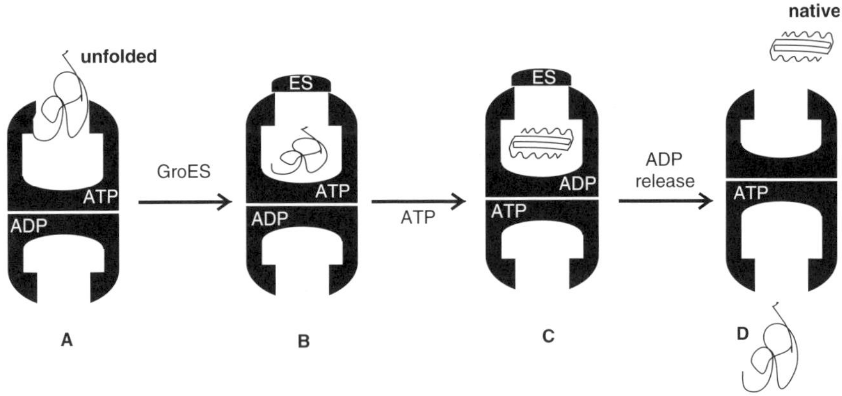

FIGURE 8.6. Schematic of the "cis" GroEL–GroES reaction cycle (77). (**A**) Upon binding of substrate peptide followed by (**B**) capping of the cavity by GroES, (**C**) cis-ATP hydrolysis triggers a conformational change within the cavity shifting the interior surface from hydrophobic to hydrophilic in nature. This promotes protein refolding. (**D**) ATP binding to the opposite side of the GroEL machine triggers release of GroES and folded substrate and prepares the opposite ring for the next peptide binding event.

different stages during the cycle and then reconstructing pictures of each state by identifying and overlaying images of individual states. While these do not show the kinetic events directly, they allow a quite detailed model of the process to be constructed.

A variety of other biophysical techniques have been applied to study the chaperone reaction cycle, for instance, fluorescence resonance energy transfer (FRET). By placing two fluorescent probes at specific points in the tertiary structure, the energy transfer between them upon excitation acts as a measure of the physical distance between them (see Chapter 2). Since GroEL and GroES contain no intrinsic fluorophores, these may be introduced by engineering cysteine residues by mutagenesis and then attaching fluorescent labels. By monitoring fluorescence changes throughout the reaction cycle, the change in distance between the donor and acceptor pair can be measured with high sensitivity. This was elegantly demonstrated by Rye et al. (54, 55), who placed the donor and acceptor pair on GroEL and GroES, respectively. As GroES binds to form the active complex during the cycle, FRET can be measured to determine the kinetics of association and dissociation. In principle, this type of methodology may be applied intramolecularly also if a large enough conformational change occurs to alter the energy transfer efficiency (56).

Much is now known about the structural changes and dynamics that occur within the chaperone macromolecular assembly itself during the reaction cycle that assists in protein folding, but significantly less is understood about how it self-assembles into the functional tetradecameric form, or the conformations adopted by the substrate polypeptide that binds within the cavity. GroEL evolved to accept a wide variety of substrates so it may be that there are no specific preferences, but it is certainly of interest to understand how folded the species are that bind and what conformational changes occur to the bound protein before it is released back into the cellular environment. Since this process is highly dynamic, it is difficult to obtain this kind of information by crystallographic measurement. The entire complex is far too large to be studied by NMR, although transferred NOE experiments of model peptides have given some significant insight into conformational preferences (57, 58). This method works if there is a significant difference in correlation time between free and bound peptide, and only if the on and off rates are just right to allow efficient transfer of NOE information. Thus, this is a very powerful technique but somewhat restricted in its usage. Other optical methods may be of use in determining the hydrophobicity of the cavity environment and for pinpointing specific binding sites.

To determine the conformational state of an entire protein bound within the GroEL cavity, MS once again has been demonstrated to be a powerful complementary technique. If the chaperone-bound polypeptide is allowed to undergo hydrogen–deuterium exchange, then this can give some valuable information as to whether the bound substrate is folded, completely unfolded, or adopts some conformational ensemble that is intermediate between the two. Upon desolvation into the gas phase, the proteins can be dissociated from one another but the hydrogen exchange information will be retained (Figure 8.7). Exchange experiments

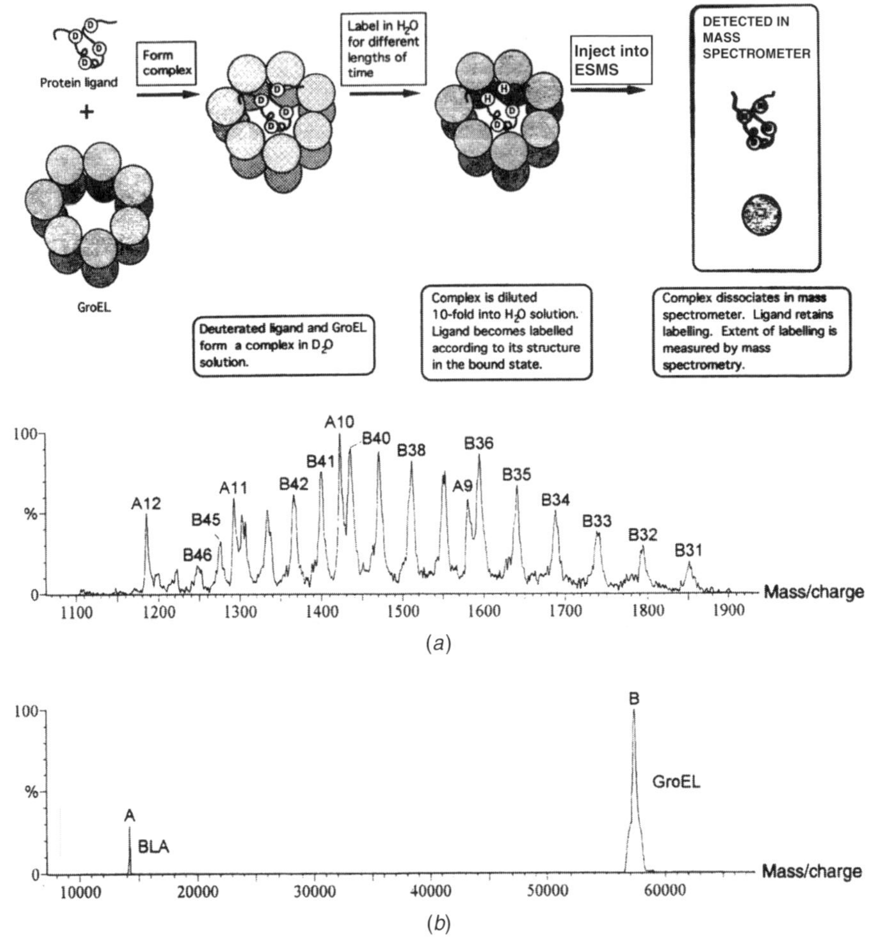

FIGURE 8.7. Schematic of the measurement of protection of partially folded states of proteins bound to GroEL. The complex is formed between deuterated protein and GroEL and then labeled for varying periods of time in solution with H_2O. Upon introduction into the mass spectrometer, the complex dissociates and hydrogen exchange protection in the ligand can be determined directly. Part (*a*) shows the raw mass spectral data with contributions from α-lactalbumin (A) and GroEL monomer (B) charge states. (*b*) Deconvoluted spectrum from which hydrogen exchange protection can be measured. Reproduced with permission from (59). © Nature Publishing Group.

performed on a three-disulfide derivative of α-lactalbumin showed that binding of this protein to GroEL caused the protein to adopt a conformation that was at least somewhat protective against exchange (59), reminiscent of the molten globule state known to be an equilibrium unfolding intermediate for this protein (14). In these experiments three complementary pieces of information could be obtained. From the mass of the substrate protein the degree of protection against exchange could be measured, indicative of the conformation of the bound peptide. Additionally, analysis of the charge state distribution of the substrate protein suggested a partially folded species. Finally, by measuring the width of the exchange peaks, some idea of the conformational heterogeneity was obtained. If the partially folded state adopted a wide range of different conformations, then this would manifest itself as a very broad distribution of masses, whereas what was actually observed was a quite narrow mass peak, indicating that the GroEL-bound species, although only partially folded, has persistent secondary structure (60).

Similar studies by NMR and MS using the protein dihydrofolate reductase (DHFR) suggested that in some cases the conformational states bound to GroEL may even be quite highly native-like (61, 62). These studies could distinguish no difference between protection of the substrate protein as it bound in the initial state and in subsequent cycles. This suggests the possibility that GroEL assists refolding by forcing subtle structural changes of the partially folded substrate protein during iterative binding and release, as opposed to a wholesale unfolding of the substrate upon binding. As mentioned earlier, when using hen lysozyme as a substrate no change in folding mechanism was observed, although the overall rate of folding for this protein was accelerated in the presence of chaperone (25).

Plückthun and co-workers have carried out extensive work to elucidate the effect of GroEL on the folding and unfolding pathway of β-lactamase. They observed very high protection against HDX of β-lactamase in the GroEL-bound form (63). Additionally, the unfolding rate from the bound state was measured to be identical to that of the native state (the unfolding of the equilibrium molten globule state occurs too rapidly to be measurable). The authors concluded that different proteins are recognized by GroEL in very different states, ranging from totally unfolded to native-like, depending on the accessibility of exposed hydrophobic patches. Reversible binding of native-like states may be an important property of GroEL in protection of proteins against aggregation under heat shock conditions. Additional studies by this group used limited proteolysis at elevated temperature to identify the binding sites on β-lactamase that are recognized by GroEL (64), and also used different denaturing conditions to identify different conformations of the protein that could be bound to the chaperone (65).

Further insight into the action of GroEL was gained from a kinetic study of the folding of malate dehydrogenase (MDH) in the cavity of a single ring variant of GroEL (SR1) in the presence of the cochaperone GroES (66). MDH does not fold efficiently in the absence of chaperones and apparently requires several rounds of binding and release from the GroEL–GroES cavity for complete recovery of native structure. The authors investigated the conformation of non-native MDH bound to SR1 alone using HDX-MS, which showed a core-like structure that was significantly protected against exchange even in this non-native state. Some loss of protection was observed upon addition of GroES and ATP, but this could be attributed to the breaking of hydrogen bonds between substrate and the walls of the chaperone cavity rather than a forced unfolding reaction. This is consistent with the notion that the binding of cochaperone induces a conformational change in the cavity that releases the unfolded polypeptide into a hydrophilic environment that promotes protein refolding. It does not support the alternative hypothesis of a mechanical unfolding process prior to release.

MS has been applied to studies of the assembly/disassembly pathway of the GroEL macromolecular complex (67). Using the methodologies described in previous chapters, Chen and co-workers used hydrogen exchange in combination with proteolytic digest and MS to investigate the equilibrium unfolding pathway of the GroEL complex. These data were combined with light scattering and size exclusion chromatography (both of which give a macroscopic view of the hydrodynamic radius of the protein). The results demonstrated, as suggested by the crystal structure, that the flexible apical and intermediate domains can be unfolded while retaining the quaternary interactions of the tetradecameric assembly. Only when the equatorial domain is denatured does the macromolecule disassemble into monomers, as might be expected since this domain contains the majority of quaternary contacts.

It still remains uncertain what the exact function of the chaperone complex is. Perhaps it can recognize proteins that are kinetically trapped or that may have a high tendency toward aggregation, so that only if the polypeptide is on the wrong track toward productive folding will it be actively unfolded in the cavity. Many more questions remain to be answered about the action of this complex molecular machine, including the exact mechanistic details of how the folding free energy landscape is altered within the cavity, or how the GroE assembly assists folding of proteins too large to be accommodated completely within the cavity (68). Other studies of interest on different chaperone systems include a substantial amount of work on lens crystallin proteins (69–73), and the trigger factor protein σ_{32} (74). This is a field of continuing active research in which MS promises to play its part as a synergistic biophysical method. With the advent of cold spray and nanospray sources it is also possible that further mechanistic details may be gleaned from observation of intact chaperone–substrate complexes (75).

Here we have demonstrated just two examples of synergy between the more traditional biophysical methods and modern MS. A combination of several techniques is clearly a powerful and indeed essential approach to obtain a complete understanding of biological systems. MS can provide information that is complementary and confirmatory of other methods, but more importantly it can also provide unique structural information when combined, for instance, with HDX. As technology continually improves both in terms of ionization and mass detection, MS is poised to take its place alongside other methodologies as a full fledged method for studying protein structure and dynamics.

REFERENCES

1. Dobson C. M., Evans P. A., Radford S. E. (1994). Understanding how proteins fold: the lysozyme story so far. *Trends Biochem. Sci.* **19**: 31–37.
2. Radford S. E., Dobson C. M. (1995). Insights into protein folding using physical techniques: studies of lysozyme and α-lactalbumin. *Philos. Trans. R. Soc. London B* **348**: 17–25.

3. Blake C. C. F., Koenig D. F., Mair G. A., North A. C. T., Phillips D. C., Sarma V. R. (1965). Structure of hen egg white lysozyme. A three dimensional Fourier synthesis at 2 Å resolution. *Nature* **206**: 757–761.
4. Imoto T., Johnson L. N., North A. C. T., Philips D. C., Rupley J. A. (1972). Vertebrate lysozymes. In: *The Enzymes*, ed. P. D. Boyer, pp. 665–864. New York: Academic Press.
5. Blake C. C. F., Johnson L. N., Mair G. A., North A. C. T., Phillips D. C., Sarma V. R. (1967). Crystallographic studies on the activity of hen egg white lysozyme. *Proc. R. Soc. London B* **167**: 378–388.
6. Saxena V. P., Wetlaufer D. B. (1970). Formation of three-dimensional structure in proteins. I. Rapid nonenzymic reactivation of reduced lysozyme. *Biochemistry* **9**: 5015–5023.
7. Anderson W. L., Wetlaufer D. B. (1976). The folding pathway of reduced lysozyme. *J. Biol. Chem.* **251**: 3147–3153.
8. Perraudin J.-P., Torchia T. E., Wetlaufer D. B. (1983). Multiple parameter kinetic studies of the oxidative folding of reduced lysozyme. *J. Biol. Chem.* **258**: 11834–11839.
9. Acharya A. S., Taniuchi H. (1976). A study of renaturation of reduced hen egg white lysozyme. *J. Biol. Chem.* **251**: 6934–6946.
10. Roux P., Ruoppolo M., Chaffotte A.-F., Goldberg M. E. (1999). Comparison of the kinetics of S—S bond, secondary structure, and active site formation during refolding of reduced denatured hen egg white lysozyme. *Protein Sci.* **8**: 2751–2760.
11. van den Berg P., Chung E. W., Robinson C. V., Mateo P. L., Dobson C. M. (1999). The oxidative refolding of hen lysozyme and its catalysis by protein disulfide isomerase. *EMBO J.* **18**: 4794–4803.
12. Radford S. E., Dobson C. M., Evans P. A. (1992). The folding of hen lysozyme involves partially structured intermediates and multiple pathways. *Nature* **358**: 302–307.
13. Itzhaki L. S., Evans P. A., Dobson C. M., Radford S. E. (1994). Tertiary interactions in the folding pathway of hen lysozyme: kinetic studies using fluorescent probes. *Biochemistry* **33**: 5212–5220.
14. Ptitsyn O. B., Pain R. H., Semisotnov G. V., Zerovnik E., Razgulyaev O. I. (1990). Evidence for a molten globule state as a general intermediate in protein folding. *FEBS Lett.* **262**: 20–24.
15. Clark P. L., Liu Z.-P., Zhang J., Gierasch L. M. (1996). Intrinsic tryptophans of CRABP I as probes of structure and folding. *Protein Sci.* **5**: 1108–1117.
16. Clark P. L., Weston B. F., Gierasch L. M. (1998). Probing the folding pathway of a β-clam protein with single-tryptophan constructs. *Folding and Design* **3**: 401–412.
17. Chen L., Wildegger G., Kiefhaber T., Hodgson K. O., Doniach S. (1998). Kinetics of lysozyme refolding: structural characterization of a non-specifically collapsed state using time-resolved X-ray scattering. *J. Mol. Biol.* **276**: 225–237.
18. Chen L., Hodgson K. O., Doniach S. (1996). A lysozyme folding intermediate revealed by solution X-ray scattering. *J. Mol. Biol.* **261**: 658–671.
19. Miranker A., Radford S. E., Karplus M., Dobson C. M. (1991). Demonstration by NMR of folding domains in lysozyme. *Nature* **349**: 633–636.

20. Miranker A., Robinson C. V., Radford S. E., Aplin R. T., Dobson C. M. (1993). Detection of transient protein folding populations by mass spectrometry. *Science* **262**: 896–900.

21. Miranker A., Robinson C. V., Radford S. E., Dobson C. M. (1996). Investigation of protein folding by mass spectrometry. *FASEB J.* **10**: 93–101.

22. Hooke S. D., Eyles S. J., Miranker A., Radford S. E., Robinson C. V., Dobson C. M. (1995). Cooperative elements in protein folding monitored by electrospray ionisation mass spectrometry. *J. Am. Chem. Soc.* **117**: 7548–7549.

23. Eyles S. J., Radford S. E., Robinson C. V., Dobson C. M. (1994). Kinetic consequences of the removal of a disulfide bridge on the folding of hen lysozyme. *Biochemistry* **33**: 13038–13048.

24. Matagne A., Chung E. W., Ball L. J., Radford S. E., Robinson C. V., Dobson C. M. (1998). The origin of the α-domain intermediate in the folding of hen lysozyme. *J. Mol. Biol.* **277**: 977–1005.

25. Coyle J. E., Texter F. L., Ashcroft A. E., Masselos D., Robinson C. V., Radford S. E. (1999). GroEL accelerates the refolding of hen lysozyme without changing its folding mechanism. *Nature Struct. Biol.* **6**: 683–690.

26. Pepys M. B., Hawkins P. N., Booth D. R., Vigushin D. M., Tennent G. A., Soutar A. K., Totty N., Nguyen O., Blake C. C. F., Terry C. J., Feest T. G., Zalin A. M., Hsuan J. J. (1993). Human lysozyme gene mutations cause hereditary systemic amyloidosis. *Nature* **362**: 553–557.

27. Booth D. R., Sunde M., Bellotti V., Robinson C. V., Hutchinson W. L., Fraser P. E., Hawkins P. N., Dobson C. M., Radford S. E., Blake C. C. F., Pepys M. B. (1997). Instability, unfolding and aggregation of human lysozyme variants underlying amyloid fibrillogenesis. *Nature* **385**: 787–793.

28. Canet D., Sunde M., Last A. M., Miranker A., Spencer A., Robinson C. V., Dobson C. M. (1999). Mechanistic studies of the folding of human lysozyme and the origin of amyloidogenic behavior in its disease-related variants. *Biochemistry* **38**: 6419–6427.

29. Koshland D. E. (1953). Stereochemistry and mechanism of enzymatic reactions. *Biol. Rev.* **28**: 416–436.

30. Phillips D. C. (1967). The hen egg white lysozyme molecule. *Proc. Natl. Acad. Sci. U.S.A.* **57**: 484–495.

31. Eshdat Y., McKelvy J. F., Sharon N. (1973). Identification of aspartic acid 52 as point of attachment of an affinity label in hen egg white lysozyme. *J. Biol. Chem.* **248**: 5892–5898.

32. Lin T. Y., Koshland D. E. (1969). Carboxyl group modification and activity of lysozyme. *J. Biol. Chem.* **244**: 505–508.

33. Moult J., Eshdat Y., Sharon N. (1973). Identification by X-ray crystallography of site of attachment of an affinity label to hen egg white lysozyme. *J. Mol. Biol.* **75**: 1–4.

34. Parsons S. M., Raftery M. A. (1969). Identification of aspartic acid residue 52 as being critical to lysozyme activity. *Biochemistry* **8**: 4199–4205.

35. Holler E., Rupley J. A., Hess G. P. (1975). Productive and unproductive lysozyme–chitosaccharide complexes: kinetic investigations. *Biochemistry* **14**: 2377–2385.

36. Fink A. L., Homer R., Weber J. P. (1980). Resolution of the individual steps in the reaction of lysozyme with the trimer and hexamer of *N*-acetyl-D-glucosamine at subzero temperatures. *Biochemistry* **19**: 811–820.

37. Hadfield A. T., Harvey D. J., Archer D. B., Mackenzie D. A., Jeenes D. J., Radford S. E., Lowe G., Dobson C. M., Johnson L. N. (1994). Crystal structure of the mutant D52S hen egg white lysozyme with an oligosaccharide product. *J. Mol. Biol.* **243**: 856–872.
38. Lumb K. J., Aplin R. T., Radford S. E., Archer D. B., Jeenes D. J., Lambert N., Mackenzie D. A., Dobson C. M., Lowe G. (1992). A study of D52S hen lysozyme–GlcNAc oligosaccharide complexes by NMR-spectroscopy and electrospray mass spectrometry. *FEBS Lett.* **296**: 153–157.
39. Ganem B., Li Y.-T., Henion J. D. (1991). Observation of noncovalent enzyme–substrate and enzyme–product complexes by ion spray mass spectrometry. *J. Am. Chem. Soc.* **113**: 7818–7819.
40. Vocadlo D. J., Davies G. J., Laine R., Withers S. G. (2001). Catalysis by hen egg-white lysozyme proceeds via a covalent intermediate. *Nature* **412**: 835–838.
41. Houry W. A., Frishman D., Eckerskorn C., Lottspeich F., Hartl F. U. (1999). Identification of in vivo substrates of the chaperonin GroEL. *Nature* **402**: 147–154.
42. Hendrix R. W. (1979). Purification and properties of GroE, a host protein involved in bacteriophage assembly. *J. Mol. Biol.* **129**: 375–392.
43. Hohn T., Hohn B., Engel A., Wurtz M., Smith P. R. (1979). Isolation and characterization of the host protein GroE involved in bacteriophage λ assembly. *J. Mol. Biol.* **129**: 359–373.
44. Saibil H., Dong Z., Wood S., Aufdermauer A. (1991). Binding of chaperonins. *Nature* **353**: 25–26.
45. Langer T., Pfeifer G., Martin J., Baumeister W., Hartl F. U. (1992). Chaperonin-mediated protein folding—GroES binds to one end of the GroEL cylinder, which accommodates the protein substrate within its central cavity. *EMBO J.* **11**: 4757–4765.
46. Azem A., Kessel M., Goloubinoff P. (1994). Characterization of a functional $GroEL_{14}(GroES_7)_2$ chaperonin hetero-oligomer. *Science* **265**: 653–656.
47. Schmidt M., Rutkat K., Rachel R., Pfeifer G., Jaenicke R., Viitanen P., Lorimer G., Buchner J. (1994). Symmetrical complexes of GroE chaperonins as part of the functional cycle. *Science* **265**: 656–659.
48. Braig K., Simon M., Furuya F., Hainfeld J. F., Horwich A. L. (1993). A polypeptide bound by the chaperonin GroEL is localized within a central cavity. *Proc. Natl. Acad. Sci. U.S.A.* **90**: 3978–3982.
49. Chen S., Roseman A. M., Hunter A. S., Wood S. P., Burston S. G., Ranson N. A., Clarke A. R., Saibil H. R. (1994). Location of a folding protein and shape changes in GroEL–GroES complexes imaged by cryoelectron microscopy. *Nature* **371**: 261–264.
50. Braig K., Otwinowski Z., Hegde R., Boisvert D. C., Joachimiak A., Horwich A. L., Sigler P. B. (1994). The crystal structure of the bacterial chaperonin GroEL at 2.8 Å. *Nature* **371**: 578–586.
51. Fenton W. A., Kashi Y., Furtak K., Horwich A. L. (1994). Residues in chaperonin GroEL required for polypeptide binding and release. *Nature* **371**: 614–619.
52. Xu Z. H., Horwich A. L., Sigler P. B. (1997). The crystal structure of the asymmetric GroEL–GroES–(ADP) (7) chaperonin complex. *Nature* **388**: 741–750.
53. Sigler P. B., Xu Z. H., Rye H. S., Burston S. G., Fenton W. A., Horwich A. L. (1998). Structure and function in GroEL-mediated protein folding. *Annu. Rev. Biochem.* **67**: 581–608.

54. Rye H. S., Roseman A. M., Chen S. X., Furtak K., Fenton W. A., Saibil H. R., Horwich A. L. (1999). GroEL–GroES cycling: ATP and nonnative polypeptide direct alternation of folding-active rings. *Cell* **97**: 325–338.

55. Rye H. S. (2001). Application of fluorescence resonance energy transfer to the GroEL–GroES chaperonin reaction. *Methods* **24**: 278–288.

56. Lakowicz J. R. (1999). *Principles of Fluorescence Spectroscopy*. New York: Kluwer Academic/Plenum.

57. Landry S. J., Gierasch L. M. (1991). The chaperonin GroEL binds a polypeptide in an alpha-helical conformation. *Biochemistry* **30**: 7359–7362.

58. Wang Z. L., Feng H. P., Landry S. J., Maxwell J., Gierasch L. M. (1999). Basis of substrate binding by the chaperonin GroEL. *Biochemistry* **38**: 12537–12546.

59. Robinson C. V., Groβ M., Eyles S. J., Ewbank J. J., Mayhew M., Hartl F. U., Dobson C. M., Radford S. E. (1994). Conformation of GroEL-bound α-lactalbumin probed by mass spectrometry. *Nature* **372**: 646–651.

60. Robinson C. V., Groβ M., Radford S. E. (1998). Probing conformations of GroEL-bound substrate proteins by mass spectrometry. *Methods Enzymol.* **290**: 296–313.

61. Groβ M., Robinson C. V., Mayhew M., Hartl F. U., Radford S. E. (1996). Significant hydrogen exchange protection in GroEL-bound DHFR is maintained during iterative rounds of substrate cycling. *Protein Sci.* **5**: 2506–2513.

62. Goldberg M. S., Zhang J., Sondek S., Matthews C. R., Fox R. O., Horwich A. L. (1997). Native-like structure of a protein-folding intermediate bound to the chaperonin GroEL. *Proc. Natl. Acad. Sci. U.S.A.* **94**: 1080–1085.

63. Gervasoni P., Staudenmann W., James P., Gehrig P., Plückthun A. (1996). Beta-lactamase binds to GroEL in a conformation highly protected against hydrogen/deuterium exchange. *Proc. Natl. Acad. Sci. U.S.A.* **93**: 12189–12194.

64. Gervasoni P., Staudenmann W., James P., Plückthun A. (1998). Identification of the binding surface on beta-lactamase for GroEL by limited proteolysis and MALDI mass spectrometry. *Biochemistry* **37**: 11660–11669.

65. Gervasoni P., Gehrig P., Plückthun A. (1998). Two conformational states of beta-lactamase bound to GroEL: A biophysical characterization. *J. Mol. Biol.* **275**: 663–675.

66. Chen J. W., Walter S., Horwich A. L., Smith D. L. (2001). Folding of malate dehydrogenase inside the GroEL–GroES cavity. *Nature Struct. Biol.* **8**: 721–728.

67. Chen J. W., Smith D. L. (2000). Unfolding and disassembly of the chaperonin GroEL occurs via a tetradecameric intermediate with a folded equatorial domain. *Biochemistry* **39**: 4250–4258.

68. Weissman J. S. (2001). The ins and outs of GroEL-mediated protein folding. *Molecular Cell* **8**: 730–732.

69. Wen Q., Smith J. B., Smith D. L., Edmonds C. G. (1992). Mass spectrometric analysis of the structure of gamma-II bovine lens crystallin. *Exp. Eye Res.* **54**: 23–32.

70. Smith D. L., Liu Y. (1993). High order structural features of lens α-crystallin by mass spectrometry. *Invest. Ophthalmol. Vis. Sci.* **34**: 1340–1340.

71. Smith J. B., Liu Y. Q., Smith D. L. (1996). Identification of possible regions of chaperone activity in lens alpha-crystallin. *Exp. Eye Res.* **63**: 125–127.

72. Smith J. B., Hasan A., Yu J., Smith D. L. (2000). Determination of the conformation of human alpha-crystallin by hydrogen–deuterium exchange/mass spectrometry. *Invest. Ophthalmol. Vis. Sci.* **41**: S582–S582.

73. Hasan A., Smith D. L., Smith J. B. (2002). Alpha-crystallin regions affected by adenosine 5′-triphosphate identified by hydrogen–deuterium exchange. *Biochemistry* **41**: 15876–15882.

74. Rist W., Jorgensen T. J. D., Roepstorff P., Bukau B., Mayer M. P. (2003). Mapping temperature-induced conformational changes in the *Escherichia coli* heat shock transcription factor sigma(32) by amide hydrogen exchange. *J. Biol. Chem.* **278**: 51415–51421.

75. Rostom A. A., Robinson C. V. (1999). Detection of the intact GroEL chaperonin assembly by mass spectrometry. *J. Am. Chem. Soc.* **121**: 4718–4719.

76. Ranson N. A., Farr G. W., Roseman A. M., Gowen B., Fenton W. A., Horwich A. L., Saibil H. R. (2001). ATP-bound states of GroEL captured by cryo-electron microscopy. *Cell* **107**: 869–879.

77. Weissman J. S., Hohl C. M., Kovalenko O., Kashi Y., Chen S. X., Braig K., Saibil H. R., Fenton W. A., Horwich A. L. (1995). Mechanism of GroEL action—productive release of polypeptide from a sequestered position under GroES. *Cell* **83**: 577–587.

9

OTHER BIOPOLYMERS AND SYNTHETIC POLYMERS OF BIOLOGICAL INTEREST

So far, the majority of applications of MS methods to study biomolecular structure and dynamics have focused on proteins. However, oligonucleotides and carbohydrates are also becoming targets of such studies. Unfortunately, many methods developed for proteins cannot be applied directly to study dynamics of glycans and oligonucleotides (e.g., due to much higher rates of labile hydrogen exchange). In this chapter we focus primarily on several experimental methods developed specifically to probe the behavior of nonprotein biopolymers. This discussion will be extended to consider uses of MS to study the behavior of polymers of nonbiological origin.

9.1. DNA

As the Human Genome Project began to get underway, many suggested that MS would overtake traditional sequencing methodologies to become the method of choice for genomics studies (1–4). The ability of MS to distinguish species by mass at high resolution, the rapid acquisition time, and the excellent sensitivity of detection promised the possibility of reading off DNA sequences directly in a matter of seconds, compared with the many hours required for electrophoretic separation. Short oligonucleotides were used as test molecules in the early developmental stages of both MALDI and ESI (5, 6). Although it is not as simple to achieve good data as with peptides or proteins, it appeared that MS would soon

Mass Spectrometry in Biophysics: Conformation and Dynamics of Biomolecules
By Igor A. Kaltashov and Stephen J. Eyles
ISBN 0-471-45602-0

surpass traditional sequencing methods purely because of its speed. However, even fifteen years after the initial experiments this has not proved to be the case, for a variety of reasons.

The techniques for chemical sequencing of DNA are mature and highly sensitive (7, 8). Since Sanger developed the technique for DNA sequencing using dideoxynucleotide chain termination (8), the process has become almost entirely automated. Originally, the method involved mixing template DNA with a short ^{32}P radiolabeled primer sequence then adding DNA polymerase and nucleoside triphosphates (dNTP). Alternatively, $\gamma(^{35}S)$-ATP was employed to provide the radioactive tracer with the remaining three unlabeled NTPs to create a new strand complementary to the target sequence. After allowing time for chain extension, dideoxynucleotides (ddNTPs) were added to terminate extension of the DNA strand. This led to a mixture of DNA oligomers of varying lengths that contained radioactive label. The chains could then be separated by molecular weight using gel electrophoresis and the gel bands visualized by exposure to photographic film (localized radioactive species would give rise to dark spots on the film from which the DNA sequence could be read). More recently, this technique has been almost completely superseded in favor of automated sequencers using colored dye terminators that can be visualized fluorometrically with high sensitivity. Using dideoxynucleotides with four different colored fluorophores, the DNA sequence of up to 1000 bases can be read directly from a single lane of a gel in a matter of several hours (Figure 9.1).

The above methods rely on duplication of the original DNA template and amplifying it using a DNA polymerase enzyme. While the fidelity of these enzymes is extremely high, there is the finite possibility of incorrect incorporation of nucleotides during synthesis that could lead to an incorrect (or at least ambiguous) sequence. On the other hand, mass spectrometric techniques can potentially be applied with high sensitivity to the original DNA template without the need for replication, so this could remove the possibility of errors. One option is to analyze a mixture of products from the Sanger reaction by mass spectrometry and read off the sequence in that manner, since acquisition of mass spectra is much more rapid than electrophoresis. An alternative and more attractive possibility is to measure intact DNA and obtain sequence information directly by MS/MS experiments (Figure 9.2).

Analysis of DNA by MS presents a number of complexities, not the least of which is sample preparation. The phosphodiester backbone of the molecule readily forms adducts with alkali and alkaline earth metal cations that remain intact in the gas phase, potentially leading to very broad peaks in the mass spectrum. This can be circumvented by buffer exchange into volatile ammonium salts that displace the metal cations and leave only the protonated (due to loss of neutral NH_3) species upon evaporation. A recent study showed that use of biotinylated dideoxynucleotides (9) served a dual purpose of allowing purification and desalting by immobilization of sequencing products on streptavidin, but also of removing incorrectly terminated products (those without ddNTP termination) (Figure 9.3).

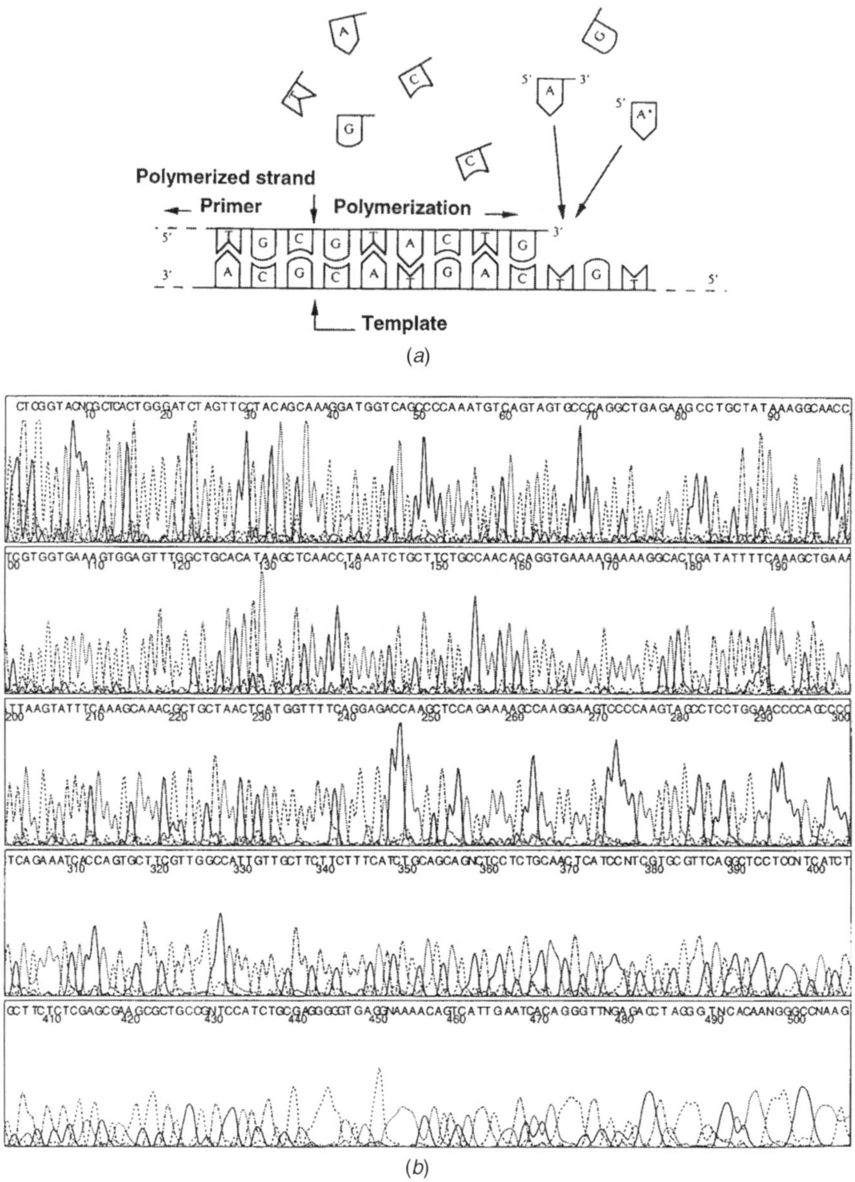

FIGURE 9.1. DNA sequencing by Sanger methods. (**A**) A short oligonucleotide primer is annealed to the template DNA strand and then extended by polymerase chain reaction in the presence of dNTPs and a low concentration of fluorescently labeled ddNTPs (different fluorophores for each of the four nucleotides). (**B**) At the end of the reaction all possible lengths of polymerized strand are randomly produced, each terminated with a fluorescent dideoxynucleotide. These are separated by electrophoresis and the sequence can be read with high fidelity by scanning the fluorescence signature of the gel. Reproduced with permission from (106). © 2000 The American Physical Society.

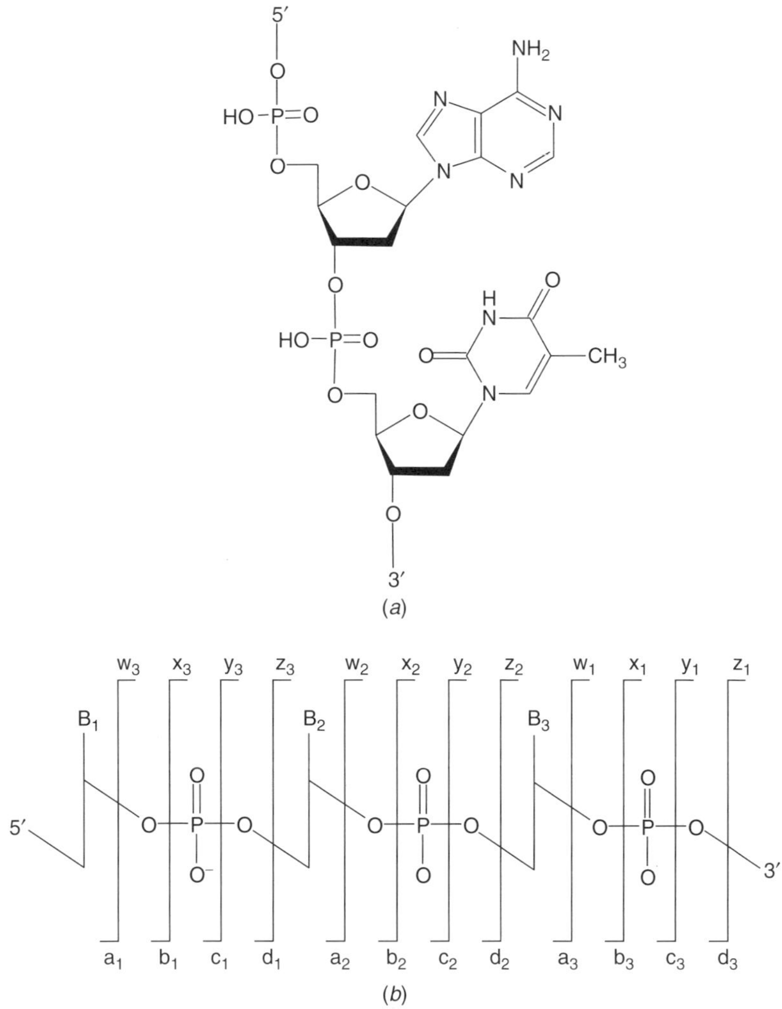

FIGURE 9.2. (**A**) Structure of nucleotides showing the ribose (RNA) or deoxyribose (DNA) backbone with attached purine and pyrimidine nucleobases. (**B**) The MS/MS fragmentation nomenclature for nucleotides of McLuckey et al. (107).

Another problem is the ease of fragmentation along the phosphodiester backbone that can lead to loss of signal from the intact DNA ion (10, 11). This has the advantage, of course, of one being able to read off limited sequence information (e.g., see Figure 3.10), but prevention of a complete sequence read is possible if there is insufficient ion intensity from intact species for subsequent tandem MS measurements. Other more disastrous fragmentation pathways (at least in terms of sequencing) involve removal of the base by 1,2-trans elimination, which renders any fragmentation information meaningless. This may be circumvented by

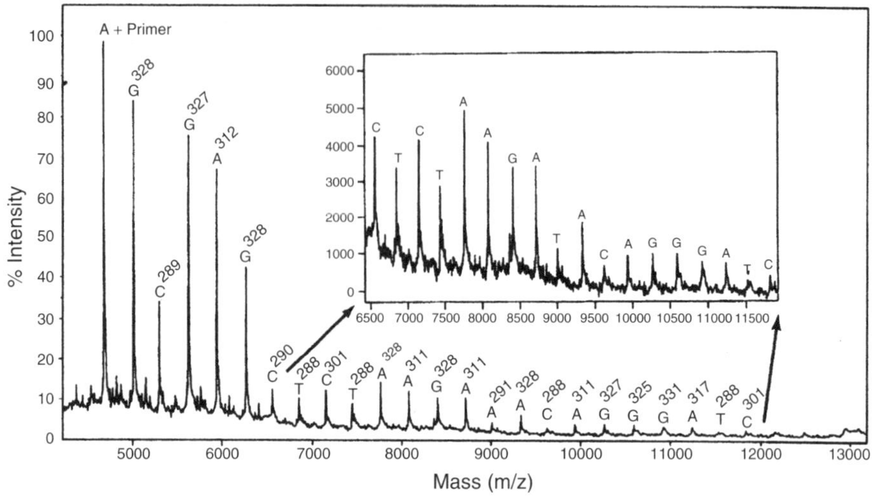

FIGURE 9.3. One of the major causes of compromised sequence data from nucleotides by MALDI TOF is formation of metal cation adducts. Biotinylation of oligonucleotides allows for rapid and efficient cleanup of PCR sequencing products and a good sequence read out to 12,000 m/z. Reproduced with permission from (9). © 2001 Oxford University Press.

using a variety of chemical modifications of the DNA molecule prior to analysis in an attempt to reduce fragmentation effects (12–15).

Given these limitations, however, there have been significant advances in analysis of DNA by MS (16). Positive ion spectra have been obtained by MALDI of DNA molecules up to several hundred bases. It may be possible to minimize fragmentation caused by the UV laser irradiation and improve these mass limits by instead employing MALDI with lasers in the visible or IR wavelength range using ice as the matrix (17–19). This produces a much more gentle ionization that may assist transportation of intact DNA into the gas phase. A more significant limitation of MALDI TOF mass spectrometry for sequencing, however, is probably the mass discrimination effects and loss of resolution at higher m/z. Without a means to obtain a linear response through the high mass range and improved resolution, it is difficult to distinguish the identity of each base. The latter problem has been investigated and determined to be primarily a problem of divergence of the ion beam for high mass ions, something that possibly may be solved by improvements in instrument design (20). However, currently it is only practically possible to obtain unambiguous sequence information for relatively short oligonucleotides where high resolution is readily achievable.

Negative ion ESI has also shown promise for the analysis of oligonucleotides. The multiple charging of the ESI process brings ions into the mass spectral region where high resolution can readily be obtained. However, each species present produces many different charge states that can lead to significant complication of the spectra. Since a single species can give rise to a broad distribution of peaks, fragmentation either in source or in tandem MS experiments would potentially give rise to a large number of peaks from various charge state fragments that may be difficult to interpret. FT ICR shows some significant promise in this respect due to its high resolving power that enables direct measurement of the charge state of fragment ions. In addition, there is strong evidence that with carefully controlled source conditions large DNA molecules can be effectively transferred intact into the gas phase for subsequent mass spectral analysis. Cheng et al. (21) were able to observe intact plasmid* DNA using ESI FT ICR measurement, and to obtain mass measurement of a 1.95 MDa plasmid molecule within 0.2% mass accuracy by employing charge reduction techniques using gas phase reaction with acetic acid (Figure 9.4). Although this method may not be readily applicable to subsequent DNA sequencing, it demonstrates that MS is extremely powerful for the analysis of polynucleotides, and further technological developments may yet bring MS to the forefront of DNA analysis.

Although MS may not be able to compete with the more traditional gel sequencing of DNA, it may excel at other applications when it comes to more high throughput screening. One of these is the detection of single nucleotide polymorphisms in genomes (22–24). Although the bulk of the DNA sequence of an organism is highly similar from one individual to another, there are cases where disease states may be caused by a single mutation at the genetic level. Higher organisms carry chromosomes in pairs within the cellular nuclei that duplicate the genetic information. While it is expected that the DNA sequence of the two paired chromosomes should be mostly identical, mutations occur at a rate of about one base per thousand. In many instances, due to parentage, a single nucleotide will differ between the two allelic sequences in an individual (known as heterozygous). If a single nucleotide is variable to a significant extent within a population, it is known as a single nucleotide polymorphism (SNP).† If this region of the DNA sequence is a coding region for a protein, then it will possibly lead to an amino acid mutation that may alter the structure and/or compromise function of the gene product, potentially leading to disease states such as anemia, diabetes, or amyloidosis. Thus, the identification and analysis of SNPs has become a highly valued extension to the Human Genome Project, and it is here that MS may excel due to the speed of analysis and ease of automation (25).

Various approaches have been developed to investigate SNPs by mass spectrometric methods. These involve either PCR amplifications of small stretches of DNA that contain the polymorphism of interest, using Sanger type sequencing reactions (see above) to determine the sequence in the desired region (Figure 9.5), or hybridization of the target sequence to peptide nucleic acids (PNAs).‡ The latter involves annealing a selection of PNA sequences to the DNA region of

*A plasmid is a circularized DNA molecule commonly used in molecular biology for cloning and protein expression in bacterial cells.

†Generally, the position is found to vary between just two of the four possible nucleotides.

‡These synthetically produced DNA analogues contain a peptide backbone to which the purine and pyrimidine bases are attached instead of the phosphodiester backbone of oligonucleotides. The former is less prone to gas phase fragmentation, making it more tractable for MS purposes.

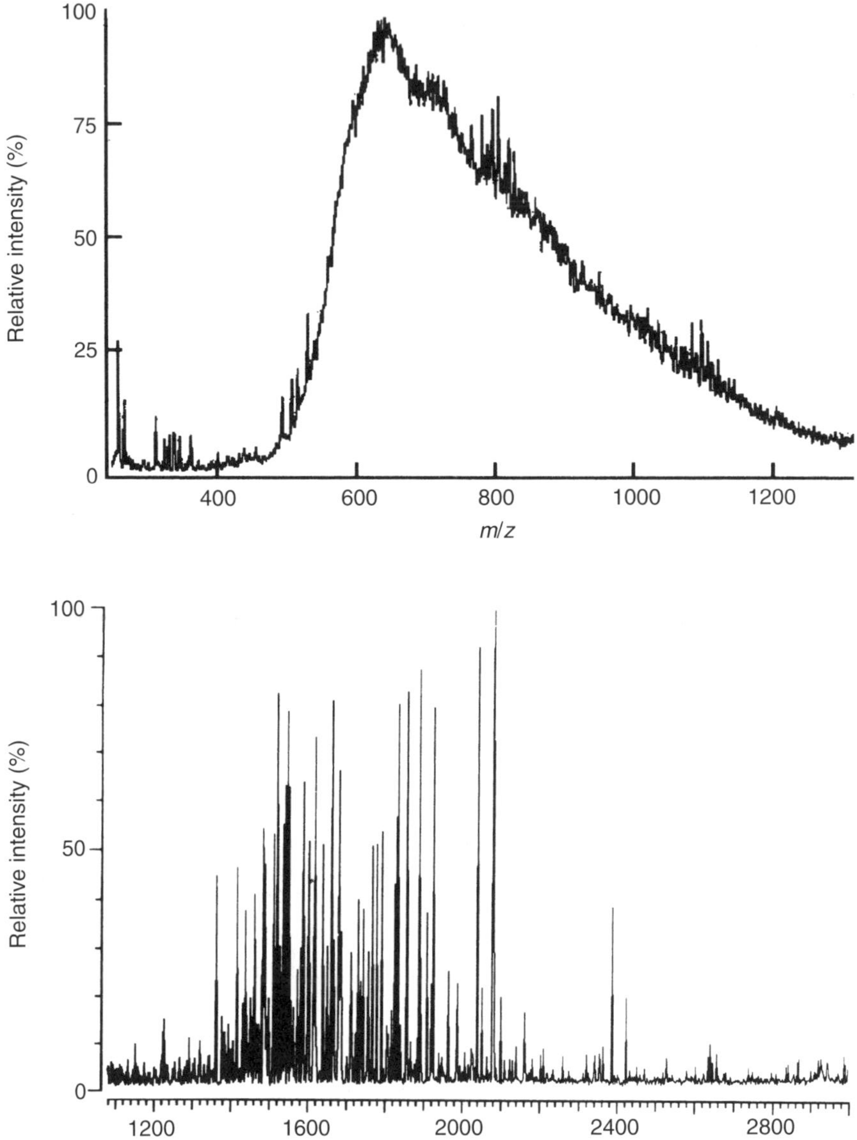

FIGURE 9.4. Mass spectrometry of intact plasmid DNA. (**A**) Negative ion ESI MS acquired using a standard quadrupole mass spectrometer shows a broad featureless hump due to overlapping charge states. (**B**) Using sequential charge reduction techniques by gas phase reactions with acetic acid in the FT ICR cell, and ion population trimming, individual charge states can be resolved and the mass of the intact plasmid measured to within 0.2% of the calculated mass. Reproduced with permission from (21). © 1996 Oxford University Press.

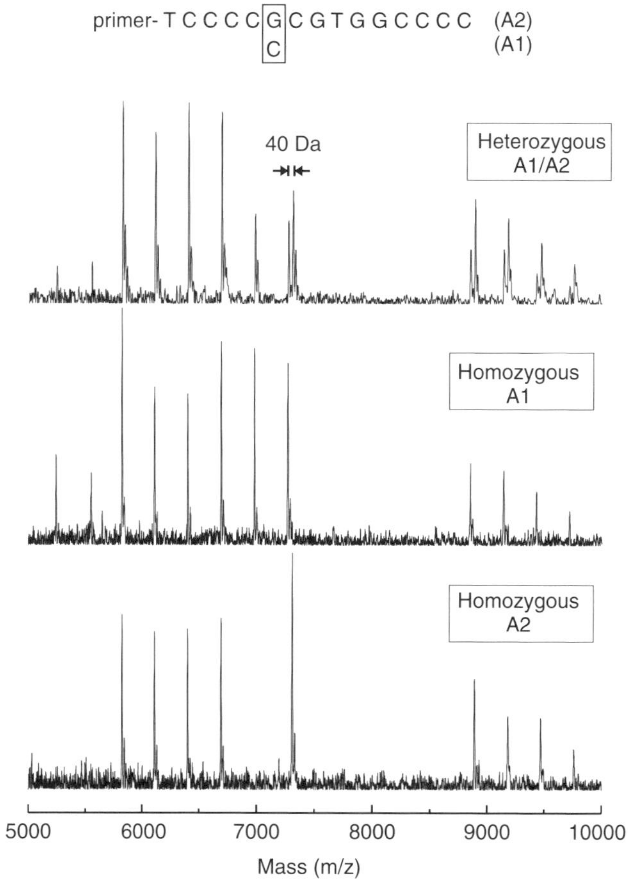

FIGURE 9.5. Identification of single nucleotide polymorphisms by MALDI TOF. Ladder sequence information can be read from the mass spectrum with split peaks arising from the polymorphic site. Reproduced with permission from (108). © 1998 Nature Publishing Group.

interest (26, 27). Only those sequences that match exactly will remain attached upon stringent washing, and the correctly hybridized PNA sequences can then be analyzed by MALDI TOF to derive the sequence of the template DNA strand. This has considerable potential for further usage with the relatively recent development of genetic chip technology, allowing a large number of PNAs to be arrayed for rapid screening purposes. For an excellent review of MALDI TOF methods to analyze DNA sequences see Jurinke et al. (28).

Other than sequencing there has been substantially less interest, at least in terms of MS, in structural characterization of DNA molecules. Unlike proteins, or indeed RNA (see below), DNA molecules can only adopt relatively few favored conformations. The majority of naturally occurring DNA contain the double stranded dimers arranged in the classic Watson–Crick double helix. Additionally, most chromosomal DNA is further supercoiled with the aid of histone proteins to form compact structures that protect the genetic information maintained in them from being compromised by nucleases. There are some instances, however, where different quaternary structure is adopted by DNA, for instance, the G-quartet motif that involves bringing together G-rich sequences to form a four-stranded structure, which is thought to be an important structural feature in telomeric repeat sequences at the end of chromosomes (29). Such structures have been observed by MS as tetramers of short oligonucleotide sequences using electrospray ionization (30). More recently, cold spray ionization (31, 32; see also Chapter 3), a modification of ESI that employs low temperature spray conditions (−80 °C to 10 °C), has demonstrated the existence of G-quartet structures in 2′-deoxyguanosine stabilized by the presence of Na^+ cations (Figure 9.6). This technique has also allowed the observation of unstable noncovalent complexes

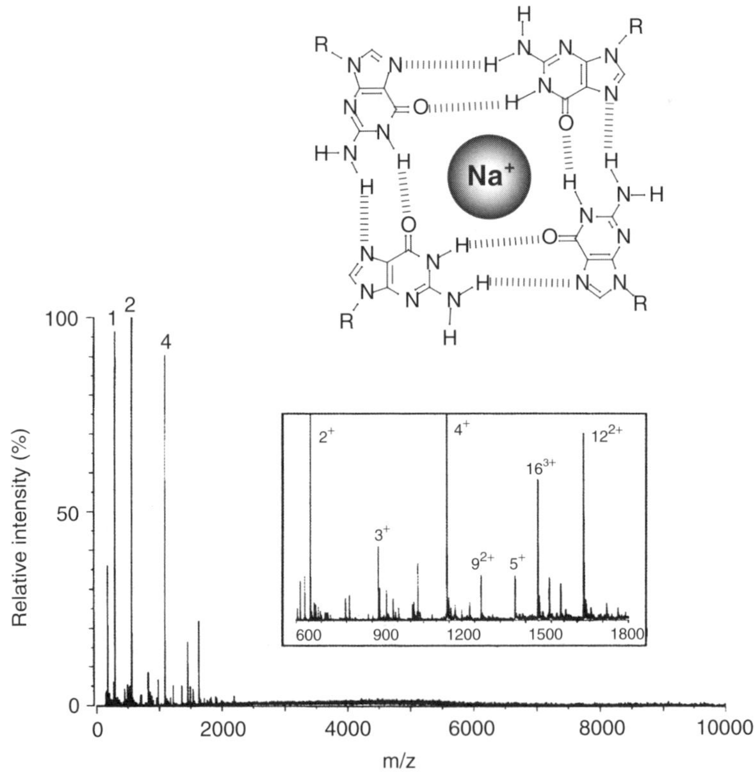

FIGURE 9.6. Noncovalent G-quartets in the presence of Na^+ ions observed by cold spray ionization MS. Monomer and dimer species are also observed, with minor contributions of trimer, dodecamer, and hexadecamer (*inset*). Reproduced with permission from (32). © 2003 Royal Society of Chemistry.

and duplex oligonucleotides (33) and promises to be of great utility for the detection of species that are unstable to standard ionization methods (34).

There have also been reported several studies of HDX measurements in DNA using NMR (35–37) and changes in optical absorbance (38, 39). Hydrogen bonding between base pairs gives rise to protection against exchange of the amino and imino proteins on the bases. As with proteins, this protection can be measured using MS and may be of use to detect regions that vary from the standard double helix structure. Several DNA–protein interactions are thought to involve a significant bending of the DNA strand, and polymerase enzymes must break apart the structure in order to replicate the strands. The techniques developed for measurement of protein fluctuations by HDX MS may also be applicable to the study of dynamics in DNA. Certainly in the gas phase, there is evidence that labile protons are more protected in duplex structures than in the absence of hydrogen bonding (40).

9.2. RNA

The first part of the genetic code is generally considered to be DNA, from which messenger RNA is transcribed and subsequently translated into gene products—proteins. This is certainly true for higher organisms, but many viruses maintain all of their genetic code in the form of RNA. Upon injection of this RNA molecule into a host cell, the viral ribonucleotide sequence is then treated like messenger RNA and translated into proteins that allow replication of the virus. Perhaps the most infamous of these RNA-only viruses are the class of retroviruses, including the human immunodeficiency virus (HIV), that carries not only RNA but also copies of the enzyme reverse transcriptase. Upon host cell infection, the viral RNA is transcribed back into DNA and incorporated into the host genome, thus allowing the cellular machinery to be hijacked and produce large numbers of daughter viruses.

More recently, it has been discovered that RNA is not merely a passive carrier of genetic information. The ribosome itself, the multiprotein complex machinery that produces proteins, also contains a number of RNA subunits that are functionally important. Additionally, several catalytic RNA sequences, known as ribozymes, have been identified. These are responsible for such tasks as splicing RNA sections together to modify the resulting gene product, and also for modifying tRNA. Still other small RNA molecules may be responsible for regulating gene expression and a variety of other intracellular processes. Consequently, there has been much recent interest in identifying the structural and functional nature of these RNA molecules.

It was noticed early in the history of MALDI MS that RNA was more stable against gas phase fragmentation than its DNA counterparts, presumably due to the presence of hydroxyl groups on the glycosidic ring that stabilize the N-linkage to the purine or pyrimidine base (41, 42). MALDI MS of in vitro transcribed RNA molecules up to 150 kDa (Figure 9.7) were demonstrated by Kirpekar et al. (43)

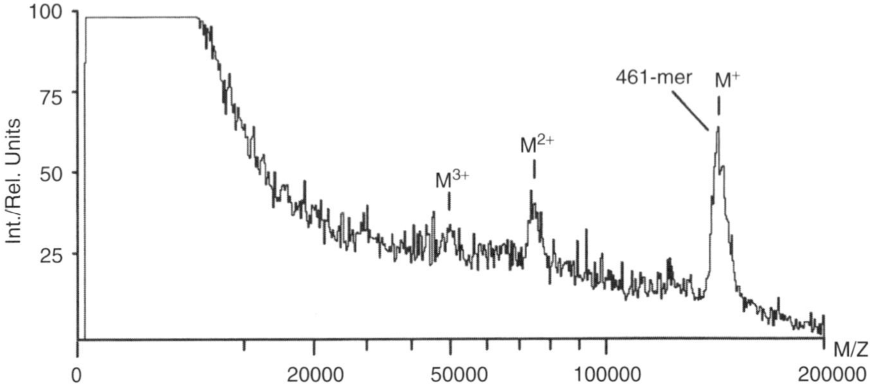

FIGURE 9.7. Positive ion MALDI TOF spectrum of an intact 461 nucleotide RNA molecule (150 kDa) after treatment with phosphatase to remove the terminal phosphate group. This demonstrates the ability of RNA molecules (unlike DNA which readily dissociates) to remain intact in the gas phase. Reproduced with permission from (43). © 1994 Oxford University Press.

in the early 1990s. The observed resolution was significantly higher than for DNA, presumably because of the reduced fragmentation, but other problems arose due to the poor fidelity of RNA polymerase transcription, and inhomogeneities at the termini of the polymer.

The same mass spectrometric techniques used for DNA sequencing (i.e. tandem MS) can, in principle, be equally applied to determine the sequence of RNA molecules. Unfortunately, one complication is caused by the similarity in masses between the nucleosides uridine (306.2 Da) and cytidine (305.2 Da). Thus, a significantly higher mass resolving power is required to analyze longer RNA by standard MS methods. One way around this takes advantage of the differences in kinetics of phosphodiester hydrolysis by exonuclease enzymes* for the two different bases, enabling ambiguities to be resolved by noting the different ion intensities (44). Bovine $5' \rightarrow 3'$ phosphodiesterase hydrolyzes bonds adjacent to cytidine bases significantly more slowly than uridine, so this manifests itself in mass spectra as an increased ion intensity from fragments terminated with the former (Figure 9.8). Thus, sequence information could be obtained from enzymatic fragments of RNA generated by this enzyme. Of course, with high-resolution instrumentation, resolving this small mass difference is quite possible; thus RNA sequencing by tandem MS measurements has since been achieved for small oligonucleotides (45). Additionally, IRMPD favors dissociation of oligonucleotides over peptides and proteins, so this method has proved useful for sequencing in FT ICR MS (46).

*Exonucleases sequentially remove bases from the ends of oligonucleotide polymers by hydrolysis of the phosphodiester bond. Enzymes are known that specifically digest from the 5′ end ($5' \rightarrow 3'$) of the molecule or vice versa ($3' \rightarrow 5'$) (cf. endonucleases, which recognize a specific internal sequence of nucleotides and hydrolyze a single bond at this site).

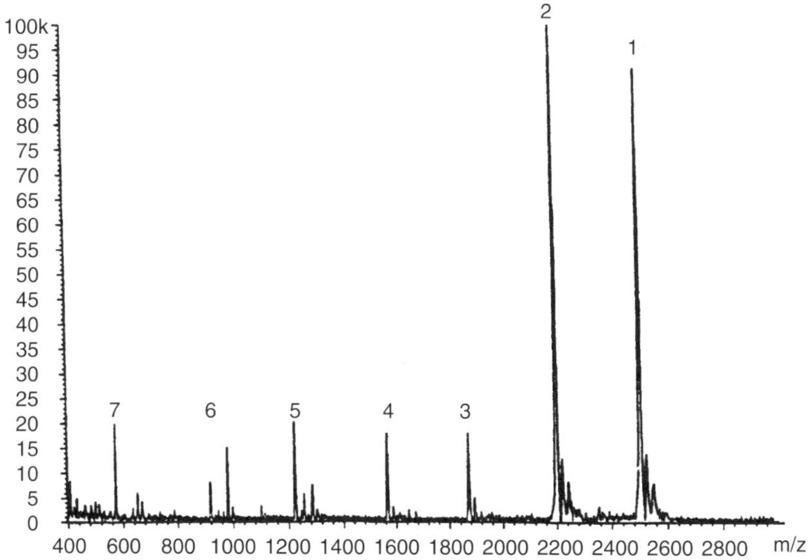

(*a*) Digestion with 5′ → 3′ phosphodiesterase from bovine spleen after 6 minutes incubation

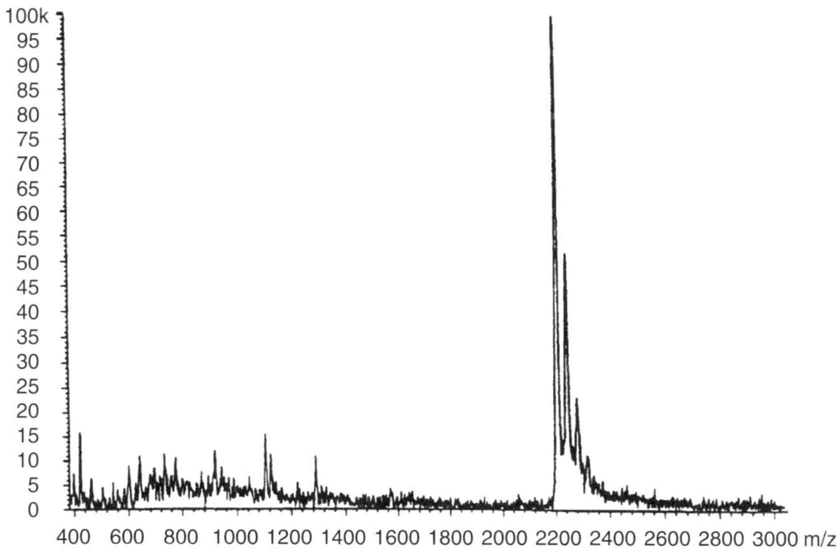

(*b*) Digestion with *RN*ase *CL*3 from *chicken liver* after 10 minutes incubation

FIGURE 9.8. Sequencing of oligoribonucleotides by taking advantage of the differential hydrolysis rates of C versus U. Since these nucleotides differ in mass by only 1 Da, they are difficult to unambiguously assign by MALDI TOF. However, different relative abundances are generated upon hydrolysis by 5′ → 3′ phosphodiesterase, giving rise to more intense peaks from nucleotides containing 5′ terminal cytidines. Reproduced with permission from (44). © 1997 American Chemical Society.

Unlike DNA, described above, which adopts relatively few different structures, a far greater range of structural features are available to RNA molecules. They are known to adopt a rich variety of secondary and tertiary interactions that make them extremely versatile, in spite of being composed of only four bases (cf. 20 different amino acids in proteins). This structural diversity is key to the wide range of functions that RNA is now known to perform. Ribozymes, the analogue of enzymes in the RNA world, must adopt conformations that enable them to specifically perform their catalytic function. Similarly, regulatory sequences must recognize their binding sites with high specificity. A whole class of small RNA molecules have been identified, known as RNA interference (RNAi), that can inhibit translation of other RNA molecules into protein, and still others that bind to chromosomal DNA and alter the genes that are exposed for translation. Consequently, there is a great deal of interest in determining the structural features of these RNA species and, in parallel with the protein folding world, how these molecules attain their active conformations.

The biophysical tools for the study of RNA structure are somewhat less mature than those for studies of proteins. For one thing, up until relatively recently it was much more difficult to obtain pure RNA in significant quantity. Whereas proteins can readily be overexpressed in high yield in bacterial cells, harnessing cells to produce large amounts of RNA is somewhat more complicated. In vitro transcription methods are now available that can produce the desired sequence of RNA in sufficient quantity for subsequent analysis.

As with oligosaccharides, described below, RNA has been the subject of intense analysis by NMR. Unfortunately, unlike proteins, there is not the same chemical diversity in oligonucleotides. The presence of aromatic nucleobases does, however, produce ring current shifts* so that there is reduced degeneracy of resonances in NMR spectra. Nevertheless, the protons are all attached either to the glycosidic backbone or else to the purine or pyrimidine base so there is generally much less dispersion than for a protein ^{1}H NMR spectrum, resonances instead being quite closely grouped in small regions of the spectrum (47–49). An additional problem is that nucleotides are much larger than amino acids, so relatively short RNA sequences can have molecular weights and consequently NMR correlation times reminiscent of large proteins, leading to unfavorable relaxation times and poor quality NMR spectra.

For similar reasons, high-resolution crystal structures have generally been reported only for relatively short RNA sequences, representing only domains of larger molecules. A major problem with RNA crystallography is the difficulty of producing isomorphous heavy atom derivatives for structure refinement (see Chapter 2). On the other hand, the crystal structure of the entire 70S ribosome was recently reported at 5.5 Å resolution (50), and of the separate subunits at still higher resolution (51, 52). However, these are painstaking and time consuming studies; thus other biophysical methods again emerge as key players to investigate structural features of RNA molecules. Once again, as with proteins,

*NMR chemical shifts are influenced by the proximity of aromatic moieties that can either shield or deshield the nuclei and affect their resonant frequencies.

the static structure is informative but still begs the question of how RNA folds into its biologically relevant structure, and what dynamic events are important for its function.

Hydrogen exchange measurements have been employed to investigate structure in RNA, using NMR (53, 54), Raman spectroscopy (55), and $^3H/^1H$ exchange with radioactivity detection (56). Although the glycosidic hydrogens are rapidly exchanging, it is possible to measure protection of the base amino and imino protons that are involved in structure. This gives valuable information about base pairing as opposed to bases that are involved in single stranded regions and/or bulges.

In Chapter 4 we described a technique using hydroxyl radical modification to probe protein structure. This method complements hydrogen exchange studies in proteins, but for the investigation of oligonucleotide structure and folding it is one of the few labeling methods that is really effective. Exposure of folded RNA molecules to synchrotron radiation induces hydrolysis of the solvent-exposed backbone regions, thus cleaving regions that are on the outer surface of the molecule. Subsequent analysis by electrophoresis yields a ladder of fragments punctuated by missed cleavages that correspond to regions of structure protected against hydrolysis. This technique is known as hydroxyl radical footprinting (57, 58). By measuring the degree of hydrolysis at equilibrium, structural changes such as the folding of the *Tetrahymena* ribozyme as a function of Mg^{2+} concentration could be determined (59). A further extension of this technique enabled measurement of the kinetics of RNA folding in response to addition of divalent cations, in a manner analogous to pulse labeling experiments (60). Folding molecules are exposed to synchrotron radiation after a period of refolding, such that protection can be monitored as a function of folding time. In principle, mass spectrometric detection may prove useful to determine the sites of cleavage. This methodology complements other low-resolution techniques to study RNA folding, such as small angle X-ray scattering.

Other chemical modifications can be employed to investigate RNA structure in a similar manner. A variety of reagents are available that act as solvent accessibility probes since they cannot penetrate to modify, for instance, bases that are involved in base-pairing, stacking, or other tertiary interactions. Using high-resolution FT ICR MS, Yu and Fabris demonstrated that these reagents could be used effectively to probe RNA structure and RNA–protein interactions (61). The extent of chemical labeling of the structure was monitored mass spectrometrically. Subsequent digestion with ribonuclease and analysis of the resulting fragments enabled sites of modification to be localized (Figure 9.9). In the case of multiple alkylation sites in a single fragment, these could be further analyzed by tandem MS. These methods can be applied both to the study of structure in RNA molecules and also to interactions of oligonucleotides with RNA binding proteins since the binding event protects a previously exposed surface of the molecule against further chemical modification.

The kinetics of protein–RNA reactions can also be usefully monitored by MS. One example of this is the binding of ricin A to a short stem-loop hairpin segment

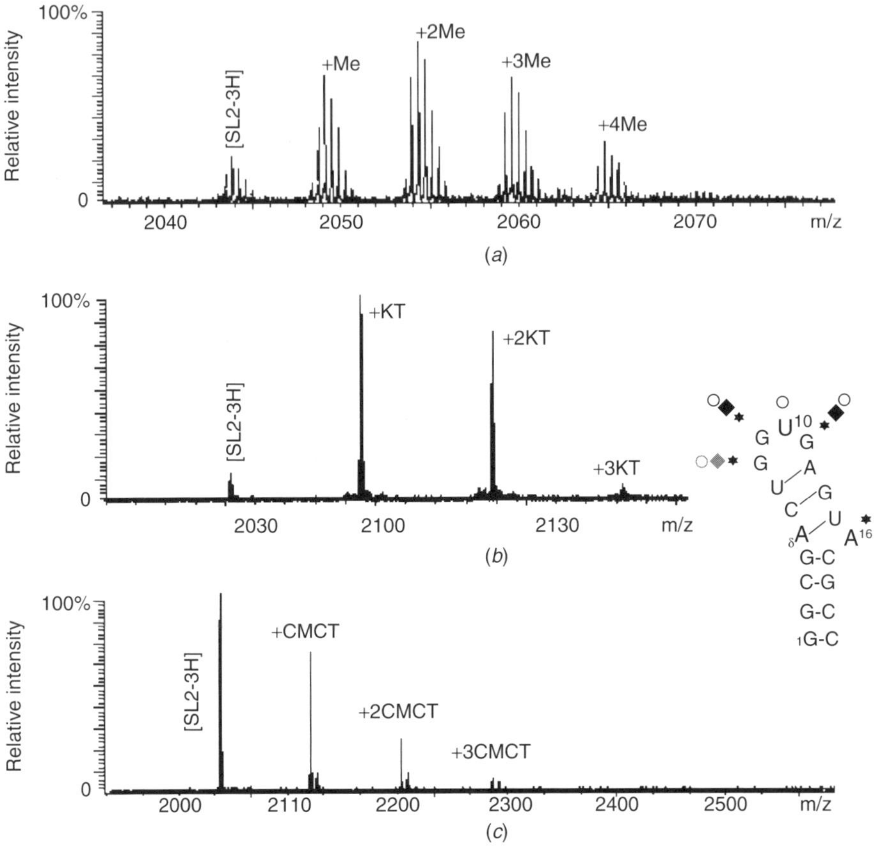

FIGURE 9.9. Structure in an RNA hairpin molecule (SL2 of the HIV-1 packaging signal) deduced from differential chemical modification. Only nucleobases that are not involved in intramolecular hydrogen bonding are available for modification by (a) methylation (Me) by dimethyl sulfate, (b) reaction with kethoxal (KT), or (c) reaction with 1-cyclohexyl-3-(2-morpholinoethyl)-carbodiimide metho-*p*-toluenesulfonate (CMCT). Reproduced with permission from (61). © 2003 Elsevier Science Ltd.

of HIV ψ-RNA. The former is a cytotoxic protein that hydrolyzes a specific N-glycosidic linkage, causing depurination at an adenine base in the sequence. The kinetics of depurination of the RNA molecule were measured by direct infusion ESI MS (62).

Of course, MS is also extremely powerful for measuring the stoichiometries of protein–DNA and protein–RNA interactions. Since binding of these molecules is predominantly an electrostatically driven process (positively charged proteins to negatively charged nucleotides), these interactions will often remain intact in the gas phase, as opposed to the more fragile hydrophobic interactions between proteins. As with the study of protein–protein interactions, there has been a great deal of interest in characterizing protein–DNA, protein–RNA, and indeed complexes of oligonucleotides with small molecules of pharmaceutical interest. MS clearly provides a relatively rapid screening method that can at least somewhat quantitatively assess ligand binding affinities. Some of these techniques are reviewed elsewhere (40, 63). Cross-linking reactions (discussed in Chapter 4) can also be applied to the study of protein–nucleotide interactions in a manner similar to studies of protein structure and protein–protein interactions. Mass spectrometric detection of the cross-linked products provides a convenient method to determine the sites of interaction between nucleotides and their protein interaction partners (64).

9.3. OLIGOSACCHARIDES

Until recently, the importance of carbohydrates has been somewhat overlooked relative to proteins, partly because their analysis is a significantly more complex task. Oligosaccharides can exist alone as polymers, such as cellulose or starch, but are also commonly found attached at various sites on the surface of proteins. In the wake of proteomics, the relatively new field of *glycomics*, the study of protein glycosylation, has developed to understand the functional implications of carbohydrate expression.

Naturally occurring oligosaccharides attached as glycoconjugates to protein molecules generally consist of hexose sugar units (Figure 9.10). These monomers can be linked to each other via any of their hydroxyl groups to form complex linear or branched polymers. Linkages are notated according to the numbering of the carbon atoms in the ring that constitute the cross-link, and the configuration at the anomeric carbon (α or β). For instance, the disaccharide maltose consists two D-glucose molecules linked via an $\alpha(1 \rightarrow 4)$ linkage (Figure 9.10). Additionally, a number of other saccharide units occur naturally, most notably the pentose sugar ribose that forms the backbone of nucleic acids in genetic material.

The majority of proteins that are either secreted or found on the surface of cells become decorated with carbohydrates after translation. The exact purpose of these glycans is not yet fully understood, but they have been implicated in structure, folding, quality control, as signals for protein export, and as recognition sites for binding of other proteins in macromolecular assemblies. While the sites of glycosylation within a protein are generally highly specific, the nature of the carbohydrate is often less well defined. The primary amino acid sequence for a given protein is always the same, but post-translational modifications that add sugars to the surface of protein can lead to some polydispersity in the final molecule. Presumably this results from less tight control of the biosynthetic pathways for oligosaccharides, which may be a design feature allowing individual glycoprotein molecules to present similar but slightly different surface architectures for recognition. The variable nature of these sugars can lead, however, to obvious problems in characterization since standard biochemical techniques of separation such as chromatography or gel electrophoresis result in broad peaks

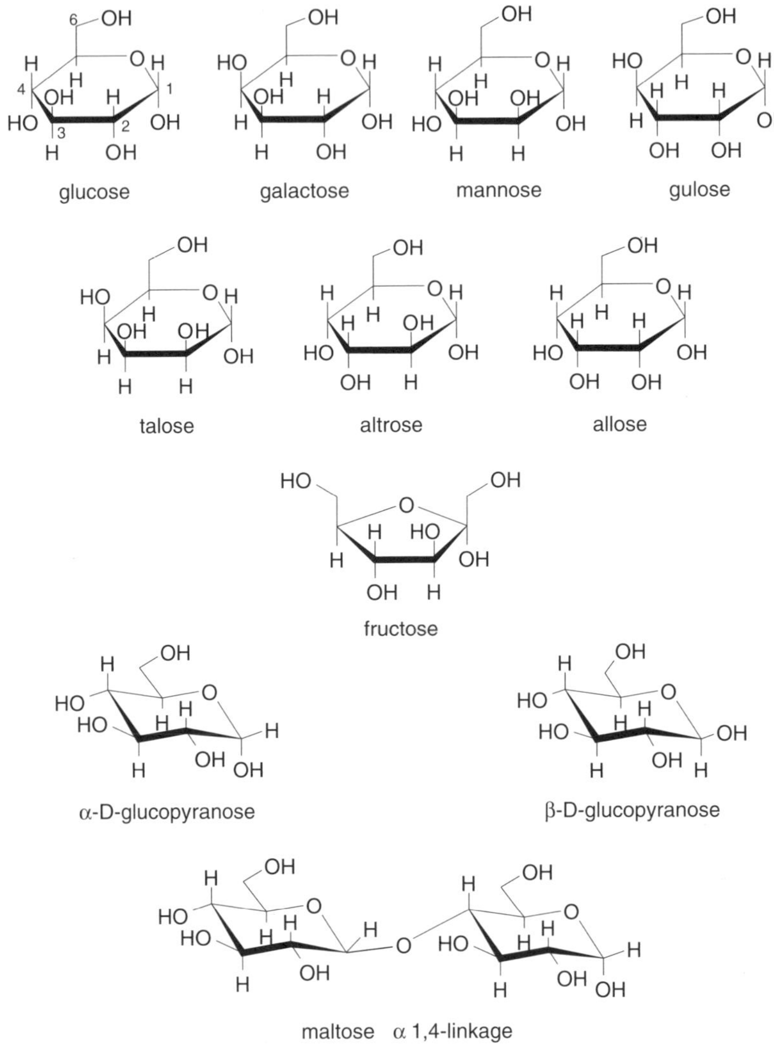

FIGURE 9.10. Structures of common saccharides. Most hexose sugars normally exist in the ribose form although fructose occurs naturally in the pyranose form. The α- and β-forms of glucose differ by their orientation of the anomeric hydroxyl group. Linkages in oligosaccharides are designated by the anomeric configuration and the carbons linked at each monosaccharide unit.

or bands from the inherent variability in molecular weight. Such polydispersity lends the subject of glycosylation to study by MS.

NMR has become a popular method for the study of carbohydrate structure, but unlike proteins there are no aromatic groups to give nuclear shielding in oligosaccharides, so resonance dispersion is generally quite poor. Additionally, whereas proteins are linear chain polymers, sugars often contain significant degrees of branching, making the analysis more complicated still (65). Nevertheless, the combination of NMR methods with X-ray crystallography and molecular modeling has enabled significant structural information to be obtained for highly purified oligosaccharides available in sufficient quantity (66). To obtain detailed structural information about the nature of the sugars and the mode of attachment to proteins, MS has emerged as a key technique (67, 68). Most methodologies involve first cleaving the oligosaccharides from the protein to simplify analysis. This can be achieved by enzymatic deglycosylation using enzymes that may be specific for N-linked or O-linked oligosaccharides.

The methodologies for analysis of proteins and peptides, at least in terms of sequence determination by tandem MS, are now quite mature relative to the analysis of carbohydrate structure. As we have seen in Chapter 3, polypeptide fragmentation tends to follow very well-defined pathways, primarily occurring along the peptide backbone, which conveniently leads to fragments from which the sequence of intact peptide can be derived (69). This is due to the inherently easier fission of peptide bonds by collisional activation relative to that of aliphatic side chain linkages. By contrast, many of the bonds in carbohydrate molecules are chemically similar, leading to much more complex fragmentation pathways. Bond fission commonly occurs not only between saccharides but also across the glycosidic ring. In addition, loss of water occurs quite trivially and multiple rearrangement pathways are available to activated species that render analysis of tandem MS data extremely complex.

Early studies of oligosaccharides by MS used FAB ionization as a means to desorb/ionize these macromolecules. This technique was limited, however, by poor ionization efficiency of neutral and basic sugars, although reasonable signals could be observed of negative ions produced from acidic carbohydrates. Permethylated or peracetylated derivatives were found to produce much stronger signals and fragmentation products from the energetic FAB ionization that could be used to obtain structural information (70–72). However, this method required a highly homogeneous preparation of oligosaccharides since fragmentation of mixtures in this manner would lead to uninterpretable results. The nomenclature for designation of product ions produced by fragmentation of glycoconjugates is analogous to that used for peptide fragmentation and was developed by Domon and Costello (73), as shown in Figure 9.11.

Analysis of glycoconjugates is significantly aided by the use of tandem MS techniques (67, 74). The ability to select a single molecular ion for fragmentation enables complex mixtures to be analyzed without prior separation. Although FAB was initially used in these types of experiments, as with proteins, ESI and MALDI ionization methods produce minimal fragmentation in the initial ionization event so these are a more prudent choice for tandem experiments. By using ESI it is possible to obtain ionized molecules from underivatized oligosaccharides either as protonated, deprotonated, or alkali metal adduct ions. Under low-energy CID conditions, the most common bond cleavages occur between the glycosidic rings, whereas high-energy dissociation can be employed to effect cross-ring

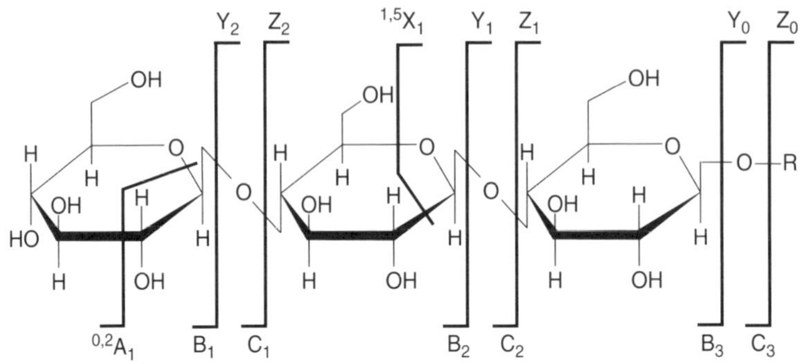

FIGURE 9.11. Fragmentation nomenclature for oligosaccharides of Domon and Costello (73).

cleavages, involving breaking two bonds, to yield further structural information. Differences in branching can be detected, often due to the altered bond energies and steric effects that influence fragmentation pathways. Additionally, fragmentation is strongly affected by the nature of the parent ion: alkali metal cationized species exhibit different fragmentation patterns from their protonated analogues, so a suite of different electrospray conditions can be employed to achieve more detailed data. Yet more information can be gained by using various derivatization techniques. One can also take advantage of the wealth of rearrangement reactions that occur and extrapolate back the original structures that produced the observed product ions. If a parent ion produces a large number of fragments, then it can become difficult to obtain good ion statistics from the detection of product ions using scanning analyzers. Therefore, mass analyzers with high duty cycles such as time of flight and the various trapping analyzers have proved very important to investigate low abundance species. Additionally, hybrid instruments with increased resolving power and mass accuracy for fragment ions will clearly be of great benefit.

Glycopeptides are another area of great interest, especially with the modern demands for high-throughput analysis. The methods described above were developed for the analysis of purified oligosaccharides or mixtures of sugars. A more attractive approach in the realm of proteomics is to be able to derive the structural and sequence information of the sugar moiety attached in situ to its host protein. Thus, from a tryptic digest of a glycosylated protein, one could in theory determine not only the exact site of glycoside attachment on the protein under scrutiny, but also the nature of the oligosaccharide in a single experiment.

MS-based methods to determine sites of glycosylation in short peptides have been available for some time. The enzyme PNGase F converts N-glycosylated Asn residues to Asp upon cleavage of the sugar moiety, enabling the site to be identified by MS/MS (75). Alternatively, enzymatic cleavage of N-linked glycans in ^{18}O-labeled water can serve as an identification strategy due to introduction of an isotopic tag on the peptide upon hydrolysis (76). Another method that shows some promise is neutral loss scanning in cases where a consistent oligosaccharide motif is present. For instance, any glycopeptide containing a readily cleavable hexose unit could be identified using tandem MS techniques by monitoring those ions that undergo neutral loss of the hexose group, and hence could in principle identify peptides in a tryptic digest mixture that are glycosylated (77). Current software allows data-dependent tandem MS experiments to be performed that may allow further analysis of the glycosylation site and the identity of the sugar.

The area of glycomics is a vast and rapidly developing field that we cannot hope to cover in sufficient detail in this text. We would point interested readers to recent excellent reviews of current methodology for oligosaccharide identification (66, 68, 78).

9.4. "PASSIVE" POLYMERS OF BIOTIC AND ABIOTIC ORIGIN

A common feature of the three classes of biopolymers considered so far is the unique sequence of the building blocks (e.g., amino acid sequence of proteins) that defines their structure, dynamic behavior, and functional properties. However, there is also a variety of biopolymers that do not display biological activity on their own, but instead play *passive* (mostly structural) roles in living organisms. Structural integrity and cohesion (from the organelle to the organism level) are vital for all living creatures, hence the importance of passive biopolymers. Since the function of such biopolymers is usually limited to structural reinforcement, they often display significant sequence redundancy (Figure 9.12). Perhaps the most illustrious example of structural proteins is fibroin, a fibrous protein that makes up most silks. For example, the heavy chain of *Bombyx mori* silk fibroin consists of 5263 amino acid residues, half of which are glycine and one-third of which are alanine residues (79). The sequence of this protein contains 433 repeat units GAGAGS that have high β-sheet propensity and form the structural basis for the remarkable physical and mechanical properties of silk (80, 81). Another example of a "structural" protein is collagen (82), a family of large proteins (>1000 amino acid residues per single chain) that make up almost a third of the total protein mass in multicellular organisms. The fibril-forming types of collagen are rich in GPX (most often, X is a 3-hydroxyproline) repeat units (Figure 9.12), a structural element that forces collagen polypeptide chains to adopt an unusual left-handed helical conformation, three of which twist together to a right-handed superhelical structure (83). Other examples of passive biopolymers are presented by various homopolysaccharides, most of which are structural polymers, while others are energy-storing polymers. Cellulose (Figure 9.12) is a typical example of a structural polysaccharide and is the most abundant organic substance on the planet.* A single cellulose molecule can contain as much as 15,000 D-glucose residues linked by $\beta(1 \rightarrow 4)$ glycosidic bonds (like most other polysaccharides discussed in the preceding section of this chapter, cellulose does not have a well-defined size). The exceptional strength of cellulose fibers is due to the formation

*It is estimated that approximately 10^{11} tons of cellulose is produced photosynthetically each year.

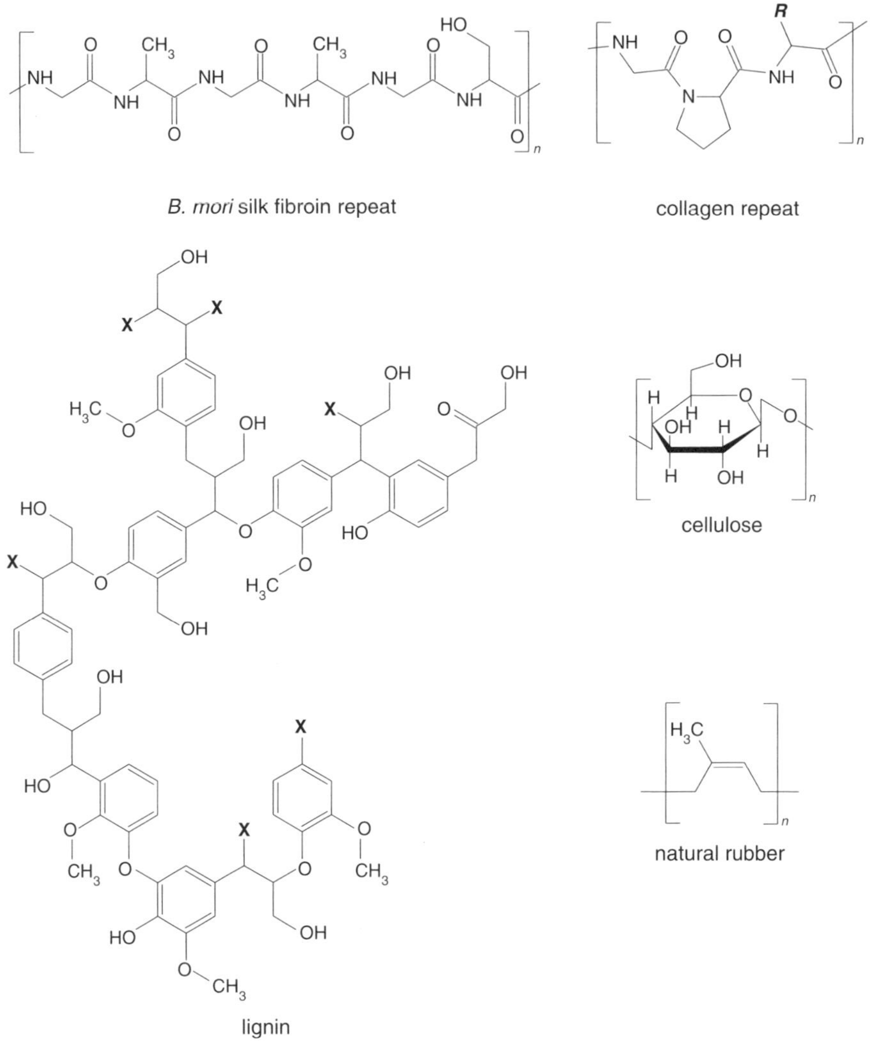

FIGURE 9.12. Examples of structural biopolymers of biological origin.

of multiple strong inter- and intramolecular hydrogen bonds within the cellulose fibers (84, 85). Additional structural reinforcement of cellulose fibers in wood is provided by a cementing matrix, whose major component is lignin, a plastic-like phenolic polymer (see Figure 9.12).

The examples of biopolymers considered above represent several major classes of polymer molecules. When a polymer chain is made by linking only one type of repeat unit, such as a D-glucose residue in cellulose, it is called a *homopolymer*. Natural rubber, or polyisoprene (see Figure 9.12), is another example of a homopolymer. When two (or more) different types of repeat units are joined in the

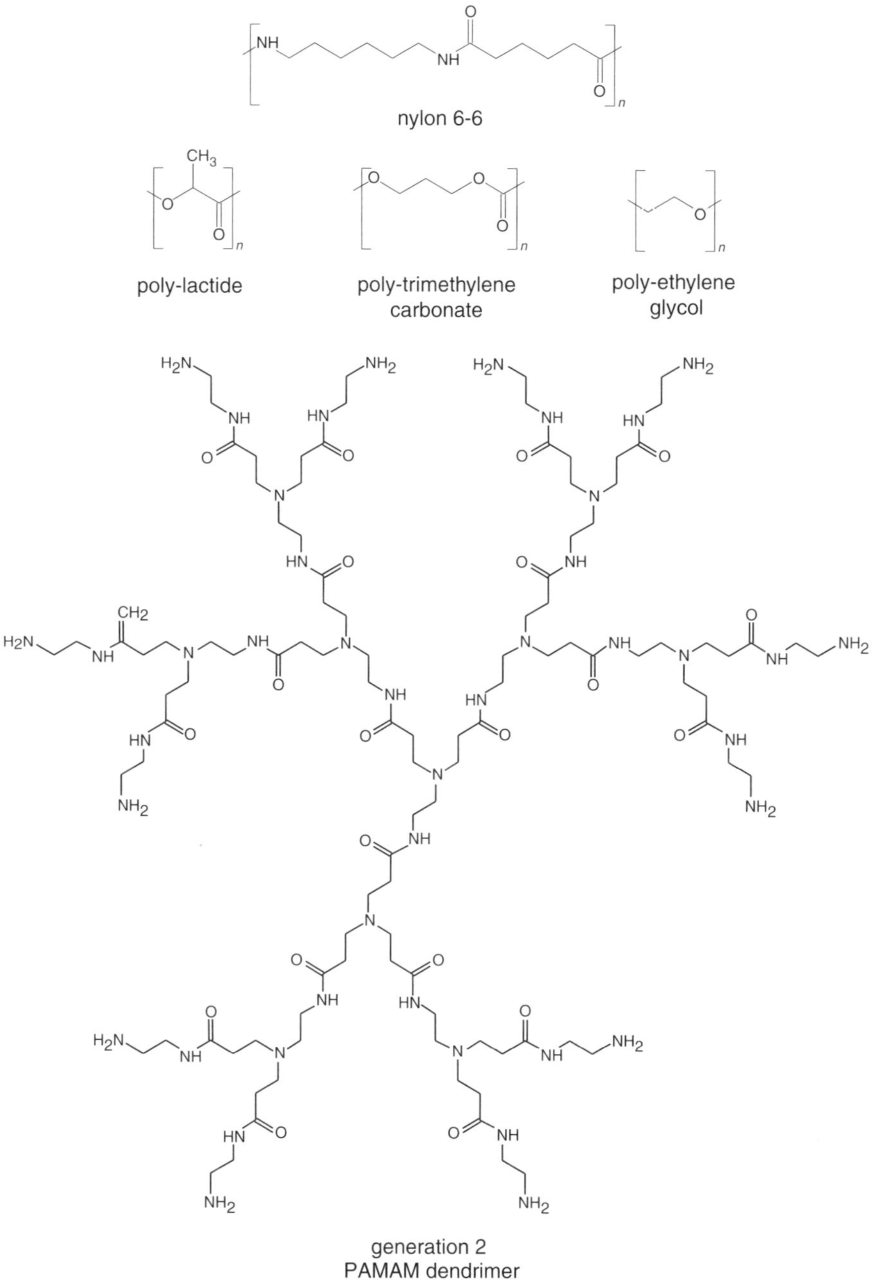

FIGURE 9.13. Examples of synthetic polymers.

same polymer chain, the polymer is called a *copolymer*. Fibroin and collagen fall into this category, as they contain several building blocks (amino acid residues). Unlike fibroin and cellulose, lignin structure is not linear, but rather represents a *random three-dimensional network polymer*.

All of the aforementioned polymers are naturally occurring and of biological origin. Of great interest is the possibility of producing polymeric materials synthetically that either mimic their biological analogues or indeed improve on their properties in some way. Following the pioneering work by Staudinger (86, 87) and Carothers et al. (88), the macromolecular nature of synthetic polymers has become a commonly accepted concept. Although a detailed consideration of synthetic polymers is beyond the scope of this book, we briefly discuss several classes of polymers whose behavior is important from the biophysical point of view. The first class is a rapidly growing group of "bioinspired" polymers that are designed to possess certain structural or functional features of biological macromolecules (89–91). Fibroin mentioned earlier has been a particularly popular target for such design due to its superior mechanical properties (80), with nylon (Figure 9.13) representing perhaps the first successful attempt to imitate the biopolymer. While the design of synthetic analogues of silk proteins remains the focal point of extensive research efforts (92), strategies that employ genetic engineering to mix the modules of the natural protein in specific proportions to attain the desired properties* are rapidly gaining popularity (93). An important issue here is how the mechanical properties of the polymers are modulated by conformational transitions. An example of such conformational change is a helix-to-sheet transition in the amorphous segments of the dragline silk proteins, which is often invoked to explain its superior expandability characteristics. Understanding the general rules of construction of silk proteins and the analysis of the relationship between their conformational and mechanical properties will undoubtedly provide a guide to achieving the desired mechanical properties in synthetic materials by designing controllable conformational switches (94).

Another class of synthetic polymers whose behavior in solution is beginning to attract the attention of biophysicists is comprised of several hydrophilic and amphiphilic macromolecules that are utilized in the design and therapeutic utilization of polymer-conjugated proteins (95–97). Conjugation of therapeutically active proteins with polyethylene glycol (PEG, Figure 9.13) or various other polymers improves their solubility, extends lifetime, and often modulates release. It had been almost always implicitly assumed in the past that the *polymer tail* is fully unstructured in solution and does not interact with the protein. This notion, however, is now being increasingly challenged. Indeed, polymer chains can collapse under certain conditions. Furthermore, electrostatic interaction can exist between the polymer tail and the protein, potentially interfering with the function of the latter. Clearly, the ability to monitor both large-scale dynamics and

*For example, spiders make webs and perform a range of tasks using several different types of silk fiber that are composed of different peptide modules and, as a result, possess distinct mechanical properties.

interaction in this system will provide valuable insight into the behavior of the PEG-conjugated proteins.

One striking difference between synthetic polymers and biopolymers whose production is genetically controlled (i.e., proteins and oligonucleotides) is the intrinsic heterogeneity of the former.* To illustrate this point, we will consider MALDI mass spectra of a 5.5 kDa polypeptide insulin and a synthetic polymer of similar mass, PEG5000 (Figure 9.14). Clearly, the monodisperse (the ratio of *weight average* to *molar average* masses is less than 1.05) PEG5000 sample actually consists of over a dozen oligomers of different length. Such heterogeneity presents a significant challenge for polymer analysis, particularly when ESI MS is used, as the total ion signal is split not only among different charge states (as is the case with essentially monodisperse proteins), but also among chains differing in the number of blocks (Figure 9.15). An additional complication for higher-MW polymers may result from the overlap of peak clusters corresponding to different charge states (as is the case for higher charge states of PEG5000 in Figure 9.15). Nonetheless, it is possible to obtain high-quality ESI MS data with clearly resolved charge states on PEG samples whose MW

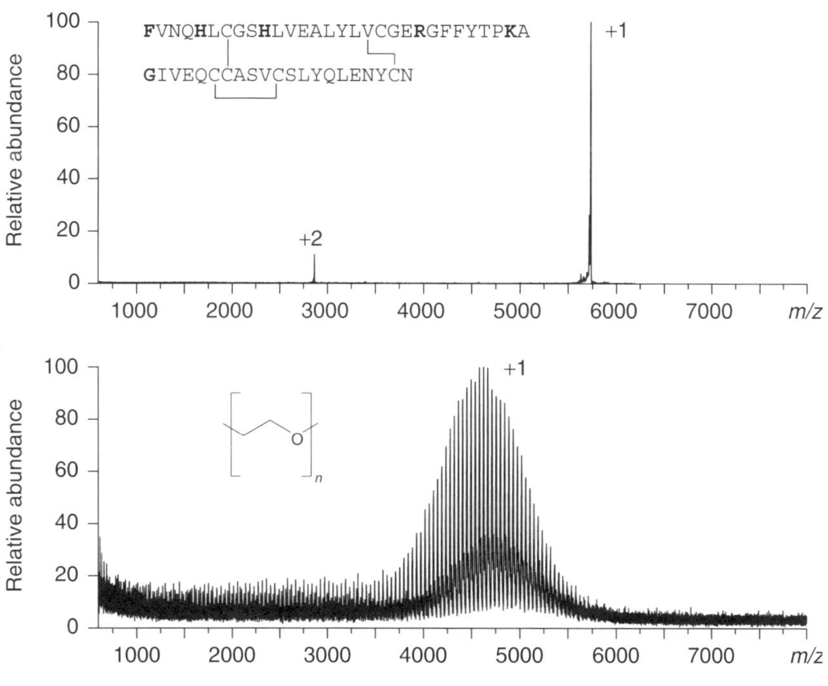

FIGURE 9.14. MALDI mass spectra of a 5.5 kDa polypeptide insulin (*top*) and a monodisperse synthetic polymer PEG5000 (*bottom*).

*Structural biopolymers whose synthesis is not genetically controlled (e.g., polysaccharides) are also heterogeneous.

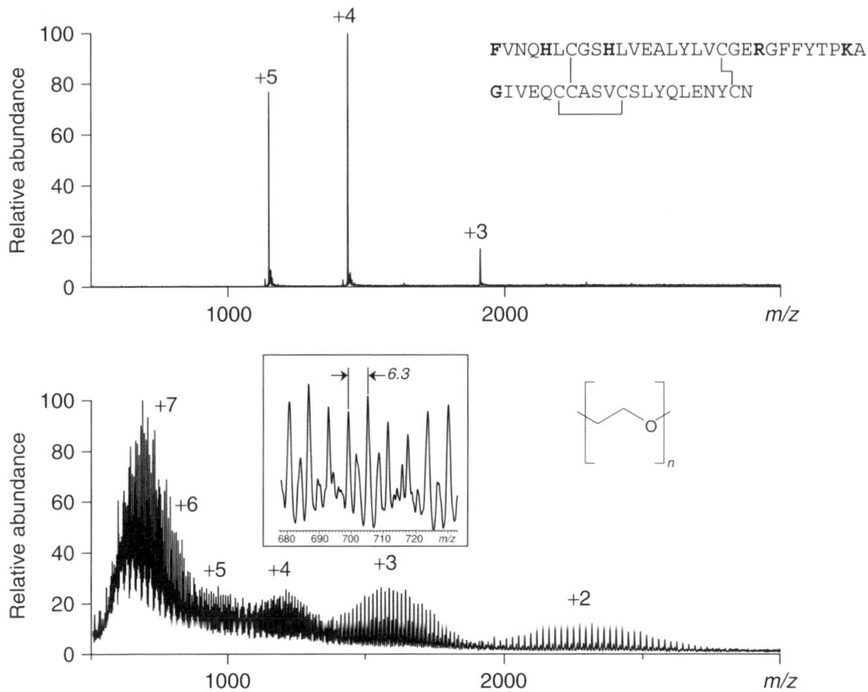

FIGURE 9.15. ESI mass spectra of insulin and PEG5000 in aqueous solutions (10 mM ammonium acetate). Molecules of PEG5000 accumulate more charges, despite the lack of basic residues and a lower mass.

approaches 10 kDa (98). ESI MS analysis of proteins conjugated with as many as five 5 kDa PEG chains (i.e., total PEG content as high as 25 kDa) has also been reported (99).

Among the many MS-based techniques considered in the previous chapters of this book, one appears to be particularly promising as far as the analysis of conformational dynamics of water-soluble synthetic polymers. A comparison of the charge state distributions of insulin and PEG5000 indicates that PEG molecules accumulate significantly more positive charges, despite the lack of basic residues (insulin contains six basic sites). The reason for such efficient charge acquisition by PEG molecules is the lack of compact structure in solution, while insulin is likely to be folded under these conditions* and, as a result, accommodates fewer charges. A detailed discussion of macromolecular ion charge state distributions in ESI MS was given in Chapter 5. However, it remains to be seen if ESI MS can be used to distinguish between the different conformations of PEG in solution (e.g., extended and collapsed forms of the polymer). If it is indeed possible to make such a distinction, charge state

*Conformational freedom of insulin is also restricted by three disulfide bonds.

distributions of polymer ions can be used to estimate surface areas of collapsed polymer chains.

A variety of synthetic polymers are known to have stable secondary structures in solution (89, 100), which are often maintained by elaborate networks of hydrogen bonds. Although in principle such networks can be characterized by measuring the kinetics of hydrogen exchange (similar to protein HDX discussed in Chapters 4 and 5), there are very few studies that actually employ this technique as a tool to probe higher-order structure of synthetic polymers (101). This is certainly an area where MS may make a significant contribution in the near future. One type of synthetic polymers that appears to be particularly suited for HDX studies is amide-based dendrimers, such as a polyamidoamine (PAMAM) dendrimer presented in Figure 9.13. Although dendrimeric structure is usually presented in a highly symmetrical fashion, these branched polymers may actually exhibit a variety of conformations. Since the dendrimer architecture is an important determinant of its physical properties, numerous studies have aimed at elucidating the structure of dendrimers (particularly amphiphilic dendrimers) in various environments, often revealing quite unexpected conformational features of these macromolecules (102). At high ionic strength, backfolding of the end groups takes place, leading to formation of a so-called *dense core* structure, while low ionic strength forces the dendrimer to stretch and adopt a *dense shell* structure (102). HDX rates of some alanine-based dendrimers have been measured by ^{1}H NMR spectroscopy in polar solvents favoring the dense shell conformation (103). The dense core–dense shell transition certainly changes solvent accessibility of the amide groups, which should be detectable by HDX MS.

An alternative approach to the structural characterization of dendrimers involves using tandem MS techniques. In a recent report a combination of MALDI and ESI MS/MS techniques was employed to investigate the fragmentation pathways of a poly(propylene imine) dendrimer (104). Depending on the polarity of the solvent, different dissociation patterns were observed: in polar protic solvents an extended conformation was adopted due to favorable solvent–solute interactions. In this case terminal branches were readily cleaved in gas phase dissociation events, whereas the compact structure with intramolecular hydrogen bonding adopted in nonpolar solvents precluded such fragmentations.

Dendrimers such as polyamidoamine (PAMAM) dendrimers have shown significant potential as efficient nonviral vehicles for delivering genetic material into cells. They have been shown to be as efficient or more so than either cationic liposomes or other cationic polymers (e.g., polyethylenimine, polylysine) for in vitro gene transfer (105). MS may be of use in determining how these molecules package genetic material and enable its introduction into cells, potentially directing further synthetic strategies.

As we have seen, the full power of MS in the non-protein biological arena has yet to be wholly exploited. Techniques that already exist and those that are still under development show great promise for the analysis of structure and dynamics in oligosaccharides, oligonucleotides, and even synthetic polymers.

REFERENCES

1. Jacobson K. B., Arlinghaus H. F., Buchanan M. V., Chen C. H., Glish G. L., Hettich R. L., McLuckey S. A. (1991). Applications of mass spectrometry to DNA sequencing. *Genet. Anal. Biomol. Eng.* **8**: 223–229.
2. Smith L. M. (1993). The future of DNA sequencing. *Science* **262**: 530–532.
3. Murray K. K. (1996). DNA sequencing by mass spectrometry. *J. Mass Spectrom.* **31**: 1203–1215.
4. Smith L. M. (1996). Sequence from spectrometry: a realistic prospect? *Nat. Biotechnol.* **14**: 1084–1088.
5. Spengler B., Pan Y., Cotter R. J., Kan L. S. (1990). Molecular weight determination of underivatized oligodeoxyribonucleotides by positive ion matrix assisted ultraviolet laser desorption mass spectrometry. *Rapid Commun. Mass Spectrom.* **4**: 99–102.
6. Covey T. R., Bonner R. F., Shushan B. I., Henion J. (1988). The determination of protein, oligonucleotide and peptide molecular weights by ion-spray mass spectrometry. *Rapid Commun. Mass Spectrom.* **2**: 249–256.
7. Maxam A. M., Gilbert W. (1977). A new method for sequencing DNA. *Proc. Natl. Acad. Sci. U.S.A.* **74**: 560–564.
8. Sanger F., Nicklen S., Coulson A. R. (1977). DNA sequencing with chain-terminating inhibitors. *Proc. Natl. Acad. Sci. U.S.A.* **74**: 5463–5467.
9. Edwards J. R., Itagaki Y., Ju J. (2001). DNA sequencing using biotinylated dideoxynucleotides and mass spectrometry. *Nucleic Acids Res.* **29**: e104.
10. Zhu L., Parr G. R., Fitzgerald M. C., Nelson C. M., Smith L. M. (1995). Oligodeoxynucleotide fragmentation in MALDI/TOF mass spectrometry using 355 nm radiation. *J. Am. Chem. Soc.* **117**: 6048–6056.
11. Nordhoff E., Karas M., Cramer R., Hahner S., Hillenkamp F., Kirpekar F., Lezius A., Muth J., Meier C., Engels J. W. (1995). Direct mass spectrometric sequencing of low picomole amounts of oligodeoxynucleotides with up to 21 bases by matrix assisted laser desorption ionization mass spectrometry. *J. Mass Spectrom.* **30**: 99–112.
12. Ono T., Scalf M., Smith L. M. (1997). 2′-Fluoro modified nucleic acids: polymerase-directed synthesis, properties and stability to analysis by matrix-assisted laser desorption/ionization mass spectrometry. *Nucleic Acids Res.* **25**: 4581–4588.
13. Schneider K., Chait B. T. (1995). Increased stability of nucleic acids containing 7-deaza-guanosine and 7-deaza-adenosine may enable rapid DNA sequencing by matrix assisted laser desorption mass spectrometry. *Nucleic Acids Res.* **23**: 1570–1575.
14. Tang W., Zhu L., Smith L. M. (1997). Controlling DNA fragmentation in MALDI-MS by chemical modification. *Anal. Chem.* **69**: 302–312.
15. Kirpekar F., Nordhoff E., Kristiansen K., Roepstorff P., Hahner S., Hillenkamp F. (1995). 7-Deaza purine bases offer a higher ion stability in the analysis of DNA by matrix assisted laser desorption/ionization mass spectrometry. *Rapid Commun. Mass Spectrom.* **9**: 525–531.
16. Köster H., Tang K., Fu D. J., Braun A., vandenBoom D., Smith C. L., Cotter R. J., Cantor C. R. (1996). A strategy for rapid and efficient DNA sequencing by mass spectrometry. *Nat. Biotechnol.* **14**: 1123–1128.

17. Schieltz D. M., Chou C. W., Luo C. W., Thomas R. M., Williams P. (1992). Mass spectrometry of DNA mixtures by laser ablation from frozen aqueous solution. *Rapid Commun. Mass Spectrom.* **6**: 631–636.

18. Williams P. (1994). Time of flight mass spectrometry of DNA laser ablated from frozen aqueous solutions—applications to the Human Genome Project. *Int. J. Mass Spectrom. Ion Processes* **131**: 335–344.

19. Sheffer J. D., Murray K. K. (2000). Infrared matrix-assisted laser desorption/ionization using a frozen alcohol matrix. *J. Mass Spectrom.* **35**: 95–97.

20. Chen X. Y., Westphall M. S., Smith L. M. (2003). Mass spectrometric analysis of DNA mixtures: instrumental effects responsible for decreased sensitivity with increasing mass. *Anal. Chem.* **75**: 5944–5952.

21. Cheng X. H., Camp D. G., Wu Q. Y., Bakhtiar R., Springer D. L., Morris B. J., Bruce J. E., Anderson G. A., Edmonds C. G., Smith R. D. (1996). Molecular weight determination of plasmid DNA using electrospray ionization mass spectrometry. *Nucleic Acids Res.* **24**: 2183–2189.

22. Schafer A. J., Hawkins J. R. (1998). DNA variations and the future of human genetics. *Nat. Biotechnol.* **16**: 33–39.

23. Wang D. G., Fan J.-B., Siao C.-J., Berno A., Young P., Sapolsky R., Ghandour G., Perkins N., Winchester E., Spencer J., Kruglyak L., Stein L., Hsie L., Topaloglou T., Hubbell E., Robinson E., Mittmann M., Morris M. S., Shen N., Kilburn D., Rioux J., Nusbaum C., Rozen S., Hudson T. J., Lipshutz R., Chee M., Lander E. S. (1998). Large scale identification, mapping, and genotyping of single nucleotide polymorphisms in the human genome. *Science* **280**: 1077–1082.

24. Brookes A. J. (1999). The essence of SNPs. *Gene* **234**: 177–186.

25. Griffin T. J., Smith L. M. (2000). Single-nucleotide polymorphism analysis by MALDI-TOF mass spectrometry. *Trends Biotechnol.* **18**: 77–84.

26. Griffin T. J., Tang W., Smith L. M. (1997). Genetic analysis by peptide nucleic acid affinity MALDI-TOF mass spectrometry. *Nat. Biotechnol.* **15**: 1368–1372.

27. Ross P. L., Lee K., Belgrader P. (1997). Discrimination of single-nucleotide polymorphisms in human DNA using peptide nucleic acid probes detected by MALDI-TOF mass spectrometry. *Anal. Chem.* **69**: 4197–4202.

28. Jurinke C., Oeth P., van der Boom D. (2004). MALDI-TOF mass spectrometry: a versatile tool for high performance DNA analysis. *Mol. Biotechnol.* **26**: 147–163.

29. Han F. X. G., Wheelhouse R. T., Hurley L. H. (1999). Interactions of TMPyP4 and TMPyP2 with quadruplex DNA. Structural basis for the differential effects on telomerase inhibition. *J. Am. Chem. Soc.* **121**: 3561–3570.

30. Goodlett D. R., Camp D. G., Hardin C. C., Corregan M., Smith R. D. (1993). Direct observation of a DNA quadruplex by electrospray ionization mass spectrometry. *Biol. Mass Spectrom.* **22**: 181–183.

31. Sakamoto S., Fujita M., Kim K., Yamaguchi K. (2000). Characterization of self-assembling nano-sized structures by means of coldspray ionization mass spectrometry. *Tetrahedron* **56**: 955–964.

32. Sakamoto S., Nakatani K., Saito I., Yamaguchi K. (2003). Formation and destruction of the guanine quartet in solution observed by cold-spray ionization mass spectrometry. *Chem. Commun.* **6**: 788–789.

33. Sakamoto S., Yamaguchi K. (2003). Low T_m DNA duplexes observed by cold-spray ionization mass spectrometry. *Tetrahedron Lett.* **44**: 3341–3344.

34. Yamaguchi K. (2003). Cold-spray ionization mass spectrometry: principle and applications. *J. Mass Spectrom.* **38**: 473–490.

35. Leroy J. L., Broseta D., Gueron M. (1985). Proton exchange and base pair kinetics of poly(Ra).poly(Ru) and poly(Ri).poly(Rc). *J. Mol. Biol.* **184**: 165–178.

36. Bendel P. (1987). A proton NMR spin-lattice relaxation study of the imino proton exchange kinetics in calf thymus DNA. *Biopolymers* **26**: 573–590.

37. Leroy J. L., Kochoyan M., Huynhdinh T., Gueron M. (1988). Characterization of base pair opening in deoxynucleotide duplexes using catalyzed exchange of the imino proton. *J. Mol. Biol.* **200**: 223–238.

38. Cross D. G., Brown A., Fisher H. F. (1975). Hydrogen–deuterium exchange in nucleosides and nucleotides—mechanism for exchange of exocyclic amino hydrogens of adenosine. *Biochemistry* **14**: 2745–2749.

39. Basu H. S., Shafer R. H., Marton L. J. (1987). A stopped flow H-D exchange kinetic study of spermine polynucleotide interactions. *Nucleic Acids Res.* **15**: 5873–5886.

40. Hofstadler S. A., Griffey R. H. (2001). Analysis of noncovalent complexes of DNA and RNA by mass spectrometry. *Chem. Rev.* **101**: 377–390.

41. Zoltewicz J. A., Clark D. F., Sharpless T. W., Grahe G. (1970). Kinetics and mechanism of acid catalyzed hydrolysis of some purine nucleosides. *J. Am. Chem. Soc.* **92**: 1741–1750.

42. Nordhoff E., Cramer R., Karas M., Hillenkamp F., Kirpekar F., Kristiansen K., Roepstorff P. (1993). Ion stability of nucleic-acids in infrared matrix-assisted laser-desorption ionization mass-spectrometry. *Nucleic Acids Res.* **21**: 3347–3357.

43. Kirpekar F., Nordhoff E., Kristiansen K., Roepstorff P., Lezius A., Hahner S., Karas M., Hillenkamp F. (1994). Matrix assisted laser desorption ionization mass spectrometry of enzymatically synthesized RNA up to 150 kDa. *Nucleic Acids Res.* **22**: 3866–3870.

44. Faulstich K., Worner K., Brill H., Engels J. W. (1997). A sequencing method for RNA oligonucleotides based on mass spectrometry. *Anal. Chem.* **69**: 4349–4353.

45. Little D. P., Thannhauser T. W., McLafferty F. W. (1995). Verification of 50-mer to 100-mer DNA and RNA sequences with high resolution mass spectrometry. *Proc. Natl. Acad. Sci. U.S.A.* **92**: 2318–2322.

46. Hofstadler S. A., Griffey R. H., Pasa-Tolic L., Smith R. D. (1998). The use of a stable internal mass standard for accurate mass measurements of oligonucleotide fragment ions using electrospray ionization Fourier transform ion cyclotron resonance mass spectrometry with infrared multiphoton dissociation. *Rapid Commun. Mass Spectrom.* **12**: 1400–1404.

47. Puglisi J. D., Tan R. Y., Calnan B. J., Frankel A. D., Williamson J. R. (1992). Conformation of the tar RNA–arginine complex by NMR-spectroscopy. *Science* **257**: 76–80.

48. Batey R. T., Inada M., Kujawinski E., Puglisi J. D., Williamson J. R. (1992). Preparation of isotopically labeled ribonucleotides for multidimensional NMR-spectroscopy of RNA. *Nucleic Acids Res.* **20**: 4515–4523.

49. Fürtig B., Richter C., Wöhnert J., Schwalbe H. (2003). NMR spectroscopy of RNA. *ChemBioChem* **4**: 936–962.

50. Yusupov M. M., Yusupova G. Z., Baucom A., Lieberman K., Earnest T. N., Cate J. H. D., Noller H. F. (2001). Crystal structure of the ribosome at 5.5 angstrom resolution. *Science* **292**: 883–896.

51. Ban N., Nissen P., Hansen J., Moore P. B., Steitz T. A. (2000). The complete atomic structure of the large ribosomal subunit at 2.4 angstrom resolution. *Science* **289**: 905–920.

52. Wimberly B. T., Brodersen D. E., Clemons W. M., Morgan-Warren R. J., Carter A. P., Vonrhein C., Hartsch T., Ramakrishnan V. (2000). Structure of the 30S ribosomal subunit. *Nature* **407**: 327–339.

53. Gmeiner W. H., Sahasrabudhe P., Pon R. T. (1995). Use of shaped pulses for semi-selective excitation of imino H-1 resonances in duplex DNA and RNA. *Magn. Reson. Chem.* **33**: 449–452.

54. Nonin S., Jiang F., Patel D. J. (1997). Imino proton exchange and base-pair kinetics in the AMP–RNA aptamer complex. *J. Mol. Biol.* **268**: 359–374.

55. Li T. S., Johnson J. E., Thomas G. J. (1993). Raman dynamic probe of hydrogen-exchange in bean pod mottle virus—base-specific retardation of exchange in packaged Ssrna. *Biophys. J.* **65**: 1963–1972.

56. Ramstein J., Erdmann V. A. (1981). A hydrogen-exchange study of *Escherichia-coli* 5s RNA. *Nucleic Acids Res.* **9**: 4081–4086.

57. Dixon W. J., Hayes J. J., Levin J. R., Weidner M. F., Dombroski B. A., Tullius T. D. (1991). Hydroxyl radical footprinting. *Methods Enzymol.* **280**: 380–413.

58. Brenowitz M., Chance M. R., Dhavan G., Takamoto K. (2002). Probing the structural dynamics of nucleic acids by quantitative time-resolved and equilibrium hydroxyl radical "footprinting." *Curr. Opin. Struct. Biol.* **12**: 648–653.

59. Celander D. W., Cech T. R. (1991). Visualizing the higher order folding of a catalytic RNA molecule. *Science* **251**: 401–407.

60. Sclavi B., Woodson S., Sullivan M., Chance M. R., Brenowitz M. (1998). Time-resolved synchrotron X-ray "footprinting", a new approach to the study of nucleic acid structure and function: application to protein–DNA interactions and RNA folding. *J. Mol. Biol.* **266**: 144–159.

61. Yu E., Fabris D. (2003). Direct probing of RNA structures and RNA–protein interactions in the HIV-1 packaging signal by chemical modification and electrospray ionization Fourier transform mass spectrometry. *J. Mol. Biol.* **330**: 211–223.

62. Fabris D. (2000). Steady-state kinetics of ricin A-chain reaction with the sarcin–ricin loop and with HIV-1 Psi-RNA hairpins evaluated by direct infusion electrospray ionization mass spectrometry. *J. Am. Chem. Soc.* **122**: 8779–8780.

63. Loo J. A. (1997). Studying noncovalent protein complexes by electrospray ionization mass spectrometry. *Mass Spectrom. Rev.* **16**: 1–23.

64. Steen H., Jensen O. N. (2002). Analysis of protein–nucleic acid interactions by photochemical cross-linking and mass spectrometry. *Mass Spectrom. Rev.* **21**: 163–182.

65. Duus J. Ø., Gotfredsen C. H., Bock K. (2000). Carbohydrate structural determination by NMR spectroscopy: modern methods and limitations. *Chem. Rev.* **100**: 4589–4614.

66. Wormald M. R., Petrescu A. J., Pao Y.-L., Glithero A., Elliott T., Dwek R. A. (2002). Conformational studies of oligosaccharides and glycopeptides: complementarity of NMR, X-ray crystallography, and molecular modeling. *Chem. Rev.* **102**: 371–386.

67. Costello C. E., Perrault H., Ngoka L. C. (1996). Mass spectrometric and tandem mass spectrometric approaches to the analysis of carbohydrates. In: *Mass*

Spectrometry in the Biological Sciences, eds. A. L. Burlingame, S. A. Carr, pp. 365–384. Totowa, NJ: Humana Press.

68. Zaia J. (2004). Mass spectrometry of oligosaccharides. *Mass Spectrom. Rev.* **23**: 161–227.
69. Biemann K., Martin S. A. (1987). Mass spectrometric determination of the amino acid sequence of peptides and proteins. *Mass Spectrom. Rev.* **6**: 1–75.
70. Dell A., Morris H. R., Egge H., Vonnicolai H., Strecker G. (1983). Fast atom bombardment mass spectrometry for carbohydrate structure determination. *Carbohydr. Res.* **115**: 41–52.
71. Carr S. A., Reinhold V. N. (1984). Structural characterization of sulfated glycosaminoglycans by fast atom bombardment mass spectrometry—application to chondroitin sulfate. *J. Carbohydr. Chem.* **3**: 381–401.
72. Egge H., Peterkatalinic J. (1987). Fast atom bombardment mass spectrometry for structural elucidation of glycoconjugates. *Mass Spectrom. Rev.* **6**: 331–393.
73. Domon B., Costello C. E. (1988). A systematic nomenclature for carbohydrate fragmentations in FAB-MS/MS spectra of glycoconjugates. *Glycoconjugate J.* **5**: 397–409.
74. Harvey D. J., Küster B., Wheeler S. F., Hunter A. P., Bateman R. H., Dwek R. A. (2000). Matrix assisted laser desorption/ionization mass spectrometry of N-linked carbohydrates and related compounds. In: *Mass Spectrometry in Biology & Medicine*, eds. A. L. Burlingame, S. A. Carr, M. A. Baldwin, pp. 403–438. Totowa, NJ: Humana Press.
75. Carr S. A., Hemling M. E., Folenawasserman G., Sweet R. W., Anumula K., Barr J. R., Huddleston M. J., Taylor P. (1989). Protein and carbohydrate structural analysis of a recombinant soluble CD4 receptor by mass spectrometry. *J. Biol. Chem.* **264**: 21286–21295.
76. Küster B., Mann M. (1999). ^{18}O labeling of N-glycosylation sites to improve the identification of gel-separated glycoproteins using peptide mass mapping and database searching. *Anal. Chem.* **71**: 1431–1440.
77. Carr S. A., Huddleston M. J., Bean M. F. (1993). Selective identification and differentiation of N-linked and O-linked oligosaccharides in glycoproteins by liquid chromatography mass spectrometry. *Protein Sci.* **2**: 183–196.
78. Küster B., Krogh T. N., Mørtz E., Harvey D. J. (2001). Glycosylation analysis of gel-separated proteins. *Proteomics* **1**: 350–361.
79. Zhou C. Z., Confalonieri F., Jacquet M., Perasso R., Li Z. G., Janin J. (2001). Silk fibroin: structural implications of a remarkable amino acid sequence. *Proteins* **44**: 119–122.
80. Gosline J. M., Guerette P. A., Ortlepp C. S., Savage K. N. (1999). The mechanical design of spider silks: from fibroin sequence to mechanical function. *J. Exp. Biol.* **202**: 3295–3303.
81. Valluzzi R., Winkler S., Wilson D., Kaplan D. L. (2002). Silk: molecular organization and control of assembly. *Philos. Trans. R. Soc. London B Biol. Sci.* **357**: 165–167.
82. Gelse K., Poschl E., Aigner T. (2003). Collagens—structure, function, and biosynthesis. *Adv. Drug Del. Rev.* **55**: 1531–1546.
83. Ottani V., Martini D., Franchi M., Ruggeri A., Raspanti M. (2002). Hierarchical structures in fibrillar collagens. *Micron* **33**: 587–596.

84. O'Sullivan A. C. (1997). Cellulose: the structure slowly unravels. *Cellulose* **4**: 173–207.

85. Kadla J. F., Gilbert R. D. (2000). Cellulose structure: a review. *Cell. Chem. Technol.* **34**: 197–216.

86. Staudinger H. (1932). *Die Hochmolekularen organischen Verbindungen, Kautschuk und Cellulose*. Berlin: Springer.

87. Staudinger H. (1947). *Makromolekulare Chemie und Biologie*. Basel: Wepf.

88. Carothers W. H., Mark H. F., Whitby G. S. (1940). *Collected Papers by Wallace Hume Carothers on High Polymeric Substances*. New York: Interscience Publishers.

89. Barron A. E., Zuckerman R. N. (1999). Bioinspired polymeric materials: in-between proteins and plastics. *Curr. Opin. Chem. Biol.* **3**: 681–687.

90. Cunliffe D., Pennadam S., Alexander C. (2004). Synthetic and biological polymers—merging the interface. *Eur. Polym. J.* **40**: 5–25.

91. van Hest J. C. M., Tirrell D. A. (2001). Protein-based materials, toward a new level of structural control. *Chem. Commun.* **2001**: 1897–1904.

92. O'Brien J. P., Fahnestock S. R., Termonia Y., Gardner K. C. H. (1998). Nylons from nature: synthetic analogs to spider silk. *Adv. Mater.* **10**: 1185–1195.

93. Hinman M. B., Jones J. A., Lewis R. V. (2000). Synthetic spider silk: a modular fiber. *Trends Biotechnol.* **18**: 374–379.

94. Bini E., Knight D. P., Kaplan D. L. (2004). Mapping domain structures in silks from insects and spiders related to protein assembly. *J. Mol. Biol.* **335**: 27–40.

95. Roberts M. J., Bentley M. D., Harris J. M. (2002). Chemistry for peptide and protein PEGylation. *Adv. Drug Del. Rev.* **54**: 459–476.

96. Greenwald R. B., Choe Y. H., McGuire J., Conover C. D. (2003). Effective drug delivery by PEGylated drug conjugates. *Adv. Drug Del. Rev.* **55**: 217–250.

97. Matthews C. N. (1994). Hardware and software in biology—simultaneous origin of proteins and nucleic-acids via hydrogen-cyanide polymers. *J. Biol. Phys.* **20**: 275–281.

98. Lord G. A., Cai H., Luo J. L., Lim C. K. (2000). HPLC-electrospray mass spectrometry of temoporfin-poly(ethylene glycol) conjugates. *Analyst* **125**: 605–608.

99. Gioacchini A. M., Carrea G., Secundo F., Baraldini M., Roda A. (1997). Electrospray mass spectrometric analysis of poly(ethylene glycol)–protein conjugates. *Rapid Commun. Mass Spectrom.* **11**: 1219–1222.

100. Nakano T., Okamoto Y. (2001). Synthetic helical polymers: conformation and function. *Chem. Rev.* **101**: 4013–4038.

101. Zagar E., Grdadolnik J. (2003). An infrared spectroscopic study of H-bond network in hyperbranched polyester polyol. *J. Mol. Struct.* **658**: 143–152.

102. Bosman A. W., Janssen H. M., Meijer E. W. (1999). About dendrimers: structure, physical properties, and applications. *Chem. Rev.* **99**: 1665–1688.

103. Mong T. K. K., Niu A. Z., Chow H. F., Wu C., Li L., Chen R. (2001). Beta-alanine-based dendritic beta-peptides: dendrimers possessing unusually strong binding ability towards protic solvents and their self-assembly into nanoscale aggregates through hydrogen-bond interactions. *Chem. Eur. J.* **7**: 686–699.

104. Adhiya A., Wesdemiotis C. (2002). Poly(propylene imine) dendrimer conformations in the gas phase: a tandem mass spectrometry study. *Int. J. Mass Spectrom.* **214**: 75–88.

105. Eichman J. D., Bielinska A. U., Kukowska-Latallo J. F., Baker J. R. Jr (2000). The use of PAMAM dendrimers in the efficient transfer of genetic material into cells. *Pharm. Sci. Technol. Today* **3**: 232–245.
106. Viovy J.-L. (2000). Electrophoresis of DNA and other polyelectrolytes: physical mechanisms. *Rev. Modern Phys.* **72**: 813–872.
107. McLuckey S. A., Van Berkel G. J., Glish G. L. (1992). Tandem mass spectrometry of small, multiply charged oligonucleotides. *J. Am. Soc. Mass Spectrom.* **3**: 60–70.
108. Fu D.-J., Tang K., Braun A., Reuter D., Darnhofer-Demar B., Little D. P., O'Donnell M. J., Cantor C. R., Köster H. (1998). Sequencing exons 5 to 8 of the p53 gene by MALDI-TOF mass spectrometry. *Nat. Biotechnol.* **16**: 381–384.

10

BIOMOLECULAR IONS IN A SOLVENT-FREE ENVIRONMENT

One of the most intriguing features of biological MS, which is also its major point of distinction from other biophysical techniques, is the environment in which the biopolymers are studied. Solvent removal from the macromolecular ion is usually a requirement for high-quality MS measurement; however, it certainly introduces rather dramatic changes in the behavior of the biopolymer. Such alteration of the protein environment and consequent changes of the macromolecular properties present a great challenge for biophysicists. At the same time, the ability to generate and manipulate solvent-free biopolymers also presents a unique opportunity to study intrinsic properties of biomolecules unaffected by the solvent. In this chapter we briefly discuss the role played by the solvent molecules in determining the structural and dynamic characteristics of biopolymers and consider the consequences of solvent elimination from the biomolecular environment. This is followed by a brief discussion of several MS-based techniques that are commonly used to study the behavior of biomolecular ions in a solvent-free environment. The chapter is not meant to be an exhaustive discussion of the current state of the field of the gas phase structure and dynamics of macromolecular ions. Instead, emphasis is placed on the differences between the solution and gas phase structures of biopolymers and their implications for the studies of biomolecular properties in solution using MS-based methods.

Mass Spectrometry in Biophysics: Conformation and Dynamics of Biomolecules
By Igor A. Kaltashov and Stephen J. Eyles
ISBN 0-471-45602-0

10.1. GENERAL CONSIDERATIONS: ROLE OF SOLVENT IN MAINTAINING BIOMOLECULAR STRUCTURE AND MODULATING ITS DYNAMICS

It is commonly accepted that water is not simply a medium in which biomolecules happen to exist, interact, and carry out their functions. Indeed, almost every aspect of biomolecular chemistry and physics is influenced by the properties of the liquid milieu. Interactions between the protein and water molecules are key to protein folding, integrity of the higher-order structure, and, ultimately, their diverse biological functions. Water is essential for the very existence of the hydrophobic interactions, one of the major forces driving protein folding toward their native structures. At the same time, the aqueous environment is an important modulator of the electrostatic interactions, another class of inter- and intramolecular forces that are important players in maintaining macromolecular architecture. While the importance of the structural aspects of protein–water interaction has been recognized for a long time, recent years have brought the realization that water is also an important determinant and modulator of the dynamic aspects of biopolymer behavior.

Recent high-resolution X-ray crystallographic studies, as well as neutron-scattering experiments and molecular dynamics simulation studies, have contributed to the view that water plays a crucial functional role in mediating protein dynamics at the molecular level (1). As a result, most cellular functions are driven by incremental changes in the biomolecular environment, such as pH and ionic strength (2). Macromolecules themselves also exert significant influence over the solvent molecules; hence the physical properties of the water molecules at the interface are different from those of the "bulk" water molecules. For example, the water density is thought to increase by as much as 50% within the first few angstroms of the shell around the protein (3). Although such "interfacial" water molecules can be visualized in some cases with high-resolution X-ray crystallography at low temperatures, it probably is more appropriate to think of local hydration patterns in terms of a distribution of, rather than individual, water molecules (2). The interfacial water molecules do not form a rigid shell around a biological macromolecule; rather, they create a fluctuating cloud of molecules that are thermodynamically affected by the biopolymer (4, 5).

Removal of solvent is certainly going to affect both the structure and dynamic properties of biomolecules, although not all such changes will have negative impact on the integrity of the higher-order structure. As pointed out by Wolynes, the stability of the internal hydrogen bonding network should not be compromised in a solvent-free environment (6). In fact, the intramolecular network of hydrogen bonds should become more stable in the absence of solvent molecules, which act as both donors and acceptors for transient *inter*molecular hydrogen bonds and normally compete with the *intra*molecular hydrogen bonds that maintain stable secondary and tertiary structure. Electrostatic interactions (such as salt bridges) should also become significantly stronger in the absence of the solvent due to a dramatic reduction of the dielectric constant. It is important to note that the increased strength of the electrostatic interactions does not necessarily have a stabilizing influence on the native-like (or even simply compact) protein ion structure in the gas phase. Since the ESI-generated macromolecular ions bear several charges of the same polarity, the repulsion between these "unshielded" charges is expected to favor the extended conformations in the gas phase and indeed lower the barriers for the dissociation of the covalent bonds (7). In contrast to the electrostatic interactions, contributions of the hydrophobic interactions to the biopolymer stability in the gas phase are expected to be dramatically reduced due to the removal of the polar solvent. It has been suggested that the vacuum can be viewed as an apolar medium, akin to the hydrophobic interiors of the membranes (6). It follows then that a protein ion may turn "inside out" in a solvent-free environment, thereby exposing its hydrophobic interior and burying the polar residues. This scenario, however, seems rather unlikely for several reasons. First, the van der Waals interaction (which is counted as a portion of the hydrophobic force) will still be present in the gas phase, favoring compact conformations.* Second, the intact network of hydrogen bonds and the strengthened electrostatic interactions† will provide significant reinforcement of the compact, native-like conformation in the absence of solvent. Indeed, molecular dynamics simulations carried out in the absence of solvent do not show signs of the proteins turning inside out at least on a nanosecond time scale. Therefore, it seems conceivable that the native-like conformations may be at least metastable in the gas phase, although it is not a priori clear if they do represent single global energy minima (6).

The ability of biopolymers to maintain specific conformations in the gas phase, transition between such conformations, and their relevance to the native structures and dynamics in aqueous solutions have been the focus of extensive studies since the advent of ESI MS. Although answers to a wide range of questions are sought in these studies, two specific issues are often seen as a major driving force and motivation behind this challenging experimental work. First, it is often argued that characterization of biomolecular behavior in the gas phase may allow one to better understand the role played by solvent in determining biomolecular structure, function, and dynamics. For example, Weinkauf and co-workers have recently pointed out that the studies of biopolymer behavior unaffected by the solvent provide unique information on individual molecules that is usually lost due to the averaging of molecular motions, distortions caused by thermal motions, conformational and structural inhomogeneity of the entire ensemble of biopolymers, and so on when these objects are studied in solution (8). Many intrinsic physicochemical properties‡ are either unavailable or not known precisely even for small biomolecules, such as single amino acids and nucleotides. Studying the behavior of individual biological molecules in the absence of solvent may allow

*The desired specificity in this case will be provided by the "packing" interactions.

†It must be mentioned, however, that the biopolymer ions carry multiple charges, and the electrostatic repulsion between these charges will disfavor the compact conformation.

‡These include ionization energies and electron affinities, dipole moments, photodynamics, structural flexibility, and strengths of non-covalent bonds to other biomolecules.

the intrinsic and externally imposed properties to be distinguished. Furthermore, the experimental results obtained under carefully controlled conditions in vacuo can be compared directly to theoretical calculations, providing a direct test and calibration for the theory (8). Investigation of biopolymer structure and behavior in vacuo may also have important implications for our understanding of processes related to the origin, evolution, and propagation of life (9–11).

Second, methods of modern MS provide a variety of unique capabilities to determine the composition of biomolecular associations ranging from small non-covalent protein–ligand complexes to large megadalton assemblies, such as intact ribosomes and viral particles (12–14). Examples of intact ribosome and viral particle characterization by MS will be presented in Chapter 11. The unique ability of MS to manipulate such complexes separately from each other based on the differences in their masses may provide a means to obtain ligand-specific information on the binding affinity, for example, by using the methods of tandem MS (15). Although these methods may evolve into a powerful analytical tool with a wide range of applications, meaningful data would only be obtained if the gas phase structures of biopolymers and their complexes bear significant similarity to those in solution. Recent studies by several groups suggest that solvent removal may introduce not only significant quantitative changes to the behavior of protein assemblies (e.g., binding energetics), but also qualitative ones, for example, by dramatic alteration of the dissociation pathways of protein complexes in the gas phase, as compared to those in solution (16).

10.2. EXPERIMENTAL METHODS TO STUDY BIOMOLECULAR STRUCTURE IN VACUO

A variety of MS-based experimental methods to study various aspects of biomolecular higher-order structure and dynamics in the gas phase have been developed over the past decade or so. This section provides a brief survey of such experimental methods. A more detailed discussion of the experimental MS-based techniques can be found in excellent reviews by Clemmer and co-workers (17) and Jarrold (18). An overview of other (non-MS) experimental methods that are popular in the studies of biomolecules in the absence of solvent is also available (19).

10.2.1. Hydrogen–Deuterium Exchange in the Gas Phase as a Probe of the Protein Ion Structure

We have already considered the use of hydrogen–deuterium exchange (HDX) reactions in solution to probe native protein structures (Chapter 4), structures of non-native equilibrium (Chapter 5) and kinetic states (Chapters 6), as well as the applications of this technique to elucidate various mechanistic aspects of macromolecular function (Chapter 7). While the exchange reactions in all these experiments are carried out in solution, the initial HDX studies to probe macromolecular structure were actually carried out in the gas phase (20). In a typical experiment, ^2H_2O vapor was allowed to pass through the bulk of the biopolymer sample of known mass and composition. The extent of exchange was determined by either measuring the change in the isotopic makeup of the solvent [e.g., by measuring the change in the solvent density (20)] or the mass change of the biopolymer sample (21).

The advent of ESI MS allowed more elegant schemes for gas phase HDX measurements to be implemented, with the progress of the exchange reactions monitored as a mass change of the protein and peptide ions. Smith and co-workers demonstrated that gas phase HDX reactions can be used to probe higher-order structure of biomolecular ions in a solvent-free environment (22). McLafferty and co-workers have applied this technique to study the gas phase behavior of several protein ions (23). ESI-generated multiply charged protein ions showed slow exchange reactions pseudo-first-order kinetics with ^2H_2O molecules at low ($\sim 10^{-7}$ torr) pressure. The extent of exchange in compact (disulfide-linked) RNase A ions was significantly lower compared to that of the extended (denatured and reduced) form of the protein. The interpretation of the exchange data suggested that the latter form of the protein may actually exist in two different conformations in the gas phase. At least three different conformations were detected for multiply charged cytochrome *c* ions (23). Later work by the same group provided evidence that interconversion between these conformations can be induced in the gas phase by collisional or infrared (IR) heating, as well as charge stripping (24, 25). Attempts to characterize local structure of the putative conformers of cytochrome *c* ions in a solvent-free environment using methods of tandem MS have been only partially successful due to extensive hydrogen scrambling* (25). Protein ion conformations have been shown to depend on the charge state (26, 27), with the gas phase ions generally exhibiting much larger structural variety compared to their counterparts in solution (27). For example, at least half a dozen various conformations of ubiquitin have been postulated to exist in the gas phase based on the results of HDX measurements (27), while only three different states of this protein exist in solution (28). Another interesting example of using HDX reactions in the gas phase to probe protein ion structure has been reported recently by Marshall and co-workers, who evaluated conformations of troponin C ions carrying different numbers of charges (29). While the ^2H uptake levels exhibited by the higher charge state protein ions were largely independent of the solvent conditions used to produce the ions, exchange of the lower charge state ions was sensitive to the choice of solvent. This phenomenon was interpreted in terms of partial retention of the solution phase structure by the ESI-generated protein ions (29). Several groups have also begun evaluating the utility of gas phase HDX reactions to probe higher-order structure of oligonucleotide ions in a solvent-free environment (30–32).

*The extensive hydrogen scrambling reported by McLafferty and co-workers is most likely due to the slow heating rates afforded by the SORI CID utilized as a method to obtain fragment ions. See Chapter 5 for a more detailed discussion of issues related to the occurrence and extent of hydrogen scrambling.

A basic premise of the experiments discussed in the preceding paragraph is that different conformers can be distinguished based on the observed difference of ^{2}H uptake levels and kinetics. It must be mentioned, however, that higher-order structure of a biopolymer ion is not the only determinant of its exchange extent and kinetics in the gas phase. Lebrilla and co-workers pointed out that the exchange rates of protonated peptides critically depend on their proton affinities, with histidine and lysine-containing peptides exhibiting greater reactivity with deuterium-substituted alcohols (33). These phenomena, as well as the limited exchange exhibited by arginine-containing peptide ions (34), have led to a suggestion that the exchange reactions proceed via proton-bridged intermediates, whose formation is facilitated by the presence of lysine residues in peptide and protein ions (35). A systematic study of the gas phase HDX reactions between relatively small peptide ions and deuterated reactants conducted by Beauchamp and co-workers has concluded that the exchange can proceed through several different mechanisms (Figure 10.1) depending on the difference in proton affinities of the reactants (36, 37). The authors of this study concluded that the gas phase HDX processes involving even small model peptide ions are very complicated; therefore, assignment of gas phase structures based on the results of measurements of hydrogen–deuterium exchange rates should be approached with caution (37). Finally, Wyttenbach and Bowers used molecular dynamics simulations to model gas phase collisions between a protonated peptide and water molecules (38). The lifetimes of water–peptide ion collision complexes appear to be long enough to allow the water molecules to explore the peptide surface. Since the energetic constraints used in the calculations required that the exchange reactions proceed through a relay mechanism (Figure 10.1), the exchangeable hydrogen atoms not only had to be accessible from the surface but also had to allow formation of hydrogen bonds from ^2H_2O to both protonated and basic sites within the peptide ion. The results of calculations based on this model compared favorably with the experimental data (38).

10.2.2. Electrostatics as a Structural Probe: Kinetic Energy Release in Metastable Ion Dissociation and Proton Transfer Reaction in the Gas Phase

The geometry of multiply charged polypeptide ions in the gas phase can also be probed in some cases by measuring the intercharge distances if the identity of the charge-bearing residues within the macromolecular ion is known. Calculation of such distances can be made in some favorable cases based on measured electrostatic repulsion energy (using a Coulomb law). Fenselau and co-workers used mass-analyzed ion kinetic energy spectrometry (MIKES) to measure electrostatic repulsion within metastable multiply charged polypeptide ions undergoing unimolecular dissociation (39). Electrostatic repulsion between the charged fragment ions manifests itself as a kinetic energy release (KER), which can be accurately measured in electrostatic analyzers of magnetic sector mass spectrometers (40). The results of evaluating intercharge distances in multiply charged ions of an α-helical polypeptide melittin (41) and model β-sheet

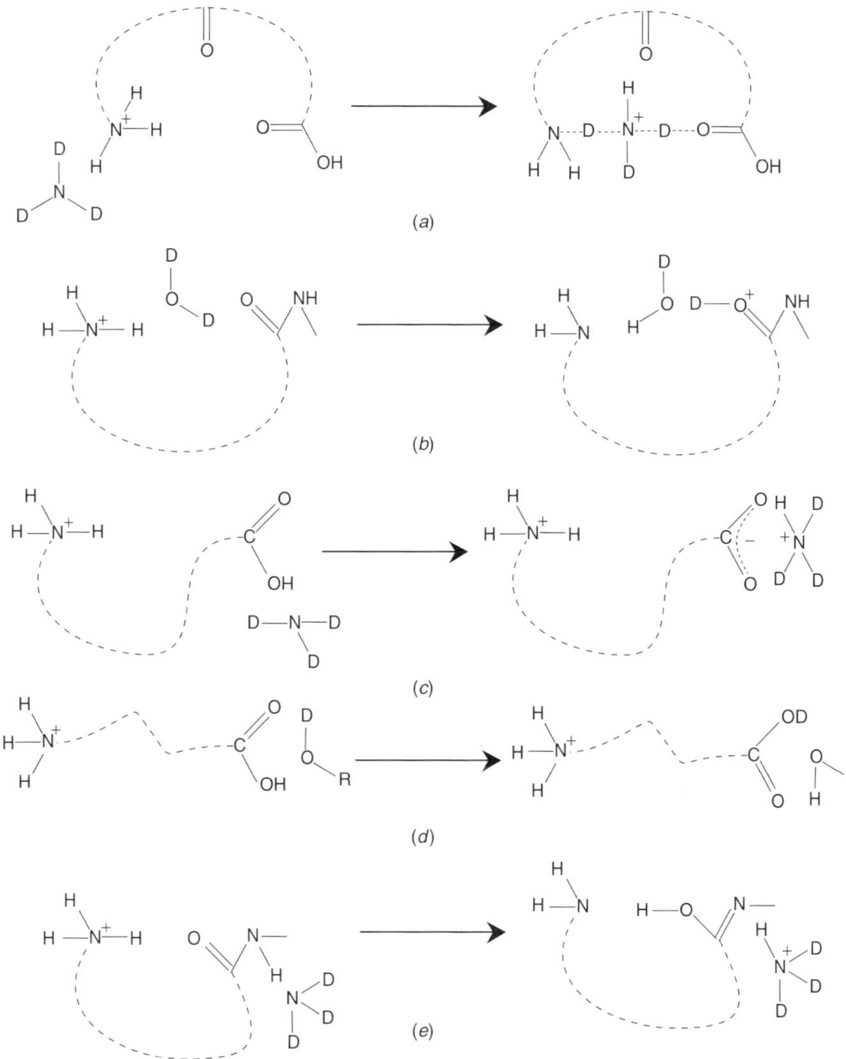

FIGURE 10.1. Schematic representation of the mechanisms of the gas phase hydrogen–deuterium exchange reactions of peptide ions proposed by Beauchamp and co-workers (37). The onium ion mechanism (**A**) allows the endothermic proton transfer (e.g., from the N-terminus of a protonated polyglycine) to occur by simultaneous solvation of the resultant ion. A relay mechanism (**B**) is favored in the case of less basic reactants (e.g., $^{2}H_2O$), whose proton affinity is too low to overcome the endothermicity of the proton transfer from the peptide ion. Exchange of the C-terminal hydrogen atoms occurs via the salt-bridge mechanism (**C**) in the case of basic reagents and the flip-flop mechanism (**D**) in the case of less basic reagents. A tautomer mechanism (**E**) is highly favorable for high proton affinity reactants (e.g., $N^{2}H_3$) and the amide protons of the peptide ion.

forming peptides (42) using MIKES were consistent with preservation of the secondary structure of these polypeptides in the absence of solvent. An alternative method for evaluation of electrostatic repulsion within multiply charged peptide and protein ions was introduced by Williams and co-workers (43). The electrostatic repulsion energy is assigned as the difference between the intrinsic gas phase basicity of the protonation sites and the apparent gas phase basicity of the polypeptide ion, whose measurements are based on the outcome of proton transfer reactions in the gas phase using a set of bases with varying proton affinities (44).

10.2.3. Ion Mobility Measurement and Biomolecular Shapes in the Gas Phase

Perhaps the most reliable evaluations of macromolecular geometry in the absence of solvent are deduced from measurements of ion mobility (or collisional cross sections) in the gas phase (45–47). The basic premise of this experimental technique is that macromolecular ions with different three-dimensional organization can be distinguished in the gas phase based on the observed differences in their collisional cross sections (and, therefore, their mobility). The experiment is typically carried out by first injecting ions into a static gas during a short time interval. A weak electrostatic field is applied across the ion drift region, forcing ions to separate according to their mobility. Measured drift velocities are used to calculate the mobility constants, from which the averaged (over all orientations of the ion) collisional cross sections can be deduced (46). Such experimentally obtained cross sections are compared to the cross sections calculated for "trial conformers," which are also averaged across all possible orientations.* The cross sections of macromolecular ions are calculated by treating these molecules as collections of hard spheres (46). This model (usually referred to as hard spheres approximation) can be extended to incorporate long-range interactions (48, 49). The hard spheres approximation ignores the details of the scattering process, such as the momentum transfer, which in some cases can lead to a significant underestimation of the collisional cross section (50). An additional complication arises due to the possibility of inelastic collisions between the macromolecular ions and the buffer gas molecules. Clemmer and Jarrold pointed out that the "evaluation of mobilities for comparison with experimental data is not [yet] a solved problem" (46). Nevertheless, in many cases ion mobility measurements do produce unique information on biopolymer ion geometry in the gas phase and have the capacity to clearly resolve different conformers (Figure 10.2).

*It is assumed that under weak field conditions ions do not align in the drift tube and sample all possible orientations throughout the measurement.

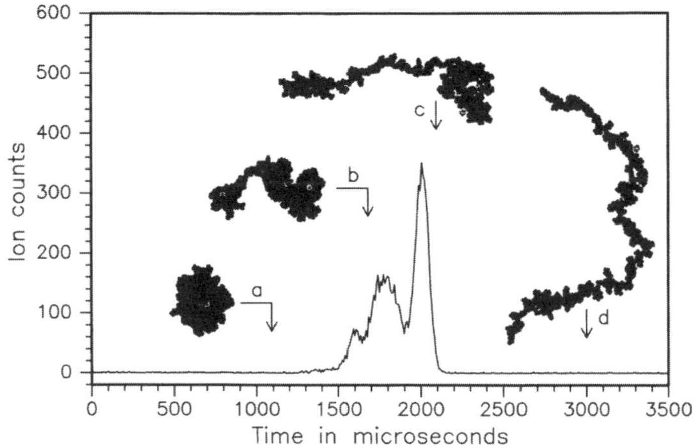

FIGURE 10.2. Drift time distribution for cytochrome *c* ions (charge state +7). Arrows indicate drift times calculated for several trial conformations of the protein: 1100 μs, crystal structure (**A**); 1680 μs, partially unfolded structure with an open heme crevice (**B**); 2070 μs, unfolded structure with retained helices (**C**); 2880 μs, typical random coil structure with no retained secondary or tertiary structural elements (**D**); and 3425 μs, fully extended (near-linear) conformation (**E**). Reproduced with permission from (86). © 1995 American Chemical Society.

10.3. PROTEIN AND PEPTIDE ION BEHAVIOR IN A SOLVENT-FREE ENVIRONMENT

10.3.1. Gas Phase Structures of Macromolecular Ions and Their Relevance to Conformations in Solution

There is a growing consensus that many biological macromolecules do retain their native-like structures when transferred from the solution to the gas phase by means of ESI. Perhaps the most convincing evidence is provided by ion mobility measurements carried out by Clemmer on ESI-generated ubiquitin ions (charge states +6 to +13) under various conditions in solution and the ESI interface (51). Three general conformer types were observed, including compact forms (favored for the +6 and +7 charge states) and partially folded states (favored for the +8 and +9 ions), as well as unfolded structures (favored for the +10 to +13 charge states). These observations are in good agreement with the charge state assignment of ubiquitin ions discussed in Chapter 5 (after taking into account instrumental effects). Interestingly, the populations of different conformational isomers were found to be highly sensitive to both solvent composition and the temperature in the ESI interface (51). For example, transitions from compact and partially folded states of the protein to a fully unfolded state could by induced by increasing the temperature in the ESI interface (Figure 10.3).

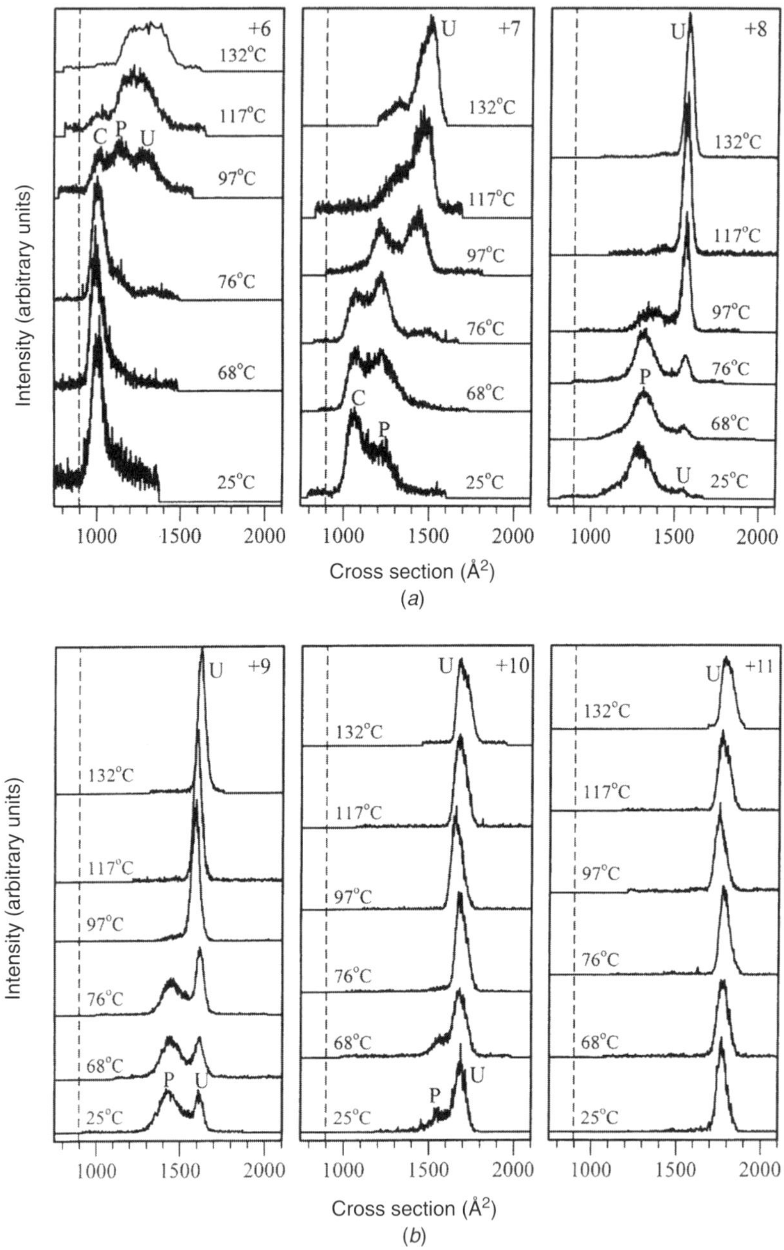

FIGURE 10.3. Cross-section distributions measured at different temperatures in the ESI interface region (capillary) for ubiquitin ions [charge states +6 to +8 (**A**); and +9 to +11, (**B**)]. Ubiquitin ions were generated by ESI from a water/acetonitrile/acetic acid solution (49:49:2, v:v:v). The labels C, P, and U on the panels denote compact, partially folded, and unfolded states, respectively. The dashed line shows the calculated cross section for the crystal structure of ubiquitin. Reproduced with permission from (51). © 1997 Elsevier.

In recent years, there has been considerable interest in questions related to putative formation of various elements of protein and polypeptide higher-order structure in the absence of solvent, with particular attention paid to the survival of the α-helices and β-sheets in the gas phase. As has already been mentioned, these elements of secondary structure are expected to maintain their integrity in the gas phase, a notion that had been confirmed by early MIKES measurements of the intercharge distances in model helical (41) and β-hairpin peptides (42). It is not clear, however, if such stable structures can actually be formed in the absence of solvent. Factors affecting formation, stability, and disruption of α-helices in the gas phase have been studied extensively by Jarrold and co-workers (52, 53), who established a thermodynamic scale of the helix propensities of different amino acids in the solvent-free environment.

A very interesting observation regarding formation of helical structures in the absence of solvent was made recently by Russell and co-workers (54), who used ion mobility measurements to study conformational preferences of tryptic peptide ions derived from the helical proteins hemoglobin and myoglobin. A particularly interesting aspect of this work is that the authors used MALDI to generate peptide ions, an ionization process that is usually expected to disrupt all native inter- and intramolecular interactions (important exceptions have been discussed in Chapter 4). Nevertheless, at least one peptide ion has a cross section, whose value is consistent with the notion that it maintains a helical conformation in the gas phase, while most other peptide ions lack any stable secondary structure (Figure 10.4). The peptides exhibiting anomalous behavior in the gas phase are suggested to be candidate "autonomous folding subunits," which have strong predisposition toward certain structure and may be important players in the protein folding processes (54).

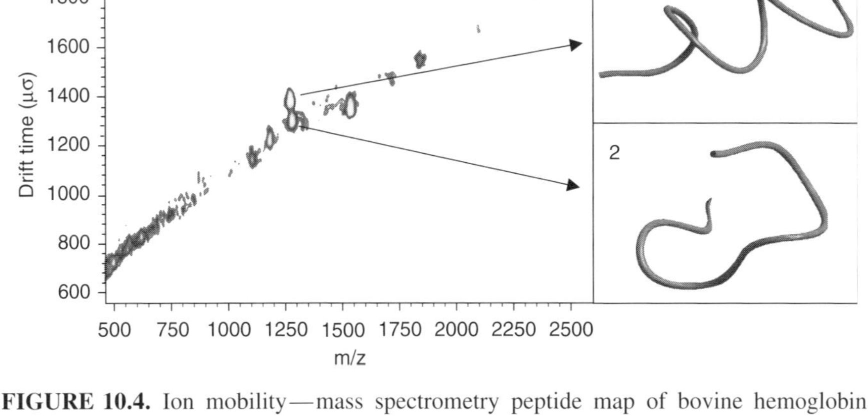

FIGURE 10.4. Ion mobility—mass spectrometry peptide map of bovine hemoglobin. The two low-energy structures (molecular dynamics calculations) are assigned to the peptides LLGNVLVVVLAR (*top right*) and LLVVYPWTQR (*bottom right*). Reproduced with permission from (54). © 2002 American Chemical Society.

10.3.2. Interaction of Protein Ions in the Gas Phase

Formation and stability of noncovalent complexes in the gas phase is another aspect of biopolymer ion behavior in the solvent-free environment that has enjoyed considerable attention in recent years. Although the issues related to the gas phase stability of complexes formed in solution are perhaps of greater practical importance, very intriguing questions also arise when one considers the rules governing the recognition and interaction processes in the gas phase (55). One particularly exciting aspect of gas phase recognition relates to chirality (56, 57), a phenomenon that is obviously very important in molecular biophysics. Pioneering work of Tao and Cooks clearly demonstrated the enormous potential of MS as a tool for the analysis of chiral molecules (58). The method exploits the differential recognition and binding between the chiral biomolecular ion and a set of enantiomeric molecules. Another interesting example of using MS to study chiral recognition in the gas phase was recently reported by Lebrilla and co-workers, who examined chiral guest-exchange reactions in the gas phase complexes of amino acids and β-cyclodextrin (59, 60). The experimental results were supported by the molecular modeling calculations, which indicated that enantioselectivity is governed by the differences in the binding interaction between the amino acid host and the β-cyclodextrin guest (Figure 10.5). A more detailed account of various MS based techniques that are currently used to study enantioselective processes can be found in an excellent review by Filippi et al. (61).

Until very recently, all studies of protein complexes in the solvent-free environment targeted macromolecular associations that were initially formed in solution and then transferred to the gas phase (usually by means of ESI). Formation of biopolymer associations in the gas phase (e.g., via ion–molecule reactions) was not feasible due to the impossibility of keeping sufficient quantities of neutral macromolecules in the gas phase. At the same time, gas phase ion–ion associations (e.g., within the high-pressure region of the ESI interface) are inhibited by strong electrostatic repulsion between the like charges. Recently, McLuckey and co-workers suggested a very elegant solution to this problem by the simultaneous trapping of ions of different polarities (62). Reactions of oppositely charged protein ions in the gas phase give rise to the formation of protein–protein complexes (Figure 10.6). However, formation of the protein–protein complexes competes with varying degrees of proton transfer (63). From the thermodynamic point of view, full neutralization of the ion of lowest charge state (either positive or negative) should be the energetically preferred outcome. Therefore, incomplete proton transfer observed by McLuckey and co-workers is rather intriguing, suggesting that the processes giving rise to the product ion distribution are under kinetic control. Of particular interest is the question of what are the key properties of the protein ions (other than their charge states) that play roles in determining the relative contributions of ion–ion complex formation versus simple proton transfer. While several different models of ion–ion interaction have been considered, the experimental results seem to have been best rationalized in terms of an orbiting ion–ion interaction model. In this case the overall rate of the ion–ion reaction is determined by the formation of an electrostatically bound complex,

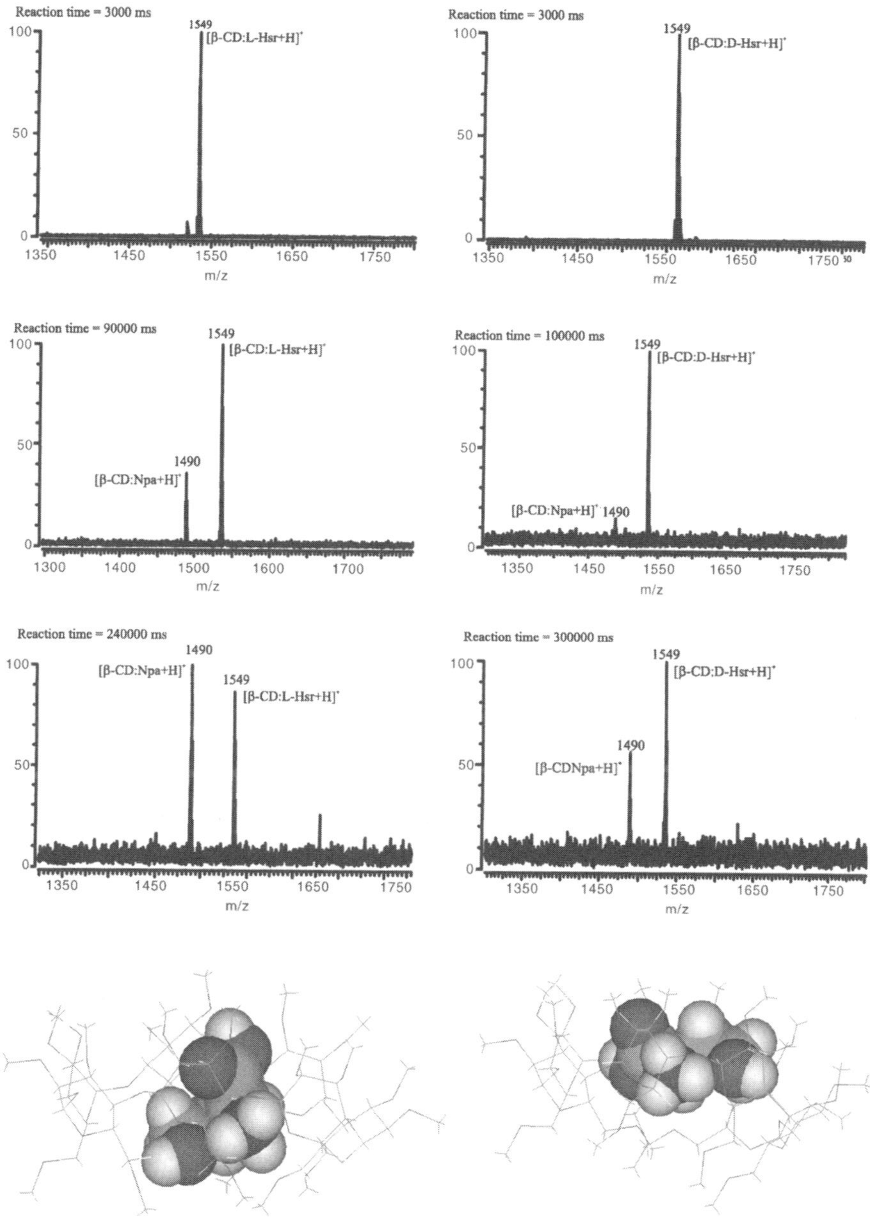

FIGURE 10.5. Representative mass spectra of the reaction of the β-cyclodextrin/L-homoserine (*left*) and β-cyclodextrin/D-homoserine (*right*) complexes with n-propyl amine in the gas phase. The bottom diagram shows lowest energy structures for the two β-cyclodextrin/homoserine complexes obtained from the molecular dynamics simulations. Adapted with permission from (60). © 2003 Elsevier.

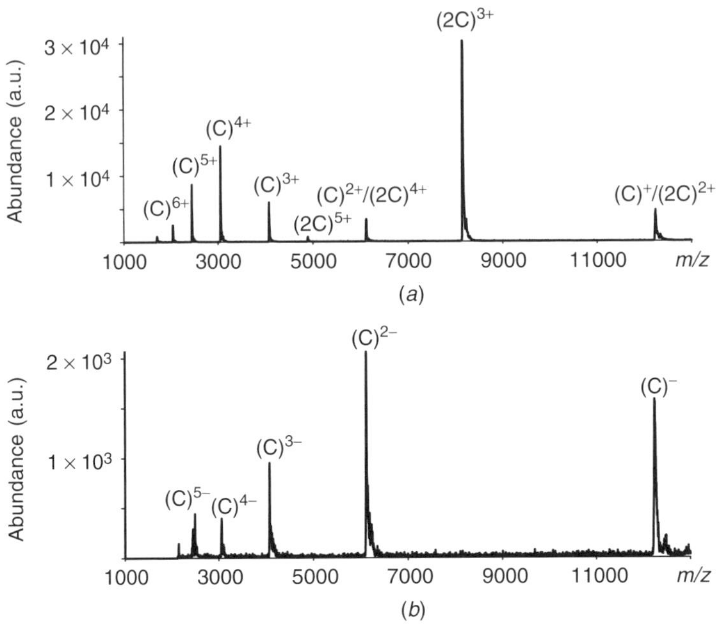

FIGURE 10.6. Mass spectra of the products of ion–ion reaction between the +8 and −5 ions of cytochrome *c* acquired in the positive (**A**) and negative (**B**) ion modes following 100 ms of reaction time. Reproduced with permission from (63). © 2003 American Chemical Society.

within which both proton transfer reactions and a physical collision between the partners can occur (63). Interestingly, the relatively limited set of observations of the dissociation of protein complexes formed in the gas phase indicates similar behavior to those formed in solution,* although the gas phase complexes can be synthesized in various ways and the dissociation behavior can retain memory of how the complex is produced (62).

10.3.3. Physical Properties of Biomolecular Ions in the Gas Phase: Spectroscopic Measurements in a Solvent-Free Environment

While most studies of macromolecular ion behavior in the gas phase employ various ion–molecule reactions and MS identification of products as an interrogation tool, there is a growing interest in the characterization of biomolecular ions using other experimental techniques that do not rely on MS detection. Prospects of the spectroscopic characterization of biomolecular ions in the gas phase seem to be particularly promising (19), given the great success of this technique in the studies of molecules of nonbiological origin, particularly inorganic cluster ions (64).

*This should not be confused with the protein complex dissociation pathways in solution, which are generally quite different from the gas phase dissociation mechanism, as noted earlier in this chapter.

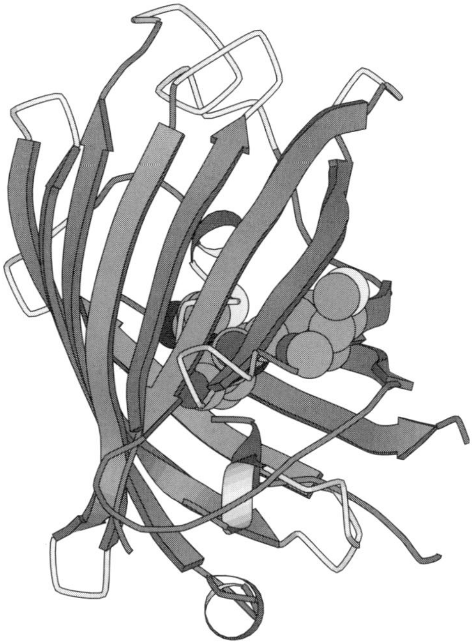

FIGURE 10.7. Crystal structure (ribbon diagram) of green fluorescent protein from *Aequorea Victoria* (PDB ID 1EMA). The chromophore is represented with a space-fill model.

The synergistic effect of combining ESI MS and optical spectroscopy became evident after the successful application of this combined technique to study various physicochemical processes in the ESI interface. Van Berkel and co-workers used absorption spectroscopy to study transformations of biomolecular ions during the ESI process (65). The focus of this study was a model porphyrin, inoctaethylporphyrin, which exists in aqueous solution predominantly as the doubly charged, diprotonated molecule, while the most abundant ionic species observed in the ESI mass spectrum is the singly charged ion. The direct optical spectroscopic measurements of the ions in solution (absorption spectra) and in the electrospray plume (fluorescence excitation spectra) were correlated with the ion distribution observed in the gas phase in order to determine how and when the transformation from the di-cation to the mono-cation occurs. The conversion of the doubly protonated porphyrin species to the singly protonated species was found to occur relatively late in the ESI process via the loss of a charged solvent molecule/cluster (65). More recently, Zhou and Cook used laser-induced fluorescence spectroscopy to profile solvent fractionation in an electrospray plume by using a solvatochromic dye (for which spectral features are sensitive to the solvent polarity) (66). Clemmer and co-workers used laser-induced fluorescence to track conformational changes within ESI-generated cytochrome *c* ions associated with the loss of solvent (67).

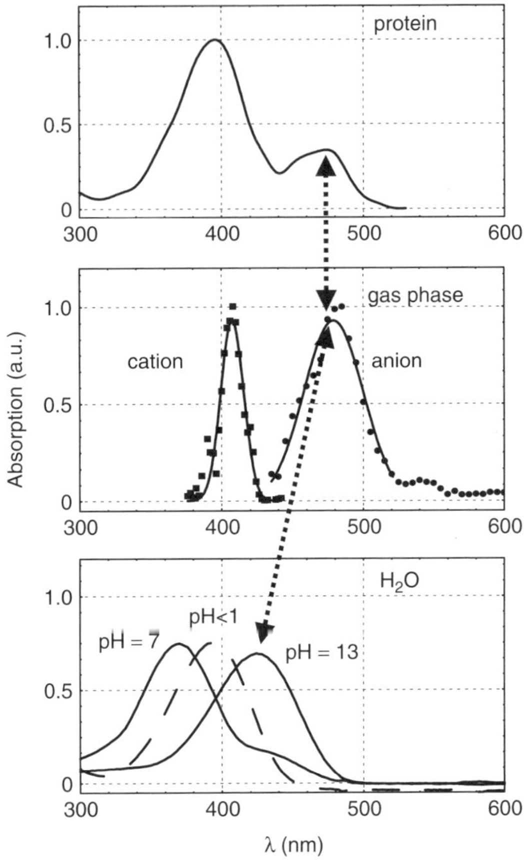

FIGURE 10.8. (*Top*) Absorption spectrum of a wild-type green fluorescent protein from *Aequorea Victoria*. The two peaks are ascribed to the neutral and the negatively charged states of the chromophore in the protein. (*Middle*) Absorption spectra of the model chromophore in vacuo (solid lines are Gaussian fits for the experimental data). (*Bottom*) Absorption spectra of the model chromophore in aqueous solutions acquired at different pH values. Reproduced with permission from (74). © 2002 Springer-Verlag GmbH.

While the studies discussed above focus on the processes occurring within the ESI interface and take advantage of the relatively high concentration of biomolecular ions in that region, spectroscopic studies of macromolecular ions in vacuo require some means to accumulate sufficient quantities of the solvent-free ions to afford spectroscopic measurements. This can be achieved by combining an ESI source (as an external ion generation device) and an ion storing device (which is also used as a "gas phase" optical cell), such as electrostatic storage rings (68, 69) or ion trapping cells of FT ICR (70, 71) and quadrupole ion trap (72) mass spectrometers. One particularly interesting study that we will mention here focuses on the intrinsic absorption characteristics of a chromophore from the green fluorescent protein (GFP) from jellyfish *Aequorea victoria* (73, 74). GFP (Figure 10.7) has been enjoying a great deal of attention recently due to its amazing ability to generate a highly visible, efficiently emitting internal fluorophore (75). GFP exhibits two absorption maxima at 395 nm and 477 nm (Figure 10.8), which are attributed to the neutral and anionic forms of the chromophore (*p*-hydroxybenzylidene-imidazoline). The spectroscopic characteristics of this chromophore in solution have been found to be very sensitive to its microenvironment, hence the interest in establishing the intrinsic properties of this molecule. The absorption spectrum of the model chromophore acquired in the absence of solvent (Figure 10.8) clearly indicates that one of the absorption bands (the anionic band) is determined almost completely by the intrinsic chemical properties of the chromophore, rather than resulting from the interaction with the solvent or the amino acid side chains of GFP (73, 74).

10.4. PROTEIN HYDRATION IN THE GAS PHASE: BRIDGING "MICRO" AND "MACRO"

Perhaps one of the most exciting opportunities in the studies of the structure and properties of biomolecular ions in the gas phase relates to the behavior of partially solvated macromolecules. In the beginning we discussed the importance of solvent as a potent modulator of biomolecular structure, dynamics, and function, in particular the neighboring layers of solvent molecules forming a transient solvation shell. Williams and co-workers demonstrated that under mild ESI conditions partial solvation around the peptide ions may be preserved (76). Hydrated ions of a small peptide gramicidin S carrying up to 60 water molecules have been observed, while the extent of the partial solvation with less polar solvent molecules is not as significant (Figure 10.9). The extent of peptide hydration that can be achieved in such experiments is significant enough to allow hydrogen bonding networks to be formed along the surface of the peptide (77). The distribution of the hydrated ions was found to be consistent with their formation via two different mechanisms. The first involves solvent condensation during adiabatic expansion in the ESI interface region. Alternatively, the hydrated ions can be formed as a result of solvent evaporation from small droplets or extensively hydrated peptide ions (77).

Fenn and Zhan produced hydrated peptide ions by dispersing them at atmospheric pressure into nitrogen containing precisely controlled concentrations of water vapor. The resulting gas–ion mixture was passed through a supersonic free jet expansion into a vacuum chamber containing a quadrupole mass analyzer. Subsequent mass spectral detection of the product ions revealed sequences of peaks due to ions of the parent peptide with varying numbers of adduct water molecules, with the extent of hydration dictated by the hydrophobicity of the peptide (78).

Regardless of the mechanism, the ability to generate partially hydrated macromolecular ions in the gas phase offers tremendous opportunities for the study of

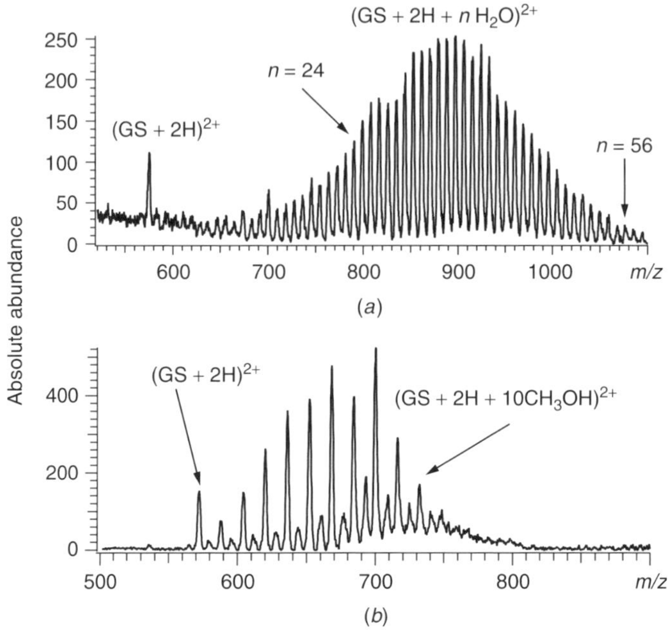

FIGURE 10.9. ESI mass spectra of partially solvated gramicidin S ions obtained from aqueous (**a**) and methanol (**b**) solutions. Adapted with permission from (77). © 1999 American Society for Mass Spectrometry.

protein–solvent interactions, a field that has been attracting significant attention in recent years and is ripe with controversies (79–81). The exciting opportunities for studying solute–solvent interaction processes at the molecular level can be illustrated by several recent studies focusing on various aspects of biopolymer hydration. Beauchamp and co-workers have used water evaporation from very large clusters to attain low temperature (estimated as 130–150 K) within the "surviving" clusters. One particularly intriguing aspect of this work is the observation of specific solvation for some (but not all) peptides, that is, the "magic numbers" of solvent molecules corresponding to very stable clusters (82). Uneven distribution of the number of water molecules in partially solvated peptide ions undergoing "evaporation" (Figure 10.10) has been also observed by Williams and co-workers (77) and is likely to reflect preferential solvent molecule distributions within the solvation shell.

Kohtani and Jarrold obtained equilibrium constants for the adsorption of the first water molecule on to a variety of unsolvated alanine-based peptides and determined both enthalpic and entropic contributions (83). These studies were designed to examine the effects of conformation, charge, and composition on the propensity for peptides to bind water. Water adsorption was found to occur more readily on the globular peptides than on helical ones, as several of the singly charged helical peptides were not observed to adsorb a water molecule even at

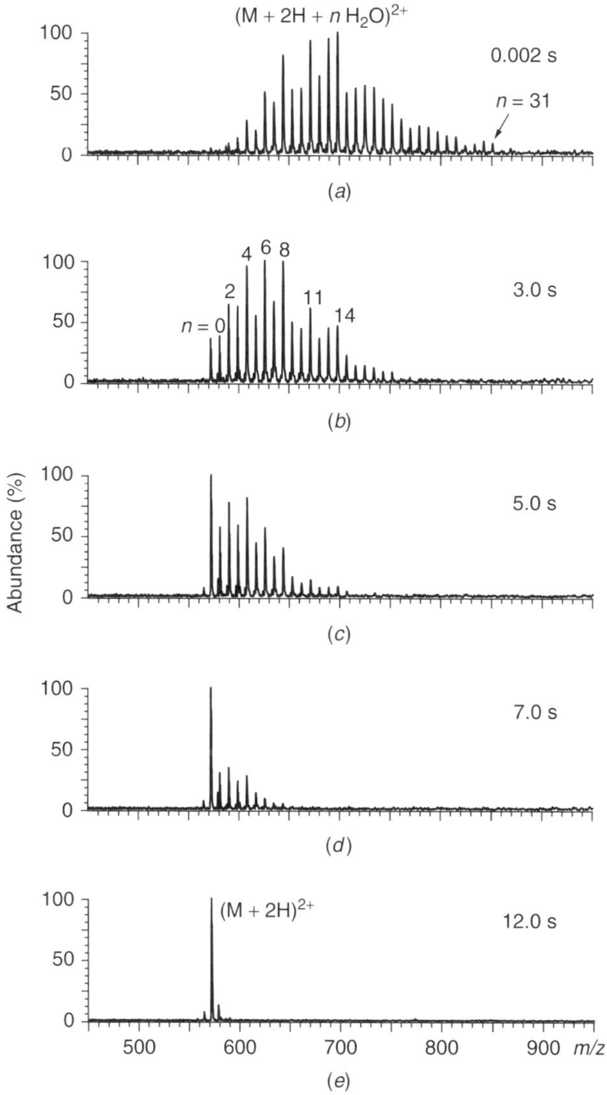

FIGURE 10.10. ESI spectra of gramicidin S obtained from an aqueous solution. Hydrated states of the doubly charged ion shown in the spectra correspond to different ion storage times. Reproduced with permission from (77). © 1999 American Society for Mass Spectrometry.

−50 °C. The ability to establish a network of hydrogen bonds to several different hydrogen bonding partners emerges as a critical factor for strong binding of the water molecule to the peptide ion (83). Bowers and co-workers measured the thermodynamics of the sequential hydration of protonated peptides (84). The binding energies were determined to be in the range of 7–15 kcal/mol, with the highest values corresponding to the binding of the first water molecules. The molecular mechanics calculations indicated that in the case of small peptide ions the first few water molecules bind preferentially to the protonation site, unless the charge is extensively solvated in the solvent-free structure. However, in the case of larger peptide ions, the charge remote groups appear to compete effectively for the binding of the first few water molecules (84). Finally, the ability to produce and manipulate hydrated peptide and protein ions in the gas phase is likely to advance our understanding of the mechanisms of hydrogen exchange both in solution and in the solvent-free environment (85).

REFERENCES

1. Mattos C. (2002). Protein–water interactions in a dynamic world. *Trends Biochem. Sci.* **27**: 203–208.
2. Makarov V., Pettitt B. M., Feig M. (2002). Solvation and hydration of proteins and nucleic acids: a theoretical view of simulation and experiment. *Acc. Chem. Res.* **35**: 376–384.
3. Bellissent-Funel M. C. (2000). Hydration in protein dynamics and function. *J. Mol. Liq.* **84**: 39–52.
4. Makarov V. A., Andrews B. K., Pettitt B. M. (1998). Reconstructing the protein–water interface. *Biopolymers* **45**: 469–478.
5. Timasheff S. N. (2002). Protein hydration, thermodynamic binding, and preferential hydration. *Biochemistry* **41**: 13473–13482.
6. Wolynes P. G. (1995). Biomolecular folding in vacuo? *Proc. Natl. Acad. Sci. U.S.A.* **92**: 2426–2427.
7. Rockwood A. L., Busman M., Smith R. D. (1991). Coulombic effects in the dissociation of large highly charged ions. *Int. J. Mass Spectrom. Ion Proc.* **111**: 103–129.
8. Weinkauf R., Schermann J. P., de Vries M. S., Kleinermanns K. (2002). Molecular physics of building blocks of life under isolated or defined conditions. *Eur. Phys. J. D* **20**: 309–316.
9. Raulin-Cerceau F., Maurel M. C., Schneider J. (1998). From Panspermia to bioastronomy, the evolution of the hypothesis of universal life. *Origins Life Evol. B.* **28**: 597–612.
10. Ehrenfreund P., Irvine W., Becker L., Blank J., Brucato J. R., Colangeli L., Derenne S., Despois D., Dutrey A., Fraaije H., Lazcano A., Owen T., Robert F. (2002). Astrophysical and astrochemical insights into the origin of life. *Rep. Prog. Phys.* **65**: 1427–1487.
11. Charnley S., Ehrenfreund P., Kuan Y. J. (2003). Molecules in space. *Phys. World* **16**: 35–38.
12. Loo J. A. (1997). Studying noncovalent protein complexes by electrospray ionization mass spectrometry. *Mass Spectrom. Rev.* **16**: 1–23.
13. Loo J. A. (2000). Electrospray ionization mass spectrometry: a technology for studying noncovalent macromolecular complexes. *Int. J. Mass Spectrom.* **200**: 175–186.

14. Veenstra T. D. (1999). Electrospray ionization mass spectrometry in the study of biomolecular noncovalent interactions. *Biophys. Chem.* **79**: 63–79.

15. Nesatyy V. J. (2002). Mass spectrometry evaluation of the solution and gas-phase binding properties of noncovalent protein complexes. *Int. J. Mass Spectrom.* **221**: 147–161.

16. Benesch J. L., Sobott F., Robinson C. V. (2003). Thermal dissociation of multimeric protein complexes by using nanoelectrospray mass spectrometry. *Anal. Chem.* **75**: 2208–2214.

17. Hoaglund-Hyzer C. S., Counterman A. E., Clemmer D. E. (1999). Anhydrous protein ions. *Chem. Rev.* **99**: 3037–3080.

18. Jarrold M. F. (2000). Peptides and proteins in the vapor phase. *Annu. Rev. Phys. Chem.* **51**: 179–207.

19. Robertson E. G., Simons J. P. (2001). Getting into shape: conformational and supramolecular landscapes in small biomolecules and their hydrated clusters. *Phys. Chem. Chem. Phys.* **3**: 1–18.

20. Champetier G., Viallard R. (1938). Reaction d'echange de la cellulose et de l'eau lourde. Hydratation de la cellulose. *Bull. Soc. Chim. Fr.* **33**: 1042–1048.

21. Burley R. W., Nicholls C. H., Speakman J. B. (1955). The crystalline/amorphous ratio of keratin fibres. Part II. The hydrogen–deuterium exchange reaction. *J. Text. Inst.* **46**: T427–T432.

22. Winger B. E., Lightwahl K. J., Rockwood A. L., Smith R. D. (1992). Probing qualitative conformation differences of multiply protonated gas-phase proteins via H/D isotopic exchange with D_2O. *J. Am. Chem. Soc.* **114**: 5897–5898.

23. Suckau D., Shi Y., Beu S. C., Senko M. W., Quinn J. P., Wampler F. M., McLafferty F. W. (1993). Coexisting stable conformations of gaseous protein ions. *Proc. Natl. Acad. Sci. U.S.A.* **90**: 790–793.

24. Wood T. D., Chorush R. A., Wampler F. M., Little D. P., O'Connor P. B., McLafferty F. W. (1995). Gas-phase folding and unfolding of cytochrome *c* cations. *Proc. Natl. Acad. Sci. U.S.A.* **92**: 2451–2454.

25. McLafferty F. W., Guan Z. Q., Haupts U., Wood T. D., Kelleher N. L. (1998). Gaseous conformational structures of cytochrome *c*. *J. Am. Chem. Soc.* **120**: 4732–4740.

26. Breuker K., Oh H. B., Horn D. M., Cerda B. A., McLafferty F. W. (2002). Detailed unfolding and folding of gaseous ubiquitin ions characterized by electron capture dissociation. *J. Am. Chem. Soc.* **124**: 6407–6420.

27. Oh H., Breuker K., Sze S. K., Ge Y., Carpenter B. K., McLafferty F. W. (2002). Secondary and tertiary structures of gaseous protein ions characterized by electron capture dissociation mass spectrometry and photofragment spectroscopy. *Proc. Natl. Acad. Sci. U.S.A.* **99**: 15863–15868.

28. Mohimen A., Dobo A., Hoerner J. K., Kaltashov I. A. (2003). A chemometric approach to detection and characterization of multiple protein conformers in solution using electrospray ionization mass spectrometry. *Anal. Chem.* **75**: 4139–4147.

29. Wang F., Freitas M. A., Marshall A. G., Sykes B. D. (1999). Gas-phase memory of solution-phase protein conformation: H/D exchange and Fourier transform ion cyclotron resonance mass spectrometry of the N-terminal domain of cardiac troponin C. *Int. J. Mass Spectrom.* **192**: 319–325.

30. Robinson J. M., Greig M. J., Griffey R. H., Mohan V., Laude D. A. (1998). Hydrogen/deuterium exchange of nucleotides in the gas phase. *Anal. Chem.* **70**: 3566–3571.
31. Griffey R. H., Greig M. J., Robinson J. M., Laude D. A. (1999). Gas-phase hydrogen–deuterium exchange in phosphorothioate d(GTCAG) and d(TCGAT). *Rapid Commun. Mass Spectrom.* **13**: 113–117.
32. Freitas M. A., Marshall A. G. (2001). Gas phase RNA and DNA ions 2. Conformational dependence of the gas-phase H/D exchange of nucleotide 5′-monophosphates. *J. Am. Soc. Mass Spectrom.* **12**: 780–785.
33. Green M. K., Gard E., Bregar J., Lebrilla C. B. (1995). H-D exchange kinetics of alcohols and protonated peptides—effects of structure and proton affinity. *J. Mass Spectrom.* **30**: 1103–1110.
34. Dookeran N. N., Harrison A. G. (1995). Gas-phase H-D exchange-reactions of protonated amino-acids and peptides with ND_3. *J. Mass Spectrom.* **30**: 666–674.
35. Green M. K., Lebrilla C. B. (1998). The role of proton-bridged intermediates in promoting hydrogen–deuterium exchange in gas-phase protonated diamines, peptides and proteins. *Int. J. Mass Spectrom.* **175**: 15–26.
36. Campbell S., Rodgers M. T., Marzluff E. M., Beauchamp J. L. (1994). Structural and energetic constraints on gas-phase hydrogen–deuterium exchange-reactions of protonated peptides with D_2O, CD_3OD, CD_3CO_2D, and ND_3. *J. Am. Chem. Soc.* **116**: 9765–9766.
37. Campbell S., Rodgers M. T., Marzluff E. M., Beauchamp J. L. (1995). Deuterium exchange reactions as a probe of biomolecule structure. Fundamental studies of gas phase H/D exchange reactions of protonated glycine oligomers with D_2O, CD_3OD, CD_3CO_2D, and ND_3. *J. Am. Chem. Soc.* **117**: 12840–12854.
38. Wyttenbach T., Bowers M. T. (1999). Gas phase conformations of biological molecules: the hydrogen/deuterium exchange mechanism. *J. Am. Soc. Mass Spectrom.* **10**: 9–14.
39. Kaltashov I. A., Li A., Szilagyi Z., Vekey K., Fenselau C. (2000). Secondary structure of peptide ions in the gas phase evaluated by MIKE spectrometry. Relevance to native conformations. *Methods Mol. Biol.* **146**: 133–146.
40. Szilagyi Z., Drahos L., Vekey K. (1997). Conformation of doubly protonated peptides studied by charge-separation reactions in mass spectrometry. *J. Mass Spectrom.* **32**: 689–696.
41. Kaltashov I. A., Fenselau C. (1997). Stability of secondary structural elements in a solvent-free environment: the alpha helix. *Proteins* **27**: 165–170.
42. Li A., Fenselau C., Kaltashov I. A. (1998). Stability of secondary structural elements in a solvent-free environment. II: the β-pleated sheets. *Proteins* Suppl. **2**: 22–27.
43. Gross D. S., Schnier P. D., Rodriges-Cruz S. E., Hagerquist C., Williams E. R. (1996). Conformations and folding of lysozyme ions in vacuo. *Proc. Natl. Acad. Sci. U.S.A.* **93**: 3143–3148.
44. Williams E. R. (1996). Proton transfer reactivity of large multiply charged ions. *J. Mass Spectrom.* **31**: 831–842.
45. Vonhelden G., Wyttenbach T., Bowers M. T. (1995). Conformation of macromolecules in the gas phase. *Science* **267**: 1483–1485.
46. Clemmer D. E., Jarrold M. F. (1997). Ion mobility measurements and their applications to clusters and biomolecules. *J. Mass Spectrom.* **32**: 577–592.

47. Wyttenbach T., Bowers M. T. (2003). Gas-phase conformations: the ion mobility/ion chromatography method. *Mod. Mass Spectrom.* **225**: 207–232.
48. Mesleh M. F., Hunter J. M., Shvartsburg A. A., Schatz G. C., Jarrold M. F. (1996). Structural information from ion mobility measurements: effects of the long-range potential. *J. Phys. Chem.* **100**: 16082–16086.
49. Wyttenbach T., vonHelden G., Batka J. J., Carlat D., Bowers M. T. (1997). Effect of the long-range potential on ion mobility measurements. *J. Am. Soc. Mass Spectrom.* **8**: 275–282.
50. Shvartsburg A. A., Jarrold M. F. (1996). An exact hard-spheres scattering model for the mobilities of polyatomic ions. *Chem. Phys. Lett.* **261**: 86–91.
51. Li J., Taraszka J. A., Counterman A. E., Clemmer D. E. (1999). Influence of solvent composition and capillary temperature on the conformations of electrosprayed ions: unfolding of compact ubiquitin conformers from pseudonative and denatured solutions. *Int. J. Mass Spectrom.* **185–187**: 37–47.
52. Hudgins R. R., Ratner M. A., Jarrold M. F. (1998). Design of helices that are stable in vacuo. *J. Am. Chem. Soc.* **120**: 12974–12975.
53. Breaux G. A., Jarrold M. F. (2003). Probing helix formation in unsolvated peptides. *J. Am. Chem. Soc.* **125**: 10740–10747.
54. Ruotolo B. T., Verbeck G. F., Thomson L. M., Gillig K. J., Russell D. H. (2002). Observation of conserved solution-phase secondary structure in gas-phase tryptic peptides. *J. Am. Chem. Soc.* **124**: 4214–4215.
55. Schalley C. A. (2001). Molecular recognition and supramolecular chemistry in the gas phase. *Mass Spectrom. Rev.* **20**: 253–309.
56. Podlech J. (2001). Origin of organic molecules and biomolecular homochirality. *Cell. Mol. Life Sci.* **58**: 44–60.
57. Compton R. N., Pagni R. M. (2002). The chirality of biomolecules. *Adv. Atom. Mol. Opt. Phys.* **48**: 219–261.
58. Tao W. A., Cooks R. G. (2003). Chiral analysis by MS. *Anal. Chem.* **75**: 25A–231A.
59. Grigorean G., Gronert S., Lebrilla C. B. (2002). Enantioselective gas-phase ion–molecule reactions in a quadrupole ion trap. *Int. J. Mass Spectrom.* **219**: 79–87.
60. Gal J. F., Stone M., Lebrilla C. B. (2003). Chiral recognition of non-natural α-amino acids. *Int. J. Mass Spectrom.* **222**: 259–267.
61. Filippi A., Giardini A., Piccirillo S., Speranza M. (2000). Gas-phase enantioselectivity. *Int. J. Mass Spectrom.* **198**: 137–163.
62. Wells J. M., Chrisman P. A., McLuckey S. A. (2001). Formation of protein–protein complexes in vacuo. *J. Am. Chem. Soc.* **123**: 12428–12429.
63. Wells J. M., Chrisman P. A., McLuckey S. A. (2003). Formation and characterization of protein–protein complexes in vacuo. *J. Am. Chem. Soc.* **125**: 7238–7249.
64. Bieske E. J., Dopfer O. (2000). High-resolution spectroscopy of cluster ions. *Chem. Rev.* **100**: 3963–3998.
65. Chillier X. F. D., Monnier A., Bill H., Gulacar F. O., Buchs A., McLuckey S. A., Van Berkel G. J. (1996). A mass spectrometry and optical spectroscopy investigation of gas-phase ion formation in electrospray. *Rapid Commun. Mass Spectrom.* **10**: 299–304.
66. Zhou S., Cook K. D. (2000). Probing solvent fractionation in electrospray droplets with laser-induced fluorescence of a solvatochromic dye. *Anal. Chem.* **72**: 963–969.

67. Ideue S., Sakamoto K., Honma K., Clemmer D. E. (2001). Conformational change of electrosprayed cytochrome *c* studied by laser-induced fluorescence. *Chem. Phys. Lett.* **337**: 79–84.

68. Andersen J. U., Hvelplund P., Nielsen S. B., Tomita S., Wahlgreen H., Moller S. P., Pedersen U. V., Forster J. S., Jorgensen T. J. D. (2002). The combination of an electrospray ion source and an electrostatic storage ring for lifetime and spectroscopy experiments on biomolecules. *Rev. Sci. Instrum.* **73**: 1284–1287.

69. Tanabe T., Noda K. (2003). Storage of bio-molecular ions in the electrostatic storage ring. *Nucl. Instrum. Meth. A* **496**: 233–237.

70. Wang Y., Hendrickson C. L., Marshall A. G. (2001). Direct optical spectroscopy of gas-phase molecular ions trapped and mass-selected by ion cyclotron resonance: laser-induced fluorescence excitation spectrum of hexafluorobenzene ($C_6F_6{}^+$). *Chem. Phys. Lett.* **334**: 69–75.

71. Cage B., McFarland M. A., Hendrickson C. L., Dalal N. S., Marshall A. G. (2002). Resolution of individual component fluorescence lifetimes from a mixture of trapped ions by laser-induced fluorescence/ion cyclotron resonance. *J. Phys. Chem. A* **106**: 10033–10036.

72. Khoury J. T., Rodriguez-Cruz S. E., Parks J. H. (2002). Pulsed fluorescence measurements of trapped molecular ions with zero background detection. *J. Am. Soc. Mass Spectrom.* **13**: 696–708.

73. Nielsen S. B., Lapierre A., Andersen J. U., Pedersen U. V., Tomita S., Anderson L. H. (2001). Absorption spectrum of the green fluorescent protein chromophore anion in vacuo. *Phys. Rev. Lett.* **87**: 2281021–2281024.

74. Andersen L. H., Lapierre A., Nielsen S. B., Nielsen I. B., Pedersen S. U., Pedersen U. V., Tomita S. (2002). Chromophores of the green fluorescent protein studied in the gas phase. *Eur. Phys. J. D* **20**: 597–600.

75. Tsien R. Y. (1998). The green fluorescent protein. *Annu. Rev. Biochem.* **67**: 509–544.

76. Rodriguez-Cruz S. E., Klassen J. S., Williams E. R. (1997). Hydration of gas-phase gramicidin S $(M+2H)^{2+}$ ions formed by electrospray: the transition from solution to gas-phase structure. *J. Am. Soc. Mass Spectrom.* **8**: 565–568.

77. Rodriguez-Cruz S. E., Klassen J. S., Williams E. R. (1999). Hydration of gas-phase ions formed by electrospray ionization. *J. Am. Soc. Mass Spectrom.* **10**: 958–968.

78. Zhan D., Fenn J. B. (2002). Gas phase hydration of electrospray ions from small peptides. *Int. J. Mass Spectrom.* **219**: 1–10.

79. Timasheff S. N. (1998). In disperse solution, "osmotic stress" is a restricted case of preferential interactions. *Proc. Natl. Acad. Sci. U.S.A.* **95**: 7363–7367.

80. Parsegian V. A., Rand R. P., Rau D. C. (2000). Osmotic stress, crowding, preferential hydration, and binding: a comparison of perspectives. *Proc. Natl. Acad. Sci. U.S.A.* **97**: 3987–3992.

81. Timasheff S. N. (2002). Protein–solvent preferential interactions, protein hydration, and the modulation of biochemical reactions by solvent components. *Proc. Natl. Acad. Sci. U.S.A.* **99**: 9721–9726.

82. Lee S. W., Freivogel P., Schindler T., Beauchamp J. L. (1998). Freeze-dried biomolecules: FT-ICR studies of the specific solvation of functional groups and clathrate formation observed by the slow evaporation of water from hydrated peptides and model compounds in the gas phase. *J. Am. Chem. Soc.* **120**: 11758–11765.

83. Kohtani M., Jarrold M. F. (2002). The initial steps in the hydration of unsolvated peptides: water molecule adsorption on alanine-based helices and globules. *J. Am. Chem. Soc.* **124**: 11148–11158.

84. Liu D. F., Wyttenbach T., Barran P. E., Bowers M. T. (2003). Sequential hydration of small protonated peptides. *J. Am. Chem. Soc.* **125**: 8458–8464.

85. Wyttenbach T., Paizs B., Barran P., Breci L., Liu D. F., Suhai S., Wysocki V. H., Bowers M. T. (2003). The effect of the initial water of hydration on the energetics, structures, and H/D exchange mechanism of a family of pentapeptides: an experimental and theoretical study. *J. Am. Chem. Soc.* **125**: 13768–13775.

86. Clemmer D. E., Hudgins R. R., Jarrold M. F. (1995). Naked protein conformations—cytochrome *c* in the gas phase. *J. Am. Chem. Soc.* **117**: 10141–10142.

11

MASS SPECTROMETRY ON THE MARCH: WHERE NEXT? FROM MOLECULAR BIOPHYSICS TO STRUCTURAL BIOLOGY, PERSPECTIVES AND CHALLENGES

Despite its recent entry into the field, MS has already made visible inroads in diverse areas of molecular biophysics and structural biology. Numerous examples presented in the previous chapters of this book attest to the highly visible (and widely appreciated) role currently played by MS as a potent biophysical tool. As the appreciation for the technique grows, so do the expectations. Increasingly, MS is being applied to systems that were never considered to be amenable to mass spectrometric analysis. While the range of such applications continues to expand at great speed, this chapter attempts to focus on several areas where MS is currently making a debut. We open with a discussion of novel uses of MS aiming at understanding "orderly" protein oligomerization processes, followed by consideration of the formation of "dysfunctional" oligomers, such as amyloid. We then proceed to consider experimental strategies aiming at the detection and characterization of subcellular structures (intact ribosomes). After considering these "relatively modest" protein complexes (in the megadalton range), we move up the "ladder of complexity" to discuss emerging experimental strategies that target entire organisms. We then proceed to discuss MS-based strategies that target structure and dynamics of a notoriously difficult class of biopolymers—membrane proteins. This discussion is followed by a brief overview of several transmembrane processes related to signaling, trafficking, and transport. We conclude the chapter with a general discussion of the relevance of in vitro studies and reductionist models to the processes occurring in vivo.

Mass Spectrometry in Biophysics: Conformation and Dynamics of Biomolecules
By Igor A. Kaltashov and Stephen J. Eyles
ISBN 0-471-45602-0

11.1. ASSEMBLY AND FUNCTION OF LARGE MACROMOLECULAR COMPLEXES: FROM OLIGOMERS TO SUBCELLULAR STRUCTURES TO ... ORGANISMS?

In Chapter 4 of this book we considered various MS-based strategies that are currently used to gain information on the composition, topology, and "topography" of multimeric protein assemblies. So far, the major focus of our discussion has been the structure and the dynamic properties of such complexes, and we now turn our attention to the assembly process itself. Oligomeric proteins are ubiquitous in nature (1), but so are various types of catastrophic aggregation and polymerization of proteins, processes implicated in a variety of pathological conditions in living organisms (2–5). What factors govern protein assembly? Why do some proteins stay in the monomeric form and interact with other macromolecules only transiently, while other proteins form stable and highly ordered complexes? What triggers spontaneous aggregation of macromolecules leading to polymerization or amyloid formation? How can such processes be arrested and reversed? Obviously, answers to these questions would be important not only for advancing our basic understanding of the fundamental principles of macromolecular binding, but also for the design of effective therapeutic strategies to combat a wide array of "conformational diseases" caused by abnormal unfolding and then aggregation of an underlying protein (6–8).

11.1.1. Formation of Protein Complexes: Ordered Self-Assembly Versus Random Oligomerization

The protein binding processes propelling efficient and highly ordered formation of a multimeric assembly are often viewed in terms of *molecular recognition* and *shape complementarity*, concepts whose origins can be traced back to the century-old "lock-and-key" model (9, 10), which has dominated thinking about interactions between proteins. However, the limitations of so-called *docking techniques* (11) as far as the adequate description of protein–ligand complex formation processes are now becoming apparent, and the problems that are encountered frequently in rigid body docking have paved the way for a comeback of the "flexible approaches" (12–15). A consensus is beginning to emerge that the dynamic nature of biopolymers is an important (although still frequently underestimated) determinant of their function in general, and binding properties in particular (16–23). Indeed, in many cases the static structure alone does not provide adequate description of function, and the dynamic aspects of biopolymer behavior have to be considered in order to understand the mechanisms of processes as diverse as enzyme catalysis, recognition, transport, and signaling. It is the delicate balance between biopolymer structure (*order*) and flexibility (*chaos*) that is indeed an ultimate determinant and a modulator of the protein function and behavior throughout its life cycle from synthesis to degradation. It appears that understanding the intimate interplay between order and the chaos is the key to deciphering the molecular mechanisms of biopolymer interaction.

We illustrate this point using hemoglobin (Hb) as an example. While the Hb molecules in nonvertebrates exhibit a wide spectrum of quaternary structures ranging from monomers (e.g., leghemoglobins in plants) to giant megadalton assemblies (earthworms) (24), Hb of higher vertebrates* is universally tetrameric. Functional mammalian Hb molecules are composed of two pairs of α and β subunits,† each containing a noncovalently bound prosthetic heme group (Figure 11.1). The heterotetrameric structure [$(\alpha^*\beta^*)_2$, where $*$ denotes a noncovalently bound heme group] of Hb is essential for the efficient delivery of oxygen from lungs to tissue (by enabling the highly cooperative ligand binding) and is a direct result of a duplication of a single ancestral globin gene, from which the α- and β-globin gene variants diverged approximately 450 million years ago (25). However, presence of the two different globin chains (encoded by different chromosomes in mammals) at very high concentration (Hb concentration inside mature erythrocytes is on the order of 5 mM) is potentially disastrous, as they may provoke random binding and aggregation. Indeed, what makes the globin monomers form heterotetrameric structures as opposed to random oligomers? Furthermore, formation of homotetrameric Hb (so-called Hb H) is possible under the extreme conditions of significant overexpression of the β-globin‡ (25, 26). The crystal structure of Hb H (27) indicates that the tetramer is

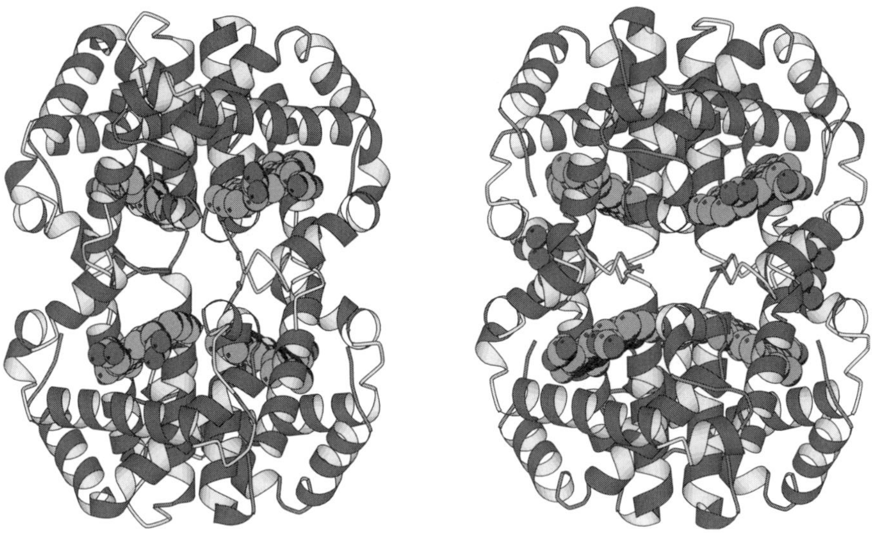

FIGURE 11.1. Crystal structures of human hemoglobin tetramer Hb A, $(\alpha^*\beta^*)_2$ (deoxy form, PDB 2HHB, *left*) and homotetramer Hb H, β_4^* (deoxy form, PDB 1CBL, *right*).

*Cyclostomes (lamprey) are the most "evolutionary advanced" vertebrates that still have homodimeric Hb.

†These two subunits of bovine Hb contain 141 and 146 amino acid residues, respectively, have very similar tertiary folds, and are highly homologous.

‡These tetramers, however, are not functional, since the molecular basis of cooperative ligand binding is eliminated.

locked in a quaternary structure closely resembling the R-state of a "normal" Hb A tetramer (Figure 11.1). Contrary to that, formation of the homotetrameric Hb molecules containing the α-globin units has never been observed. This is quite remarkable, given the near-identical tertiary folds of the two globin chains. Clues for explaining this striking difference in the behavior of the α- and β-globins (as well as γ-globin, closely related to β-globin) were previously sought in the subtle structural differences between the two globins (28).

Our recent investigation of the molecular basis of such directed oligomerization has demonstrated that structural plasticity is an essential characteristic of ordered Hb self-assembly, as it both guides the assembly process and provides a vital safeguard against random oligomerization (29). The dynamics of mammalian Hb assembly was investigated in this study by monitoring monomer/oligomer equilibria in dilute Hb solutions with ESI MS (Figure 11.2). The shifts in these equilibria were correlated to changes in flexibility of the monomeric species assessed by monitoring the charge state distributions of the protein ions in ESI mass spectra. (See Chapter 5 for a detailed discussion of using charge state distributions of protein ions to monitor large-scale dynamic events in solution.) The results of this study provide strong evidence that the monomeric β-chain exhibits a high degree of structural disorder and indeed fails to bind the prosthetic heme group on its own.

Binding of such an unstructured apo-β-chain to a tightly folded holo-α-chain to form a heme-deficient dimer ($\alpha^*\beta$) is the initial step in Hb assembly. This binding event "locks" the β-chain in a highly ordered conformation, which enables it to acquire a heme group, followed by docking of two hemoglobin dimers to form a tetrameric form of the protein, $(\alpha^*\beta^*)_2$ (29). The asymmetry of the roles of the two globin chains in the assembly process (α-globin serving as a "rigid" template to which a "flexible" β-globin can easily adapt) is rather surprising, given the high sequence homology (~43%). Such "role differentiation" is certainly advantageous, as it effectively inhibits the formation of homodimers during the first step of Hb assembly and, therefore, appears to be a rather intriguing consequence of molecular evolution. The importance of protein flexibility for binding is also demonstrated by our experiments with isolated and heme-reconstituted α-chains, which are tightly folded and monomeric (Figure 11.2). On the other hand, isolated β-chains mixed with a (molar) excess of heme do show the ability to form tetramers, as well as dimers ($\beta^*\beta^*$ and $\beta^*\beta$), although the most abundant species are loosely structured monomeric species (both β and β^*). Careful examination of the charge state distributions of the β-globin ions indicates that a small fraction of the protein populate a highly structured conformation (29), which appears to be able to serve as a template for binding, leading to homodimer and, eventually, homotetramer formation (Figure 11.2). As we have already mentioned, production of a homotetrameric Hb H molecule is a well-documented clinical consequence of significant overexpression of β-globin in α-thalassemic disorder (so-called hemoglobin H disease) (25, 26).

The ability of flexible β-globins to undergo ordered oligomerization (as opposed to aggregation) highlights the importance of the large-scale polypeptide

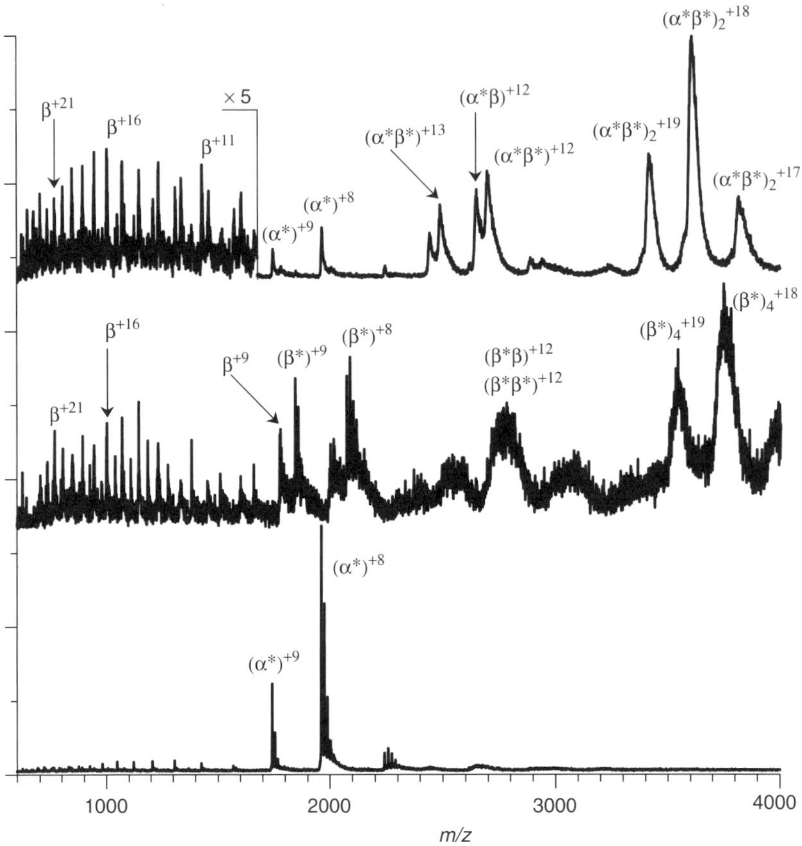

FIGURE 11.2. ESI mass spectra of dilute bovine Hb (*top trace*) and isolated β- (*middle trace*) and α- (*bottom trace*) globins. The spectra were acquired from aqueous medium-salt solutions maintained at neutral pH (10 mM $CH_3CO_2NH_4$). Soft desolvation conditions in the ESI source were used in order to avoid dissociation of noncovalent complexes in the gas phase [control experiments are described in (29)]. The spectra of the isolated globins were acquired in the presence of a fivefold molar excess of prosthetic heme group. Asterisks indicate a heme group bound to a globin chain.

chain dynamics for protein assembly. The intriguing asymmetry in the globin interaction reported in (29) may provide a vital safeguard, which (in addition to recently reported α-globin chaperonins) (30, 31) may inhibit random oligomerization, thus directing the assembly process along a "correct" pathway.

11.1.2. Protein Oligomerization as a Chain Reaction: Catastrophic Aggregation and "Ordered" Polymerization

Characterization of the catastrophic oligomerization of proteins leading to their aggregation and amyloidosis (8) is a very challenging experimental task

Normally, only the "endpoints" of the amyloidosis process can be observed directly, namely, the free proteins in solution ("reactants") and the amyloid fibrils ("products"). The structure of the amyloid fibrils and the conformation of individual proteins within such aggregates can be visualized by methods of cryo-electron microscopy (Figure 11.3); however, very little information is available on the structure and composition of the intermediate states of protein aggregates. In some circumstances the initial steps of the aggregation process can be observed directly using ESI MS (32); however, in most cases indirect methods have to be used to probe amyloid structure and formation kinetics.

Miranker and co-workers designed a method for monitoring fibrillogenesis in a quantitative fashion and applied it to characterize the aggregation of islet amyloid polypeptide (IAPP), an amyloidogenic peptide (33). The approach, based on ESI, is complementary to existing assays of fibril formation as it monitors directly the population of the precursor rather than the product molecules. Fiber formation is monitored in two modes: a quenched mode in which fibril formation

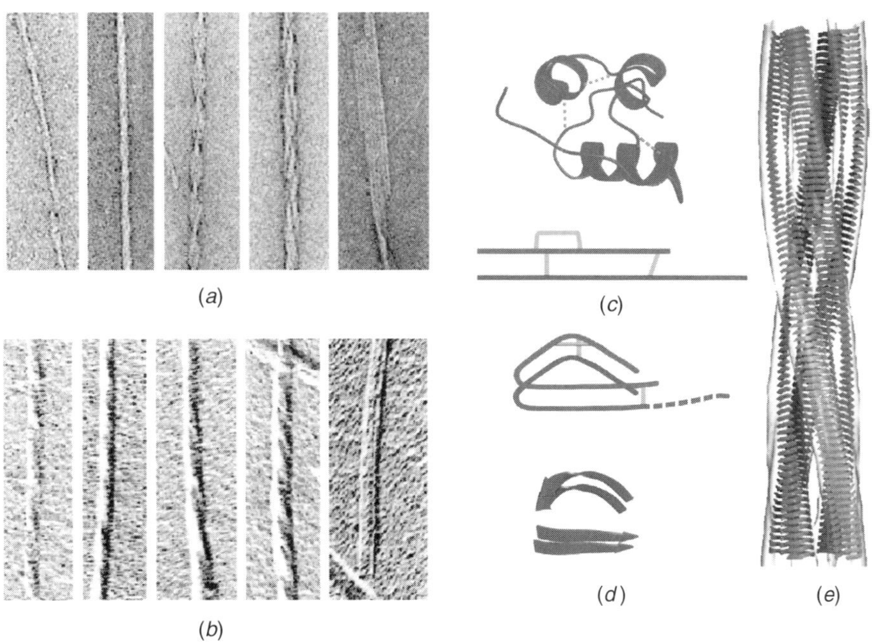

FIGURE 11.3. Gallery of negatively stained (**A**) and shadowed (**B**) insulin fibrils, showing the diversity of fibril structures. (**C**) Insulin structure showing the three native disulfide bonds (*top*) and a topology diagram of insulin. (**D**) Proposed topology for the amyloid protofilament (orientations of the termini and disulfide bonds within the curved structure are arbitrary; the C-terminus of chain B shown as a dashed line is not required for amyloid fibril formation). (**E**) A proposed β-strand model docked into the cryo-electron microscopy density of the compact fibril (translucent surface). Adapted with permission from (198). © 2002 National Academy of Science, USA.

is halted by dilution into denaturant and a real-time mode in which fibril formation is conducted within the capillary of the ESI source. The fibrillar IAPP does not compromise the ionization of the monomeric form of the peptide. Furthermore, under mild ionization conditions, fibrillar IAPP does not dissociate and subsequently contribute to the monomeric signal. Quantitation was made possible through the introduction of an internal standard, IAPP from another animal species.* Applied to the IAPP fibrillogenesis, the technique revealed that the precursor consumption in seeded reactions obeys first-order kinetics, while a consistent level of monomer persists in both seeded and unseeded experiments after the fibril formation is complete (33).

Serrano and co-workers developed a MS-based procedure to monitor conformational changes that precede the formation of fibrils (34). Aggregation of a model protein ADA2h (activation domain of human procarboxypeptidase A2) was tracked by monitoring the disappearance of protein monomers and generation of resistance to proteolysis. Limited proteolysis of amyloid fibrils monitored by MS was also employed by Pucci and co-workers, who studied the structural determinants of amyloid nucleation (35). Seven proteases with different degrees of specificity were used to map solvent-exposed regions within the native, ligand-bound and amyloidogenic states of muscle acylphosphatase (AcP), a protein that forms amyloid fibrils in the presence of trifluoroethanol. The results of this study suggest that a considerable degree of solvent exposure is an important structural feature that initiates the protein aggregation process (35). Limited proteolysis monitored by MALDI or ESI MS has been used to characterize the structure of amyloid fibrils in a number of other studies as well (36, 37).

Inhibition of amyloid formation can be studied using ESI MS to observe the formation of stable protein–ligand complexes that prevent protein aggregation. This approach was recently used by Robinson and co-workers to evaluate a small library of ligands proposed as inhibitors of transthyretin aggregation (38). Tetrameric transthyretin is involved in the transport of thyroxine and the metabolism of vitamin A (through its interactions with retinol binding protein). Dissociation of these structures is considered to be a first step in the formation of transthyretin amyloid fibrils. Binding of certain ligands to the transthyretin tetramer (a process that can be observed directly using ESI MS) stabilizes the tetrameric structure, effectively arresting amyloidosis (38). A similar approach has been used by Skribanek and co-workers to study the interaction between a synthetic β-amyloid peptide $A\beta(1-40)$ and its aggregation inhibitors (39).

Several research groups have used HDX MS to probe stability and structure of various amyloid formations (32, 40–47). In a typical experiment, amide protection patterns are acquired for the free protein in solution and also in the amyloid form. A comparison of the two exchange patterns often provides valuable information on the structural changes that accompany amyloid formation.

*The standard (rat IAPP) is structurally close to the human IAPP, so that the ionization efficiencies of the two peptides are identical. At the same time, the masses of the two peptides are slightly different, so that they do not interfere with each other in the ESI mass spectrum

Robinson and co-workers employed HDX MS to monitor backbone protection of insulin molecules within relatively small aggregates in an oligomer-specific fashion (32). ESI mass spectra were obtained from solutions of insulin at millimolar concentrations at pH 2.0, conditions known to promote insulin aggregation. Clusters containing up to 12 insulin molecules could be detected in the gas phase. HDX MS measurements show that in solution these higher oligomers are in rapid equilibrium with monomeric insulin. At elevated temperatures, under conditions where insulin rapidly forms amyloid fibrils, the concentration of soluble higher oligomers was found to decrease with time, yielding insoluble high molecular weight aggregates and then fibrils (32). Further studies of insulin aggregation by the Robinson group utilized a combination of HDX and peptic digestion (performed after quenching the exchange) to obtain site-specific information on amide protection within insulin molecules derived from the fibers (41).

Nazabal and colleagues used HDX MS to study the conformational transition occurring upon amyloid aggregation of HET-s prion protein, an infectious element of the filamentous fungus *Podospora anserina* (47). HET-s is a 289 amino acid residue protein that aggregates into amyloid fibers in vitro. Such fibers produced in vitro are infectious, indicating that the HET-s prion can propagate as a self-perpetuating amyloid aggregate of the HET-s protein. HDX of the soluble form of HET-s monitored using MALDI MS reveals that the C-terminal region of the protein (240–289) displays a high solvent accessibility. However, solvent accessibility is drastically reduced in this region in the amyloid form of the protein. The HDX pattern in the N-terminal part of the protein (residues 1–220) is not significantly affected by protein aggregation, suggesting that aggregation of HET-s involves a conformational transition only in the C-terminal part of the protein (47).

Lansbury and co-workers have used HDX MS to characterize the secondary structure of protofibrils, heterogeneous transient structures observed during the in vitro formation of mature amyloid fibrils that have been implicated as the toxic species responsible for cell dysfunction and neuronal loss in Alzheimer's disease (48). Populations of protofibrils of different sizes were isolated chromatographically from amyloid assembly reactions of Aβ(1–40) (amyloid β-peptide, residues 1–40). HDX MS was shown to be able to distinguish among unstructured monomer, protofibrils, and fibrils based on their different amide protection patterns (Figure 11.4). About 40% of the backbone amides of Aβ(1–40) protofibrils were highly resistant to exchange even after 2 days of incubation in aqueous deuterated buffer, implying a very stable core structure. This was in contrast to the "mature" amyloid fibrils, whose equally stable structure protects about 60% of the backbone amide hydrogens over the same time period. A surprising degree of specificity was observed in amyloid assembly, with the wild type Aβ(1–40) being preferentially excluded from both protofibrils and fibrils grown from an equimolar mixture of the wild-type and the mutant peptides (48). Fernandez and co-workers used HDX followed by proteolytic (peptic) digest of Aβ(1–40) carried out under slow exchange conditions to probe the structure of amyloid fibrils (43). These experiments revealed differential local protection patterns within

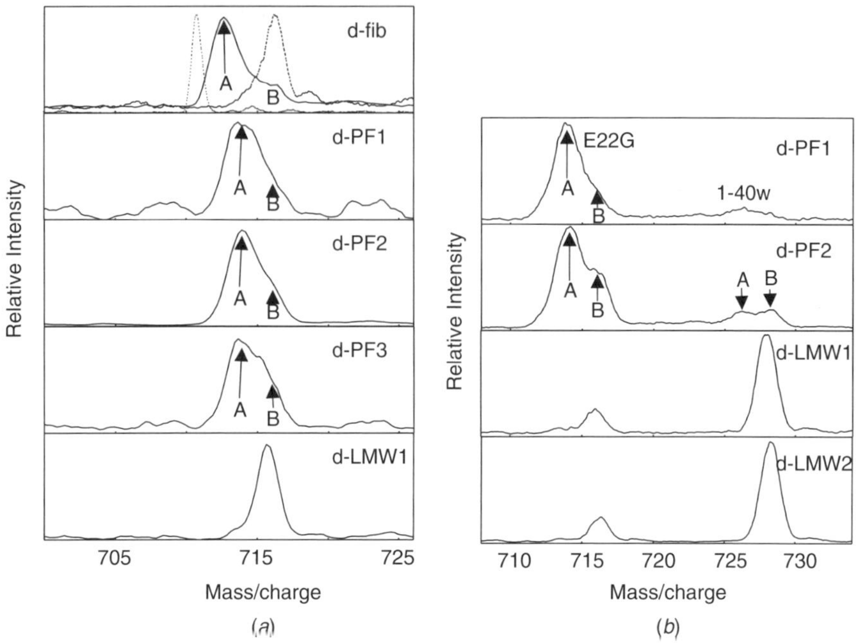

FIGURE 11.4. ESI mass spectra of partially deuterated fibrils and protofibrils (fractions PF1 through PF3 and low molecular weight fraction, LMW 1) of a mutant β-amyloid peptide Aβ_{ARC} (Arctic type) acquired following a 2 hour incubation in a deuterated buffer (**A**). The peaks shown in the spectra correspond to a +6 charge state of the β-amyloid peptide Aβ_{ARC}(1–40). Peak shapes of protonated and fully deuterated Aβ_{ARC} (1–40) monomer are shown as dashed lines on the top panel. (**B**) ESI mass spectra of a 1:1 mixture of the mutant and wild type Aβ (1–40) acquired after 2 hours of incubation in a deuterated buffer. Marks on the spectra denote partially resolved protected (A) and unprotected (B) peptide populations. Reproduced with permission from (48). © 2003 American Chemical Society.

Aβ(1–40), with the N-terminus (residues 1–4) being completely unprotected, the C-terminal segment exhibiting intermediate levels of protection [approximately 35% of the amides in this segment were protected from the exchange in the fibril containing samples, as opposed to the monomeric Aβ(1–40) species prepared in DMSO/H_2O], and the fragment containing residues 5–19 of the peptide exhibiting very high protection in the fibril-containing samples (over 50% of amides within this segment were protected from the exchange) (43).

Site-specific information on amyloid fibril structure can also be obtained using a combination of HDX in solution and peptide ion fragmentation in the gas phase (49), an emerging technique that was discussed in detail in Chapter 5. Kraus and co-workers used this technique* to investigate the behavior of synthetic β-amyloid peptides. In contrast to the C-terminally truncated peptides Aβ(1–40)

*Fragmentation of β-amyloid peptide ions was carried out using collision-induced dissociation, inlet fragmentation, and postsource decay in a time-of-flight mass analyzer.

and Aβ(1–36) showing lack of protection, Aβ(1–42) and the pyroglutamyl peptide pAβ(3–42) exhibited more complex exchange patterns resulting from the formation of β-sheet structured oligomers with 18–20 highly protected amides. Peptide ion fragmentation in the gas phase indicated that the protected region of Aβ(1–42) is located in the central part of the chain (residues 8–23). The results of this study confirm that the hydrophobic binding domain LVFFA is important for Aβ aggregation (44).

Structure of amyloid fibrils can also be probed by chemical cross-linking. Egnaczyk and co-workers used photoaffinity cross-linking to determine structural constraints for Aβ monomer arrangement within the amyloid fibrils. A photoreactive Aβ(1–40) was synthesized by substituting L-*p*-benzoyl-phenylalanine (Bpa) for Phe at position 4 (Aβ(1–40) Phe4Bpa) and incorporated into synthetic amyloid fibrils, followed by UV irradiation of the peptide aggregate (50). SDS-PAGE of dissolved fibrils revealed the light-dependent formation of a covalent Aβ dimer. MS analysis of the products of enzymatic cleavage of the dimers suggests that Bpa-4 is covalently attached to the C-terminal segment of a β-amyloid peptide [tryptic fragment (29–40)]. MS/MS experiments and further chemical modifications of the cross-linked dimer allowed the precise locations of the chemical modification to be established as the Met-35 side chain. The Bpa-4 $\rightarrow$ Met-35 intermolecular cross-link is consistent with an antiparallel alignment of β-amyloid peptides within the amyloid fibrils (50).

Although the examples of catastrophic protein aggregation and polymerization considered so far lead to pathological conditions with grave physiological manifestations, controlled large-scale protein polymerization processes are crucial for normal cell functioning. One particularly fascinating example of such controlled polymerization is self-organization of tubulin to form microtubule cytoskeleton (51). The microtubules are cylindrical aggregates that are present in the cytoplasm of all eukaryotic cells and carry out a diverse set of functions vital for cell morphogenesis and organization, organelle transport, and cell motility and division (52). The microtubule building block is a heterodimer formed by α- and β-tubulin molecules.* Tubulin dimers (Figure 11.5A) aggregate through longitudinal interactions into protofilaments, which are then associated laterally to form cylindrical structure (Figure 11.5B). (These cylindrical structures are 25 nm in diameter and may span the entire length of the cell.) Individual polymers exhibit spontaneous length fluctuations involving polymer assembly and disassembly, a feature that appears to be vital for many of its functions (51). Assembly of microtubule fibers from tubulin involves both microtubule nucleation and elongation. While the elongation and disassembly of microtubules have been studied extensively using methods of light and electron microscopy, the detailed understanding of its nucleation is still lacking due to the difficulties associated with detection and characterization of transient intermediates between tubulin dimers and microtubules (53). One of the important players in the regulation of microtubule assembly at the molecular level is stathmin, a protein (more correctly, a

*The α- and β-tubulins are 50 kDa (~450 amino acid residues each) proteins that are highly homologous (~50% sequence identity).

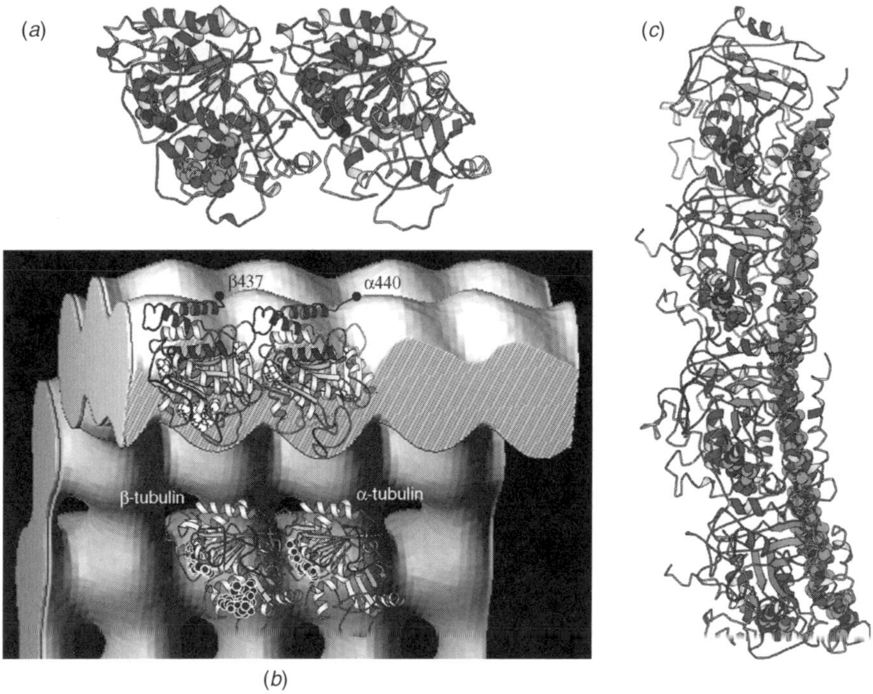

FIGURE 11.5. (**A**) Three-dimensional structure of tubulin $\alpha\beta$ heterodimer (PDB 1TUB). (**B**) A low-resolution model of microtubule structure showing positions and orientation of tubulin $\alpha\beta$ heterodimers [adapted with permission from (199)]. (**C**) A three-dimensional structure of stathmin–tubulin complex (PDB 1FFX). Stathmin is shown as a space-fill model.

family of proteins) that interacts with tubulin to form a complex with two tubulin $\alpha\beta$ heterodimers (54). The X-ray structure of crystals of the complex reveals a head-to-tail arrangement of two tubulin heterodimers, which are connected by a long stathmin α-helix (Figure 11.5C). The curved geometry of the complex is thought to prevent its incorporation into microtubules, thereby arresting the tubulin polymerization process (55). Mass spectrometry has been used recently with great success to probe tubulin–stathmin interaction (56, 57). Curmi and co-workers used a multifaceted strategy (limited proteolysis, size exclusion chromatography, and MALDI MS) to probe the native structure of stathmin and to delineate its minimal region able to interact with tubulin (56). A hypothesis of stathmin folding was formulated based on the detection of four structured domains within the native protein. Analysis of the interaction of various stathmin proteolytic fragments with tubulin allowed a "minimal core region" of stathmin* to be identified (56). Further insight into the molecular aspects of the

*This core region was actually confined to the predicted α-helical "interaction region" of stathmin and a short N- or C-terminal extension.

tubulin–stathmin interaction was provided by Wallon and co-workers, who used chemical cross-linking to locate the binding sites with higher precision (57). MALDI MS analysis of the cross-links* of tubulin and stathmin or its fragments was used to identify short protein segments from the tubulin–stathmin interface region. The nonglobular nature of stathmin was shown to be a major factor allowing this protein to interact with more than one tubulin monomer and bind to structurally noncontiguous regions of α-tubulin (57), giving rise to a "bent" conformation of the stathmin–tubulin complex.

We are confident that the vast arsenal of MS-based strategies of detection and characterization of transient intermediate states (discussed in Chapters 5 and 6 of this book) can also be applied to study other early events in tubulin polymerization. Furthermore, it seems MS has potential to further advance our understanding of the molecular mechanisms of the dynamics of mature microtubules. In addition to its obvious importance in the variety of cellular processes, understanding microtubule dynamics will certainly have profound fundamental implications by shedding light on physical aspects of biological phenomena related to nonequilibrium organization (58–61) and the balance between self-assembly and self-organization (51).

11.1.3. Subcellular Structures: Ribosomes

Ribosomes are cellular components that execute the last step in the gene expression pathway by catalyzing mRNA-directed protein synthesis (62, 63). Elucidation of the mechanism of protein synthesis has been a central problem in structural biology for several decades. While the crystal structures of bacterial ribosomes have recently been obtained [see (64) for a historical account], many questions still remain as far as the functioning of this formidable cellular apparatus in both prokaryotic and eukaryotic organisms (65, 66). The general features of ribosomal structure and functional mechanism (which, in addition to amino acid coupling by synthesizing a peptide bond, also includes a selection of tRNA molecule carrying a "correct" amino acid at each coupling step) are conserved in all cellular forms of life. Among all ribosomes, the prokaryotic ones are the smallest ("only" 2.5 MDa) and best studied. They are on the order of 250 Å and are composed of two subunits, 30S and 50S† (the latter about twice the size of the former). The 30S subunit consists of a single RNA molecule (16S rRNA), which is approximately 1500 nucleotides in length, and single copies of each of about 20 different protein molecules (Figure 11.6). The 50S subunit consists of a large RNA molecule (23S rRNA, which contains approximately 2900 nucleotides), a smaller molecule of RNA (5S rRNA, ~120 nucleotides), and approximately 30 different proteins (Figure 11.6). Although their architectural organization is very similar, eukaryotic ribosomes are larger as compared to

*A zero-length cross-linker ECD used in this study allows identification of pairs of acidic and basic residues that are in direct contact (see Chapter 4 for more detail).

†Subunits are denoted by their apparent sedimentation coefficient expressed in svedberg units (intact prokaryotic ribosome sediments at 70 S; see Chapter 2 for a detailed discussion of sedimentation experiments).

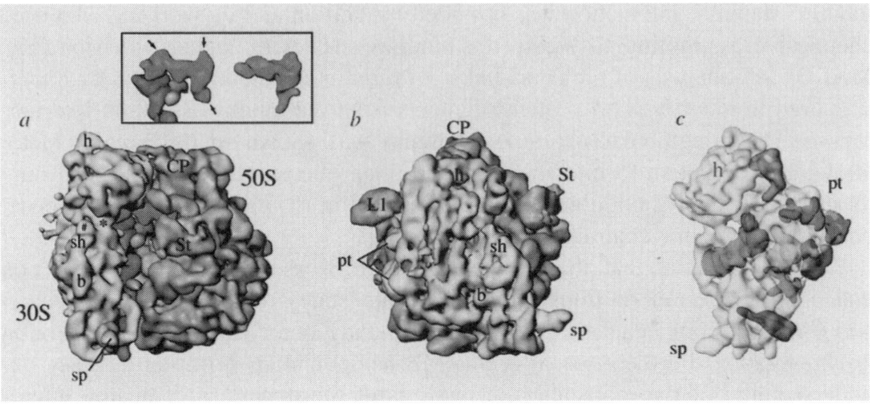

FIGURE 11.6. *Escherichia coli* initiation complex, with fMet-tRNA bound to the ribosome in the presence of mRNA with an AUG codon in its middle. (See color insert, where the 30S subunit is shown in yellow, the 50S subunit in blue, the helix 44 of 16S rRNA in red, and the tRNA in green.) (*a*, *b*) Two views of the 70S ribosome. Inset in (*a*) shows on the left the experimental mass attributed to tRNA in contact with the decoding center, and on the right, atomic structure of initiator tRNA, filtered to 10 Å resolution. Arrows point to contacts between tRNA and ribosome. (*c*) The 30S portion of the map, with fitted X-ray structures of proteins. Adapted with permission from (200) as presented in (201).

those of prokaryotic organisms. (Generally, they contain more components; on average, each component's size is larger than that of its prokaryotic analogue.)

The initial attempt to use MS as a structural tool in the study of ribosomes was made by Robinson, who used nano-ESI to transfer an intact *E. coli* ribosome* from solution to the gas phase with a consequent mass analysis of its constituents (67). Although gas phase dissociation of the intact ribosome could not be avoided in this study, the partial disassembly processes were remarkably selective, showing strong correlation with the predicted features of ribosomal protein–protein and protein–RNA interactions. In addition, the observation of distinct peaks corresponding to individual protein components in the ESI mass spectra provided an opportunity to probe the conformational and dynamic properties of these proteins. Many of these proteins are known to be at least somewhat unfolded when removed from the ribosome, and very little is known about their dynamics within the context of the intact particle. HDX measurements carried out on 50S subunits indicated that many proteins are highly protected. For example, a very high protection was exhibited by L11 protein, which is consistent with a highly folded conformation assumed by this protein within the ribosome. In fact, the extent of L11 protection was so high for a protein of its size (14 kDa), that it matched the protection levels characteristic of similarly sized proteins in crystals, rather than in solution (67). Intriguingly, earlier NMR measurements indicated

*Products of ribosome dissociation in solution were removed by ultrafiltration and buffer exchange prior to ESI MS experiments.

that a C-terminal fragment of protein L11 (hosting the RNA binding domain) is significantly destabilized in the absence of rRNA (68). The remarkable difference between the two states of L11 suggests that tight packing within the intact ribosome restrains the structural fluctuations of otherwise flexible segments of the protein (67). This initial study demonstrated the power of ESI MS as a tool to probe not only the "topological features" of ribosomal assemblies, but also the dynamic behavior of its constituents within the context of the intact ribosomal particle.

The absence of intact ions of the ribosome or either of its subunits in the ESI spectra of *E. coli* ribosome (67) was initially interpreted by many as a proof that such large complexes cannot survive the transition from solution to the gas phase. However, our earlier discussion of the relationship between the protein size and the number of charges it accumulates in the gas phase offers a different explanation (see Chapter 4). The number of charges expected to be accumulated on a ~250 Å particle is not significant enough to produce ions within the m/z range used in the initial study (67). Spectacular advances made by both ESI and time-of-flight technologies in the late 1990s led to a dramatic expansion of the m/z range within which protein ions can be detected. This allowed the Robinson group to observe ions corresponding to the intact 30S subunit of an *E. coli* ribosome (69). The spectrum (Figure 11.7) features broad overlapping peaks corresponding to different charge states, making charge state assignment a difficult task. However, the mass deduced from the spectrum was only 5.5 kDa higher than that calculated based on the sequences of the 21 proteins and the 16S RNA molecule (846,7 kDa). Mass spectra of an intact subunit 50S and intact ribosome (70S) were acquired as well (stabilized in 10 mM magnesium acetate, pH 6.4), however, peak overlap prevented charge assignment to be done in a manner similar to that shown in Figure 11.7. Instead, the average charges of both ionic species were calculated based on the known protein and RNA composition of the ribosome. The average charges were calculated as +90 (for 50S) and +105 (70S), leading the authors to a suggestion that a significant proportion of the subunit surface is buried upon their association to form the 70S particle (69).

Once the feasibility of intact ribosome detection was demonstrated, Robinson and co-workers focused their attention on investigating various factors that govern the loss of proteins from ribosomes in the mass spectrometer (70). Such information can be used to probe dynamic events in the ribosome induced by binding of various cofactors (such as the elongation factor G, whose binding triggers translocation). MS was used to identify proteins that are released in the gas phase from *E. coli* ribosomes in response to varying solution conditions and cofactor binding. Protein components that dissociate most readily from the ribosome were identified and such dissociation patterns were analyzed against the available structural information for ribosomes.* The results of such analysis strongly suggested that a high

*The high-resolution atomic structures of a ribosomal subunit 50S (from *H. marismortui*) and intact 70S (*T. thermophilus*) were used to infer the structural details of the protein–RNA interactions in *E. coli* ribosome, an approach justified by a high degree of structural conservation across bacterial ribosomes. The "interaction strength" between individual proteins and RNA were quantified using changes in the solvent accessible surface area (ΔSASA).

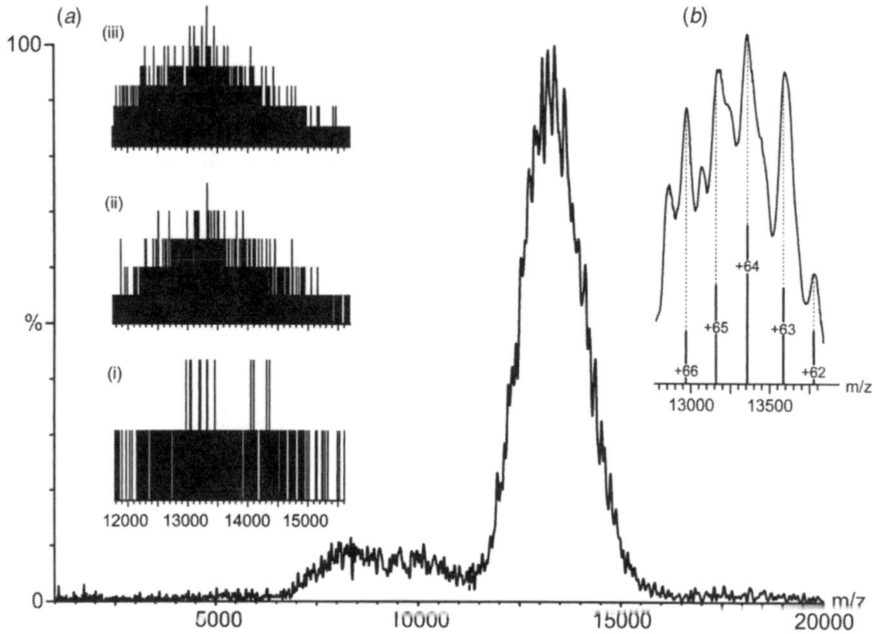

FIGURE 11.7. Nano-ESI mass spectrum of the 30S subunit of the *E. coli* ribosome. The ribosome sample was buffer exchanged in 10 mM $CH_3CO_2NH_4$ (pH 6.4) prior to MS measurements. Inserts show the raw data from single scans, as well as summations of ion signals from five and ten scans (**A**) and an expansion of the apex region of the cluster of partially resolved ion peaks of the intact assembly with assigned charge states (**B**). Charge state assignment for poorly resolved peaks was carried out using an algorithm described in (72). Reproduced with permission from (69). © 2000 National Academy of Sciences, USA.

interaction surface area of the protein with RNA (ΔSASA) was the major factor in preventing dissociation. Gas-phase dissociation was enhanced by "loosening" the structure of the intact ribosome in solution resulting from pH changes, metal ion content variation, and so on. This relationship between the protein–RNA ΔSASA and the propensity for gas phase dissociation was used to interpret the release of the protein component from the ribosome, which was complexed with the elongation factor G (EFG). Two inhibitors of elongation were used in these experiments, fusidic acid and thiostrepton, freezing the ribosome in complexes analogous to post- and pretranslocation states, respectively. Complexes formed between ribosomes and EFG undergo major conformational changes depending on the particular translocation stage inhibited by a given antibiotic. ESI mass spectra of the ribosome–EFG complex in the presence of each antibiotic are shown on Figure 11.8. While the presence of fusidic acid in solution does not result in significant alteration of the ribosome dissociation pattern, thiostrepton

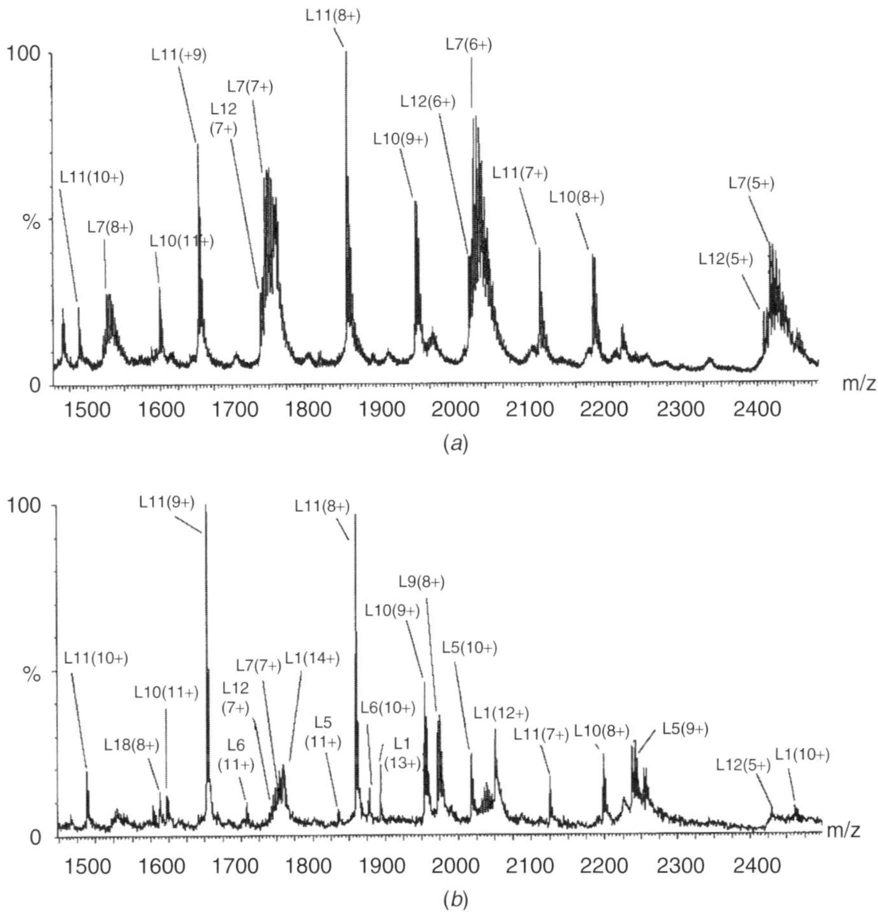

FIGURE 11.8. Nano-ESI mass spectra (low m/z region) of the ribosome–EFG complex in the presence of fusidic acid (**A**) and thiostrepton (**B**). The mass spectrum recorded in the presence of fusidic acid (**A**) is similar to that observed for ribosomes in the absence of EFG. By contrast, the complex inhibited by thiostrepton (**B**) demonstrates the absence of L7/L12 and the presence of additional proteins L5, L6, and L18. L5 and L18 have the largest surface area of interaction with the 5S RNA and, therefore, their dissociation from the ribosome is consistent with conformational changes of 5S RNA (which reduce the surface area of interaction between the RNA and both proteins). Adapted with permission from (70).

does induce major perturbations in protein–RNA interactions, as suggested by dramatic changes in the ribosomal dissociation pathways. The results of these experiments clearly demonstrate that the two ribosome–EFG complexes are significantly different and are indicative of conformational changes occurring during translocation (70).

11.1.4. Mass Spectrometry at the Organism Level?

Perhaps some of the most impressive examples of the dramatically increased capabilities of MS to characterize large macromolecular complexes come from the recent reports on the analysis of whole viral particles. The MS-based methods of analyzing macromolecular interactions have been very prominent in the experimental arsenal of virology for quite some time (71). However, it was not until very recently that dramatic improvements in ESI and mass analysis of macromolecules enabled researchers to produce and detect intact viral ions in the gas phase.

Robinson and co-workers used a time-of-flight mass analyzer with an extended m/z range (up to 36,000) to detect intact virus capsid particles carrying over a hundred charges in the gas phase (72). In these experiments a recombinant MS2 coat protein was expressed in *E. coli* and the virus capsid was purified by ultrafiltration. A 5 μM capsid solution in unbuffered H_2O (pH 7.1) was sprayed using a nano-ESI source under carefully controlled conditions in the interface region. Collisional cooling of ions was used to remove excess internal energy and prevent capsid dissociation in the gas phase. Poor desolvation resulted in the appearance of broad and barely resolved ion peaks corresponding to different charge states. Nevertheless, computer-aided data processing allowed the mass of the capsid particle to be estimated as 2485 ± 25 kDa, a number that is very close (within 5%) to the mass calculated based on the known composition of the capsid particle (72).

An alternative approach to the detection and characterization of intact viral particles in the gas phase was used by Siuzdak and co-workers (73), who employed charge detection MS (74) for this purpose. This ion detection technique allows mass analysis of ESI-produced ions to be carried out virtually without any mass limit. Two viruses were examined in this study, rice yellow mottle virus, RYMV, and tobacco mosaic virus, TMV (Figure 11.9). RYMV is an icosahedral virus consisting of a single-stranded RNA surrounded by a homogenuous protein shell made up of 180 copies of a single coat protein. The calculated molecular weight of RYMV is 6.5 MDa. TMV is rod shaped and is formed from ~2140 identical protein subunits wound in a 300 nm long helix. A central hollow cylindrical core holds the viral genome—a 6395 nucleotide strand of RNA. The calculated molecular weight of TMV is 40.5 MDa.

Mass spectra were recorded using solutions of purified RYMV and TMV in unbuffered H_2O diluted to ~5 pM concentration. Coulombic repulsion limits the maximum number of charges that can be placed on the surface of a viral particle. The RYMV and TMV ions possessed charge state distributions ranging from 300 to 1000 positive unit charges with higher charge states observed for TMV. The mass spectrum of RYMV (Figure 11.9, top) shows the most abundant signal at 6–7 MDa with tailing in the high mass region, consistent with the theoretical mass of a single intact RYMV particle. The mass spectrum of TMV (Figure 11.9, bottom) provides an estimate for its mass as 39–42 MDa, which is consistent with the calculated mass of this virus (73). This study demonstrates the tremendous potential of charge detection MS as a powerful experimental tool

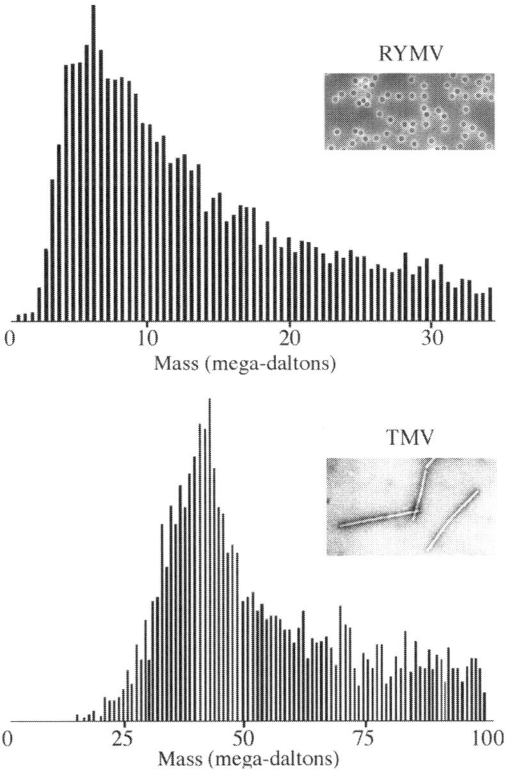

FIGURE 11.9. Mass spectra of rice yellow mottle virus (RYMV, *top*) and tobacco mosaic virus (TMV, *bottom*) particles obtained with an ESI charge detection time-of-flight mass spectrometer. Insets show the electron micrographs of the icosahedral RYMV (diameter 28.8 nm; *top*) and the cylindrical TMV (approximately 300 nm long and 17 nm in diameter; *bottom*). The calculated molecular weights of the viral particles are 6.5 MDa (RYMV) and 40.5 MDa (TMV). Reproduced with permission from (73).

to obtain information on masses of biological superassemblies that are too large to be analyzed using routine MS techniques.

11.2. STRUCTURE AND DYNAMICS OF MEMBRANE PROTEINS

Membrane proteins constitute a broad class of biopolymers (it is estimated that approximately 30% of all proteins belong to this class) whose amphiphilic nature makes them notoriously difficult as far as structural and functional analysis at the whole protein level (75–77). Structures are currently known for less than 0.2% of all membrane proteins (75, 76). Structural coverage of the entire protein repertoire representing sequenced genes is estimated to be 6–10%. In most cases, membranes are essential for the integrity of the proteins they host, although the structure and dynamic properties of membrane proteins are actually defined by a wide spectrum of intermolecular forces. These include protein interactions with the hydrophobic "bulk" of the membrane, its polar interface region, internal and external water molecules, as well as interactions between various segments of the protein itself (Figure 11.10). Such amphipathic character of membrane proteins results in their general insolubility, which makes any experimental study extremely difficult. MS is not an exception, since even sequencing of membrane proteins is often problematic (78).

Two approaches are commonly used to deal with the solubility problem. One relies on using detergents to isolate, solubilize, and manipulate membrane proteins (79), while the other utilizes lipid vesicles as a means to mimic membranes (80). Both approaches, as well as select examples of relevant studies that use MS as a biophysical tool, are briefly considered in the following sections.

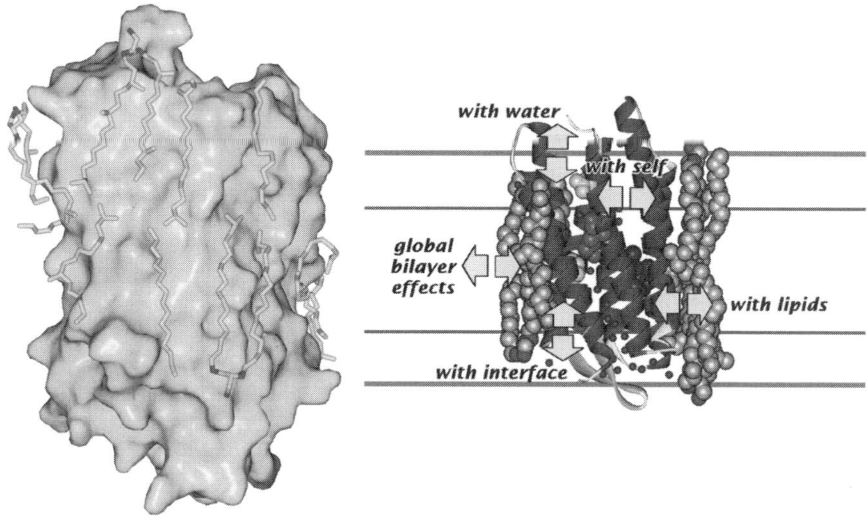

FIGURE 11.10. (*Left*) A view of membrane protein interactions with lipids. Native lipids are seen bound to the surface of bacteriorhodopsin and suggest intimate and specific interactions between the protein and lipids. Reproduced with permission from (79). (*Right*) A schematic representation of the polypeptide interactions that determine the structure and stability of membrane proteins. (See color insert, where the blue lines represent schematically the total thickness of the lipid bilayer, and the red lines the central hydrocarbon core bounded by the interfacial regions.) Besides interactions of the polypeptide chain with itself, water, neighboring lipids, and the membrane interface, the thermodynamic and electrostatic properties of the lipid bilayer itself are important. The lipid bilayer, like proteins, resides in a free energy minimum resulting from numerous interactions. This equilibrium can be disturbed by the introduction of proteins or other solutes, resulting in so-called bilayer effects, which also include solvent properties peculiar to bilayers that arise from motional anisotropy and chemical heterogeneity. Reproduced with permission from (202).

11.2.1. Structural Studies of Membrane Proteins Utilizing Detergents

Detergents are amphipathic molecules that act as surfactants and are indispensable as solubilizing agents for membrane proteins (81). Most detergents belong to one of the following groups: ionic, nonionic, and zwitterionic. Several popular detergents are listed in Table 11.1. Detergents have a natural tendency to self-associate and interact with other molecules and exhibit a wide range of behavior and assembled structures depending on their concentration, solution conditions (pH, ionic strength, etc.), and the presence of proteins and other molecules. Self-association (leading to micelle formation) occurs at a threshold concentration, usually termed critical micelle concentration (CMC). The micelle size is usually characterized by an aggregation number (the average number of detergent monomers per micelle). The shape of the micelle is determined by the packing of the monomer tails and is dependent on the aggregation number. It is usually quite nonuniform, despite

TABLE 11.1. Properties of Commonly Used Detergents[a]

Name	Monomeric Mass	Critical Micelle Concentration	Aggregation Number	Comments
Sodium dodecyl sulfate (SDS)	288	1.2–7.1 mM	62–101	Ionic detergent, strongly denaturing
3-[(3-Cholamidopropyl) dimethylammonio]-1-propanesulfonate (CHAPS)	615	3–10 mM	4–14	Steroid-based detergent, relatively mild
Dodecyldimethyl-*N*-amineoxide (DDAO)	229	2.2 mM	69–73	Zwitterionic detergent (uncharged at pH > 5)
N-Dodecylphosphocholine (DPC)	352	1.1 mM	50–60	Ionic detergent, an efficient solubilizer of hydrophobic or amphipathic α-helices. Dynamic behavior of DPC at low temperatures corresponds to that in a phosphotidylcholine membrane–water interface above its room temperature.
*p-tert*C8ØE$_{<9.5>}$ (Triton X-100)	625	0.25 mM	75–165	Polyoxyethylene glycol detergent, generally mild
C_{12} sorbitan E$_{<9.5>}$ (Tween-20)	1240	60 μM	—	Polyoxyethylene glycol detergent, generally mild

[a]Based on data compiled in: le Maire M., Champeil P., Moller J.V. (2000). Interaction of membrane proteins and lipids with solubilizing detergents. *Biochim. Biophys. Acta* **1508**: 86–111.

the commonly used notion of an ideal "spherical" micelle (79). Packing defects result in considerable contact between hydrophobic tails and water. Micelles are very dynamic structures and rapidly exchange micellar components with the solvent. A detergent is capable of solubilizing amphipathic molecules only at a concentration that exceeds the CMC. Solubilization of many integral membrane proteins actually occurs significantly above the CMC level, since the detergent molecules also interact with the hydrophobic surfaces of the membrane protein to create protein–detergent complexes, PDCs (79).

The behavior of a solubilized membrane protein is obviously affected by interactions with the detergent molecules in PDCs. There are several well-documented examples when complete solubilization of a membrane protein cannot be achieved without loss of activity, even if the detergents are "mild" and do not grossly affect the conformation of the protein after solubilization (81). The exact mechanism of protein–detergent interactions remains a subject of intense debate. For example, it is still unclear whether solubilization of the hydrophobic segments of membrane proteins proceeds via their being engulfed by micellar-like structures or by forming a detergent monolayer on the protein surface.

A major problem that is often encountered when MS is used to characterize detergent-solubilized membrane proteins is the suppressive effect of detergents (82). Sometimes it is possible to substitute detergents with organic solvent, in which the protein would not precipitate (83–85). Typically, this is accomplished by protein precipitation, followed by the removal of the detergent and resolubilization of the protein in a suitable nonpolar solvent. Separation of membrane proteins from detergents can also be accomplished in one step using HPLC, which is compatible with direct mass spectrometric analysis, for example, online HPLC/ESI MS (86). Reverse-phase chromatography is normally expected to denature proteins; however, there is substantial evidence that this is not always the case (86). Another method to obtain high-quality mass spectra of detergent-solubilized membrane proteins avoids the protein precipitation and resolubilization steps by utilizing direct protein extraction into a nonpolar solvent (87). Finally, a group of "electrospray-friendly" surfactants (such as perfluorooctanoic acid, PFOA, and perfluorooctanesulfonic acid, PFOSA) were reported recently (88). The presence of such surfactants in solution does result in some decrease of the protein ion signal in the ESI mass spectra. However, such signal loss could be fully recovered as the surfactant could easily be removed by evaporation (88).

There are several methods of detergent removal particularly suitable for MALDI MS that make use of phase separation during the crystallization step (89). It is commonly accepted that only relatively mild (nonionic or zwitterionic) detergents should be used in the analyses of membrane proteins by MS. However, there are reports of MALDI MS analyses of protein samples containing strong (ionic) detergents, such as sodium dodecyl sulfate, SDS (90), as well as its "milder" analogue, ammonium dodecyl sulfate, ADS (91).

The most common MS-based method of probing higher-order structure of detergent-solubilized membrane proteins is selective chemical modification (see

Chapter 4 for a detailed discussion of technical aspects of selective chemical labeling). A very interesting example of such studies has been presented recently by Whitelegge and co-workers in a series of reports aiming at elucidating details of protein–substrate interactions for a 47 kDa transmembrane protein, lactose permease from *E. coli* (92, 93). Initially, protein modification by *N*-ethylmaleimide (targeting thiol groups) followed by ESI MS detection and identification of the reaction products was used to probe changes of the microenvironments of cysteine residues induced by substrate binding (92). Treatment of the native protein solubilized in detergent micelles reveals only two reactive thiol groups out of eight Cys residues. Both Cys residues become protected in the presence of D-galactopyranosyl β-D-thiogalactopyranoside, a substrate analogue. Interestingly, labeling of one of these two Cys residues (which is a component of the substrate binding site according to the model of the protein–substrate interaction, see Figure 11.11) is inhibited completely in the presence of the substrate analogue. Significantly reduced (but not completely eliminated) alkylation of the second reactive Cys residue in the presence of the substrate analogue reflects a long-range conformational change caused by binding of the substrate (92). The molecular model of lactose permease (94) also predicts that the substrate interacts with Glu^{269} via a hydroxyl group in the galactopyranosyl ring (Figure 11.11C). Covalent modification of carboxyl groups with carbodiimides followed by identification of the reaction products with ESI MS provides strong evidence that the substrate protects the protein against carbodiimide reactivity (93). A significant proportion of the decrease in reactivity occurs specifically in a nonapeptide containing Glu^{269}, while the reactivity of a E269D mutant (which exhibits significantly lower affinity toward the substrate) is unaffected by the substrate (93). Monitoring the ability of different substrate analogues to protect against carbodiimide modification of Glu^{269} provided evidence that the C-3 hydroxyl group of the galactopyranosyl ring (Figure 11.11C) plays an important role in specificity, possibly by H-bonding with Glu^{269} (93).

A very creative approach to probing contact topology of assemblies of membrane proteins was presented by Przybylski and co-workers, who used proteolysis in the presence of strong detergents (sodium dodecyl sulfate) to identify protein–protein contact regions (95). A complex formed by an ion channel protein (mitochondrial porin) and its ligand (adenine nucleotide translocator) was separated from its constituents with SDS-PAGE. In situ digestion was performed on the bands corresponding to the complex and the unbound form of the porin, followed by MALDI MS analysis and identification of the fragment peptides. The proteolytic peptide patterns of the two bands were significantly different due to the different accessibility of the cleavage sites located in the interface region (95).

11.2.2. Detergent-Free Analysis of Membrane Proteins

In many cases the structural integrity of a membrane protein is absolutely reliant on the presence of a lipid bilayer (96, 97), or else it is impossible to identify a

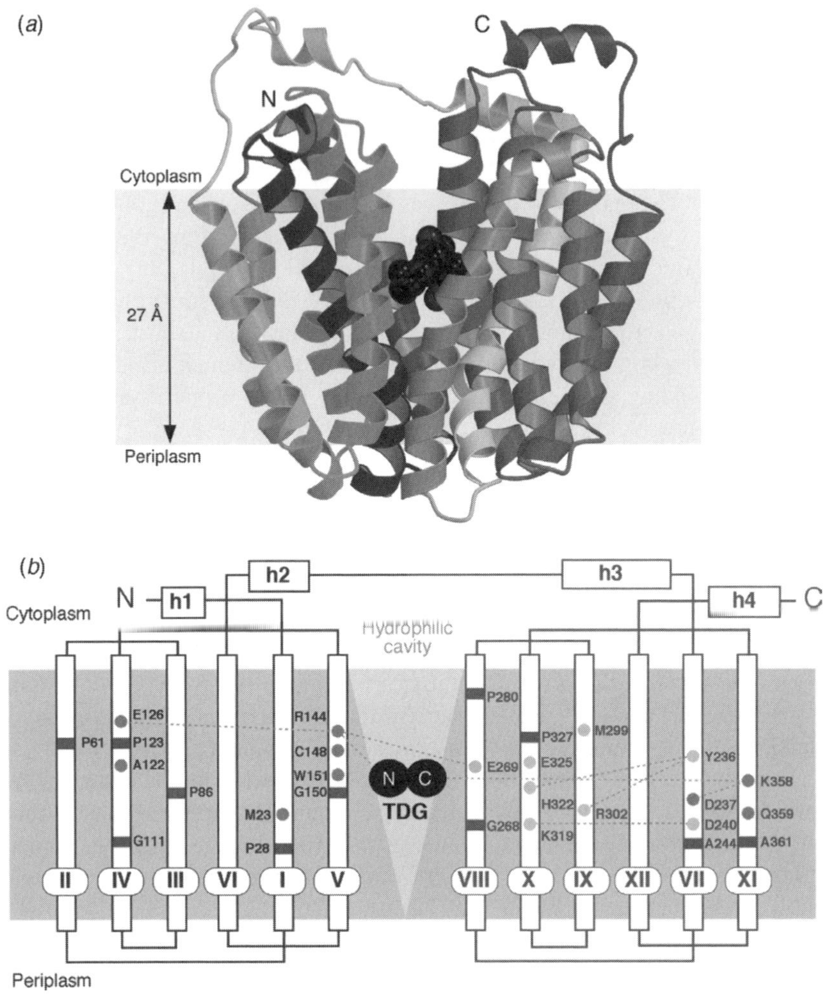

FIGURE 11.11. Overall structure of lactose permease from *E. coli* (LacY) with a bound substrate homologue D-galactopyranosyl-β-D-thio-galactopyranoside (TDG). (**A**) Ribbon representation of LacY viewed parallel to the membrane. (See color insert, where the twelve transmembrane helices are colored from the N-terminus in purple to the C-terminus in pink and TDG is represented by black spheres.) (**B**) Secondary structure schematic of LacY. (See color insert, where the N- and C-terminal domains of the transporter are colored blue and red, respectively; residues marked with green and yellow circles are involved in substrate binding and proton translocation, respectively; residue Glu^{269} marked by a light blue circle, is involved in both substrate binding and proton transfer; and the hydrophilic cavity is designated by a light blue triangle.) h1 to h4 denote surface helices. (**C**) Substrate-binding site of LacY. (See color insert, where transmembrane helices in the N- and C-terminal domains are colored blue and red, respectively.) Reproduced with permission from (94). © 2003 AAAS.

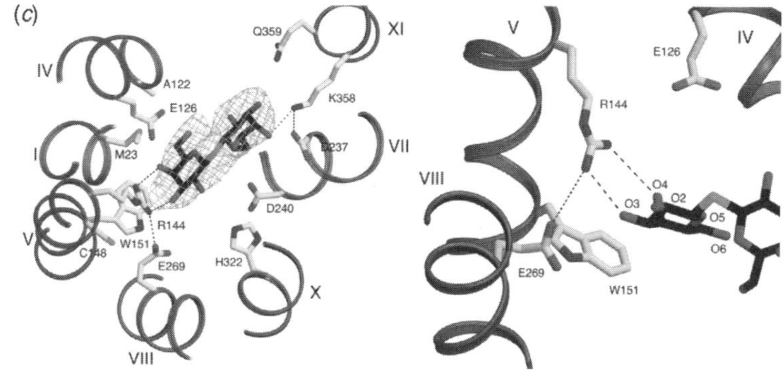

FIGURE 11.11. *(Continued)*

detergent that would extract such a protein from a membrane while preventing its aggregation and precipitation. In such cases protein structure and dynamic behavior can be analyzed directly within the context of a membrane. For quite some time now NMR spectroscopy has been used to probe the structure and dynamics of membrane-bound proteins (98, 99) and has recently been joined by X-ray crystallography (100). Although MS is a relative newcomer in this field, we have seen already several important contributions, and their number will certainly continue to grow. Several experimental approaches are currently being used to probe the structure and behavior of membrane-bound proteins using mass spectrometric methods of detection. One group includes an array of methods that use proteolytic degradation to identify membrane-bound protein segments. Another utilizes hydrophobic probes to obtain topological information on such segments. Finally, hydrogen–deuterium exchange can be used to provide information on interfacial positioning and stability of transmembrane polypeptides in lipid bilayers. These methods are discussed briefly in this section and are illustrated with selected examples from recent literature.

Producing peptide maps of membrane-bound proteins may yield information on their topological arrangement by identifying the segments that are confined to a membrane and, therefore, are protected from proteolytic enzymes. Yates suggested that proteinase K, a relatively nonspecific enzyme, can be used to cleave selectively soluble domains of membrane proteins, followed by MS-assisted identification of the fragment peptides (78). Proteolysis of the "solvent-exposed" soluble domains can be temporally separated from the "protected" segments of these domains, providing further structural differentiation (78). Smith and co-workers described another method of analysis of membrane proteins that specifically targets the transmembrane segments (101).

Selective chemical modification is perhaps among the most popular techniques that are currently used to probe the structural arrangements of various domains of membrane-bound proteins. An array of existing hydrophobic probes would only modify protein segments confined to the membrane, while hydrophilic probes would only label solvent-exposed regions of the protein. Hydrophobic photoreactive probes [particularly benzophenone photophores (102)] have seen a particular surge in popularity in recent years. Benzophenone (BP) -containing photoreagents can be manipulated at ambient light. Activation of BP-based photoprobes (at 350–360 nm) does not cause protein damage (other than selective chemical modification), while providing a means to control the extent of covalent modification. Importantly, BP-based photoprobes react preferentially with inert C—H bonds* (Figure 11.12).

Leite and co-workers have recently presented a particularly intriguing example of how such photoreagents can be used to characterize structural changes within

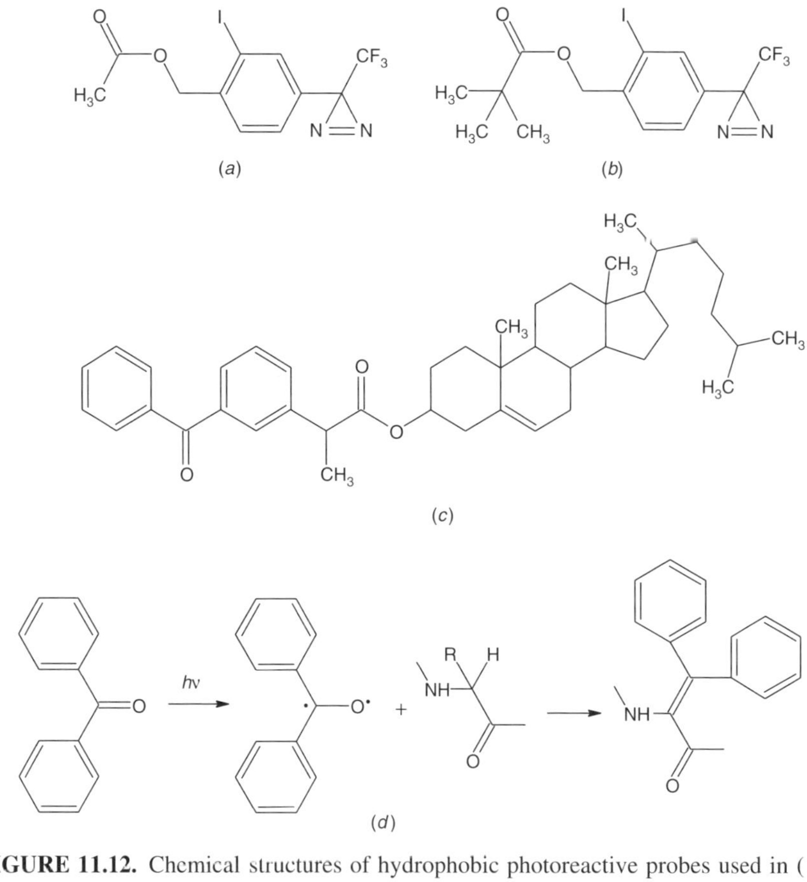

FIGURE 11.12. Chemical structures of hydrophobic photoreactive probes used in (103) to probe topology of a membrane protein: acetate (**A**) and trimethylacetate (**B**) derivatives of 3-trifluoromethyl-3-*m*-iodophenyl diazirine and cholesteryl benzoylphenyl propionate (**C**). A schematic diagram depicting the mechanism of photolabeling by benzophenone is depicted in (**D**).

*The reactivity order of C—H bonds in various groups is: NCH_x > SCH_x > methine > $C{=}CCH_2$ > CH_2 > CH_3.

the membrane protein induced by physiologically relevant variations of the physical parameters of the membrane (103). Several hydrophobic photoreactive agents (Figure 11.12) were used to characterize the differential accessibility of the nicotinic acetylcholine receptor $\alpha 1$ subunit in the open, closed, and desensitized states. Photoactivation of the probes was carried out by UV irradiation during pulses of voltage across the membrane ranging from +40 mV (producing closed states) to −140 mV (producing a approximately 1:1 mixture of open and closed states). Labeling defined the lipid-exposed parts of the transmembrane segments of the protein. More importantly, the results of such experiments helped to identify protein segments that are involved in gating-dependent conformational shifts (103).

Knapp used site-specific cleavages of the cytoplasm-exposed loops of rhodopsin induced upon activation of a Cu-phenanthroline tethered cleavage reagent attached to the protein (104). The cleavage site was identified by mass analyzing the CNBr fragments of the protein, which provided an unbiased mapping of rhodopsin (105).

A very promising approach to mapping transmembrane regions and characterizing their dynamics using amide hydrogen exchange (HDX) (see Chapter 5 for a detailed discussion of HDX measurements with mass spectrometric detection) and electrospray MS was recently introduced by Heck and co-workers (106, 107). Short model transmembrane peptides WALP16(+10) were reconstituted in fully hydrated dispersed phospholipid bilayers, and the entire mixture was analyzed by nano-ESI MS. The flow rate was actually higher than that achieved in a typical nano-ESI source; a special procedure was used to manufacture needles with relatively large (tens of μm) tip openings. The ESI spectra of the suspension of vesicles (Figure 11.13) clearly show that the lipid bilayer structures decompose once removed from the aqueous environment and transferred to the gas phase. This should not be surprising in light of our discussion of the stability of various types of noncovalent associations in a solvent-free environment. The stability of lipid bilayers in aqueous environments is mostly due to hydrophobic interaction; once water is removed, the bilayer disintegrates into its components and releases the transmembrane peptides. Although the spectrum is dominated by the ionic signal corresponding to lipid monomers and small aggregates, the peptide ion peaks can also be clearly seen (Figure 11.13). Hydrogen exchange was initiated by diluting the suspensions in deuterated ammonium acetate buffer and the progress of HDX was measured by monitoring the mass gain of the peptide ions. Approximately ten hydrogen atoms remained protected from exchange, indicating effective solvent shielding by the lipid bilayer from the $(LeuAla)_5$ repeat.* The combination of HDX in solution and peptide ion fragmentation in the gas phase provided even more definitive proof that this repeat is embedded in the hydrophobic portion of the membrane and is effectively shielded from the solvent (Figure 11.14). The local deuterium content levels

*The length of this repeat is close to the hydrophobic thickness of the DMPC bilayer (~23 Å in the fluid state), which suggests that this hydrophobic segment is completely embedded in the hydrophobic region of the bilayer.

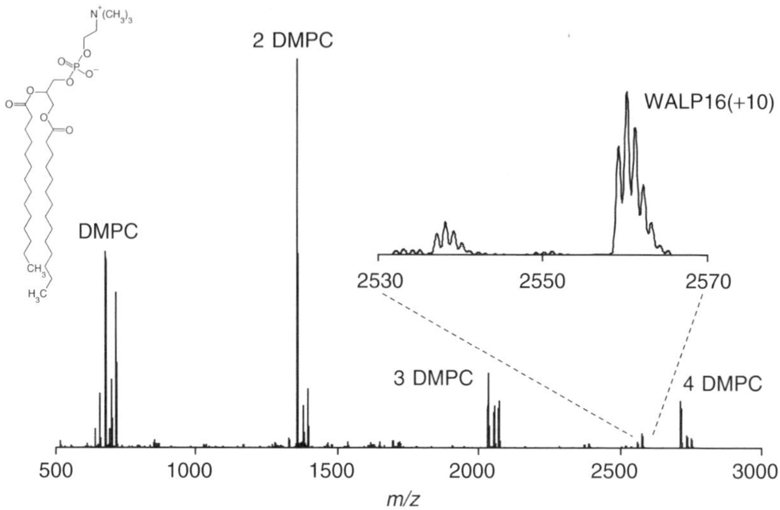

FIGURE 11.13. Nano-ESI mass spectrum of lipid vesicles composed of 1,2-dimyristoyl-*sn*-glycero-3-phosphocholine (DMPC) reconstituted with model transmembrane peptides acetyl-$(GA)_3WW(LA)_5WW(AG)_3$-ethanolamine [WALP16(+10)] at a 1:25 ratio [WALP16(+10) : DMPC]. Abundance of the largest aggregate of DMPC detected in this experiment (16-mer, not shown in the spectrum) was four orders of magnitude below that of the DMPC monomer. The peaks labeled as WALP16 are singly charged Na^+ adducts of the peptide (the unlabeled low-abundance isotopic cluster shown in the *insets* corresponds to a protonated form of the peptide). Approximately 200 nL of a vesicle suspension (equivalent of 20 μM peptide concentration) was used to acquire the spectrum. Adapted with permission from (106). © 2000 National Academy of Sciences, USA.

measured as mass gains clearly indicated that the amide exchange within the central region of the peptide is extremely slow as compared to the terminal (solvent-exposed) segments (compare mass increases of the $a_{10}{}^+$ and $a_{18}{}^+$ in Figure 11.14). A detailed analysis of the local exchange patterns showed some (but rather slow) deuterium incorporation at the ends of the transmembrane segment (106), which can be attributed to limited hydrogen scrambling within the peptide ions prior to their fragmentation (108). Chapter 5 contains a more detailed discussion of issues related to internal proton mobility and hydrogen scrambling within gas phase protein and peptide ions. Alternatively, local dynamic fraying can be invoked to explain such behavior. Indeed, lipid bilayers are known to be dynamic systems and allow significant water penetration into the hydrophobic "bulk" (109). Hydrogen exchange in the transmembrane segments due to such penetration would be slow for several reasons. The "penetration" events are rather rare (109) and the hydrophobic environment may impose severe steric constraints on the water molecule. Furthermore, the dielectric constant in the membrane interior is much lower than that of water itself, which would also adversely affect the exchange rate.

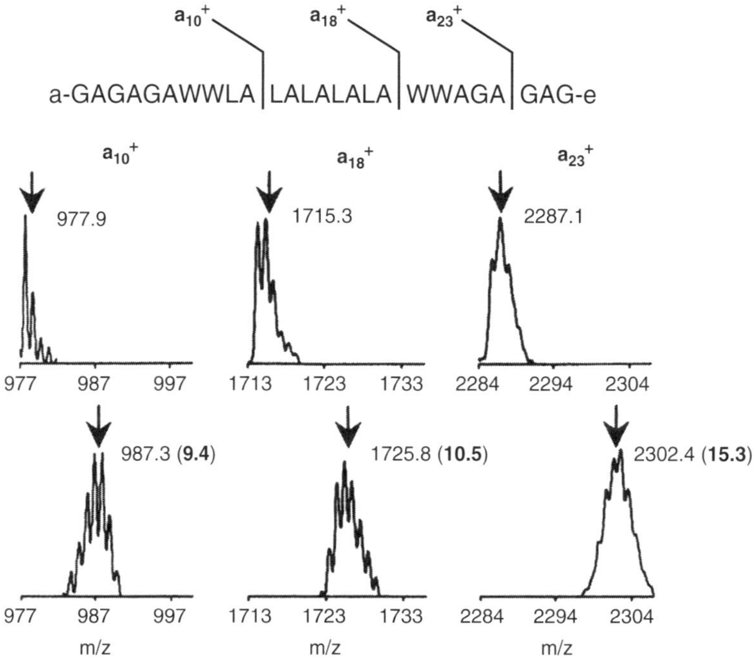

FIGURE 11.14. Isotopic clusters of $a_{10}{}^+$, $a_{18}{}^+$, and $a_{23}{}^+$ fragment ions of WALP16(+10) desorbed from a suspension of the peptide-reconstituted DMPC vesicles hydrated in protiated buffer (*top*) and in deuterated buffer (*bottom*). The numbers in parentheses indicate the mass difference due to ^{2}H incorporation. The positions of the backbone cleavages giving rise to the fragment ions discussed are marked on the amino acid sequence of WALP16(+10). Adapted with permission from (106). © 2000 National Academy of Sciences, USA.

In subsequent studies, Heck and co-workers used this approach to explore the effects of transmembrane segment length and composition on the local protection patterns (108). Surprisingly, longer peptide analogues were found to be more protected than was expected on the basis of their length with respect to bilayer thickness. This phenomenon was explained by an increased protection from the bilayer environment, because of stretching of the lipid acyl chains and/or tilting of the longer peptides. Next, the role of the flanking tryptophan residues was investigated. The length of the transmembrane segment that shows very slow HDX was also found to depend on the exact position of the tryptophan residues in the peptide sequence Ac-GWW(LA)$_n$L$_x$WWA-Etn ($n = 5$ to 9, $x = 0$ or 1), suggesting that these residues act as strong determinants for positioning of proteins at the bilayer–water interface (108). The influence of putative helix breakers on the dynamics of membrane-bound proteins was also studied. The insertion of a proline residue in the middle of the transmembrane segment resulted in a significant increase of the exchange rates, while incorporation of a glycine residue did not have any noticeable effect on the stability of the membrane-bound segment of the peptide.* More importantly, site-specific measurements of deuterium incorporation into Pro-containing membrane-bound peptides (Figure 11.15) clearly showed increased conformational flexibility of the transmembrane segments and enhanced water penetration (108).

A similar approach was recently used by Broadhurst and co-workers to study the properties of a transmembrane fragment of the M2 protein of Influenza A incorporated into lipid vesicles and detergent micelles (110). Interestingly, the HDX rates of this protein were significantly lower when the peptide was incorporated into aqueous DMPC vesicles, as compared to the HDX rates of the peptide in the presence of a large excess of the detergent.† In turn, the HDX rate of the peptide incorporated into the detergent micelles was significantly lower than that of the denatured peptide in methanol. The results of this study emphasize that the dynamics and solvent accessibility of the "transmembrane" segments are greatly affected by the properties of the model bilayer systems (110). Another important observation reported by the authors relates to the optimal sample preparation procedure for direct ESI MS analysis of lipid vesicle-bound peptides. Electron microscopy experiments provided evidence that vesicle concentration by centrifugation induced aggregation, leading to the formation of large multimicellar aggregates, which could not be analyzed by ESI MS (Figure 11.16). However, the liposome size profile could be maintained when lyophilization was used, followed by thawing above the liquid crystal transition temperature of the lipid component (110).

The initial studies briefly summarized in the preceding paragraphs clearly demonstrate the great potential of HDX MS experiments as a tool to probe both topology and dynamics of membrane-bound polypeptides. It remains to be seen if these methods can be applied to larger systems, such as integral membrane proteins and membrane-bound proteins. Although there has been significant progress in understanding the mechanisms of folding of water-soluble proteins, the factors governing correct folding and dynamic behavior of integral membrane proteins remain largely unknown (96). Information on the dynamics of membrane proteins within their native environment would certainly be indispensable for providing answers to an array of important fundamental questions related to "lipid-assisted" protein folding.

Membranes themselves are not static structures, and their dynamic properties are greatly affected by their protein components (111). An important question that still awaits an answer is whether MS will be useful in understanding how proteins influence the mechanical properties of membranes at the molecular level. Deciphering the mechanisms of such processes will obviously have a profound effect on our understanding of a variety of biological phenomena ranging from endocytosis (112) to cell lysis (113–115) to viral entry into the host cell (116, 117). An intriguing example of using HDX CAD MS to study polypeptide interaction

*Earlier studies have also shown that glycine does not disrupt α-helical structure of transmembrane segments.

†Triton X-100 was used as a detergent.

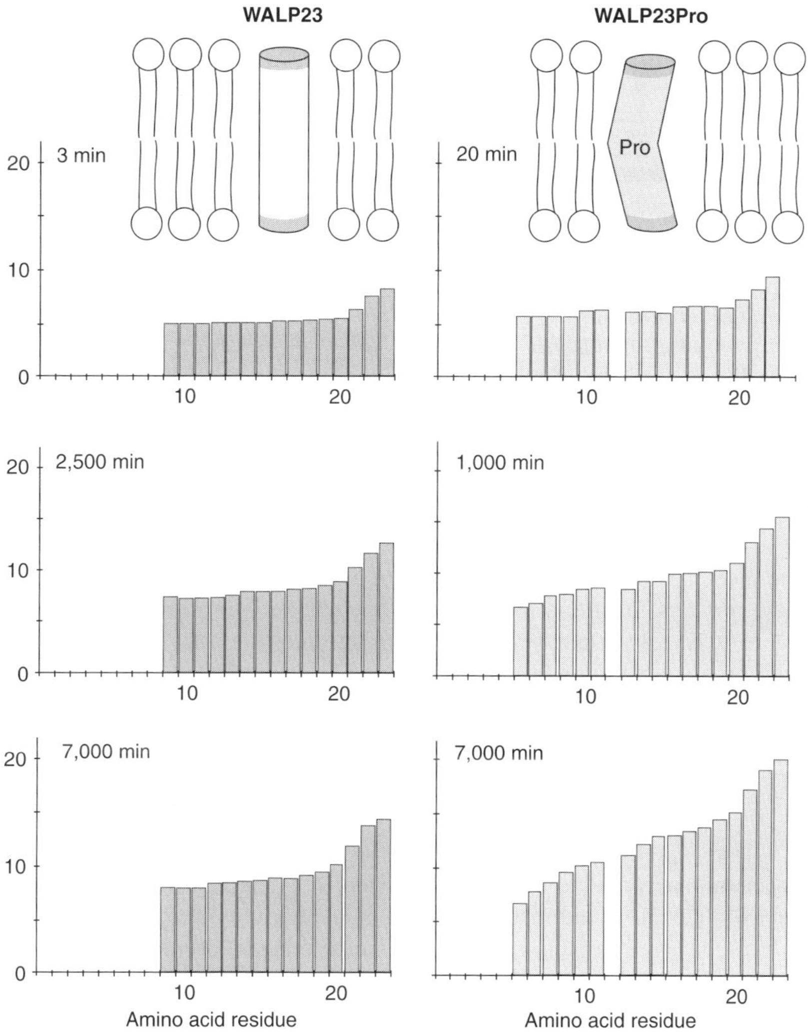

FIGURE 11.15. Bar diagrams representing deuterium levels of $a_n{}^+$ fragment ions derived from a model peptide Ac-GWW(LA)$_8$LWWA-Etn (*left*) and Ac-GWW(LA)$_4$P(AL)$_4$WWA-Etn (*right*) using collision-induced dissociation of Na^+-cationized molecular ions of the peptides that were incorporated in DMPC bilayers and incubated in deuterated buffers for the time periods indicated on the graphs. Models for the incorporation of the peptides in a bilayer are presented on the insets. Fast-exchanging segments of the peptides are indicated in gray. Adapted with permission from (107).

with a phospholipid micelle was recently reported by Akashi and Takio (118). Measurements of the amide protection of melittin, a peptide from bee venom with hemolytic activity, was carried out in the absence and in the presence of micelles formed by dodecylphosphocholine (DPC). Although amide exchange of

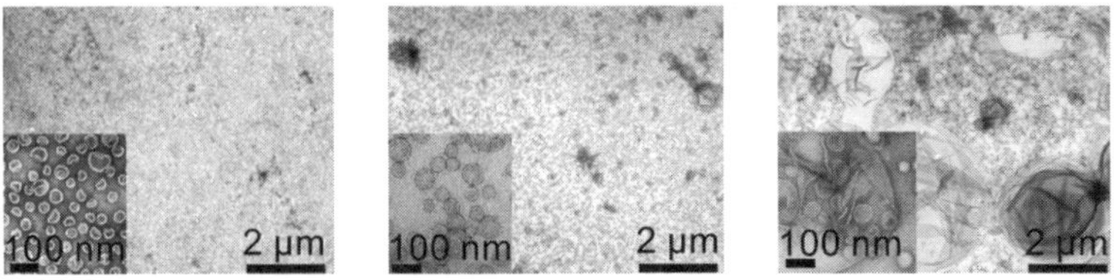

(*a*) (*b*) (*c*)

FIGURE 11.16. Images of mixtures of DMPC liposomes incorporating hydrophobic peptides prepared by dialysis of mixed micelles containing the peptide, phospholipid, and the detergent (**A**); concentrated by lyophilization after the dialysis (**B**); and concentrated by centrifugation after the dialysis (**C**). Reproduced with permission from (110). © 2002 American Society for Mass Spectrometry.

melittin in aqueous solution was very fast, the presence of DPC micelles resulted in a significant decrease of the exchange rate. Site-specific measurements of deuterium incorporation (carried out using collisional activation of melittin ions in the ESI interface) indicated that only one short segment of the peptide located next to the helix-breaking proline residue remained relatively flexible. It remains to be seen if this feature is related to the membrane-lytic properties of melittin, as recent studies by Dempsey and co-workers indicate that lipid perturbations induced by monomeric melittin are very modest, unlike those caused by melittin dimers under identical conditions (119). It appears that a carefully crafted experimental strategy that involves both HDX and chemical cross-linking may actually provide a detailed account of the melittin–membrane interaction.

11.2.3. Organic Solvent Mixtures

Many membrane proteins can be solubilized in organic solvent mixtures. The concept of "naked" membrane proteins in such mixtures is appealing for structural analysis, since organic solvents, unlike detergents, do not adversely affect spectral quality. However, it is probably unrealistic to expect that the behavior and structure of complex membrane proteins would not be affected by membrane removal (99). While secondary structure may often be retained, a significant proportion of the protein tertiary structure will likely be perturbed or fully lost (99). Dobson and co-workers used a small transmembrane channel peptide gramicidin A to explore the influence of the environment on its conformation (120). The "channel conformation" of the peptide adopted in lipid membranes and SDS micelles is a right-handed $\beta^{6.3}$ helix dimer, although other conformations have been reported as well (Figure 11.17). The exact conformation assumed by the peptide in lipid bilayers remains a subject of controversy (121, 122). Furthermore, it still is unclear what conformation is adopted by gramicidin A in polar solvents and whether or not it retains its dimer-forming ability in such solutions. Mass spectral data provide strong evidence that gramicidin A remains a monomer in trifluoroethanol (TFE) and dimethylsulfoxide (DMSO) solutions, while the existence of the dimeric species becomes evident in ethanol solutions (120). Despite

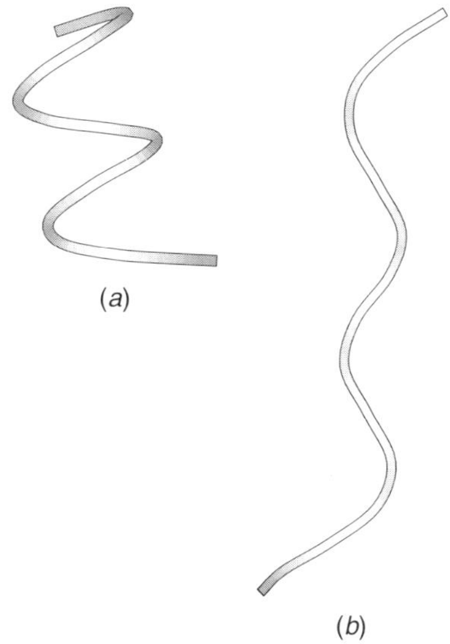

FIGURE 11.17. Conformations adopted by gramicidin A in SDS micelles (PDB code 1JNO) and methanol (1MIC).

its inability to form a dimer in TFE, the peptide is apparently highly structured in this environment, as suggested by significant backbone amide protection revealed by HDX MS measurements (120). These measurements were rather unusual, as the exchange was carried out in anhydrous solvent. HDX reactions in TFE were initiated by diluting a concentrated TFE (CF_3CH_2OH) solution of gramicidin A in d_1-TFE (CF_3CH_2OD) 1:99 (v:v). Exchange of a reference tripeptide (Ala_3) was also monitored under these conditions to provide a measure of the intrinsic exchange rate in TFE.

In a more recent study, Chitta and Gross used ESI MS to characterize gramicidin A dimerization in solutions ranging from relatively hydrophilic (TFE) to very hydrophobic (*n*-propanol) (123). The degree of dimerization (as deduced from MS measurements, Figure 11.18) is clearly correlated with the dielectric constant of the solvent. Careful evaluation of the ESI MS data led the authors to a conclusion that up to 70% noncovalently bound dimers formed in solution could be preserved in the gas phase under the most favorable conditions.

Fluorinated alcohols (TFE and hexafluoroisopropanol, HFIP) were also used by Waring and co-workers to examine structure of a viral fusion peptide (N-terminal peptide FP of HIV-1 gp41) under conditions mimicking near-membrane environments (124). The amide protection pattern was evaluated in a site-specific fashion using HDX in solution (50% TFE or 70% HFIP) and peptide ion fragmentation in the gas phase. The resulting map of the fusion peptide secondary structure compared favorably with earlier NMR studies carried out in similar environments mimicking membranes (124).

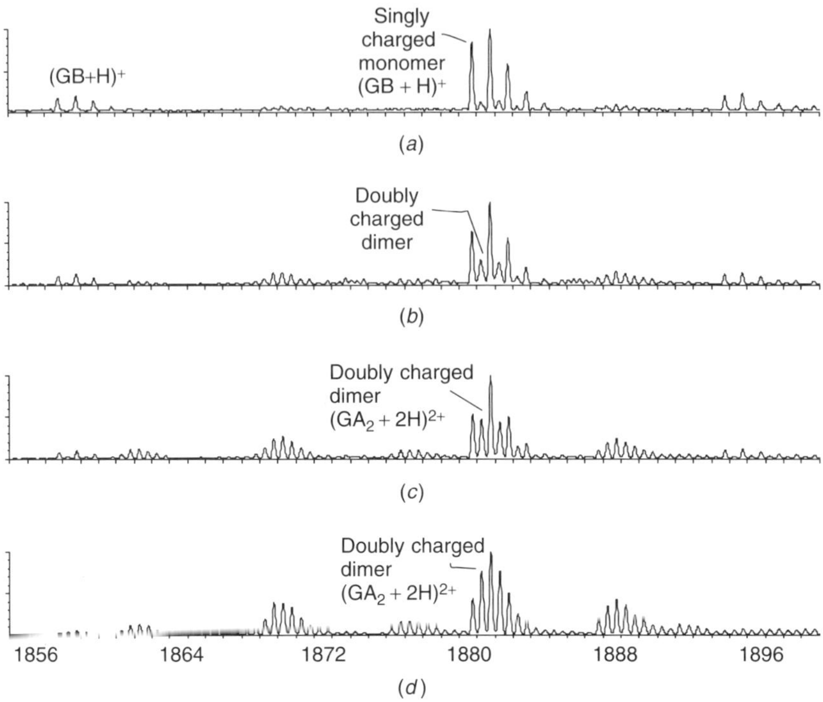

FIGURE 11.18. ESI mass spectra (molecular ion regions) of gramicidin A in TFE (**A**), methanol (**B**), ethanol (**C**), and *n*-propanol (**D**). Peaks of singly charges ions at m/z 1880 correspond to gramicidin A monomers, while those of doubly charged ions correspond to the noncovalently bound dimers of the peptide. Reproduced with permission from (123). © 2004 Biophysical Society.

11.2.4. Noncovalent Interaction by MS

Conventional wisdom holds that hydrophobic interactions are all but eliminated in the gas phase and, therefore, MS is hardly suitable for direct observation and characterization of noncovalent complexes that thrive in lipophilic environments. However, in the last several years this notion is being increasingly challenged and several examples have been presented where noncovalent protein–lipid interactions are preserved upon transition from solution to the gas phase. Wirtz and co-workers have recently used nano-ESI MS to detect intact complexes formed by phosphatidylinositol transfer protein α (PI-TPα) and a range of lipids, as well as phosphatidylcholine transfer protein (PC-TP) and different phosphatidylcholine (PC) species (125). The stability of the protein–lipid complex in the gas phase was clearly determined by the electrostatic interactions. Although PI-TPα from mammalian cells normally carries either a PC or a PI molecule (126), the complexes formed by these zwitterionic lipids and PI-TPα were much less stable in the gas phase than those formed by PI-TPα complex with a negatively charged phospatidylglycerol (PG). Examination of the PI-TPα structure (Figure 11.19)

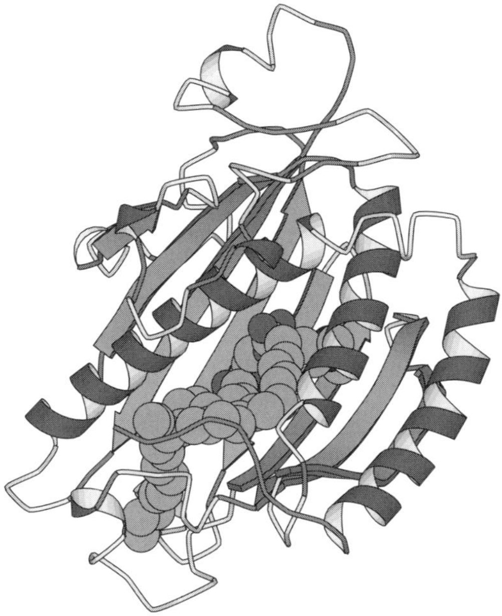

FIGURE 11.19. Crystal structure of phosphatidylcholine transfer protein (PC-TP) complexed to phosphatidylcholine (PC), PDB code 1FVZ.

shows the presence of an array of lysine and arginine residues located at the entrance to the lipid-binding cavity. Many of these residues are likely to carry a positive charge in the gas phase and will interact strongly with the negatively charged head-group of PG. Indeed, probing the strength of interaction between PI-TPα and PG in the gas phase indicates that it approaches that of a peptide bond (125). At the same time, the stability of the protein–lipid complex was not affected by the length of the fatty acid tails of the lipids, indicating little or no contribution to the complex stability from the lipophilic segments of the ligands (125).

Another study of specific phospholipid binding to a soluble plasma protein (apolipoprotein C-II*) using ESI MS was recently reported by Robinson and coworkers (127). Mass spectra of phospholipid suspensions acquired under very gentle conditions in the ESI interface and ion guide contained a series of peaks corresponding to multiply charged phospholipid associations containing nearly a hundred molecules. Addition of the protein to the suspension resulted in the appearance of multiple peaks corresponding to a range of complexes formed by apoC-II and phospholipids in addition to the free protein signal. Binding specificity was confirmed by limited proteolysis followed by ESI MS analysis of proteolytic fragments. Lipid-bound peptides present in the mass spectra

*ApoC-II is a 9 kDa protein that belongs to a family of apolipoproteins involved in lipid transport in plasma; apoC-II binds in vitro to a range of synthetic and natural lipid surfaces.

encompass the proposed lipid-bound region of apoC-II (127). Observation of a noncovalent association between a short soluble peptide (viral fusion peptide) and phospholipids was also reported by Cole and co-workers (128).

An example of a noncovalent interaction between a membrane protein and a lipid that survives transition to the gas phase was recently reported by Demmers et al. (129). In this study ESI MS was used to investigate the interaction between bacterial K^+ channel KcsA* and membrane phospholipids. Binding of several different lipids to the protein was detected, when the vesicle dispersion was mixed with TFE prior to ESI MS measurements. However, the protein was present in the complex ion in the monomeric form, despite the well-characterized tetrameric structure of KcsA in lipid bilayers (129). Nevertheless, the authors provide some evidence that the stability of the protein monomer–lipid complex in the gas phase may in fact reflect their binding affinity within the membrane (129).

Finally, Griffiths and co-workers recently presented the first example of direct ESI MS analysis of a noncovalently bound membrane protein complex in its native state (130). The observation of the intact noncovalent trimer of a membrane-bound enzyme microsomal glutathione transferase† was revealed in a series of experiments that employed ESI MS to assess the oligomerization of this protein reconstituted in a minimal amount of detergent (130). The measured mass of the detected homotrimer was 53 kDa, indicating that the protein complex also contained one glutathione molecule. Collision-induced dissociation of the trimer complex resulted in the formation of monomer and homodimer ion species. The latter was observed in two forms, one unliganded and one carrying a glutathione molecule (130). The detergent used in this study was Triton X-100, which was apparently well tolerated by ESI MS,‡ although increase of the detergent content of the sample inevitably led to deterioration of the spectral quality.§ To avoid excessive protein–detergent adduct formation, relatively high collisional activation of ions in the ESI interface region was used to induce dissociation of adduct ions and increase signal-to-noise ratio (130). Some tetrameric ions "survived" such harsh conditions in the ESI interface, although the mass spectrum was clearly dominated by signal corresponding to the monomer ions, emphasizing marginal stability of the trimeric protein complex in the gas phase.

11.3. MACROMOLECULAR TRAFFICKING AND CELLULAR SIGNALING

Membranes separate cellular compartments from each other, allowing them a certain degree of autonomy. Still, most compartments and organelles are highly

*KcsA is an α-helical homotetrameric K^+ channel protein from *Streptomyces lividans*; each monomer contains two transmembrane segments.

†This enzyme is involved in the detoxification of xenobiotics and protects cells from oxidative stress.

‡A range of other detergents were tested by the authors of this study (MEGA-10, Thesit and Zwittergent 3–12); however, they were found to be less compatible with ESI MS compared to Triton X-100.

§"Tolerable" levels of Triton X-100 were found to be $<0.1\%$.

dependent upon each other, and normal functioning of the cell requires both sophisticated communication (signaling) and highly orchestrated exchange of biomaterial (trafficking). These two groups of processes are central to biology and both involve the transport of soluble macromolecules across the membranes. Mass spectrometry is still making first steps in solving problems related to trafficking and signaling. Although the list of accomplishments in this field is still rather short, the contributions made already are certainly very impressive. This section provides several particularly intriguing examples that illustrate the potential of MS as an analytical tool for the study of protein transport and signal transduction.

11.3.1. Trafficking Through Nuclear Pores

Proper functioning of eukaryotic cells is maintained through continuous trafficking of various proteins between the cytoplasm and the nucleus. Such import into and export from the nucleus occur through nuclear pore complexes (NPCs), which consist of at least 30 distinct components embedded in the nuclear envelope (Figure 11.20). Although many aspects of the nuclear pore trafficking machinery have been clarified in recent years, the molecular mechanisms of protein transport into and out of the nuclei are still a subject of extensive investigation (131–136). Although smaller macromolecules can pass freely through the nuclear pore, proteins exceeding ~40 kDa must be specifically transported through the NPC. Such trafficking is mediated by karyopherins (Kaps), nucleocytoplasmic shuttles that interact with NPC proteins (nucleoporins or Nups) and carry hundreds of different cargoes across the NPC. A key to understanding the inner workings of NPC regulation is to identify and localize the nucleoporins (Nups) within NPCs and to decipher their individual roles in the trafficking processes. This is an extremely

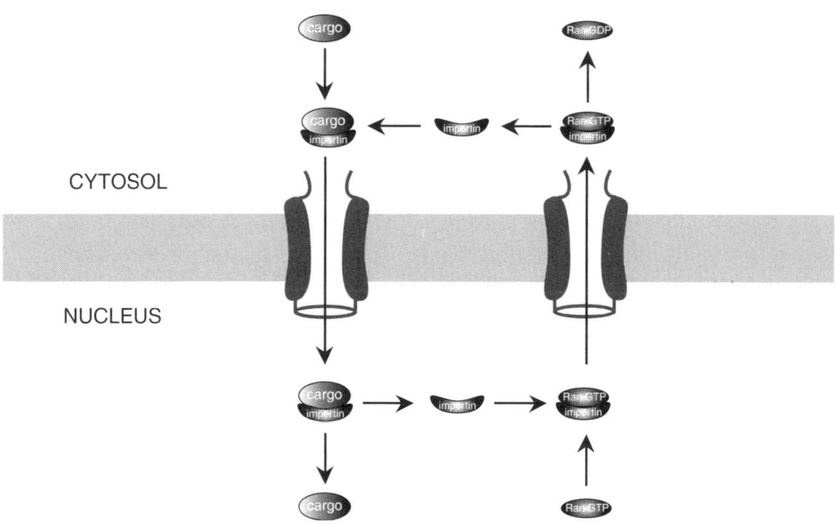

FIGURE 11.20. Schematic diagram of protein trafficking through a nuclear pore complex.

difficult task due to the enormous complexity of the NPC and the multiplicity of interactions among its components.

Until very recently, cryo-electron microscopy remained perhaps the only technique capable of providing detailed structural information on NPCs (137, 138). Rapid development of proteomic techniques, as well as the central role of MS in proteomics are certain to put MS at the center stage of investigations of NPC structure and function. Indeed, MS has already begun to experience an enormous surge in popularity in studies related to structure, dynamic behavior, and mechanistic aspects of NPC function. Burlingame and co-workers developed a strategy to identify proteins that interact with individual components of the nucleocytoplasmic transport machinery (139, 140). Further studies by this group provided a more detailed account as far as how these interactions actually occur (141). The novel interactions among the components of nuclear pore trafficking machinery uncovered in this study will certainly be invaluable for understanding the mechanisms of nucleocytoplasmic transport.

The complicated structure of the NPC was initially believed to function as an elaborate mechanochemically gated portal, which promotes a series of binding and exchange reactions. An alternative view of macromolecular transport through the NPC has recently been introduced based on the results of a series of studies in which MS has enjoyed a very prominent role (142, 143). Cataloging all protein components of a NPC using proteomic tools and MS allowed all proteins present in a highly enriched NPC fraction to be identified. Localization of all identified Nups within the NPC using other (non-MS) biophysical and biochemical methods allowed a detailed architectural map of the yeast NPC to be produced (Figure 11.21) (142). This study eventually led to the introduction of a "virtual gating" model of NPC functioning (144). The "virtual gating" model explains the mechanism of rapid and selective macromolecular trafficking through the nuclear pore by invoking the notion of passive diffusion that can be enhanced by favorable interaction of the cargo-carrying Kaps with Nups. When a macromolecule is confined to a volume encompassed by the central channel of the NPC, its mobility is limited, leading to an entropy decrease. Such "entropic price" of passage through the NPC increases as the macromolecular size increases, diminishing the probability of the passage and making the channel effectively impermeable to macromolecules above a certain size. The densely packed Nups add to this entropic price, by further constraining the free space available for diffusion. As a macromolecule must pass through this region to get from one side of the NPC to another, being within the NPC can therefore be considered as a "transition state" for translocation (Figure 11.22). In order to pass through the channel, the macromolecule has to enter such a transition state. Transport factors actually bind to the NPC, thus gaining access to the "transition state" (Figure 11.23) and eventually passing through the NPC. On the other hand, macromolecules that do not bind to the NPC would have a very low probability of accessing the "transition state" and, consequently, would not be able to pass through the NPC (144).

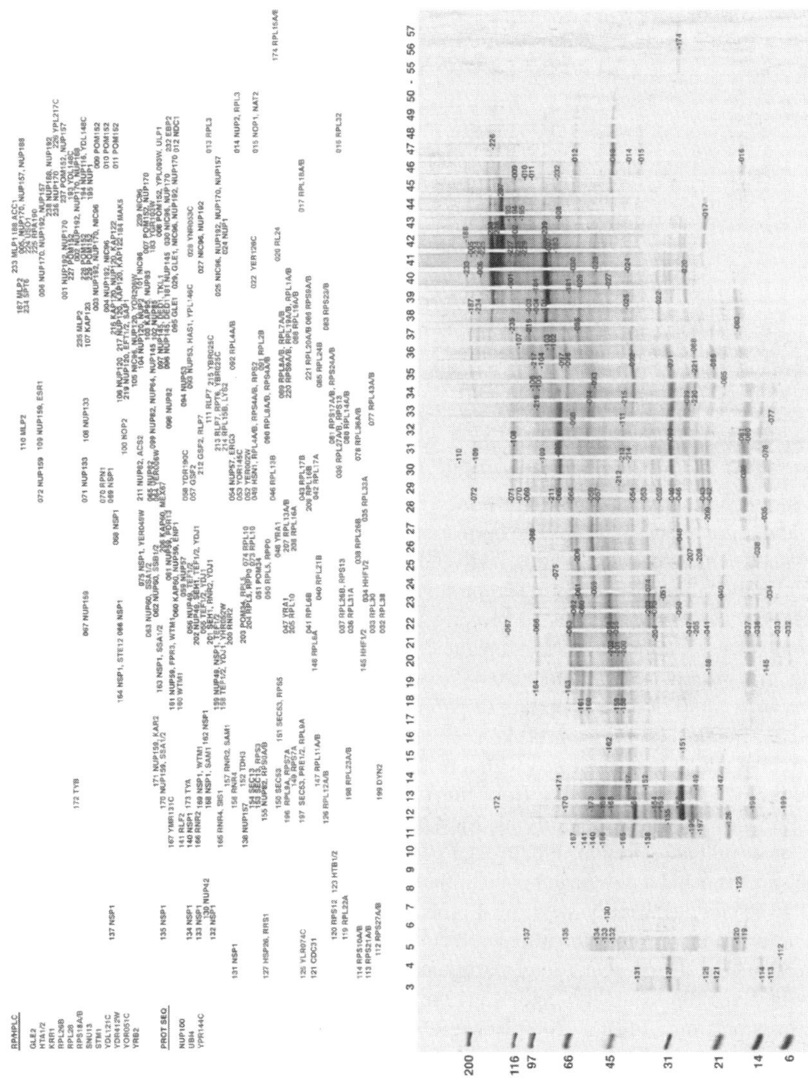

(*a*)

FIGURE 11.21

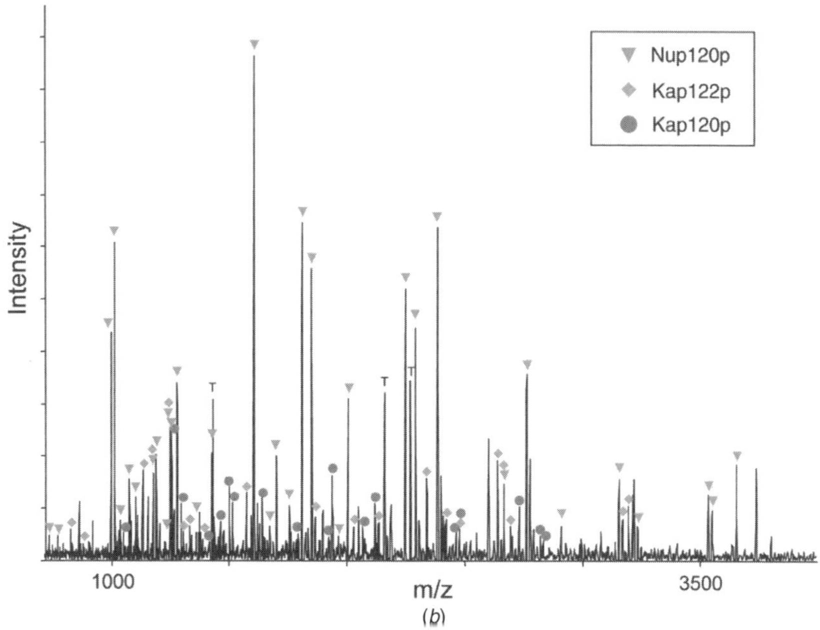

FIGURE 11.21. Identification of proteins in the yeast nuclear pore complex (NPC). Proteins in the highly enriched NPC fraction were separated by HPLC and SDS-PAGE. The number above each lane indicates the corresponding fraction number (proteins were visualized with Coomassie blue). Bands analyzed by MALDI TOF mass spectrometry are indicated by the adjacent numbers. The proteins identified in each band are shown in the top panel. On the top left are listed proteins not found in this separation but identified by MS/MS of reverse-phase HPLC-separated peptides (RP/HPLC) or peptide microsequencing. An example of a protein identification in a gel band by MS is shown at the bottom (a MALDI TOF mass spectrum of peptides produced by in-gel trypsin digestion of band no. 217, molecular mass ~100 kDa, fraction 35). The dominant protein component was Nup120p, which was identified with a probability 5×10^{22} higher than the next most probable protein. The peaks arising from Nup120p were subtracted from the spectrum and a new search was initiated, identifying Kap120p with a probability 3×10^{9} higher than the next most probable protein. Finally, the peaks arising from Kap120p were subtracted from the spectrum and a new search was initiated identifying Pdr6p/Kap122p with a probability 1×10^{8} higher than the next most probable protein. Markers indicate which peaks arise from each of the three identified proteins. The peaks labeled T arise from trypsin self-digestion. Unlabeled peaks likely originate from modified or incompletely digested peptides in Nup120p, Kap120p, and Kap122p, or from additional proteins. Reproduced with permission from (142). © 2000 Rockefeller University Press.

Translocation of proteins across biological membranes is another example of macromolecular trafficking that is mediated by an array of proteins interacting in a highly organized fashion (145–151). Very few examples have been reported so far on using MS to characterize mechanistic aspects of this complex machinery. Kipping and co-workers demonstrated a great potential of HDX MS experiments

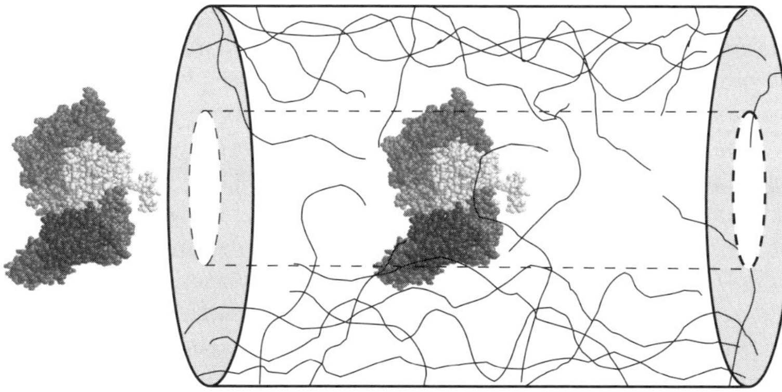

FIGURE 11.22. The oily-spaghetti model for translocation through a nuclear pore. The approximate dimensions of the pore are 45 nm (length) and 10 nm (inner diameter). Nucleoporins containing flexible FG repeats are shown as randomly oriented lines. Molecules with a diameter of <10 nm can diffuse freely through the pore. Diffusion of larger molecules is hindered by the nucleoporins. Reproduced with permission from (203). © 2001 American Society for Microbiology.

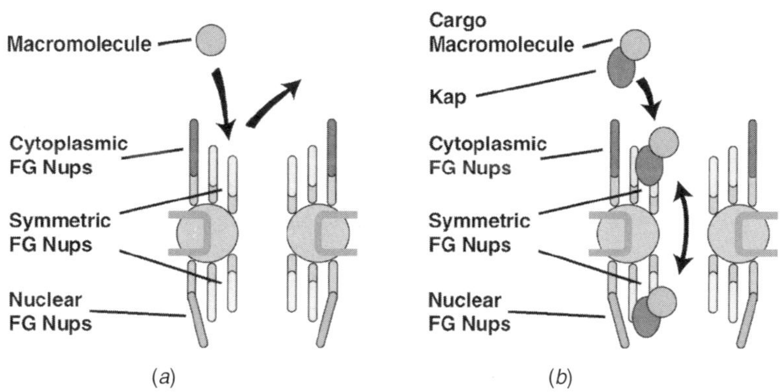

FIGURE 11.23. The Brownian affinity gate model. The narrow bore of the central tube ensures that macromolecules that do not bind to Nups find it hard to diffuse across the NPC and are thus largely excluded; the diffusive movement of the filamentous Nups may contribute to this exclusion (**A**). Macromolecules that bind to Nups increase their residence time at the entrance of the central tube, and so their diffusion across the nuclear pore complex is greatly facilitated (**B**). Reproduced with permission from (142). © 2000 Rockefeller University Press.

for characterization of structural and dynamic properties of individual components of translocation systems (152).

11.3.2. Signaling

Signaling is usually defined as a chain of biochemical events triggered by either external or internal stimuli (physical actions or chemical signals). It is now becoming clear that the localization of key signaling components is highly regulated during signal transduction. All signaling pathways are regulated by specific protein–protein interactions and often involve the assembly of large signaling complexes whose functional characteristics are difficult to discern. Even a simple list of protein complexes that can potentially form in response to a stimulus is problematic because of the number of ways signaling molecules can be modified and combined (153). The recent explosive progress of proteomics catalyzed by dramatic improvements in sensitivity and accuracy of mass spectral analysis of proteins (154) makes it possible in many cases to detect and verify suspected interactions. Mann and co-workers have recently demonstrated that these methods are indispensable for elucidating functional protein–protein interactions involved in signaling (155).

Although a detailed discussion of proteomics techniques is beyond the scope of this book, it should be mentioned that multiple strategies have been developed for characterization of various signaling pathways. Methods of quantitative (156) and functional proteomics (157, 158) have been particularly popular in the studies of signal transduction. Many signaling pathways contain enzymes, which modify high-abundance proteins that are not directly involved in the signaling. Therefore, modulation of the signaling at a certain stage would produce a "footprint" in the proteome characteristic of a specific cell phenotype (e.g., changes in protein phosphorylation in response to certain chemical signals).

Mann and co-workers have recently presented a very elegant study that utilized methods of functional proteomics to probe protein composition of "lipid rafts" and assess their relevance for signaling (159). Quantitative high-resolution MS was used to specifically detect proteins depleted from the lipid rafts by cholesterol-disrupting drugs, resulting in a set of 241 authentic lipid raft components. Detection of a large proportion of signaling molecules (highly enriched versus total membranes and detergent-resistant fractions) provided solid evidence for the connection of rafts with signaling pathways (159). A more detailed account of various uses of functional proteomics for elucidation of the signal transduction pathways can be found in several excellent recent reviews on the subject (160–163).

11.4. IN VIVO VERSUS IN VITRO BEHAVIOR OF BIOPOLYMERS

Since the goal of structural biology is to understand the mechanics of the building blocks* of life, it seems natural that the experiments should be carried out under conditions that are designed to mimic the "in vivo" environment. Very few proteins have the ability to act alone in both cellular and extracellular environments. Both transient interactions and the formation of stable assemblies of multiple proteins governed by complex signaling pathways are often required for the tasks of target recognition, binding, transport, and function. As we have seen

*We use here a very liberal definition of a "building blocks of life," which can range from single biopolymers to their complex interacting networks.

in Chapters 4–7 of this book, MS often provides a unique opportunity to detect the presence of such complexes in solution (both directly and indirectly), and to study their structure and behavior. The analysis of multiprotein systems is almost always problematic for all spectroscopic techniques due to signal interference. MS is unique in that signal interference does not pose a significant problem,* since the ionic signals corresponding to different protein species (as well as their complexes) will generally appear at different m/z ratios. Furthermore, even overlapping peaks can often be resolved (provided the masses of the two species are different) based on the differences in the number of charges carried by the two ionic species in the gas phase. The suggestion that ESI MS can be used to actually monitor protein interactions, under conditions designed to mimic their environment in the cytosole or the extracellular milieu, is indeed very appealing. However, the feasibility of this proposal is still being debated and it is not presently clear when (or even if) its implementation will actually take place. The two challenges that lie at the heart of this problem are considered in the following sections.

11.4.1. Salts and Buffers

It is not uncommon to see references in the MS literature to experiments in which the data were acquired "under native (or near-native) conditions." As a matter of fact, multiple examples of such references are scattered throughout this book. Very often this simply means that ESI mass spectra are acquired directly from aqueous solutions maintained at physiological pH. "Native" environment, of course, cannot be reduced to water buffered to neutral pH; in fact there is no such thing as a "universal" native environment. Serum is arguably one of the simplest matrices in which protein behavior can be studied; however, even its chemical make-up is extremely complex (Table 11.2). In this section we use the term "physiological matrix" to describe a solution whose composition resembles that of the milieu in which a biomolecule exists in vivo. The composition of physiological matrices can be split into three categories: (i) inorganic ions, (ii) organic molecules and small peptides, and (iii) proteins. Most major inorganic components of serum adversely affect the quality of ESI MS measurements, although there are a few "MS-friendly" components as well.† Deterioration of spectral quality may be caused by a variety of factors, most notably suppression of the protein ion signal (82). Alkali metal ions present in serum can also cause heterogeneous cationization of protein ions (formation of protein ions $[M + nNa^+ + kK^+ + pH^+]^{(n+k+p)+}$ carrying the same charge, but differing in the number of alkali metals attached to the protein M), which results in a significant decrease of the protein ion signal-to-noise ratio even in the absence of signal suppression. However, several recent developments in

*This, of course, implies that the number of the proteins in the system is "moderate" and that signal suppression does not present a significant problem.

†A good example is bicarbonate, an "electrospray-friendly" anion, which is actually widely used in buffer preparations for direct ESI MS measurements.

TABLE 11.2. Composition of Human Serum[a]

Component	Typical Concentration[b]
Solvent	
H_2O	~90% by volume
Electrolytes—anions	
HCO_3^-, CO_2	21–31 mM
Cl^-	103 mM
HPO_4^{2-}, $H_2PO_4^-$	1 mM
SO_4^{2-}	0.5 mM
F^-	0.5–10 μM
CN^-	0.2 μM
Lactate	0.5–2.2 mM
Vitamin C (L-ascorbate)	23–85 μM
Oxalate	11–27 μM
Other organic acids (total)	~5 mM
Electrolytes—cations	
Na^+	136–145 mM
K^+	3.5–5.1 mM
Ca^{2+}	1.2–1.3 mM (free), 2.5 mM (total)
Mg^{2+}	0.6–1.1 mM
NH_4^+/NH_3	11–32 μM
Cholesterol (total)	2.9–7.8 mM
Urea	2.1–7.1 mM
Phospholipids (total)	0.7–1.7 g/L
Plasma proteins (total)	64–83 g/L
Albumin	39–51 g/L
α_1-Antitrypsin	0.8–2.0 g/L
α_2-Macroglobulin	1.3–3.0 g/L
Apolipoprotein A-I	0.9–2.0 g/L
C3 (β_1 C-globulin)	0.8–1.8 g/L
C4 (β_1 E-globulin)	0.3–0.7 g/L
Fibrinogen	2–4 g/L
Haptoglobin	0.3–2.0 g/L
IgA	0.4–4.0 g/L
IgG	5.7–18 g/L
IgM	0.4–3.0 g/L
Transferrin	2.0–3.8 g/L

[a]Based on compilations in: Burtis C. A., Ashwood E. R., Tietz N. W. 1999. *Tietz Textbook of Clinical Chemistry*, p. xxxvii. Philadelphia: Saunders.
[b]Adults 20–60 years of age.

ESI technology give some hope that inorganic salt-tolerant ionization sources can become a reality in the very near future. In some cases nonvolatile buffers and salts can be removed "on-line" just prior to sample introduction into the ESI source (164). Alternatively, salt-tolerant modifications of ESI can be considered.

Karas and co-workers noted that salt tolerance of nano-ESI is significantly higher than that of conventional ESI MS (165). This was attributed to the different initial droplet sizes (micrometer range for conventional ESI, and an order of magnitude smaller for nano-ESI) and higher surface charge density for nano-ESI generated droplets. These two factors are expected to result in "early fission" without extensive evaporation for the nano-ESI generated droplets, avoiding a dramatic increase in the salt concentration. Indeed, the "conventional" ESI spectra closely resemble nano-ESI spectra acquired from solutions with about an order of magnitude higher salt concentration (165). Another recently introduced modification of ESI, fused-droplet electrospray ionization (FD ESI) (166), also shows greater tolerance to inorganic salts and nonvolatile buffers. Although the "general appearance" of FD ESI mass spectra of peptides and proteins is very close to those obtained with conventional ESI MS, the former show a very high salt tolerance. FD ESI mass spectra of proteins could be obtained from solutions containing as much as 10% NaCl (i.e., 1.7 M) or 2.5% (425 mM) NaH_2PO_4 (425 mM) (166).

11.4.2. Macromolecular Crowding Effect

Modern biological MS in general, and ESI MS in particular, are very sensitive analytical techniques. The superior sensitivity of ESI MS (compared to many other biophysical methods, such as NMR) is certainly a great advantage when only minute amounts of sample are available for the analyses. On the other hand, working with very concentrated samples can present a challenge for ESI MS (as well as MALDI MS) analysis. Such situations are extremely rare, as the emphasis in most "typical" applications is placed on detecting analytes present at very low concentrations. However, as MS is increasingly being looked at as a means of monitoring protein behavior under conditions that mimic the "in vivo environment," such a limitation may become important. This can be illustrated with our own attempts to monitor behavior (conformational dynamics and oligomerization) of hemoglobin under "near-native" conditions. ESI mass spectra of bovine hemoglobin (Hb) in dilute solutions maintained at neutral pH and medium salt show significant tetramer dissociation, as expected from the general principles of chemical equilibria. However, our attempts to acquire Hb spectra using typical concentrations of this protein inside red blood cells (typical Hb concentration in red blood cells is 2–6 mM) have failed due to precipitous deterioration of the spectral quality once the protein concentration exceeds 100 μM. This difficulty can be ameliorated to a certain degree by using nano-ESI instead of conventional ESI.

The example considered above is quite unique, as the endogenous concentrations of most proteins are several orders of magnitude lower than that of hemoglobin. However, the total concentration of macromolecules in a typical in vivo setting would still be very high, reaching levels of 50–400 mg/mL in cellular interiors (167). Since all species occupy a significant proportion of the total volume of the medium (regardless of how insignificant the contribution of any given macromolecule could be), such media are referred to as "crowded" rather than "concentrated" (167–170). Taken together, the macromolecules occupy a significant fraction of the total volume, making it physically unavailable to other macromolecules (Figure 11.24).

In such a crowded environment, nonspecific interactions between macrosolutes may exert very significant influence on their properties (the so-called *macromolecular crowding effect*). Minton points out that the concept of "nonspecific interactions" is widely misunderstood (170). These interactions are often considered as an artifact that interferes with the acquisition of meaningful data and should be reduced or totally eliminated (e.g., by extrapolating the measurement results to zero macromolecular concentration). Although such procedures may be useful for understanding the behavior of isolated macromolecules, they do not

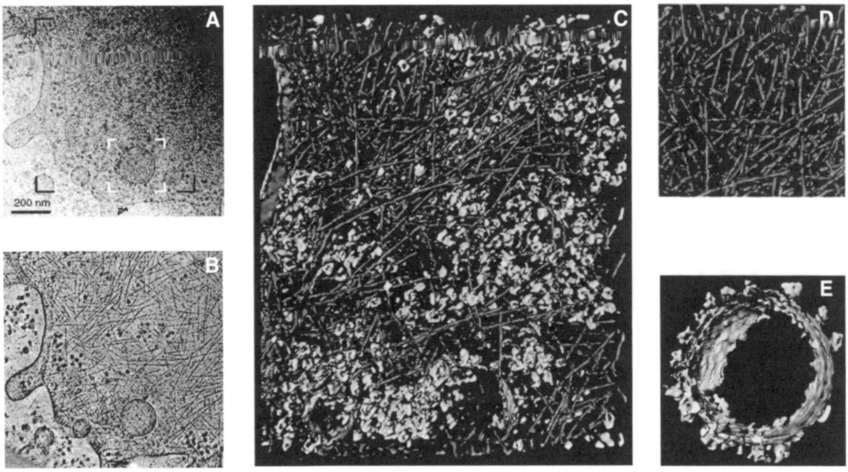

FIGURE 11.24. Macromolecular crowding of eukaryotic cytoplasm visualized by cryo-electron tomography. Shown are a peripheral region of a vitrified *Dictyostelium* imaged with conventional transmission electron microscopy, cell thickness 200–350 nm (**A**), and a representative tomographic slice of 60 nm thickness, revealing a cortical actin network (**B**). Visualization (surface rendering) of actin network, membranes, and cytoplasmic macromolecular complexes (**C**) in a $815 \times 870 \times 97$ nm volume corresponds to the area in panel (**A**) framed in black. [See color insert, where actin filaments are colored in red, other macromolecular complexes (mostly ribosomes) in green, and membranes in blue.] An image of an actin layer (**D**) taken from the upper left portion of (**C**) reveals a crowded network of branched and cross-linked filaments. Surface rendering of a ribosome-decorated portion of endoplasmic reticulum segmented into membrane and ribosomes is shown in (**E**). The densely packed structures in the lumen are omitted for clarity. Adapted with permission from (171). © 2002 AAAS.

necessarily provide results that are more meaningful in a biological context (167, 168, 172). Nonspecific interactions caused by crowding are unavoidable in vivo, as the crowding itself is an intrinsic property of most or all physiological fluid media. Therefore, many important aspects of macromolecular behavior in such media cannot be understood without taking the nonspecific interactions into account. The main effect of macromolecular crowding on biochemical equilibria is to favor the association of macromolecules. Crowding usually favors aggregation, particularly for slow-folding chains. This effect is greater for small polypeptides, since larger proteins have slower diffusion and, therefore, reduced encounter rate (172).

The effect of macromolecular crowding on protein behavior can be modeled by using high concentrations of macromolecules that mimic those found in living cells* and do not directly interact with the macromolecules being investigated, except via steric repulsion (172). Commonly used synthetic crowding agents include dextrans, Ficolls [synthetic polymers of sucrose, poly(sucrose-*co*-epichlorhydrin)], PEG, and polyvinyl alcohol. Unfortunately, utilization of any of these crowding agents in ESI MS studies at relevant concentrations ($>1\%$) seems to be problematic due to the protein ion signal suppression (as discussed earlier in this section). Another serious problem is caused by the intrinsic heterogeneity of synthetic polymers giving rise to a wide distribution of ion peaks in the ESI mass spectra (see Chapter 9 for a detailed discussion of the ESI MS of synthetic polymers), which would almost certainly interfere with the protein ion signal. This problem can be eliminated by using "inert" ubiquitous proteins (such as lysozyme, serum albumin, or hemoglobin) as crowding agents. Once again, nano-ESI may become an indispensable tool in such studies due to its increased tolerance to high concentrations of proteins in solution (165).

11.4.3. Complexity of Macromolecular Interactions in vivo

Another important aspect of macromolecular behavior in vivo is the extreme complexity of the interactions due to the very high number of participants in such interactions. Throughout most of this book, our attention has been focused on studies that use the *reductionist approach* to the problem. Each biological process is considered as a set of well-defined steps, or interactions in which the number of players is reduced to a minimum. The advantages of this approach are obvious, since it allows a huge many-body problem to be substituted with the analysis of properties of individual key biomolecules and their interactions with a limited number of key partners. The implicit assumption here is that if the function and properties of each individual component of a system are characterized, then the functioning of the entire system can be understood by a simple superposition. As pointed out by Westerhoff and co-workers, this "principle is true in theory but in practice is often false" because of a widespread nonlinearity in biological

*An ideal "crowding agent" should be fairly large (~50–200 kDa), highly water soluble, and not prone to self-aggregation. Globular (rather then extended) shapes may prevent the solutions from becoming too viscous.

systems at every level of organization (173). Nevertheless, until very recently the reductionist approach remained a commanding methodological dogma not only in molecular biophysics, but also in the entire field of structural biology.

The situation is now changing dramatically. Recent years have brought a realization of the limitations of the reductionist methodology in the life sciences (173–175). The completion of entire genome sequencing of several organisms highlights the enormous repertoire of biomolecules that make up living cells. However, simple cataloging of the DNA and protein inventory alone does not automatically generate understanding of how a live cell functions. At the same time, it is unlikely that such knowledge will be provided by characterization of the individual properties even of a large number of individual biomolecules. It appears that there is need for a synthesis of the two approaches that would take our understanding of the complex networks of cellular processes to a new level (176–179). One approach to dealing with the complexity of real living systems, that has enjoyed great popularity in the past several years, is proteomics. As can be seen from even the few examples presented earlier in this chapter, mass spectrometry has become a major player in this field. The limitations of this book do not allow us to provide a detailed discussion of various methodologies that are currently in use in proteomics. An interested reader is referred to several extensive review articles (154, 157, 180–182) and books (183–186) on this subject.

11.4.4. "Live" Macromolecules: Equilibrium Systems or Dissipative Structures?

Another implicit assumption that has not been articulated in this chapter is that the in vivo behavior of biomolecules can be properly modeled and studied under equilibrium conditions. In fact, everywhere in this book we have considered studies of various properties of biological macromolecules under equilibrium conditions. The only exception was Chapter 6, where kinetic processes were considered, but even there such processes were almost always leading from one equilibrium situation to another (e.g., protein refolding induced by changing the conditions from denaturing to near-native).

The appropriateness of equilibrium conditions in the studies of living matter was first explicitly articulated by Tanford, who wrote on the occasion of the centennial anniversary of the Gibbs papers on chemical equilibrium (187, 188): "The time now seems appropriate to use his method of thermodynamic analysis as a basis of thinking about the organization of the living cell and, beyond that, the assembly of cells into multicellular organisms" (189). Tanford suggested that understanding of the organization of the living matter can be achieved by dividing the process into two parts, biosynthesis and assembly of biosynthetic products into organized structure. The latter process should be under thermodynamic control: that is, self-organized structures represent the macromolecular arrangement providing the lowest chemical potential (which, in the absence of chemical transformations, is equal to the free energy) (189). According to this view, the cell can be thought of as a sophisticated reaction vessel, whose dynamics is governed by diffusion and random collisions of various chemical species.

However, recent decades have brought the realization that most cellular processes are far from equilibrium: live cells are polar structures, and their interiors are neither homogeneous nor isotropic, with most of the cellular functions involving directional movement and transport (190). Furthermore, every living cell is essentially a nonequilibrium system, as it relies on the constant flow of matter and energy.

The philosophical and theoretical framework for understanding how such complicated systems can assemble and function under nonequilibrium conditions has been provided by the work of the Brussels School of Physics (191–193). It resulted in the creation of nonlinear and nonequilibrium thermodynamics, which rationalizes the emergence and evolution of the *far-from-equilibrium states* (194, 195). These states often achieve very high levels of order, leading to *self-organization*,* a phenomenon that is central to the assembly and functioning of the live cell and its individual compartments (196). For example, directionality of transport mentioned earlier is achieved by utilizing various molecular motors, which overcome the randomizing effect of Brownian motion by converting chemical energy into mechanical work (61, 190, 197). So far, there have been very few examples of using MS to advance our understanding of nonequilibrium processes in vivo. Nevertheless, the tremendous potential of MS as a biophysical tool makes it certain that this technique will become a major player in removing the shroud of mystery from various nonequilibrium phenomena, thus bringing us closer to the understanding of life itself.

REFERENCES

1. D'Alessio G. (1999). The evolutionary transition from monomeric to oligomeric proteins: tools, the environment, hypotheses. *Prog. Biophys. Mol. Biol.* **72**: 271–298.
2. Caughey B., Lansbury P. T. (2003). Protofibrils, pores, fibrils, and neurodegeneration: separating the responsible protein aggregates from the innocent bystanders. *Annu. Rev. Neurosci.* **26**: 267–298.
3. Giasson B. I., Lee V. M., Trojanowski J. Q. (2003). Interactions of amyloidogenic proteins. *Neuromol. Med.* **4**: 49–58.
4. Hashimoto M., Rockenstein E., Crews L., Masliah E. (2003). Role of protein aggregation in mitochondrial dysfunction and neurodegeneration in Alzheimer's and Parkinson's diseases. *Neuromol. Med.* **4**: 21–36.
5. Aguzzi A., Haass C. (2003). Games played by rogue proteins in prion disorders and Alzheimer's disease. *Science* **302**: 814–818.
6. Horwich A. (2002). Protein aggregation in disease: a role for folding intermediates forming specific multimeric interactions. *J. Clin. Invest.* **110**: 1221–1232.
7. Carrell R. W., Lomas D. A. (1997). Conformational disease. *Lancet* **350**: 134–138.
8. Buxbaum J. N. (2003). Diseases of protein conformation: what do in vitro experiments tell us about in vivo diseases? *Trends Biochem. Sci.* **28**: 585–592.

*Self-organization is often contrasted to self-assembly: the latter is viewed as association of macromolecules into an equilibrium structure, while the former is a steady-state (nonequilibrium) process.

9. Fischer E. (1894). Einfluss der configuration auf die wirkung derenzyme. *Ber. Dtsch. Chem. Ges.* **27**: 2985–2993.
10. Behr J.-P. (1994). *The Lock-and-Key Principle: The State of the Art—100 Years on.* Chichester, England: Wiley.
11. Brooijmans N., Kuntz I. D. (2003). Molecular recognition and docking algorithms. *Annu. Rev. Biophys. Biomol. Struct.* **32**: 335–373.
12. Sundberg E. J., Mariuzza R. A. (2000). Luxury accommodations: the expanding role of structural plasticity in protein–protein interactions. *Structure Fold. Des.* **8**: R137–R142.
13. Camacho C. J., Vajda S. (2002). Protein–protein association kinetics and protein docking. *Curr. Opin. Struct. Biol.* **12**: 36–40.
14. Verkhivker G. M., Bouzida D., Gehlhaar D. K., Rejto P. A., Freer S. T., Rose P. W. (2002). Complexity and simplicity of ligand–macromolecule interactions: the energy landscape perspective. *Curr. Opin. Struct. Biol.* **12**: 197–203.
15. Carlson H. A. (2002). Protein flexibility is an important component of structure-based drug discovery. *Curr. Pharm. Des.* **8**: 1571–1578.
16. Neubig R. R., Thomsen W. J. (1989). How does a key fit a flexible lock—structure and dynamics in receptor function. *Bioessays* **11**: 136–141.
17. Miller D. W., Dill K. A. (1997). Ligand binding to proteins: the binding landscape model. *Protein Sci.* **6**: 2166–2179.
18. Tsai C.-J., Xu D., Nussinov R. (1998). Protein folding *via* binding and *vice versa*. *Folding and Design* **3**: R71–R80.
19. Tsai C.-J., Kumar S., Ma B., Nussinov R. (1999). Folding funnels, binding funnels, and protein function. *Protein Sci.* **8**: 1181–1190.
20. Kumar S., Ma B., Tsai C. J., Sinha N., Nussinov R. (2000). Folding and binding cascades: dynamic landscapes and population shifts. *Protein Sci.* **9**: 10–19.
21. Shoemaker B. A., Portman J. J., Wolynes P. G. (2000). Speeding molecular recognition by using the folding funnel: the fly-casting mechanism. *Proc. Natl. Acad. Sci. U.S.A.* **97**: 8868–8873.
22. Luque I., Leavitt S. A., Freire E. (2002). The linkage between protein folding and functional cooperativity: two sides of the same coin? *Annu. Rev. Biophys. Biomol. Struct.* **31**: 235–256.
23. Papoian G. A., Wolynes P. G. (2003). The physics and bioinformatics of binding and folding—an energy landscape perspective. *Biopolymers* **68**: 333–349.
24. Weber R. E., Vinogradov S. N. (2001). Nonvertebrate hemoglobins: functions and molecular adaptations. *Physiol. Rev.* **81**: 569–628.
25. Dickerson R. E., Geis I. (1983). *Hemoglobin: Structure, Function, Evolution, and Pathology*. Menlo Park, CA: Benjamin/Cummings.
26. Chui D. H., Fucharoen S., Chan V. (2003). Hemoglobin H disease: not necessarily a benign disorder. *Blood* **101**: 791–800.
27. Borgstahl G. E. O., Rogers P. H., Arnone A. (1994). The 1.9 Åstructure of deoxy β_4 hemoglobin: analysis of the partitioning of quaternary-associated and ligand-induced changes in tertiary structure. *J. Mol. Biol.* **236**: 831–843.
28. Kidd R. D., Baker H. M., Mathews A. J., Brittain T., Baker E. N. (2001). Oligomerization and ligand binding in a homotetrameric hemoglobin: two

high-resolution crystal structures of hemoglobin Bart's γ_4, a marker for α-thalassemia. *Protein Sci.* **10**: 1739–1749.

29. Griffith W. P., Kaltashov I. A. (2003). Highly asymmetric interactions between globin chains during hemoglobin assembly revealed by electrospray ionization mass spectrometry. *Biochemistry* **42**: 10024–10033.
30. Luzzatto L., Notaro R. (2002). Hemoglobin's chaperone. *Nature* **417**: 703–705.
31. Kihm A. J., Kong Y., Hong W., Russell J. E., Rouda S., Adachi K., Simon M. C., Blobel G. A., Weiss M. J. (2002). An abundant erythroid protein that stabilizes free a-haemoglobin. *Nature* **417**: 758–763.
32. Nettleton E. J., Tito P., Sunde M., Bouchard M., Dobson C. M., Robinson C. V. (2000). Characterization of the oligomeric states of insulin in self-assembly and amyloid fibril formation by mass spectrometry. *Biophys. J.* **79**: 1053–1065.
33. Larson J. L., Ko E., Miranker A. D. (2000). Direct measurement of islet amyloid polypeptide fibrillogenesis by mass spectrometry. *Protein Sci.* **9**: 427–431.
34. Villanueva J., Villegas V., Querol E., Aviles F. X., Serrano L. (2003). Monitoring disappearance of monomers and generation of resistance to proteolysis during the formation of the activation domain of human procarboxypeptidase A2 (ADA2h) amyloid fibrils by matrix assisted laser desorption ionization time-of-flight MS. *Biochem. J.* **374**: 489–495.
35. Monti M., Garolla di Bard B. L., Calloni G., Chiti F., Amoresano A., Ramponi G., Pucci P. (2004). The regions of the sequence most exposed to the solvent within the amyloidogenic state of a protein initiate the aggregation process. *J. Mol. Biol.* **336**: 253–262.
36. Kheterpal I., Williams A., Murphy C., Bledsoe B., Wetzel R. (2001). Structural features of the Aβ amyloid fibril elucidated by limited proteolysis. *Biochemistry* **40**: 11757–11767.
37. Monti M., Principe S., Giorgetti S., Mangione P., Merlini G., Clark A., Bellotti V., Amoresano A., Pucci P. (2002). Topological investigation of amyloid fibrils obtained from β2-microglobulin. *Protein Sci.* **11**: 2362–2369.
38. McCammon M. G., Scott D. J., Keetch C. A., Greene L. H., Purkey H. E., Petrassi H. M., Kelly J. W., Robinson C. V. (2002). Screening transthyretin amyloid fibril inhibitors: characterization of novel multiprotein, multiligand complexes by mass spectrometry. *Structure* **10**: 851–863.
39. Skribanek Z., Balaspiri L., Mak M. (2001). Interaction between synthetic amyloid-beta-peptide (1–40) and its aggregation inhibitors studied by electrospray ionization mass spectrometry. *J. Mass Spectrom.* **36**: 1226–1229.
40. Nettleton E. J., Robinson C. V. (1999). Probing conformations of amyloidogenic proteins by hydrogen exchange and mass spectrometry. *Methods Enzymol.* **309**: 633–646.
41. Tito P., Nettleton E. J., Robinson C. V. (2000). Dissecting the hydrogen exchange properties of insulin under amyloid fibril forming conditions: a site-specific investigation by mass spectrometry. *J. Mol. Biol.* **303**: 267–278.
42. Kheterpal I., Zhou S., Cook K. D., Wetzel R. (2000). Aβ amyloid fibrils possess a core structure highly resistant to hydrogen exchange. *Proc. Natl. Acad. Sci. U.S.A.* **97**: 13597–13601.

43. Wang S. S., Tobler S. A., Good T. A., Fernandez E. J. (2003). Hydrogen exchange–mass spectrometry analysis of β-amyloid peptide structure. *Biochemistry* **42**: 9507–9514.

44. Kraus M., Bienert M., Krause E. (2003). Hydrogen exchange studies on Alzheimer's amyloid-beta peptides by mass spectrometry using matrix-assisted laser desorption/ionization and electrospray ionization. *Rapid Commun. Mass Spectrom.* **17**: 222–228.

45. Kheterpal I., Wetzel R., Cook K. D. (2003). Enhanced correction methods for hydrogen exchange–mass spectrometric studies of amyloid fibrils. *Protein Sci.* **12**: 635–643.

46. Williams A. D., Portelius E., Kheterpal I., Guo J.-T., Cook K. D., Xu Y., Wetzel R. (2004). Mapping Aβ amyloid fibril secondary structure using scanning proline mutagenesis. *J. Mol. Biol.* **335**: 833–842.

47. Nazabal A., Dos Reis S., Bonneu M., Saupe S. J., Schmitter J. M. (2003). Conformational transition occurring upon amyloid aggregation of the HET-s prion protein of *Podospora anserina* analyzed by hydrogen/deuterium exchange and mass spectrometry. *Biochemistry* **42**: 8852–8861.

48. Kheterpal I., Lashuel H. A., Hartley D. M., Walz T., Lansbury P. T. Jr., Wetzel R. (2003). Aβ protofibrils possess a stable core structure resistant to hydrogen exchange. *Biochemistry* **42**: 14092–14098.

49. Kaltashov I. A., Eyles S. J. (2002). Crossing the phase boundary to study protein dynamics and function: combination of amide hydrogen exchange in solution and ion fragmentation in the gas phase. *J. Mass Spectrom.* **37**: 557–565.

50. Egnaczyk G. F., Greis K. D., Stimson E. R., Maggio J. E. (2001). Photoaffinity cross-linking of Alzheimer's disease amyloid fibrils reveals interstrand contact regions between assembled β-amyloid peptide subunits. *Biochemistry* **40**: 11706–11714.

51. Nedelec F., Surrey T., Karsenti E. (2003). Self-organisation and forces in the microtubule cytoskeleton. *Curr. Opin. Cell Biol.* **15**: 118–124.

52. Valiron O., Caudron N., Job D. (2001). Microtubule dynamics. *Cell. Mol. Life Sci.* **58**: 2069–2084.

53. Job D., Valiron O., Oakley B. (2003). Microtubule nucleation. *Curr. Opin. Cell Biol.* **15**: 111–117.

54. Gigant B., Martin-Barbey C., Curmi P. A., Sobel A., Knossow M. (2003). L'interaction stathmine-tubuline et la regulation de la dynamique des microtubules [The stathmin–tubulin interaction and the regulation of the microtubule assembly]. *Pathol. Biol.* **51**: 33–38.

55. Gigant B., Curmi P. A., Martin-Barbey C., Charbaut E., Lachkar S., Lebeau L., Siavoshian S., Sobel A., Knossow M. (2000). The 4 ÅX-ray structure of a tubulin:stathmin-like domain complex. *Cell* **102**: 809–816.

56. Redeker V., Lachkar S., Siavoshian S., Charbaut E., Rossier J., Sobel A., Curmi P. A. (2000). Probing the native structure of stathmin and its interaction domains with tubulin. Combined use of limited proteolysis, size exclusion chromatography, and mass spectrometry. *J. Biol. Chem.* **275**: 6841–6849.

57. Wallon G., Rappsilber J., Mann M., Serrano L. (2000). Model for stathmin/OP18 binding to tubulin. *EMBO J.* **19**: 213–222.

58. Tuszynski J. A., Trpisova B., Sept D., Brown J. A. (1997). Selected physical issues in the structure and function of microtubules. *J. Struct. Biol.* **118**: 94–106.

59. Tuszynski J. A., Brown J. A., Sept D. (2003). Models of the collective behavior of proteins in cells: tubulin, actin and motor proteins. *J. Biol. Phys.* **29**: 401–428.

60. Christophorov L. N., Holzwarth A. R., Kharkyanen V. N., van Mourik F. (2000). Structure–function self-organization in nonequilibrium macromolecular systems. *Chem. Phys.* **256**: 45–60.

61. Reimann P. (2002). Brownian motors: noisy transport far from equilibrium. *Phys. Rep.* **361**: 57–265.

62. Moore P. B. (2001). The ribosome at atomic resolution. *Biochemistry* **40**: 3243–3250.

63. Moore P. B., Steitz T. A. (2003). The structural basis of large ribosomal subunit function. *Annu. Rev. Biochem.* **72**: 813–850.

64. Kopylov A. M. (2002). X-ray analysis of ribosomes: the static of the dynamic. *Biochemistry (Mosc.)* **67**: 372–382.

65. Mathews M. B. (2002). Lost in translation. *Trends Biochem. Sci.* **27**: 267–269.

66. Preiss T., Hentze M. W. (2003). Starting the protein synthesis machine: eukaryotic translation initiation. *Bioessays* **25**: 1201–1211.

67. Benjamin D. R., Robinson C. V., Hendrick J. P., Hartl F. U., Dobson C. M. (1998). Mass spectrometry of ribosomes and ribosomal subunits. *Proc. Natl. Acad. Sci. U.S.A.* **95**: 7391–7395.

68. Markus M. A., Hinck A. P., Huang S., Draper D. E., Torchia D. A. (1997). High resolution solution structure of ribosomal protein L11-C76, a helical protein with a flexible loop that becomes structured upon binding to RNA. *Nat. Struct. Biol.* **4**: 70–77.

69. Rostom A. A., Fucini P., Benjamin D. R., Juenemann R., Nierhaus K. H., Hartl F. U., Dobson C. M., Robinson C. V. (2000). Detection and selective dissociation of intact ribosomes in a mass spectrometer. *Proc. Natl. Acad. Sci. U.S.A.* **97**: 5185–5190.

70. Hanson C. L., Fucini P., Ilag L. L., Nierhaus K. H., Robinson C. V. (2003). Dissociation of intact *Escherichia coli* ribosomes in a mass spectrometer. Evidence for conformational change in a ribosome elongation factor G complex. *J. Biol. Chem.* **278**: 1259–1267.

71. Trauger S. A., Junker T., Siuzdak G. (2003). Investigating viral proteins and intact viruses with mass spectrometry. *Mod. Mass Spectrom.* **225**: 265–282.

72. Tito M. A., Tars K., Valegard K., Hajdu J., Robinson C. V. (2000). Electrospray time-of-flight mass spectrometry of the intact MS2 virus capsid. *J. Am. Chem. Soc.* **122**: 3550–3551.

73. Fuerstenau S. D., Benner W. H., Thomas J. J., Brugidou C., Bothner B., Siuzdak G. (2001). Mass spectrometry of an intact virus. *Angew. Chem. Int. Ed.* **40**: 542–544.

74. Fuerstenau S. D., Benner W. H. (1995). Molecular weight determination of megadalton DNA electrospray ions using charge detection time-of-flight mass spectrometry. *Rapid Commun. Mass Spectrom.* **9**: 1528–1538.

75. Werten P. J., Remigy H. W., de Groot B. L., Fotiadis D., Philippsen A., Stahlberg H., Grubmuller H., Engel A. (2002). Progress in the analysis of membrane protein structure and function. *FEBS Lett.* **529**: 65–72.

76. Arora A., Tamm L. K. (2001). Biophysical approaches to membrane protein structure determination. *Curr. Opin. Struct. Biol.* **11**: 540–547.

77. Torres J., Stevens T. J., Samso M. (2003). Membrane proteins: the "Wild West" of structural biology. *Trends Biochem. Sci.* **28**: 137–144.

78. Wu C. C., Yates J. R. III (2003). The application of mass spectrometry to membrane proteomics. *Nat. Biotechnol.* **21**: 262–267.

79. Garavito R. M., Ferguson-Miller S. (2001). Detergents as tools in membrane biochemistry. *J. Biol. Chem.* **276**: 32403–32406.

80. Marsh D. (2003). Lipid interactions with transmembrane proteins. *Cell. Mol. Life Sci.* **60**: 1575–1580.

81. le Maire M., Champeil P., Moller J. V. (2000). Interaction of membrane proteins and lipids with solubilizing detergents. *Biochim. Biophys. Acta* **1508**: 86–111.

82. Annesley T. M. (2003). Ion suppression in mass spectrometry. *Clin. Chem.* **49**: 1041–1044.

83. Fearnley I. M., Walker J. E. (1996). Analysis of hydrophobic proteins and peptides by electrospray ionization MS. *Biochem. Soc. Trans.* **24**: 912–917.

84. Hufnagel P., Schweiger U., Eckerskorn C., Oesterhelt D. (1996). Electrospray ionization mass spectrometry of genetically and chemically modified bacteriorhodopsins. *Anal. Biochem.* **243**: 46–54.

85. Whitelegge J. P., Gundersen C. B., Faull K. F. (1998). Electrospray-ionization mass spectrometry of intact intrinsic membrane proteins. *Protein Sci.* **7**: 1423–1430.

86. Whitelegge J. P. (2004). HPLC and mass spectrometry of intrinsic membrane proteins. *Methods Mol. Biol.* **251**: 323–340.

87. Barnidge D. R., Dratz E. A., Jesaitis A. J., Sunner J. (1999). Extraction method for analysis of detergent-solubilized bacteriorhodopsin and hydrophobic peptides by electrospray ionization mass spectrometry. *Anal. Biochem.* **269**: 1–9.

88. Ishihama Y., Katayama H., Asakawa N. (2000). Surfactants usable for electrospray ionization mass spectrometry. *Anal. Biochem.* **287**: 45–54.

89. Cadene M., Chait B. T. (2000). A robust, detergent-friendly method for mass spectrometric analysis of integral membrane proteins. *Anal. Chem.* **72**: 5655–5658.

90. Zhang N., Doucette A., Li L. (2001). Two-layer sample preparation method for MALDI mass spectrometric analysis of protein and peptide samples containing sodium dodecyl sulfate. *Anal. Chem.* **73**: 2968–2975.

91. Zhang N., Li L. (2002). Ammonium dodecyl sulfate as an alternative to sodium dodecyl sulfate for protein sample preparation with improved performance in MALDI mass spectrometry. *Anal. Chem.* **74**: 1729–1736.

92. Whitelegge J. P., le Coutre J., Lee J. C., Engel C. K., Prive G. G., Faull K. F., Kaback H. R. (1999). Toward the bilayer proteome, electrospray ionization-mass spectrometry of large, intact transmembrane proteins. *Proc. Natl. Acad. Sci. U.S.A.* **96**: 10695–10698.

93. Weinglass A. B., Whitelegge J. P., Hu Y., Verner G. E., Faull K. F., Kaback H. R. (2003). Elucidation of substrate binding interactions in a membrane transport protein by mass spectrometry. *EMBO J.* **22**: 1467–1477.

94. Abramson J., Smirnova I., Kasho V., Verner G., Kaback H. R., Iwata S. (2003). Structure and mechanism of the lactose permease of *Escherichia coli*. *Science* **301**: 610–615.

95. Buhler S., Michels J., Wendt S., Ruck A., Brdiczka D., Welte W., Przybylski M. (1998). Mass spectrometric mapping of ion channel proteins (porins) and identification of their supramolecular membrane assembly. *Proteins* Suppl. **2**: 63–73.

96. Bogdanov M., Dowhan W. (1999). Lipid-assisted protein folding. *J. Biol. Chem.* **274**: 36827–36830.

97. Lee A. G. (2003). Lipid–protein interactions in biological membranes: a structural perspective. *Biochim. Biophys. Acta* **1612**: 1–40.

98. Pebay-Peyroula E., Rosenbusch J. P. (2001). High-resolution structures and dynamics of membrane protein–lipid complexes: a critique. *Curr. Opin. Struct. Biol.* **11**: 427–432.

99. Sanders C. R., Oxenoid K. (2000). Customizing model membranes and samples for NMR spectroscopic studies of complex membrane proteins. *Biochim. Biophys. Acta* **1508**: 129–145.

100. Stahlberg H., Fotiadis D., Scheuring S., Remigy H., Braun T., Mitsuoka K., Fujiyoshi Y., Engel A. (2001). Two-dimensional crystals: a powerful approach to assess structure, function and dynamics of membrane proteins. *FEBS Lett.* **504**: 166–172.

101. Blonder J., Goshe M. B., Moore R. J., Pasa-Tolic L., Masselon C. D., Lipton M. S., Smith R. D. (2002). Enrichment of integral membrane proteins for proteomic analysis using liquid chromatography–tandem mass spectrometry. *J. Proteome Res.* **1**: 351–360.

102. Dorman G., Prestwich G. D. (1994). Benzophenone photophores in biochemistry. *Biochemistry* **33**: 5661–5673.

103. Leite J. F., Blanton M. P., Shahgholi M., Dougherty D. A., Lester H. A. (2003). Conformation-dependent hydrophobic photolabeling of the nicotinic receptor: electrophysiology-coordinated photochemistry and mass spectrometry. *Proc. Natl. Acad. Sci. U.S.A.* **100**: 13054–13059.

104. Ball L. E., Oatis J. E., Dharmasiri K., Busman M., Wang J., Cowden L. B., Galijatovic A., Chen N., Crouch R. K., Knapp D. R. (1998). Mass spectrometric analysis of integral membrane proteins: application to complete mapping of bacteriorhodopsins and rhodopsin. *Protein Sci.* **7**: 758–764.

105. Gelasco A., Crouch R. K., Knapp D. R. (2000). Intrahelical arrangement in the integral membrane protein rhodopsin investigated by site-specific chemical cleavage and mass spectrometry. *Biochemistry* **39**: 4907–4914.

106. Demmers J. A., Haverkamp J., Heck A. J., Koeppe R. E. II, Killian J. A. (2000). Electrospray ionization mass spectrometry as a tool to analyze hydrogen/deuterium exchange kinetics of transmembrane peptides in lipid bilayers. *Proc. Natl. Acad. Sci. U.S.A.* **97**: 3189–3194.

107. Demmers J. A., van Duijn E., Haverkamp J., Greathouse D. V., Koeppe R. E. II, Heck A. J., Killian J. A. (2001). Interfacial positioning and stability of transmembrane peptides in lipid bilayers studied by combining hydrogen/deuterium exchange and mass spectrometry. *J. Biol. Chem.* **276**: 34501–34508.

108. Demmers J. A., Rijkers D. T., Haverkamp J., Killian J. A., Heck A. J. (2002). Factors affecting gas-phase deuterium scrambling in peptide ions and their

implications for protein structure determination. *J. Am. Chem. Soc.* **124**: 11191–11198.

109. Marsh D. (2002). Membrane water-penetration profiles from spin labels. *Eur. Biophys. J.* **31**: 559–562.
110. Hansen R. K., Broadhurst R. W., Skelton P. C., Arkin I. T. (2002). Hydrogen/deuterium exchange of hydrophobic peptides in model membranes by electrospray ionization mass spectrometry. *J. Am. Soc. Mass Spectrom.* **13**: 1376–1387.
111. Vereb G., Szollosi J., Matko J., Nagy P., Farkas T., Vigh L., Matyus L., Waldmann T. A., Damjanovich S. (2003). Dynamic, yet structured: the cell membrane three decades after the Singer–Nicolson model. *Proc. Natl. Acad. Sci. U.S.A.* **100**: 8053–8058.
112. Clague M. J. (1998). Molecular aspects of the endocytic pathway. *Biochem. J.* **336**: 271–282.
113. Young I., Wang I., Roof W. D. (2000). Phages will out: strategies of host cell lysis. *Trends Microbiol.* **8**: 120–128.
114. Bernhardt T. G., Wang I. N., Struck D. K., Young R. (2002). Breaking free: "protein antibiotics" and phage lysis. *Res. Microbiol.* **153**: 493–501.
115. Grundling A., Manson M. D., Young R. (2001). Holins kill without warning. *Proc. Natl. Acad. Sci. U.S.A.* **98**: 9348–9352.
116. Cullen B. R. (2001). Journey to the center of the cell. *Cell* **105**: 697–700.
117. Poranen M. M., Daugelavicius R., Bamford D. H. (2002). Common principles in viral entry. *Annu. Rev. Microbiol.* **56**: 521–538.
118. Akashi S., Takio K. (2001). Structure of melittin bound to phospholipid micelles studied using hydrogen–deuterium exchange and electrospray ionization Fourier transform ion cyclotron resonance mass spectrometry. *J. Am. Soc. Mass Spectrom.* **12**: 1247–1253.
119. Hristova K., Dempsey C. E., White S. H. (2001). Structure, location, and lipid perturbations of melittin at the membrane interface. *Biophys. J.* **80**: 801–811.
120. Bouchard M., Benjamin D. R., Tito P., Robinson C. V., Dobson C. M. (2000). Solvent effects on the conformation of the transmembrane peptide gramicidin A: insights from electrospray ionization mass spectrometry. *Biophys. J.* **78**: 1010–1017.
121. Cross T. A., Arseniev A., Cornell B. A., Davis J. H., Killian J. A., Koeppe R. E. II, Nicholson L. K., Separovic F., Wallace B. A. (1999). Gramicidin channel controversy—revisited. *Nat. Struct. Biol.* **6**: 610–611.
122. Burkhart B. M., Duax W. L. (1999). Gramicidin channel controversy—reply. *Nat. Struct. Biol.* **6**: 611–612.
123. Chitta R. K., Gross M. L. (2004). Electrospray ionization–mass spectrometry and tandem mass spectrometry reveal self-association and metal-ion binding of hydrophobic peptides: a study of the gramicidin dimer. *Biophys. J.* **86**: 473–479.
124. Waring A. J., Mobley P. W., Gordon L. M. (1998). Conformational mapping of a viral fusion peptide in structure-promoting solvents using circular dichroism and electrospray mass spectrometry. *Proteins* **S2**: 38–49.
125. de Brouwer A. P., Versluis C., Westerman J., Roelofsen B., Heck A. J., Wirtz K. W. (2002). Determination of the stability of the noncovalent phospholipid transfer protein–lipid complex by electrospray time-of-flight mass spectrometry. *Biochemistry* **41**: 8013–8018.

126. van Tiel C. M., Schouten A., Snoek G. T., Gros P., Wirtz K. W. (2002). The structure of phosphatidylinositol transfer protein alpha reveals sites for phospholipid binding and membrane association with major implications for its function. *FEBS Lett.* **531**: 69–73.

127. Hanson C. L., Ilag L. L., Malo J., Hatters D. M., Howlett G. J., Robinson C. V. (2003). Phospholipid complexation and association with apolipoprotein C-II: insights from mass spectrometry. *Biophys. J.* **85**: 3802–3812.

128. Li Y., Heitz F., Le Grimellec C., Cole R. B. (2004). Hydrophobic component in noncovalent binding of fusion peptides to lipids as observed by electrospray mass spectrometry. *Rapid Commun. Mass Spectrom.* **18**: 135–137.

129. Demmers J. A. A., van Dalen A., de Kruijff B., Heck A. J. R., Killian J. A. (2003). Interaction of the K^+ channel KcsA with membrane phospholipids as studied by ESI mass spectrometry. *FEBS Lett.* **541**: 28–32.

130. Lengqvist J., Svensson R., Evergren E., Morgenstern R., Griffiths W. J. (2004). Observation of an intact non-covalent homotrimer of detergent-solubilised rat microsomal glutathione transferase 1 by electrospray mass spectrometry. *J. Biol. Chem.* **279**: 13311–13316.

131. Nakielny S., Dreyfuss G. (1999). Transport of proteins and RNAs in and out of the nucleus. *Cell* **99**: 677–690.

132. Stoffler D., Fahrenkrog B., Aebi U. (1999). The nuclear pore complex: from molecular architecture to functional dynamics. *Curr. Opin. Cell Biol.* **11**: 391–401.

133. Yoneda Y. (2000). Nucleocytoplasmic protein traffic and its significance to cell function. *Genes Cells* **5**: 777–787.

134. Rout M. P., Aitchison J. D. (2001). The nuclear pore complex as a transport machine. *J. Biol. Chem.* **276**: 16593–16596.

135. Rogue P. J. (2002). Nups, Kaps, NXFs and others: making sense in the Babel of nuclear transport. *Trends Cell Biol.* **12**: 6–8.

136. Suntharalingam M., Wente S. R. (2003). Peering through the pore: nuclear pore complex structure, assembly, and function. *Dev. Cell* **4**: 775–789.

137. Yang Q., Rout M. P., Akey C. W. (1998). Three-dimensional architecture of the isolated yeast nuclear pore complex: functional and evolutionary implications. *Mol. Cell* **1**: 223–234.

138. Stoffler D., Feja B., Fahrenkrog B., Walz J., Typke D., Aebi U. (2003). Cryo-electron tomography provides novel insights into nuclear pore architecture: implications for nucleocytoplasmic transport. *J. Mol. Biol.* **328**: 119–130.

139. Allen N. P., Huang L., Burlingame A., Rexach M. (2001). Proteomic analysis of nucleoporin interacting proteins. *J. Biol. Chem.* **276**: 29268–29274.

140. Huang L., Baldwin M. A., Maltby D. A., Medzihradszky K. F., Baker P. R., Allen N., Rexach M., Edmondson R. D., Campbell J., Juhasz P., Martin S. A., Vestal M. L., Burlingame A. L. (2002). The identification of protein–protein interactions of the nuclear pore complex of *Saccharomyces cerevisiae* using high throughput matrix-assisted laser desorption ionization time-of-flight tandem mass spectrometry. *Mol. Cell. Proteomics* **1**: 434–450.

141. Allen N. P. C., Patel S. S., Huang L., Chalkley R. J., Burlingame A., Lutzmann M., Hurt E. C., Rexach M. (2002). Deciphering networks of protein interactions at the nuclear pore complex. *Mol. Cell. Proteomics* **1**: 930–946.

142. Rout M. P., Aitchison J. D., Suprapto A., Hjertaas K., Zhao Y., Chait B. T. (2000). The yeast nuclear pore complex: composition, architecture, and transport mechanism. *J. Cell Biol.* **148**: 635–652.

143. Cronshaw J. M., Krutchinsky A. N., Zhang W., Chait B. T., Matunis M. J. (2002). Proteomic analysis of the mammalian nuclear pore complex. *J. Cell Biol.* **158**: 915–927.

144. Rout M. P., Aitchison J. D., Magnasco M. O., Chait B. T. (2003). Virtual gating and nuclear transport: the hole picture. *Trends Cell Biol.* **13**: 622–628.

145. Corsi A. K., Schekman R. (1996). Mechanism of polypeptide translocation into the endoplasmic reticulum. *J. Biol. Chem.* **271**: 30299–30302.

146. Matlack K. E. S., Mothes W., Rapoport T. A. (1998). Protein translocation: tunnel vision. *Cell* **92**: 381–390.

147. Driessen A. J., Fekkes P., van der Wolk J. P. (1998). The Sec system. *Curr. Opin. Microbiol.* **1**: 216–222.

148. Pilon M., Scheckman R. (1999). Protein translocation. How Hsp70 pulls it off. *Cell* **97**: 679–682.

149. Agarraberes F. A., Dice J. F. (2001). Protein translocation across membranes. *Biochim. Biophys. Acta* **1513**: 1–24.

150. Schnell D. J., Hebert D. N. (2003). Protein translocons: multifunctional mediators of protein translocation across membranes. *Cell* **112**: 491–505.

151. Eichler J., Irihimovitch V. (2003). Move it on over: getting proteins across biological membranes. *Bioessays* **25**: 1154–1157.

152. Kipping M., Lilie H., Lindenstrauss U., Andreesen J. R., Griesinger C., Carlomagno T., Bruser T. (2003). Structural studies on a twin-arginine signal sequence. *FEBS Lett.* **550**: 18–22.

153. Hlavacek W. S., Faeder J. R., Blinov M. L., Perelson A. S., Goldstein B. (2003). The complexity of complexes in signal transduction. *Biotechnol. Bioeng.* **84**: 783–794.

154. Aebersold R., Mann M. (2003). Mass spectrometry-based proteomics. *Nature* **422**: 198–207.

155. Blagoev B., Kratchmarova I., Ong S. E., Nielsen M., Foster L. J., Mann M. (2003). A proteomics strategy to elucidate functional protein–protein interactions applied to EGF signaling. *Nat. Biotechnol.* **21**: 315–318.

156. Ong S. E., Foster L. J., Mann M. (2003). Mass spectrometric-based approaches in quantitative proteomics. *Methods* **29**: 124–130.

157. Godovac-Zimmermann J., Brown L. R. (2001). Perspectives for mass spectrometry and functional proteomics. *Mass Spectrom. Rev.* **20**: 1–57.

158. Hunter T. C., Andon N. L., Koller A., Yates I., John R., Haynes P. A. (2002). The functional proteomics toolbox: methods and applications. *J. Chromatogr. B* **782**: 165–181.

159. Foster L. J., de Hoog C. L., Mann M. (2003). Unbiased quantitative proteomics of lipid rafts reveals high specificity for signaling factors. *Proc. Natl. Acad. Sci. U.S.A.* **100**: 5813–5818.

160. Resing K. A. (2002). Analysis of signaling pathways using functional proteomics. *Ann. N.Y. Acad. Sci.* **971**: 608–614.

161. Graves P. R., Haystead T. A. J. (2003). A functional proteomics approach to signal transduction. *Recent Prog. Horm. Res.* **58**: 1–24.

162. Ping P. (2003). Identification of novel signaling complexes by functional proteomics. *Circ. Res.* **93**: 595–603.

163. Godovac-Zimmermann J., Brown L. R. (2003). Proteomics approaches to elucidation of signal transduction pathways. *Curr. Opin. Mol. Ther.* **5**: 241–249.

164. Cavanagh J., Benson L. M., Thompson R., Naylor S. (2003). In line desalting mass spectrometry for the study of noncovalent biological complexes. *Anal. Chem.* **75**: 2949–2954.

165. Juraschek R., Dulcks T., Karas M. (1999). Nanoelectrospray—more than just a minimized-flow electrospray ionization source. *J. Am. Soc. Mass Spectrom.* **10**: 300–308.

166. Chang D. Y., Lee C. C., Shiea J. (2002). Detecting large biomolecules from high-salt solutions by fused-droplet electrospray ionization mass spectrometry. *Anal. Chem.* **74**: 2465–2469.

167. Minton A. P. (2001). The influence of macromolecular crowding and macromolecular confinement on biochemical reactions in physiological media. *J. Biol. Chem.* **276**: 10577–10580.

168. Minton A. P. (2000). Implications of macromolecular crowding for protein assembly. *Curr. Opin. Struct. Biol.* **10**: 34–39.

169. Davis-Searles P. R., Saunders A. J., Erie D. A., Winzor D. J., Pielak G. J. (2001). Interpreting the effects of small uncharged solutes on protein-folding equilibria. *Annu. Rev. Biophys. Biomol. Struct.* **30**: 271–306.

170. Hall D., Minton A. P. (2003). Macromolecular crowding: qualitative and semiquantitative successes, quantitative challenges. *Biochim. Biophys. Acta* **1649**: 127–139.

171. Medalia O., Weber I., Frangakis A. S., Nicastro D., Gerisch G., Baumeister W. (2002). Macromolecular architecture in eukaryotic cells visualized by cryoelectron tomography. *Science* **298**: 1209–1213.

172. Ellis R. J. (2001). Macromolecular crowding: obvious but underappreciated. *Trends Biochem. Sci.* **26**: 597–604.

173. Bruggeman F. J., van Heeswijk W. C., Boogerd F. C., Westerhoff H. V. (2000). Macromolecular intelligence in microorganisms. *Biol. Chem.* **381**: 965–972.

174. Huang S., Ingber D. E. (2000). Shape-dependent control of cell growth, differentiation, and apoptosis: switching between attractors in cell regulatory networks. *Exp. Cell Res.* **261**: 91–103.

175. Pogun S. (2001). Are attractors "strange," or is life more complicated than the simple laws of physics? *Biosystems* **63**: 101–114.

176. Hartwell L. H., Hopfield J. J., Leibler S., Murray A. W. (1999). From molecular to modular cell biology. *Nature* **402**: C47–52.

177. Eisenberg D., Marcotte E. M., Xenarios I., Yeates T. O. (2000). Protein function in the post-genomic era. *Nature* **405**: 823–826.

178. Kitano H. (2002). Systems biology: a brief overview. *Science* **295**: 1662–1664.

179. Gavin A. C., Superti-Furga G. (2003). Protein complexes and proteome organization from yeast to man. *Curr. Opin. Chem. Biol.* **7**: 21–27.

180. Aebersold R., Goodlett D. R. (2001). Mass spectrometry in proteomics. *Chem. Rev.* **101**: 269–295.

181. Kettman J. R., Frey J. R., Lefkovits I. (2001). Proteome, transcriptome and genome: top down or bottom up analysis? *Biomol. Eng.* **18**: 207–212.

182. Patterson S. D., Aebersold R. H. (2003). Proteomics: the first decade and beyond. *Nat. Genet.* **33 Suppl**: 311–323.

183. James P. (2001). *Proteome Research: Mass Spectrometry*. Berlin: Springer.

184. Pennington S. R., Dunn M. J. (2001). *Proteomics: From Protein Sequence to Function*. New York: Springer-Verlag.

185. Liebler D. C. (2002). *Introduction to Proteomics: Tools for the New Biology*. Totowa, NJ: Humana Press.

186. Westermeier R., Naven T. (2002). *Proteomics in Practice: A Laboratory Manual of Proteome Analysis*. Weinheim: Wiley-VCH.

187. Gibbs J. W. (1876). On the equilibrium of heterogeneous substances. *Trans. Conn. Acad.* **3**: 108–248.

188. Gibbs J. W. (1878). On the equilibrium of heterogeneous substances. *Trans. Conn. Acad.* **3**: 343–524.

189. Tanford C. (1978). Hydrophobic effect and organization of living matter. *Science* **200**: 1012–1018.

190. Bustamante C., Keller D., Oster G. (2001). The physics of molecular motors. *Acc. Chem. Res.* **34**: 412–420.

191. Prigogine I., Wiame J. M. (1946). Biologie et thermodynamique des phenomenes irreversibles. *Experientia* **2**: 451–453.

192. Prigogine I. (1972). La thermodynamique de la vie. *Recherche* **3**: 547–570.

193. Prigogine I., Nicolis G., Babloyan A. (1974). Nonequilibrium problems in biological phenomena. *Ann. N.Y. Acad. Sci.* **231**: 99–105.

194. Nicolis G., Prigogine I. (1977). *Self-organization in Nonequilibrium Systems: From Dissipative Structures to Order Through Fluctuations*. Hoboken, NJ: John Wiley & Sons.

195. Kondepudi D. K., Prigogine I. (1998). *Modern Thermodynamics: From Heat Engines to Dissipative Structures*. Hoboken, NJ: John Wiley & Sons.

196. Misteli T. (2001). The concept of self-organization in cellular architecture. *J. Cell Biol.* **155**: 181–186.

197. Vale R. D., Milligan R. A. (2000). The way things move: looking under the hood of molecular motor proteins. *Science* **288**: 88–95.

198. Jimenez J. L., Nettleton E. J., Bouchard M., Robinson C. V., Dobson C. M., Saibil H. R. (2002). The protofilament structure of insulin amyloid fibrils. *Proc. Natl. Acad. Sci. U.S.A.* **99**: 9196–9201.

199. Amos L. A. (2000). Focusing-in on microtubules. *Curr. Opin. Struct. Biol.* **10**: 236–241.

200. Gabashvili I. S., Agrawal R. K., Spahn C. M., Grassucci R. A., Svergun D. I., Frank J., Penczek P. (2000). Solution structure of the *E. coli* 70S ribosome at 11.5 Åresolution. *Cell* **100**: 537–549.

201. Frank J. (2003). Electron microscopy of functional ribosome complexes. *Biopolymers* **68**: 223–233.

202. White S. H., Ladokhin A. S., Jayasinghe S., Hristova K. (2001). How membranes shape protein structure. *J. Biol. Chem.* **276**: 32395–32398.

203. Macara I. G. (2001). Transport into and out of the nucleus. *Microbiol. Mol. Biol. Rev.* **65**: 570–594.

APPENDIX

PHYSICS OF ELECTROSPRAY

Electrospray ionization is a convoluted process that involves a large number of steps, each having a profound effect on the outcome of the measurements. It occurs upon forcing a continuous flow of conducting liquid through a metal* capillary (needle), which is maintained at a high electrostatic potential (as discussed in Chapter 3). Such conditions often cause liquid spray and form aerosol particles, a phenomenon that was explored in detail and photographed by John Zeleny in 1917 (1). The electrospray process was rediscovered in the early 1950s and thoroughly investigated in the following several decades. Aerosol production may occur in a variety of functional modes, which have been reviewed by Cloupeau and Prunet-Foch (2). One of the most commonly observed modes of aerosol formation is a *dripping mode* (or "field enhanced dripping"), in which drops are formed at regular time intervals. The drops usually have a diameter exceeding that of the capillary; however, such "main drops" can sometimes be accompanied by formation of several satellite drops (2). In a *microdripping mode* the emission also occurs with single droplets (as in the conventional dripping mode), but the droplets have much smaller diameter compared to that of the capillary. A jet appears at the apex of the liquid meniscus, and liquid begins to accumulate at the end of the jet, finally resulting in the formation of a single droplet. Following the droplet separation and jet retraction, the cycle begins again (2). This regime can

*Although most ESI sources do have a metal or metal-coated capillary as a means to deliver the flow of liquid to the interface region, it is possible to use other types of capillaries as well (e.g., fused silica).

Mass Spectrometry in Biophysics: Conformation and Dynamics of Biomolecules
By Igor A. Kaltashov and Stephen J. Eyles
ISBN 0-471-45602-0

easily evolve into the *cone-jet mode*, which is usually favored at higher electric fields. The occurrence of this mode of electrospraying was first documented by Zeleny, who noted that low viscosity liquids (e.g., alcohol) were capable of forming "a cone with a fine thread of liquid coming from its apex" (1). This thread "remains intact for but a short distance ... breaking up into drops rather suddenly." The drops spread "into a more or less conical volume," leading to the appearance of a "brush spray" (1). This mode appears to be quite robust and can be maintained within a wide range of conditions. An intermittent (pulsed) cone jet is observed once the voltage at the capillary drops below a certain level. Further decrease of the electric field may result in transition to a multicone jet, which is characterized by the formation of two (or more) emissive cusps at the tip of the capillary. At very high fields, the permanent part of the spray may disappear, giving rise to totally random spraying (2).

Analytical treatment of the cone-jet regime was first attempted by Taylor, who obtained an exact equilibrium solution for the shape of the cone in the absence of liquid emission (no jet formed) (3). Under this "local equilibrium" (surface tension is balanced by electric forces) a cone is formed, whose opening angle is 98.6° (3). Taylor's exact electrostatic solution can be extended analytically to account for ejection of a jet from the cone apex* (4, 5). Consideration of these analyses extends beyond the scope of this book; however, we will mention one conclusion that is particularly important. It relates to field penetration into the liquid during the electrospray process. Even if the liquid is not perfectly conducting, its bulk will remain quasineutral, given the motion is slow enough to allow for electric relaxation (4). In this case the electric field in the liquid bulk is very small compared to the external field.

Breakup of the charged jet into a "brush spray" can be better understood by first considering disintegration of electrically neutral jets of fluids. Rayleigh considered perturbation of the infinite cylindrical flow of liquid in the form of harmonic oscillation along the spray axis z (6):

$$r = r_0 + \alpha e^{-ikz}, \tag{A-1}$$

where r and r_0 are the radii of perturbed and unperturbed jet surfaces, and $k = 2\pi/\lambda$ is a wavenumber of the disturbance. It was found that the most rapidly propagating perturbation corresponds to a wavelength λ, which is approximately 4.5 times the diameter of the jet; a more extensive discussion of this problem, including the influence of viscosity, can be found in a recent review by Eggers (7). This defines the geometrical scale of jet disintegration in the *varicose mode*, with the expected droplet diameter being almost two times that of the jet. Surprisingly, this holds even for electrically charged jets (8).

The oscillations, of course, do not have to be axisymmetric, and the perturbations can be presented in the general form as (9–11):

$$r = r_0 + \alpha e^{(\omega t + im\theta - ikz)}, \tag{A-2}$$

*The jet is modeled as being infinitely long; that is, its disintegration into a "brush spray" is ignored.

where $m\theta$ represents an angular component and ω is the growth rate of perturbations. The varicose instability occurs when $m = 0$; otherwise the jet shape (and breakup) will be dependent on the angle θ. If $m = 1$, then there will be alternating regions of θ, where r is either larger or smaller than r_0. This represents a whipping* motion of the jet (9), a phenomenon often observed at higher electrical field strength and liquid flow rates. If $m = 2$, the jet is no longer circular. An extreme example of such a jet is a *ramified jet*, which can arise due to insufficient flow rate of liquid and high enough electric field. This leads to excessive charge on the jet, and the electric stresses overcome the surface tension (9). A succession of thickened regions are formed on the main jet, each giving rise to fine jets. All three perturbations considered here ($m = 0$, 1, and 3) can arise simultaneously, and the breakup of a ramified jet is a complex mixture of all three types of perturbation (9).

An external electric field will accelerate the jet, leading to an increase of the perturbation wavelength. If such acceleration is very fast, the jet breakup may never occur. In most cases, however, the breakup will occur somewhere along the jet, leading to the formation of a droplet whose diameter in the varicose regime can be estimated as (4, 9):

$$d \sim Q^{1/2} \cdot \left(\frac{\rho\varepsilon_0}{\gamma K}\right)^{1/6}, \tag{A-3}$$

where Q is the flow rate, ρ is the density of the liquid, γ is its surface tension coefficient, and K is its conductivity.† There is an apparent disagreement in the literature as to what the average charge on the droplet is. De Juan and de la Mora suggested that the jet charge "remains nearly tied to the liquid during the break-up process," so that the charge density is nearly constant for a given spray (12):

$$q \sim d^3. \tag{A-4}$$

Others maintain that a droplet charge-to-diameter ratio should be

$$q \sim d^{3/2} \tag{A-5}$$

and present experimental evidence that the power function coefficient can actually be as low as 1.2 (9).

Formation of the main droplets is accompanied by the appearance of smaller secondary ("satellite") droplets. The "satellite" droplets likely originate from the atomization of the liquid bridges connecting the incipient "main" droplet with the rest of the jet just prior to their separation. The frequency of droplet production is typically very high (it can easily break the megahertz level), and their velocities may reach tens of meters per second (9). Electrostatic repulsion between the droplets results in their divergence from the jet axis to form a brush spray

*This type of jet instability is often termed lateral "kink-type" instability (2).

†Droplets formed in the presence of a "kink-type" instability are scaled as $d \sim Q^{1/3}$.

observed by Zeleny (1). If one ignores solvent evaporation from the droplets, as well as their deformation, the dynamics of droplet movement in the external field $\vec{E}_{ext}$ can be described simply as (10):

$$m_i\ddot{\vec{r}}_i = q_i\vec{E}_{ext}(\vec{r}_i) + \sum_{j\neq i} \frac{q_i q_j(\vec{r}_i - \vec{r}_j)}{4\pi\varepsilon_0 \cdot |\vec{r}_i - \vec{r}_j|^3} + C_D\frac{\pi d_i^2}{8}\rho\left|\dot{\vec{r}}_i - \vec{u}\right| \cdot \left(\dot{\vec{r}}_i - \vec{u}\right), \quad \text{(A-6)}$$

where m_i, q_i, and $\vec{r}_i$ are the droplet's mass, net charge, and position, respectively. The drag force coefficient, gas density, and flow rate are represented by C_D, ρ, and $\vec{u}$. The space-charge effect represented in (A-6) by the Coulombic term is quite significant, and its influence would be particularly high in the case of smaller satellite droplets. Because of its lower inertia and drag force, a smaller droplet will attain larger radial velocity. As a result, droplets are segregated according to their size in the "brush" part of the spray, with the area along the jet axis dominated by large droplets, while the smaller ones are forced to the periphery of the brush (10).

Droplet Fate and Ion Formation: Charged Residue Model (CRM). The ultimate fate of the droplet is determined by two parallel processes: solvent loss due to evaporation and electrodynamic instability. Even an uncharged conducting droplet will not maintain an ideal spherical form in an external electrostatic field. Such a droplet is likely to assume the shape of an oblate spheroid and generate uneven charge distribution on its surface (maintaining zero net charge) to maintain field-free conditions in the bulk of the liquid (13). The dynamics of conducting liquid drops in electrostatic fields was first investigated experimentally by Macky (14). He observed that given sufficient field strength, the droplets deform (elongate) and become pointed at the ends. An increase of the surface tension will be compensated for by a decrease of electrostatic energy (elongation must be accomplished by charge segregation). A further increase of the field results in the rapid development of instability, with liquid jets being ejected from the cusps (14). It appears that this process can adequately be described using the Taylor formalism we just discussed (3). The droplet does not have to carry a net charge to become unstable (in fact, the original experiments were performed with electrically neutral droplets); however, the field strength required to induce such instability is very high, approaching 10 kV/cm (14). The presence of net positive (or negative) charge on the droplet is likely to reduce the instability threshold.

Perhaps a more efficient droplet atomization process is related to the so-called Rayleigh instability and does require an external electric field (15). The electrostatic repulsion among the like charges on the droplet is offset by the cohesive action of surface tension. The potential energy of the former is inversely proportional to the radius of the droplet, while the energy of the latter is proportional to the square of the radius (assuming the droplet has a spherical form). If solvent evaporation from the droplet proceeds without loss of charge, the electrostatic component of the potential energy will increase dramatically upon droplet shrinkage, while the stabilizing role of the surface tension will be diminished.

Eventually, the droplet will become unstable with respect to high multipole oscillations, resulting in "liquid ... thrown out in fine jets, whose fineness, however, has a limit" (15). The instability criterion is (15)

$$\gamma = \frac{(Ne)^2}{8\pi\varepsilon_0 d_0^3}, \tag{A-7}$$

where Ne represents a net charge of the droplet and d_0 its diameter in the spherical shape. Numerous measurements have confirmed that droplet fission does occur at or slightly below the Rayleigh limit (16). The characteristic time of instability development can be estimated as (17, 18):

$$\tau \sim d_0^2 \cdot \sqrt{\frac{\rho}{\gamma}}, \tag{A-8}$$

which is on the order of a tenth of a second for a water droplet with a radius of ~1 mm. Since the "typical" droplet size in ESI is three orders of magnitude below that (19), the characteristic time for instability development would fall below a microsecond, significantly below the rate of droplet evaporation.

Although the original consideration of the droplet fission problem by Rayleigh predicts formation of fine jets at the droplet's ends, a notion confirmed experimentally by many groups, another outcome termed *rough fission* is also possible. In this case, the precursor droplet is fragmented into a few droplets of similar size. It seems, however, that such a mode of fission only occurs when the liquid is a poor conductor (20). Droplets composed of conducting liquids (e.g., water, alcohols, which make, of course, a more interesting case as typical ESI solvents) do decompose in a *fine fission mode* (21). Such fission removes a significant proportion of the net charge from the droplet (up to 25–33%), while the mass loss is negligible (on the order of 0.3% or even less) (16, 22). The droplets generated during jet disintegration are of similar size and close to the instability criterion (A-7), which means they undergo fission soon after separation from the main droplet. In the meantime, the main droplet relaxes to a spherical form following the ejection of the jet, as its net charge falls comfortably below the instability limit.

De la Mora considered the "fine fission" of conducting droplets charged to a Rayleigh limit (A-7) in the absence of the electric field (21). He concluded that the fission occurs via formation of a Taylor cone, followed by ejection of a fine jet from its tip. Although the lifetime of the Taylor cone is small as compared to the characteristic droplet evaporation time, the process can still be considered quasistatic in the case of conducting liquids. (The electrical relaxation time for conducting liquids is much smaller than the expected lifetime of a Taylor cone.) As a result, one can use Taylor's original formalism (3) to describe fission processes. Continuous evaporation brings the diameter of the initially stable droplet to the Rayleigh limit, resulting in a Coulombic explosion. The precursor droplet loses a significant proportion of its charge (but not volume),

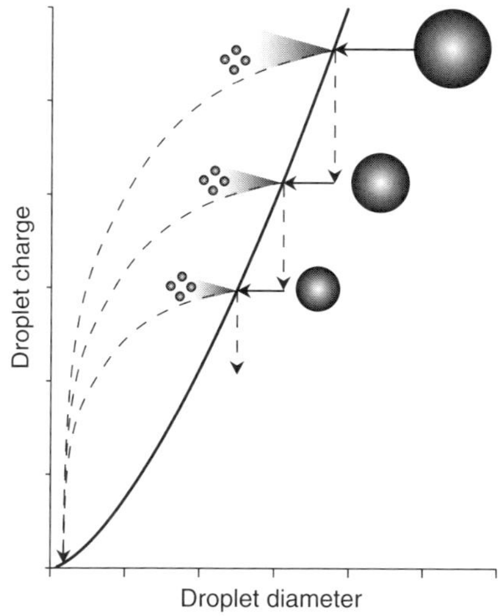

FIGURE A.1. A schematic diagram representing evolution of a conducting charged droplet. Horizontal arrows indicate solvent loss due to evaporation prior to reaching the Rayleigh instability limit (solid curve). Dashed lines indicate ejection of progeny droplets (curved lines) and loss of charge/relaxation of the precursor droplet (straight vertical lines) following ejection of the progeny droplets.

thus becoming stable again. Subsequent solvent evaporation can, of course, bring this droplet back to the unstable conditions, resulting in another fission event (Figure A.1).

Eventually, charges on both the precursor and the progeny droplets will be reduced to such levels that further solvent evaporation will no longer bring about Rayleigh instability. Complete evaporation of solvent from the droplets will, therefore, force the residual charge onto the solute molecules. The solute molecules are assumed to be large macromolecules that either lack charge in solution or else maintain electroneutrality through association with counterions. This scenario forms the foundation of the so-called charge residue model (CRM) of ion formation, which is attributed to Dole et al. (23). It should probably be mentioned here once again that despite the violent nature of the charged jet breakup and subsequent Coulombic explosions of the electrospray droplets, the bulk of the liquid may remain essentially charge- and field-free (24). Given the sufficiently high concentration of charge carrier in the liquid, a quasi-equilibrium charge layer will be formed at the liquid–gas interface. As a result, even the weakly conducting liquid would conform to a quasi-electrostatic model with the charge confined to the surface and the bulk of the liquid remaining quasi-neutral and field-free. This assertion, however, is not universally accepted (e.g., see (12)). Therefore, the solute molecules residing in the bulk of the liquid will remain "oblivious" to the harsh conditions on the droplet surface and the environment beyond it until the last of the solvent is gone. Indeed, there is solid evidence that proteins do not denature during droplet evolution in the course of ESI (25). This is a very important conclusion, since it opens up an opportunity to use ESI MS to study biological macromolecules in their "near-native" environment.

Alternative View of Ion Formation: Ion Evaporation Model (IEM). An alternative model of ion formation was offered by Iribarne and Thompson (26), who suggested that under certain conditions a single ion (or an ion cluster) can evaporate from the surface of the droplet. Removal of an ion from the droplet becomes thermodynamically favored as soon as the loss of solvation energy is offset by enthalpic gains due to the separation of the charges of the same polarity (electrostatic repulsion between the escaped ion and the remaining charges on the droplet). However, such an escape must proceed through an energy barrier created by the attraction between the ion and its electrical image on the surface of the droplet (26):

$$\Delta G^{\ddagger} = 4\pi\gamma r^2 + \frac{e^2}{2r}\left(1 - \frac{1}{\varepsilon}\right) - e^2\left[\frac{N(\chi_m + d)}{(R-d)(R+\chi_m)} + \frac{1}{4\chi_m}\right], \tag{A-9}$$

where the first two terms represent the solvation energy of an ion of radius r in a liquid whose dielectric constant is ε and surface tension coefficient is γ, and the last term represents the electrostatic interaction between the ion* and the droplet of radius R carrying N charges. The equilibrium position of the ion inside the droplet (prior to evaporation) is d, and the location of the barrier maximum ("point of no return") is χ_m:

$$\chi_m = \frac{R}{2\sqrt{N} - 1}. \tag{A-10}$$

In order for the ion evaporation process to be efficient, it must occur faster than the total evaporation of solvent from the droplet, thus setting an upper limit for $\Delta G^{\ddagger}$ of ~9 kcal/mol in the case of water droplets (26). Substitution of this value in (A-9) allows a critical droplet "radius–charge" relationship to be established. As one might expect, small ion evaporation is more efficient, compared to larger ones (Figure A.2). Importantly, droplets carrying relatively few charges have the critical size for ion evaporation exceeding that of the Rayleigh instability, indicating that there are conditions when the reduction of the droplet charge occurs through expulsion of a single ion, not through a fission process (Figure A.2).

Although the original Iribarne–Thompson work dealt with singly charged cluster ions, it was later invoked to explain production of large macromolecular ions as well (27, 28). However, it is still unclear how much is contributed by ion evaporation to the total macromolecular ion production. Reservations regarding the validity of IEM have been articulated by Röllgen and co-workers, who questioned the use of classical electrostatics to describe processes that are essentially dynamic (29). Furthermore, it was suggested that "the removal of an ion

*Evaporated ions are assumed to be singly charged; extension to multiply charged ions is rather straightforward.

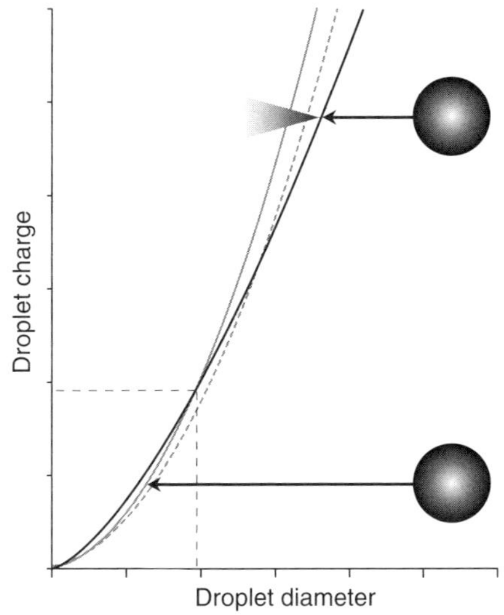

FIGURE A.2. Two scenarios of a charged conducting droplet fate in a strong electrostatic field. Depending on the initial droplet charge, solvent loss due to evaporation will lead to either Coulombic explosion (Rayleigh instability limit is indicated with a solid black curve) or ion evaporation (the critical radius for a large ion evaporation as a function of the droplet total charge is indicated with a solid gray curve). Dashed gray curve indicates critical droplet radius for evaporation of a small ion from the droplet.

with part of its solvation sphere from a charged liquid surface under field stress represents the onset of an electrohydrodynamic disintegration process on a molecular scale" (29). However, Labowsky, Fenn, and de la Mora recently pointed out that "the process of emission of a singly charged nanodrop from a larger drop is radically different from a Coulomb explosion, being in fact indistinguishable from Iribarne–Thompson ionization" (30). Such nanoscale processes (ion or nanodroplet desorption from the surface of the precursor droplet) should be facilitated by the discrete nature of the surface charge, which is shown to increase the local electrostatic pressure by at least an order of magnitude compared to the continuous charge distribution (31). Still, even though the careful analytical treatment of such processes indicates that there is a distinct possibility of small ion cluster evaporation from the charged droplet, the window of conditions that lead to ion evaporation appears to be very narrow (30). In the case of water droplets, evaporation of small ions can only occur from relatively small droplets close to the Rayleigh instability limit. The possibility of larger ion (e.g., biopolymers) desorption from electrosprayed droplets still awaits careful consideration, although the accumulated experimental evidence suggests that production of such macro-ions occurs via Dole's CRM mechanism (19, 32, 33).

A very distinct feature of ESI is the accumulation of multiple charges on a single analyte molecule during ionization. Both the CRM and IEM predict accumulation of a significant number of charges on a macromolecule, and this number is expected to increase as the physical size of the molecule increases. According to the CRM, the number of charges on the macromolecular ion will be determined by the total charge of the fission-incompetent droplet whose lower limit is set by the size of the macromolecule it encapsulates. Here we ignore gas phase processes that may result in charge transfer to/from the macromolecular ions. These ion–molecular reactions were briefly discussed in Chapter 10. The IEM scenario of macromolecular ion desorption from a charged droplet invokes the notion of equidistant charges immobilized on the surface of the droplet, a configuration that minimizes the electrostatic repulsion energy (27). The number of charges accumulated by the macromolecular ion would then be determined by its cross section upon traversing the surface of the droplet.

REFERENCES

1. Zeleny J. (1917). Instability of electrified liquid surfaces. *Phys. Rev.* **10**: 1–6.
2. Cloupeau M., Prunet-Foch B. (1994). Electrohydrodynamic spraying functioning modes: a critical review. *J. Aerosol Sci.* **25**: 1021–1036.
3. Taylor G. (1964). Disintegration of water drops in an electrical field. *Proc. R. Soc. London A* **280**: 383–397.
4. Gañán-Calvo A. M. (1997). Cone-jet analytical extension of Taylor's electrostatic solution and the asymptotic universal scaling laws in electrospraying. *Phys. Rev. Lett.* **79**: 217–220.
5. Cherney L. T. (1999). Electrohydrodynamics of electrified liquid menisci and emitted jets. *J. Aerosol Sci.* **30**: 851–862.
6. Rayleigh J. W. S. (1878). On the instability of jets. *Proc. London Math. Soc.* **11**: 4–13.
7. Eggers J. (1997). Nonlinear dynamics and breakup of free-surface flows. *Rev. Mod. Phys.* **69**: 865–929.
8. Cloupeau M., Prunetfoch B. (1989). Electrostatic spraying of liquids in cone-jet mode. *J. Electrostat.* **22**: 135–159.
9. Hartman R. P. A., Brunner D. J., Camelot D. M. A., Marijnissen J. C. M., Scarlett B. (2000). Jet break-up in electrohydrodynamic atomization in the cone-jet mode. *J. Aerosol Sci.* **31**: 65–95.
10. Hartman R. P. A., Borra J. P., Brunner D. J., Marijnissen J. C. M., Scarlett B. (1999). The evolution of electrohydrodynamic sprays produced in the cone-jet mode, a physical model. *J. Electrostat.* **47**: 143–170.
11. Hartman R. P. A., Brunner D. J., Camelot D. M. A., Marijnissen J. C. M., Scarlett B. (1999). Electrohydrodynamic atomization in the cone-jet mode physical modeling of the liquid cone and jet. *J. Aerosol Sci.* **30**: 823–849.
12. de Juan L., de la Mora J. F. (1997). Charge and size distributions of electrospray drops. *J. Colloid Interf. Sci.* **186**: 280–293.
13. Landau L. D., Lifshitz E. M., Pitaevskii L. P. (1984). *Electrodynamics of Continuous Media*. New York: Pergamon.
14. Macky W. A. (1931). Some investigations on the deformation and breaking of water drops in strong electric fields. *Proc. R. Soc. London A* **133**: 565–587.

15. Rayleigh J. W. S. (1882). On the equilibrium of liquid conducting masses charged with electricity. *Philos. Mag.* **14**: 184–186.
16. Duft D., Lebius H., Huber B. A., Guet C., Leisner T. (2002). Shape oscillations and stability of charged microdroplets. *Phys. Rev. Lett.* **89**: article no. 084503.
17. Belonozhko D. F., Grigor'ev A. I. (1999). Characteristic time for the evolution of instability of a droplet charged to the Rayleigh limit. *Tech. Phys. Lett.* **25**: 610–611.
18. Belonozhko D. F., Grigor'ev A. I. (2000). Nonlinear capillary oscillation of a charged drop. *Tech. Phys.* **45**: 1001–1008.
19. Kebarle P., Peschke M. (2000). On the mechanisms by which the charged droplets produced by electrospray lead to gas phase ions. *Anal. Chim. Acta* **406**: 11–35.
20. Richardson C. B., Pigg A. L., Hightower R. L. (1989). On the stability limit of charged droplets. *Proc. R. Soc. London A* **422**: 319–328.
21. de la Mora J. F. (1996). On the outcome of the coulombic fission of a charged isolated drop. *J. Colloid Interf. Sci.* **178**: 209–218.
22. Duft D., Achtzehn T., Muller R., Huber B. A., Leisner T. (2003). Coulomb fission—Rayleigh jets from levitated microdroplets. *Nature* **421**: 128–128.
23. Dole M., Mack L. L., Hines R. L. (1968). Molecular beams of macroions. *J. Chem. Phys.* **49**: 2240–2249.
24. Ganan-Calvo A. M. (1999). The surface charge in electrospraying: its nature and its universal scaling laws. *J. Aerosol Sci.* **30**: 863–872.
25. Mirza U. A., Chait B. T. (1997). Do proteins denature during droplet evolution in electrospray ionization? *Int. J. Mass Spectrom. Ion Proc.* **162**: 173–181.
26. Iribarne J. V., Thomson B. A. (1976). On the evaporation of small ions from charged droplets. *J. Chem. Phys.* **64**: 2287–2294.
27. Fenn J. B. (1993). Ion formation from charged droplets: roles of geometry, energy and time. *J. Am. Soc. Mass Spectrom.* **4**: 524–535.
28. Fenn J. B., Rosell J., Meng C. K. (1997). In electrospray ionization, how much pull does an ion need to escape its droplet prison? *J. Am. Soc. Mass Spectrom.* **8**: 1147–1157.
29. Röllgen F. W., Bramer-Weger E., Butfering L. (1987). Field-ion emission from liquid solutions—ion evaporation against electrohydrodynamic disintegration. *J. Phys.* **48**: 253–256.
30. Labowsky M., Fenn J. B., de la Mora J. F. (2000). A continuum model for ion evaporation from a drop: effect of curvature and charge on ion solvation energy. *Anal. Chim. Acta* **406**: 105–118.
31. Labowsky M. (1998). Discrete charge distributions in dielectric droplets. *J. Colloid Interf. Sci.* **206**: 19–28.
32. de la Mora J. F. (2000). Electrospray ionization of large multiply charged species proceeds via Dole's charged residue mechanism. *Anal. Chim. Acta* **406**: 93–104.
33. Kebarle P. (2000). A brief overview of the present status of the mechanisms involved in electrospray mass spectrometry. *J. Mass Spectrom.* **35**: 804–817.

INDEX

Mass Spectrometry in Biophysics: Conformation and Dynamics of Biomolecules
By Igor A. Kaltashov and Stephen J. Eyles
ISBN 0-471-45602-0